Nonlinear Integer Programming

Recent titles in the **INTERNATIONAL SERIES IN OPERATIONS RESEARCH & MANAGEMENT SCIENCE**

Frederick S. Hillier, Series Editor, *Stanford University*

Maros/ *COMPUTATIONAL TECHNIQUES OF THE SIMPLEX METHOD*

Harrison, Lee & Neale/ *THE PRACTICE OF SUPPLY CHAIN MANAGEMENT: Where Theory and Application Converge*

Shanthikumar, Yao & Zijm/ *STOCHASTIC MODELING AND OPTIMIZATION OF MANUFACTURING SYSTEMS AND SUPPLY CHAINS*

Nabrzyski, Schopf & Węglarz/ *GRID RESOURCE MANAGEMENT: State of the Art and Future Trends*

Thissen & Herder/ *CRITICAL INFRASTRUCTURES: State of the Art in Research and Application*

Carlsson, Fedrizzi, & Fullér/ *FUZZY LOGIC IN MANAGEMENT*

Soyer, Mazzuchi & Singpurwalla/ *MATHEMATICAL RELIABILITY: An Expository Perspective*

Chakravarty & Eliashberg/ *MANAGING BUSINESS INTERFACES: Marketing, Engineering, and Manufacturing Perspectives*

Talluri & van Ryzin/ *THE THEORY AND PRACTICE OF REVENUE MANAGEMENT*

Kavadias & Loch/*PROJECT SELECTION UNDER UNCERTAINTY: Dynamically Allocating Resources to Maximize Value*

Brandeau, Sainfort & Pierskalla/ *OPERATIONS RESEARCH AND HEALTH CARE: A Handbook of Methods and Applications*

Cooper, Seiford & Zhu/ *HANDBOOK OF DATA ENVELOPMENT ANALYSIS: Models and Methods*

Luenberger/ *LINEAR AND NONLINEAR PROGRAMMING, 2nd Ed.*

Sherbrooke/ *OPTIMAL INVENTORY MODELING OF SYSTEMS: Multi-Echelon Techniques, Second Edition*

Chu, Leung, Hui & Cheung/ *4th PARTY CYBER LOGISTICS FOR AIR CARGO*

Simchi-Levi, Wu & Shen/ *HANDBOOK OF QUANTITATIVE SUPPLY CHAIN ANALYSIS: Modeling in the E-Business Era*

Gass & Assad/ *AN ANNOTATED TIMELINE OF OPERATIONS RESEARCH: An Informal History*

Greenberg/ *TUTORIALS ON EMERGING METHODOLOGIES AND APPLICATIONS IN OPERATIONS RESEARCH*

Weber/ *UNCERTAINTY IN THE ELECTRIC POWER INDUSTRY: Methods and Models for Decision Support*

Figueira, Greco & Ehrgott/ *MULTIPLE CRITERIA DECISION ANALYSIS: State of the Art Surveys*

Reveliotis/ *REAL-TIME MANAGEMENT OF RESOURCE ALLOCATIONS SYSTEMS: A Discrete Event Systems Approach*

Kall & Mayer/ *STOCHASTIC LINEAR PROGRAMMING: Models, Theory, and Computation*

Sethi, Yan & Zhang/ *INVENTORY AND SUPPLY CHAIN MANAGEMENT WITH FORECAST UPDATES*

Cox/ *QUANTITATIVE HEALTH RISK ANALYSIS METHODS: Modeling the Human Health Impacts of Antibiotics Used in Food Animals*

Ching & Ng/ *MARKOV CHAINS: Models, Algorithms and Applications*

***** *A list of the early publications in the series is at the end of the book* *****

NONLINEAR INTEGER PROGRAMMING

DUAN LI
Department of Systems Engineering and
Engineering Management
The Chinese University of Hong Kong
Shatin, N. T.
Hong Kong

XIAOLING SUN
Department of Mathematics
Shanghai University
Baoshan, Shanghai 200444
P. R. China

Duan Li
The Chinese University of Hong Kong
P.R. China

Xiaoling Sun
Shanghai University
P.R. China

ISBN-13: 978-1-4419-3991-3

ISBN-10: 0-387-32995-1 (e-book)
ISBN-13: 978-0387-32995-6 (e-book)

Printed on acid-free paper.

9 8 7 6 5 4 3 2 1

springer.com

To my mother, Yunxiu Xue,
my wife, Xiuying Lu, and my
children, Bowen Li, Kevin
Bozhe Li, Andrew Boshi Li

To my parents, Zhiping Sun
and Cuifang Lan, my wife,
Huijue Yang, and my son,
Lei Sun

Contents

List of Figures

List of Tables

Preface

It is not an exaggeration that much of what people devote in their life resolves around optimization in one way or another. On one hand, many decision making problems in real applications naturally result in optimization problems in a form of integer programming. On the other hand, integer programming has been one of the great challenges for the optimization research community for many years, due to its computational difficulties: Exponential growth in its computational complexity with respect to the problem dimension. Since the pioneering work of R. Gomory [80] in the late 1950s, the theoretical and methodological development of integer programming has grown by leaps and bounds, mainly focusing on linear integer programming. The past few years have also witnessed certain promising theoretical and methodological achievements in nonlinear integer programming.

When the first author of this book was working on duality theory for nonconvex continuous optimization in the middle of 1990s, Prof. Douglas J. White suggested that he explore an extension of his research results to integer programming. The two authors of the book started their collaborative work on integer programming and global optimization in 1997. The more they have investigated in nonlinear integer programming, the more they need to further delve into the subject. Both authors have been greatly enjoying working in this exciting and challenging field.

Applications of nonlinear (mixed) integer programming can be found in various areas of scientific computing, engineering, management science and operations research. Its prominent applications include, for example, portfolio selection, capital budgeting, production planning, resource allocation, computer networks, reliability networks and chemical engineering. Due to nonlinearity, theory and solution methodologies for nonlinear integer programming problems are substantially different from the linear integer programming.

There are numerous books and monographs on linear integer programming [68][109][116][168] [191][188][212][222] [228]. In contrast, there is no book that comprehensively discusses theory and solution methodologies for general nonlinear integer programming, despite its importance in real world applications and its academic significance in optimization. The book by Ibaraki and Katoh [106] systematically describes solution algorithms for resource allocation problems, a special class of nonlinear integer programming problems. Sherali and Adams, in their book [196], develop a reformulation-linearization technique for constructing a convex hull of a nonconvex domain. Floudas, in his book [60], develops lower-bounding convexification schemes for nonlinear mixed integer programming problems with applications in chemical engineering. Tawarmalani and Sahinidis [213] also investigate convexification and global optimization methods for nonlinear mixed integer programming.

This book addresses the topic of general nonlinear integer programming. The overall goal of this book is to bring the state-of-the-art of the theoretical foundation and solution methods for nonlinear integer programming to readers who are interested in optimization, operations research and computer science. Of note, recent theoretical progress and innovative methodologies achieved by the authors are presented. This book systematically investigates theory and solution methodologies for general nonlinear integer programming and, at the same time, provides a timely and comprehensive summary of the theoretical and algorithmic development in the last 30 years on this topic.

We assume that readers are already familiar with some basic knowledge of *linear* integer programming. The book thus focuses on the theory and solution methodologies of *nonlinear* integer programming. The following are some features of the book:

- Duality theory for nonlinear integer programming: Investigation of the relationship between the duality gap and the perturbation function has led to the development of the novel nonlinear Lagrangian theory, thus establishing a theoretical foundation for solution methodologies of nonlinear integer programming.

- Convergent Lagrangian and cutting methods for separable nonlinear integer programming problems: Performing objective-level cut, objective contour cut or domain cut reshapes the perturbation function, thus exposing eventually an optimal solution to the convex hull of a revised perturbation function and guaranteeing a zero duality gap for a convergent Lagrangian method.

- Convexification scheme: The relationship between the monotonicity and convexity has been explored. Convexification schemes have been developed for monotone and nonconvex integer programming problems, thus extending the reach of branch-and-bound methods whose success depends on an ability to achieve a global solution of the continuous relaxation problem.

- A solution framework using global descent: The exact solution of a nonlinear integer programming problem is sought from among the local minima. A theoretical basis has been established to escape from the current minimum to a better minimum in an iterative global descent process.

- Computational implementation for large-scale nonlinear integer programming problems with dimensions up to several thousands is demonstrated for several efficient solution algorithms presented in the book.

Readers of this book can be researchers, practitioners, graduate students and senior undergraduate students in operations research and computer science. This book aims at people in academics as well as people in applied areas who already have basic knowledge of optimization and want to broaden their knowledge in integer programming. It can be used as a textbook for graduate students in the fields of operations research, management science and computer science. It can be also used as a reference book for researchers, engineers and practitioners to solve real-world application problems by nonlinear integer programming models and to design and implement sophisticated algorithms for their specific application problems.

Acknowledgments

We appreciate many individuals for their contributions to the materials presented in this book. Especially, we thank Dr. Jun Wang for his contributions in the convergent Lagrangian and objective level cut method and the revised Taha's method, Dr. Kevin Ng and Prof. Liansheng Zhang for their contributions in global descent methods, Prof. Douglas J. White for his contributions in development of the p-th power Lagrangian, Prof. Yifan Xu and Ms. Chunli Liu for their contributions in nonlinear Lagrangian theory, Prof. Ken McKinnon for his contributions in the convergent Lagrangian and domain cut method, Ms. Fenlan Wang for her contribution in the convergent Lagrangian and contour cut method, and Dr. Zhiyou Wu for her help in deriving Proposition 12.1.

We would like to express our gratitude to Dr. Kevin Ng, Mr. Hezhi Luo, Mr. Hongbo Sheng, Ms. Fei Gao, Ms. Chunli Liu, Ms. Fenlan Wang, Ms. Jianling Li, Ms. Ning Ruan and Ms. Juan Sun for their careful proofreading of the manuscript of our book.

We are grateful to our institutes, The Chinese University of Hong Kong and Shanghai University, for providing us with excellent working environments. We are grateful for the financial support to our research by the Research Grants Council of Hong Kong and the National Natural Science Foundation of China. The first author is grateful to Prof. Yacov Y. Haimes for the guidance and the all-round support during his career development in the past twenty years.

We are also very thankful for the encouragement from Prof. Frederick S. Hillier and the professional support from Mr. Gary Folven and other staff at Springer.

Finally, we owe a great debt of thanks to our families for their lasting understanding and support.

Chapter 1

INTRODUCTION

Most of the contents of this book deal with the following general class of nonlinear integer programming problems:

$$(NLIP) \qquad \min\ f(x) \\ \text{s.t. } g_i(x) \le b_i,\ i = 1, \ldots, m, \\ x \in X,$$

where f and g_i, $i = 1, \ldots, m$, are real-valued functions on $\mathbb{R}^n$, and X is a finite subset in $\mathbb{Z}^n$, the set of all integer points in $\mathbb{R}^n$.

While problem $(NLIP)$ is a *nonlinear pure integer programming* problem, Chapter 13 of this book deals with the following *mixed-integer nonlinear programming* problem

$$(MINLP) \qquad \min\ f(x, y) \\ \text{s.t. } g_i(x, y) \le b_i,\ i = 1, \ldots, m, \\ x \in X,\ y \in Y,$$

where f and g_i, $i = 1, \ldots, m$, are real-valued functions on $\mathbb{R}^{n+q}$, X is a finite subset in $\mathbb{Z}^n$, and Y is a continuous subset in $\mathbb{R}^q$.

When all functions f and g_i, $i = 1, \ldots, m$, are linear, problems $(NLIP)$ and $(MINLP)$ reduce to linear integer programming and mixed integer linear programming problems, respectively. The focus of this book is on nonlinear integer programming, which implies that at least one function of f and g_i, $i = 1, \ldots, m$, is nonlinear.

1.1 Classification of Nonlinear Integer Programming Formulations

Problem $(NLIP)$ can be classified into different subclasses according to its special structure.

Unconstrained Nonlinear Integer Programming. When the inequality constraints in $(NLIP)$ are absent, the problem is called unconstrained nonlinear integer programming problem. Two important classes of the unconstrained nonlinear integer programming problems are *unconstrained polynomial 0-1 optimization* problems and *unconstrained quadratic 0-1 optimization* problems.

Singly Constrained Nonlinear Integer Programming. When $m = 1$, i.e., there is only one constraint in $(NLIP)$, the problem is called a singly constrained nonlinear integer programming problem.

Multiply Constrained Nonlinear Integer Programming. When $m \geq 2$, problem $(NLIP)$ is called a multiply constrained nonlinear integer programming problem. It will be revealed later in this book that there exist essential differences between singly- and multiply-constrained nonlinear integer programming problems.

Convex Integer Programming. If all functions f and g_i, $i = 1, \ldots, m$, are convex on the convex hull of X in problem $(NLIP)$, problem $(NLIP)$ is called a convex integer programming problem. Note that the convexity is a sufficient condition to obtain a global solution to the continuous relaxation of $(NLIP)$.

Separable Integer Programming. When a function is of an additive form with respect to all of its variables, the function is called separable. In many situations, the objective function and the constraint functions of $(NLIP)$ are separable ([44][45]). A separable nonlinear integer programming formulation of $(NLIP)$ takes the following form:

$$
\begin{aligned}
(SIP) \qquad & \min\ f(x) = \sum_{j=1}^{n} f_j(x_j) \\
& \text{s.t. } g_i(x) = \sum_{j=1}^{n} g_{ij}(x_j) \leq b_i,\ i = 1, \ldots, m, \\
& x \in X = \{x \in \mathbb{Z}^n \mid l_j \leq x_j \leq u_j,\ j = 1, \ldots, n\}.
\end{aligned}
$$

Nonlinear resource allocation problem is a special case of (SIP) where all f_j's and g_{ij}'s are convex functions. In many nonlinear resource allocation problems, only a single constraint is presented in (SIP) with the form $g(x) = \sum_{j=1}^{n} x_j = N$ (see [106]). If all f_j's in (SIP) are of a quadratic form of x_j, $f_j(x_j) = q_j x_j^2$

+ c_jx_j, then (SIP) is classified as a *separable nonlinear integer programming problem with a quadratic objective function.*

Nonseparable Integer Programming. When at least one of the objective function and the constraint functions of $(NLIP)$ is nonseparable, problem $(NLIP)$ is a nonseparable integer programming problem. There are many real cases of nonlinear integer programming models where some of the functions involved are nonseparable. For example, in reliability optimization, the reliability function of an overall system is a multi-linear function of the reliability levels of all individual subsystems.

Nonlinear Knapsack Problem. If in a separable nonlinear integer programming problem (SIP), all f_j's are nonincreasing while all g_{ij}'s are nondecreasing, then the problem is called a *nonlinear knapsack problem*. If in a nonlinear knapsack problem, all f_j's are concave and all g_{ij}'s are linear, then the problem is called *concave knapsack problem.*

Monotone Nonlinear Integer Programming. If in a nonseparable integer programming problem $(NLIP)$, f is nonincreasing while all g_i's are nondecreasing, then the problem is called a *monotone nonlinear integer programming problem* or a *nonseparable knapsack problem.*

Nonlinear 0-1 Programming. When all the integer variables x_j's are restricted to be 0 or 1 in $(NLIP)$, problem $(NLIP)$ is called a *nonlinear 0-1 programming problem*. Theoretically, any integer programming problem can be reduced to a 0-1 integer programming problem ([31][92]). The methodologies for solving nonlinear 0-1 programming problems are inherently different from methods for other problem formulations.

Polynomial 0-1 Programming. If in a nonlinear 0-1 programming formulation, all functions f and g_i's are of a multi-linear polynomial form:

$$\sum_{j=1}^{n} c_j x_j + \sum_{k=1}^{K} q_k \prod_{i \in S(k)} x_i,$$

where $S(k)$ is an index set with $|S(k)| \geq 2$, then the problem is called a *polynomial 0-1 programming problem* or *pseudo-Boolean optimization problem*. In particular, if f and g_i's are of the following form of quadratic functions:

$$\sum_{j=1}^{n} c_j x_j + \sum_{1 \leq i < j \leq n} q_{ij} x_i x_j,$$

then the problem is called a *quadratic 0-1 programming* problem. The polynomial 0-1 programming and quadratic 0-1 programming problems have been extensively studied over the last thirty years (see [31][92][95]).

1.2 Examples of Applications

Integer programming has its root in various real applications. We present in this section some nonlinear integer programming models arising from different application areas.

1.2.1 Resource allocation in production planning

Optimal lot sizing is often sought in production planning in order to minimize the total cost via optimal resource allocation of labor and machine-hour among n different items. Let x_j denote the lot size of item j, D_j the total demand of item j, O_j the ordering cost per order of item j, h_j the holding cost per period of item j, c_j the storage requirement per item j, and C the total storage capacity. Then (i) the term $O_j D_j / x_j$ represents the total ordering cost of item j since item j is ordered D_j / x_j times; and (ii) the term $h_j x_j / 2$ gives the average holding cost of item j. The optimal lot size problem can be then formulated as

$$
\begin{aligned}
(OL) \qquad & \min \sum_{j=1}^{n} (O_j D_j / x_j + h_j x_j / 2) \\
& \text{s.t.} \sum_{j=1}^{n} c_j x_j \leq C \\
& x \in \mathbb{Z}^n_+,
\end{aligned}
$$

where $\mathbb{Z}^n_+$ denotes the set of integer points in $\mathbb{R}^n_+$. Notice that problem (OL) is a separable convex integer programming problem.

1.2.2 Portfolio selection

Portfolio selection is to seek a best allocation of wealth among a basket of securities. Quantifying the investment risk by the variance of the random return of the portfolio, the mean-variance formulation proposed by Markowitz [150] in the 1950s provides a fundamental basis for portfolio selection.

The trade practice often only allows trade of integer lots of stocks. Consider a market with n available securities where the purchasing of the securities is confined to integer number of lots. An investor with initial wealth W_0 seeks to improve his wealth status by investing his wealth into these n risky securities and into a risk-free asset (e.g., a bank account). Let X_i be the random return per lot of the i-th security ($i = 1, \ldots, n$) before deducting associated transaction costs. The mean and covariance of the returns are assumed to be known,

$$
\mu_i = \mathrm{E}(X_i), \text{ and } \sigma_{ij} = \mathrm{Cov}(X_i, X_j), \quad i, j = 1, \ldots, n.
$$

Let x_i be the integer number of lots the investor invests in the i-th security. Denote the decision vector in portfolio selection by $x = (x_1, \ldots, x_n)^T$. Then, the random return from holding securities is $P_s(x) = \sum_{i=1}^n x_i X_i$. The mean and variance of $P_s(x)$ are

$$s(x) = \mathrm{E}[P_s(x)] = \mathrm{E}[\sum_{i=1}^{n} x_i X_i] = \sum_{i=1}^{n} \mu_i x_i$$

and

$$V(x) = \mathrm{Var}(P_s(x)) = \mathrm{Var}[\sum_{i=1}^{n} x_i X_i] = \sum_{i=1}^{n}\sum_{j=1}^{n} x_i x_j \sigma_{ij} = x^T C x,$$

where $C = (\sigma_{ij})_{n\times n}$ is the covariance matrix. Let r be the interest rate of the risk-free asset. Assume that the same rate is applied when borrowing money from the risk-free asset. Let b_i be the current price of one lot of the i-th security. The balance $x_0 = W_0 - \sum_{i=1}^n b_i x_i$ is assumed to be deposited into the risk-free asset and rx_0 is the corresponding return. Note that a negative x_0 implies a debt from the risk-free asset.

The budget constraint of the investor is given by

$$b^T x \le W_0 + U_b$$

where $b = (b_1, \ldots, b_n)^T$ and U_b is the upper borrowing limit from the risk-free asset. Let $c(x) = \sum_{i=1}^n c_i(x_i)$ be the transaction cost associated with the portfolio decision $x = (x_1, \ldots, x_n)^T$. It is always assumed in the literature that each $c_i(\cdot)$ is a nondecreasing concave function.

The total expected return of portfolio decision x can be now summarized as:

$$R(x) = s(x) + rx_0 - \sum_{i=1}^{n} c_i(x_i) = \sum_{i=1}^{n} [(\mu_i - rb_i)x_i - c_i(x_i)] + rW_0.$$

Note that $R(x)$ is a convex function since each $c_i(x_i)$ is a concave function.

In most situations, an investor would like to invest his wealth only to a limited number of stocks. Thus a cardinality constraint is often necessary to be considered in portfolio selection,

$$\mathrm{supp}(x) \le K,$$

where $\mathrm{supp}(x)$ denotes the number of nonzero components in x and K is a given positive integer with $K \le n$.

By introducing n zero-one variables, y_i, $i = 1, \ldots, n$, a discrete-feature constrained mean-variance model can be formulated as follows for an investor who would like to minimize his investment risk while attaining an expected return level higher than a given value, ε, under transaction costs and a cardinality constraint:

$$
\begin{aligned}
(MV) \qquad & \min V(x) = x^T C x \\
& \text{s.t. } R(x) = \sum_{i=1}^{n} [(\mu_i - r b_i) x_i - c_i(x_i)] + r W_0 \geq \varepsilon, \\
& U(x) = b^T x \leq W_0 + U_b, \\
& \sum_{i=1}^{n} y_i \leq K, \\
& x \in X = \{x \in \mathbb{Z}^n \mid l_i y_i \leq x_i \leq u_i y_i,\ i = 1, 2, \ldots, n\}, \\
& y \in \{0, 1\}^n,
\end{aligned}
$$

where l_i and u_i are lower and upper bounds on purchasing the i-th security, respectively. A negative l_i implies that short selling is allowed. Upper bound u_i is either imposed by the investor or can be set as the largest integer number less than or equal to $\frac{W_0+U_b}{b_i}$.

Problem (MV) is of a nonlinear nonconvex integer programming formulation. Varying the value of ε, the efficient frontier in the mean-variance space can be traced out which provides a valuable decision-aid for investors.

1.2.3 Redundancy optimization in reliability networks

Systems reliability plays an important role in systems design, operation and management. Systems reliability can be improved by adding redundant components to subsystems.

Assume that there are n subsystems in a network. Let r_i $(0 < r_i < 1)$ be a fixed value of component reliability in the i-th subsystem and x_i represent the number of redundant (parallel) components in the i-th subsystem. Then, the reliability of the i-th subsystem, R_i, is given as follows:

$$R_i(x_i) = 1 - (1 - r_i)^{x_i}, \quad i = 1, \ldots, n.$$

Let $x = (x_1, \ldots, x_n)^T$ be the decision vector for the redundancy assignment. The overall system reliability, $R_s(x)$, is in general a nonlinear increasing function of $R_1(x_1)$, ..., $R_n(x_n)$. For example, if the network is the 7-link ARPA complex system given in Figure 1.1, then we have

$$R_s = R_6 R_7 + R_1 R_2 R_3 (Q_6 + R_6 Q_7) + R_1 R_4 R_7 Q_6 (Q_2 + R_2 Q_3),$$

where $Q_i = 1 - R_i$, $i = 1, \ldots, n$.

Determination of the optimal amount of redundancy among various subsystems under limited resource constraints leads to a nonseparable nonlinear

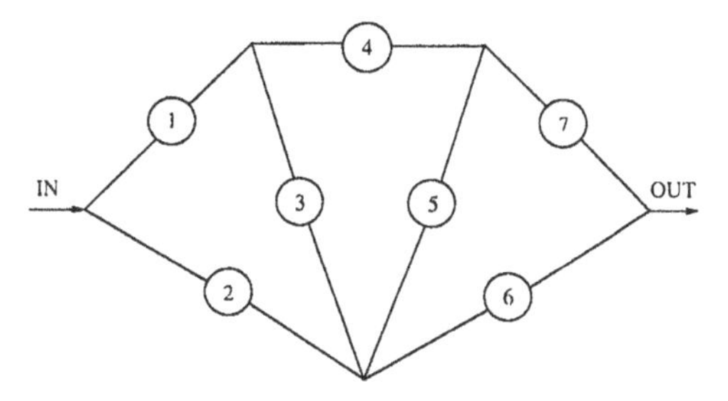

Figure 1.1. The ARPA complex system ($n = 7$).

integer programming problem,

$$(RELI) \qquad \max\ R_s(x) = f(R_1(x_1), \ldots, R_n(x_n))$$
$$\text{s.t. } g_i(x) = \sum_{j=1}^{n} g_{ij}(x_j) \le b_i,\ i = 1, \ldots, m,$$
$$x \in X = \{x \in \mathbb{Z}^n \mid 1 \le l_j \le x_j \le u_j,\ j = 1, \ldots, n\},$$

where $g_i(x)$ is the i-th resource consumed; b_i is the total available i-th resource, l_j and u_j are lower and upper integer bounds of x_j, respectively. The resource constraints often correspond to the constraints in cost, volume, and weight.

An inherent property in problem $(RELI)$ is that functions R_s and g_i's are strictly increasing with respect to each variable x_j. Thus, problem $(RELI)$ is a nonconvex nonseparable knapsack problem.

1.2.4 Chemical engineering

Chemical engineers often seek at the same time an optimal structure for a chemical process and the corresponding optimal operating parameters in order to satisfy given design specifications. This often results in nonlinear mixed integer programming formulations at the design stage. Figure 1.2 from [53] presents a superstructure of a chemical process in which all competitive alternative process configurations have been embedded. A zero-one variable y_i is attached to each process unit, while its final value will determine whether or not a process unit is in the final optimal configuration. The continuous variable x_{ij} represents a process parameter such as the flow rate of materials. The objective is to minimize a summation of fixed-charge costs and operation costs, while the constraints correspond to design specifications, topological considerations and physical conservation laws. The nonlinearities in the formulation are often caused by some intrinsic nonlinear input-output relationships of some process units.

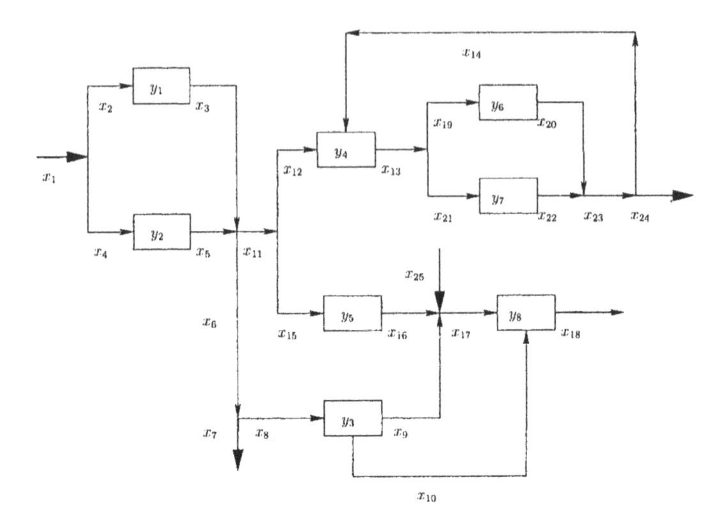

Figure 1.2. Superstructure of synthesization of a chemical process [53].

1.3 Difficulties and Challenges

There is no doubt that the first number system which mankind understood and utilized was the integer number system. More specifically, counting fingers (integer number) could be the first step humankind ever took in their long journey in advancing mathematics. In the later development of mathematics, however, focus has been primarily placed on the real number system, mainly due to powerful analytical tools that have been developed for mathematical study under the real number system. Compared to continuous optimization, discrete optimization presents more difficult research tasks, posting great long-standing challenges.

While convexity in continuous optimization guarantees that a local search offers a global solution, this is certainly not the case for discrete optimization or integer programming. To support this argument, let us consider a two-dimensional example:

EXAMPLE 1.1

$$\begin{aligned}&\min\ f(x) = (x-\bar{x})^T Q (x-\bar{x})\\ &\text{s.t. } x \in X = \{x \in \mathbb{Z}^2 \mid 0 \le x_1 \le 7,\ 0 \le x_2 \le 6\},\end{aligned}$$

where $\bar{x} = (3.1, 2.5)^T$ and $Q = \begin{pmatrix} 42.67 & -49.41 \\ -49.41 & 57.38 \end{pmatrix}$.

The global minimizer of this problem is $x_{global} = (6,5)^T$ with $f(x_{global}) = 1$. Since matrix Q is positive definite, there is only one continuous local minimizer $\bar{x} = (3.1, 2.5)^T$ which is also the global minimizer. An integer $x \in X$ is defined here as a discrete local minimizer of f over X if its function value is less than or equal to that of its 4 neighboring points $x \pm (1,0)^T$ and $x \pm (0,1)^T$ (if they are included in X). Table 1.1 lists the values of f for all integer points on X, where discrete local minima are labelled by "*" and the global minimum by "$\star$". It is clear that even for a convex function, there may exist multiple minima on an integer domain.

Table 1.1. Multiple discrete local minimizers of a convex function.

	x_1 =0	1	2	3	4	5	6	7
x_2 =0	3*	28	139	334	616	982	1434	1971
1	80	6*	18	115	297	565	918	1356
2	271	99	12	10*	93	262	516	856
3	578	306	120	20	4*	75	230	471
4	999	629	344	144	30	2*	58	200
5	1535	1066	682	384	171	43	1$\star$	44
6	2185	1617	1135	738	426	200	59	3*

Since set X is finite, people may think naturally to find out an optimal solution of $(NLIP)$ by enumerating all integer points in X, checking their feasibility and comparing their objective value. This approach of total enumeration is exact and probably efficient for small scale problems, but is definitely computationally infeasible when n is large. A modest size problem with 200 zero-one variables leads to 2^{200} or equivalently, 10^{60} possible candidates to compare. Increasing n from 200 to 201 will generate 10^{60} more points in X. Note here that the computational effort grows exponentially as the dimension n goes up.

Under certain conditions, the optimality of a continuous solution can be in general checked against Karush-Kuhn-Tucker conditions. Except for a very few cases, optimality conditions, however, have not been developed for integer programming problems. Thus, verifying the optimality of a solution essentially requires enumerating (implicitly) all the feasible solutions for an integer programming problem in most situations.

Not only is discrete optimization usually more difficult than its continuous counterpart, confining solutions to integers could bring an essential structural change in feasibility check of the problem. Fermat's Last Theorem is a wonderful example to demonstrate this point. It is obvious that there exist infinite real triples a, b and c such that $a^n + b^n = c^n$ is satisfied for any integer n greater than 2. If we confine solutions to integer numbers, however, the answer to the above statement becomes negative, astonishingly. Fermat's Last Theorem states that there do not exist nonzero integers a, b and c such that $a^n + b^n = c^n$ holds true for any integer n greater than 2. This most famous question in number theory has troubled and excited many scholars since 1637 and has only been resolved by Andrew Wiles in 1994.

It has been widely accepted that an efficient solution algorithm should have its theoretical running time (time complexity) bounded from above by $O(l^k)$ where k is a constant and l the problem size which is measured by the length of the input data to specify the problem instance in binary representation. Algorithms with a polynomial-time complexity exist for many optimization problems, such as linear program and maximum flow problem. The class of problems for which algorithms with polynomial-time complexity exist is denoted by class P. For many other optimization problems, such as linear integer program and the set covering problem, only algorithms with exponential time complexity have been developed up to now.

A problem is called a *decision* problem if it seeks a Yes or No answer. Satisfiability problem is a decision problem. An optimization problem can be always transformed into a corresponding decision problem. The set of all decision problems which can be solved in polynomial time by nondeterministic algorithms is denoted as class NP. It follows from the definition of NP that $P \subseteq NP$. S. Cook first introduced the concept of NP-completeness for a set of problems in NP and proved that the satisfiability problem is NP-complete. By the definition of NP-completeness, if any NP-complete problem belongs to P, then all NP-complete problems belong to P. The common belief is an almost-sure impossibility of this occurrence as the intention behind the notion of NP-completeness is to strongly suggest that there does not exist polynomial algorithms for NP-complete problems.

An optimization problem A is defined to be *NP-hard*, if all problems in NP can be transformed into A with a polynomial time complexity. It has been proved in the literature that 0-1 linear knapsack problem, quadratic 0-1 integer program, and redundancy optimization for series-parallel reliability networks are all NP-hard. Thus, most nonlinear integer programming problems investigated in this book are not in class P and there is almost no chance to develop solution algorithms with a polynomial-time complexity for this kind of problems.

1.4 Organization of the Book

The first part of the book, Chapters 2 to 5, provides theoretical foundation of nonlinear integer programming. Chapter 2 discusses general solution concepts for integer programming, including optimality, relaxation, and implicit enumeration schemes. Chapters 3 to 5 are devoted to the study of duality theory, including both Lagrangian duality and surrogate duality. Nonlinear Lagrangian theory, discussed in Chapter 5, has been developed to achieve the strong duality and to guarantee the success of the dual search.

The remaining chapters in the book, except Chapter 14, deal with solution methodologies for different classes of nonlinear integer programming problems. The treatment of the development evolves according to a sequence of nonlinear knapsack problems (Chapter 6), separable nonlinear integer programming (Chapter 7), nonlinear integer programming with a quadratic objective function (Chapter 8), nonseparable nonlinear integer programming (Chapter 9), polynomial 0-1 programming (Chapters 10–12), and mixed integer programming (Chapter 13). Chapter 14 discusses the global descent method which searches for a (global) optimal solution of a general nonlinear integer programming problem from among its local minima.

1.5 Notes

The reader may refer to [34][106] for further discussions about integer programming formulations in resource allocation. Further investigation about integer programming formulations for portfolio selection can be found in [140]. For more applications of (mixed) integer programming models in reliability optimization, see [217]. The interested reader may refer to [60] for more sophisticated models of mixed-integer nonlinear programming in chemical engineering. Detailed discussions about computational complexity can be found in [191].

Chapter 2

OPTIMALITY, RELAXATION AND GENERAL SOLUTION PROCEDURES

In this chapter, we discuss some fundamental concepts and basic solution frameworks for the following general nonlinear integer programming problem:

$$
\begin{aligned}
(P) \qquad & \min \ f(x) \\
& \text{s.t. } g_i(x) \le b_i, \ i = 1, \ldots, m, \\
& \qquad h_k(x) = c_k, \ k = 1, \ldots, l, \\
& \qquad x \in X \subseteq \mathbb{Z}^n,
\end{aligned}
$$

where all f, g_i's and h_k's are real-valued functions defined on $\mathbb{R}^n$ and $\mathbb{Z}^n$ is the set of integer points in $\mathbb{R}^n$.

A solution $\tilde{x} \in X$ is said to be a *feasible* solution of (P) if $g_i(\tilde{x}) \le b_i$, for all $i = 1, \ldots, m$, and $h_k(\tilde{x}) = c_k$, for all $k = 1, \ldots, l$. A feasible solution x^* is said to be an *optimal* solution of (P) if $f(x^*) \le f(x)$ for any feasible solution x of (P).

This chapter is organized as follows: We introduce the concept of an optimality condition using bounds in Section 2.1. In Section 2.2, we present a general framework of partial enumeration methods, first a general branch-and-bound method, then a backtrack partial enumeration method for 0-1 programming and its implementation in 0-1 linear integer programming. In Section 2.3, we introduce the concept of relaxation and discuss the relationship between Lagrangian relaxation and continuous relaxation. We study the relationship between continuous and integer optimal solutions of nonlinear integer programming problems in Section 2.4. In Section 2.5, we discuss how to convert a general constrained nonlinear integer programming problem into an unconstrained one by using an exact penalty function. Finally, we present in Section 2.6 optimality conditions for binary quadratic problems.

2.1 Optimality Condition via Bounds

An essential task in designing any solution algorithm for (P) is to derive an optimal condition or a stopping criterion to terminate the algorithm, i.e., to judge if the current solution is optimal to (P) or to conclude that there is no feasible solution to (P). Except for very few special cases, such as unconstrained quadratic binary problems (see Section 2.6), it is difficult to obtain an explicit optimality condition for problem (P). As in linear integer program and other discrete optimization problems, however, optimality of the nonlinear integer programming problem (P) can be verified through the convergence of a sequence of upper bounds and a sequence of lower bounds of the objective function. Let f^* be the optimal value of (P). Suppose that an algorithm generates a nonincreasing sequence of upper bounds

$$\overline{f}_1 \geq \overline{f}_2 \geq \cdots \geq \overline{f}_k \geq \cdots \geq f^*$$

and a nondecreasing sequence of lower bounds

$$\underline{f}_1 \leq \underline{f}_2 \leq \cdots \leq \underline{f}_k \leq \cdots \leq f^*,$$

where $\underline{f}_k$ and $\overline{f}_k$ are the lower and upper bounds of f^* generated at the k-th iteration, respectively. If $\overline{f}_k - \underline{f}_k \leq \epsilon$ holds for some small $\epsilon \geq 0$ at the k-th iteration, then the following is evident:

$$f^* - \epsilon \leq \underline{f}_k \leq f^*.$$

Notice that an upper bound of f^* is often associated with a feasible solution x^k to (P), since $f(x^k) \geq f^*$. A lower bound of f^* is usually achieved by solving a relaxation problem of (P) which we will discuss in later sections of this chapter. A feasible solution x^k is called an *ϵ-approximate* solution to (P) when $f(x^k) = \overline{f}_k$ and $\overline{f}_k - \underline{f}_k \leq \epsilon > 0$.

We have the following theorem.

THEOREM 2.1 *Suppose that $\{\overline{f}_k\}$ and $\{\underline{f}_k\}$ are the sequences of upper bounds and lower bounds of f^*, respectively. If $\overline{f}_k - \underline{f}_k = 0$ for some k and x^k is a feasible solution to (P) with $f(x^k) = \overline{f}_k$, then x^k is an optimal solution to (P).*

The key question is how to generate two converging sequences of upper and lower bounds of f^* in a solution process. Continuous relaxation, Lagrangian relaxation (Chapter 3) and surrogate relaxation (Chapter 4) are three typical ways of getting a lower bound of an integer programming problem. The upper bound of f^* is usually obtained via feasible solutions of problem (P).

2.2 Partial Enumeration

Although the approach of total enumeration is infeasible for large-scale integer programming problems, the idea of partial enumeration is still attractive if there is a guarantee of identifying an optimal solution of (P) without checking explicitly all the points in X. The efficiency of any partial enumeration scheme can be measured by the average reduction of the search space of integer solutions to be examined in the execution of the solution algorithm. The branch-and-bound method is one of the most widely used partial enumeration schemes.

2.2.1 Outline of the general branch-and-bound method

The branch-and-bound method has been widely adopted as a basic partial enumeration strategy for discrete optimization. In particular, it is a successful and robust method for linear integer programming when combined with linear programming techniques. The basic idea behind the branch-and-bound method is an implicit enumeration scheme that systematically discards non-promising points in X that are hopeless in achieving optimality for (P). The same idea can be applied to nonlinear integer programming problem (P). To partition the search space, we divide the integer set X into p (≥ 2) subsets: $X_1, \ldots, X_p$. A *subproblem* at *node* i, $(P(X_i))$, $i = 1, \ldots, p$, is formed from (P) by replacing X with X_i. One or more subproblems are selected from the subproblem list. For each selected node, a lower bound LB_i of the optimal value of subproblem $(P(X_i))$ is estimated. If LB_i is greater than or equal to the function value of the *incumbent*, the best feasible solution found, then the subproblem $(P(X_i))$ is removed or *fathomed* from further consideration. Otherwise, problem $(P(X_i))$ is kept in the subproblem list. The incumbent is updated whenever a better feasible solution is found. One of the unfathomed nodes, $(P(X_i))$, is selected and X_i is further divided or *branched* into smaller subsets. The process is repeated until there is no subproblem left in the list. It is convenient to use a node-tree structure to describe a branch-and-bound method in which a *node* stores the information necessary for describing and solving the corresponding subproblem. We describe the general branch-and-bound method in details as follows.

ALGORITHM 2.1 (GENERAL BRANCH-AND-BOUND METHOD FOR (P))

Step 0 (Initialization). Set the subproblem list $L = \{P(X)\}$. Set an initial feasible solution as the incumbent x^* and $v^* = f(x^*)$. If there is no feasible solution available, then set $v^* = +\infty$.

Step 1 (Node Selection). If $L = \emptyset$, stop and x^* is the optimal solution to (P). Otherwise, choose one or more nodes from L. Denote the set of k selected nodes by $L^s = \{P(X_1), \ldots, P(X_k)\}$. Let $L := L \setminus L^s$. Set $i = 1$.

Step 2 (Bounding). Compute a lower bound LB_i of subproblem $(P(X_i))$. Set $LB_i = +\infty$ if $(P(X_i))$ is infeasible. If $LB_i \geq v^*$, go to Step 5.

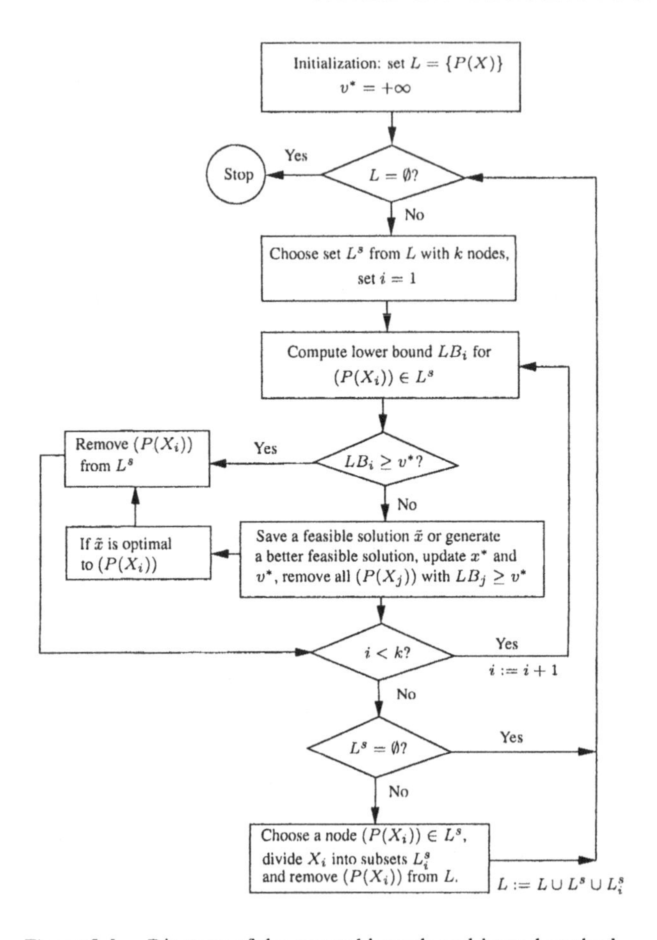

Figure 2.1. Diagram of the general branch-and-bound method.

Step 3 (Feasible solution). Save the best feasible solution found in Step 2 or generate a better feasible solution when possible by certain heuristic method. Update the incumbent x^* and v^* when needed. Remove from L^s all $(P(X_j))$ satisfying $LB_j \geq v^*$, $1 \leq j \leq i$. If $i < k$, set $i := i + 1$ and return to Step 2. Otherwise, go to Step 4.

Step 4 (Branching). If $L^s = \emptyset$, go to Step 1. Otherwise, choose a node $(P(X_i))$ from L^s. Further divide X_i into smaller subsets: $L_i^s = \{X_i^1, \ldots, X_i^p\}$. Remove $(P(X_i))$ from L^s and set $L := L \cup L^s \cup L_i^s$. Go to Step 1.

Step 5 (Fathoming). Remove $(P(X_i))$ from L^s. If $i < k$, set $i := i + 1$ and return to Step 2. Otherwise, go to Step 4.

Figure 2.1 illustrates the diagram of Algorithm 2.1.

THEOREM 2.2 *Algorithm 2.1 stops at an optimal solution to (P) within a finite number of iterations.*

Proof. Note that the fathoming procedure, either in Step 3 or Step 5 of the algorithm, will not remove any feasible solution of (P) better than the incumbent. Notice that X is finite. Thus only a finite number of branching steps can be executed. At an extreme, when X_i is a singleton, either $(P(X_i))$ is infeasible or an optimal solution to $(P(X_i))$ can be found, thus $(P(X_i))$ being fathomed in Step 5. Within a finite number of iterations, L will become empty and the optimality of the incumbent is evident. □

One key issue to develop an efficient branch-and-bound method is to get a good (high) lower bound LB_i generated by the bounding procedure in Step 2. The better the lower bound, the more subproblems can be fathomed in Steps 3 and 5 and the faster the algorithm converges. There is a trade-off, however, between the quality of the lower bounds and the associated computational efforts. For nonlinear integer programming problem (P), *continuous relaxation* and *Lagrangian relaxation* are two commonly used methods for generating lower bounds in Step 2.

2.2.2 The back-track scheme

The back-track scheme was proposed originally as a systematic way to thread a maze. Known by its different names, the back-track scheme was rediscovered from time to time in different fields. Especially, it was adopted as an efficient procedure for implicit enumeration in solving many kinds of combinatorial problems. We discuss the back-track scheme in this subsection as a powerful partial enumeration scheme for 0-1 programming problems.

Let's consider the following general nonlinear 0-1 integer programming problem:

$$(0\text{-}1P) \qquad \begin{aligned} &\min\ f(x) \\ &\text{s.t. } g_i(x) \le b_i,\ i = 1, 2, \ldots, m, \\ &\qquad x \in X = \{0, 1\}^n, \end{aligned}$$

where f is assumed to be monotonically increasing, i.e., $f(x) \ge f(y)$ if $x \ge y$. It is clear that there are at most 2^n possible candidates to be considered for achieving an optimality of problem $(0\text{-}1P)$. However, an efficient solution algorithm should be devised such that, in most situations, only a significantly small portion of the 2^n possible solutions needs to be explicitly enumerated. These possible solutions should rather be implicitly enumerated group by group.

To group the 2^n solutions, we define a *partial solution* to be an assignment of binary values to a subset of the n decision variables. Let $N = \{1, \ldots, n\}$.

At iteration t, let $J_t = \{j \text{ or } -j \mid j \in I_t \subseteq N\}$ denote the partial solution with $x_j = 1$ when $j \in J_t$ and $x_j = 0$ when $-j \in J_t$, where I_t is the index set of J_t. Only one of j or $-j$ could be included in J_t. Any variable x_j whose index j is not included in I_t is defined to be *free*. A completion of J_t is defined as a solution determined by J_t together with a binary specification of the free variables. It is clear that a k-element partial solution could determine 2^{n-k} different completions as a group. When all free variables are set to be zero, the completion is termed *typical*. Since the objective function f in problem $(0\text{-}1P)$ is monotonically increasing, the typical completion of J_t has the minimum objective function value among all completions of J_t. For example, $J_t = \{3, 5, -2\}$ with $n = 5$ specifies a partial solution of $x_3 = 1$, $x_5 = 1$ and $x_2 = 0$. J_t has two free variables (x_1 and x_4) and four possible completions, among which the one with $x_1 = x_4 = 0$ is the typical completion.

After a partial solution J_t is generated at iteration t, we need to determine if its corresponding solution group (completions) could include an optimal solution to (P). In the following two situations, J_t can be *fathomed.*

Case (i): If the typical completion of J_t is feasible in $(0\text{-}1P)$, J_t can be fathomed in this case (after updating the incumbent if the typical completion of J_t has an objective value less than the one of the incumbent), since no other completion of J_t could generate an objective value of $(0\text{-}1P)$ smaller than the objective value of the typical completion as f is monotonically increasing.

Case (ii): If the typical completion of J_t has an objective value larger than or equal to the one of the incumbent, J_t can be fathomed in this case since no other completion of J_t, including the typical completion, could do better than the incumbent.

There is only one remaining situation which fits neither Case (i) nor Case (ii): the typical completion of J_t is infeasible in $(0\text{-}1P)$ and has an objective value less than that of the incumbent. In this situation, we augment J_t by assigning values to some free variables of J_t according to some rules such that a new partial solution is generated for further fathoming.

The back-track scheme, as a systematic method, is designed to implicitly enumerate all solutions without generating any redundant partial solutions. To ensure having a new non-redundant partial solution when a partial solution is fathomed, at least one element of the partial solution has to be changed to its complement. When the chosen element is replaced by its complement, it is marked by an underline in order to prevent a turning back in the solution process. This process repeats and terminates when there is no non-underlined component in the partial solution, which implies that all possible solutions are implicitly enumerated. In the back-track procedure, we always locate in a partial solution the right-most element which is not underlined. We replace this right-most non-underlined element by its underlined complement and delete all elements to its right. If no non-underlined element exists in the partial solution,

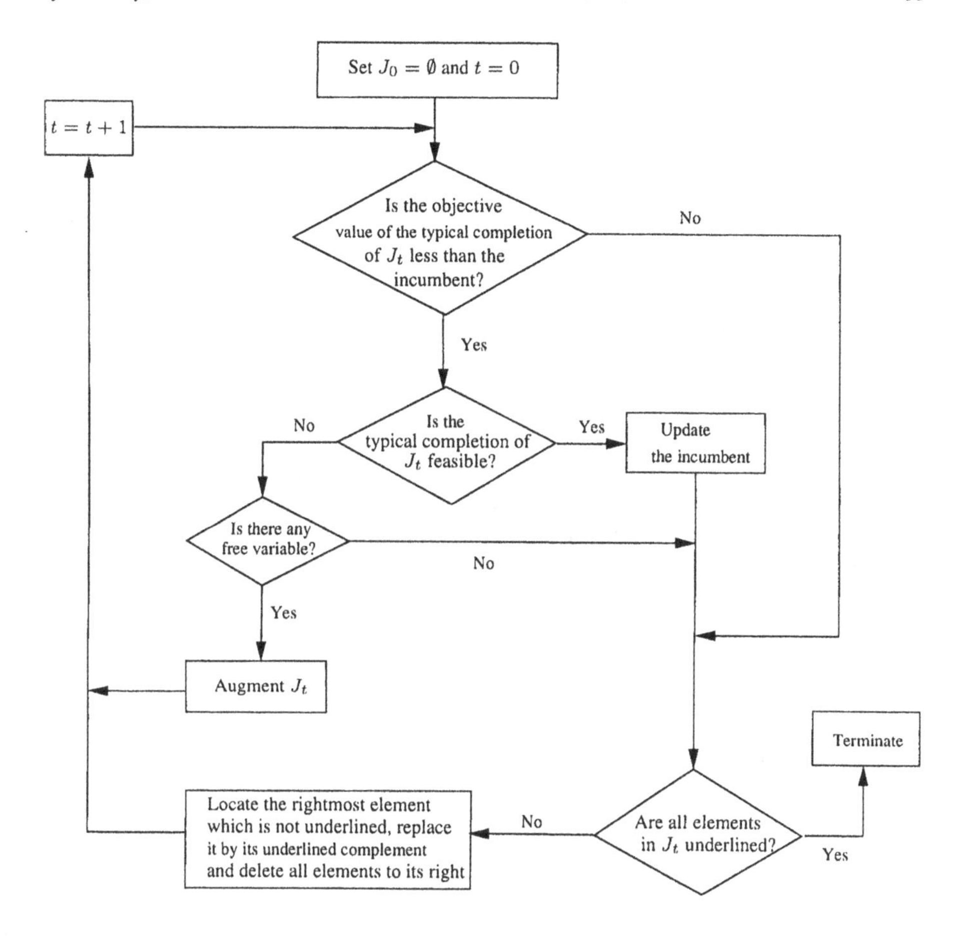

Figure 2.2. Diagram of the back-track scheme.

we can claim that all 2^n solutions have been implicitly enumerated and the solution procedure terminates. For example, if $J_t = \{3, 5, \underline{-2}\}$ is fathomed at iteration t, the new partial solution J_{t+1} is $\{3, \underline{-5}\}$ in the back-track procedure.

A diagram of the general solution framework for the back-track scheme is given in Figure 2.2. Notice that for different types of 0-1 programming problems, such as 0-1 linear programming problems and polynomial 0-1 programming problems, different fathoming and augmenting rules could be designed to explore special structures of the problems.

THEOREM 2.3 *The back-track scheme leads to a non-redundant sequence of partial solutions which terminates only when all 2^n solutions have been (implicitly) enumerated.*

Theorem 2.3 indicates that the back-track scheme is a finite algorithm. If $(0\text{-}1P)$ is feasible, the optimal solution will be in store of the incumbent at termination of the procedure.

Although we start with $J_0 = \emptyset$ in Figure 2.2, J_0 could essentially be any other partial solution without an underlined element. In addition, in the process of augmentation, we can augment more than one free variable on the right of J_t.

2.2.2.1 The additive algorithm for solving linear 0-1 programming problems

In 1965, Balas proposed an implicit enumeration method to directly solve linear zero-one programming problems [7]. Due to the fact that only addition is required as an arithmetic operation in the solution procedure, the solution procedure is called as the additive algorithm. One advantage of the additive algorithm is that there is no roundoff error. The additive algorithm is considered to be fundamental for the later development of various implicit enumeration methods for integer programming problems.

In this subsection we consider the following linear zero-one programming problem:

$$(0\text{-}1LP) \qquad \min\ f(x) = \sum_{j=1}^{n} c_j x_j,$$
$$\text{s.t. } g_i(x) = \sum_{j=1}^{n} a_{ij} x_j \le b_i,\ i \in M = \{1, 2, \ldots, m\},$$
$$x_j \in \{0, 1\},\ j \in N = \{1, 2, \ldots, n\}.$$

Without loss of generality, we assume that $c_j \ge 0$ for all $j \in N$. By introducing m slack variables, problem $(0\text{-}1LP)$ can be rewritten as follows,

$$(0\text{-}1LP_s) \qquad \min\ f(x) = \sum_{j=1}^{n} c_j x_j,$$
$$\text{s.t. } g_i(x) = \sum_{j=1}^{n} a_{ij} x_j + y_i = b_i,\ i \in M,$$
$$x_j \in \{0, 1\},\ j \in N,$$
$$y_i \ge 0,\ i \in M,$$

where y_i, $i \in M$, are nonnegative slack variables.

The additive algorithm starts with a partial solution $J_0 = \emptyset$ and an upper bound of the minimum value of the objective function, $f^* = \sum_{j=1}^{n} c_j$. At iteration t, the partial solution is J_t. Let x^t be the typical completion of J_t and $y^t \in \mathbb{R}^m$ be the corresponding vector of slack variables.

When $f(x^t) \geq f^*$, the partial solution J_t can be fathomed, no matter if x^t is feasible or not in (0-1LP), since no completion of J_t will give an objective value less than f^*. The algorithm proceeds then to the back-track procedure.

When $f(x^t) < f^*$ and $y^t \geq 0$, x^t is a better feasible solution. We update the incumbent by setting $f^* = f(x^t)$. The partial solution J_t can be fathomed, since no other completions of J_t can yield an objective value less than $f(x^t)$. The algorithm proceeds then to the back-track procedure.

When $f(x^t) < f^*$ and $y^t \ngeq 0$, the typical completion of J_t, x^t, is infeasible in (0-1LP) and we need to augment J_t with at least one free variable (if any). The principle of augmentation is to pursue a reduction in both the objective value and the degree of infeasibility. To identify a candidate of augmentation from among all free variables, a set T^t is constructed as follows,

$$\begin{aligned} T^t \; = \; & \{j \in N \setminus I_t \mid f(x^t) + c_j < f^* \text{ and there exists } i \in M \text{ such that} \\ & a_{ij} < 0 \text{ and } y_i^t < 0\}. \end{aligned}$$

It is clear that only those x_j's with j in T^t need to be considered as candidates to augment J_t on the right because assigning 1 to some free variable not in T^t would either lead to a larger lower objective value than f^* or increase the degree of the infeasibility of x^t. If T^t is empty, we know that there does not exist a feasible completion of J_t which can do better than the incumbent, and J_t is thus fathomed.

When T^t is not empty, we check further the following inequality for those $i \in M$ with $y_i^t < 0$:

$$y_i^t - \sum_{j \in T^t} \min\{0, a_{ij}\} \geq 0. \tag{2.2.1}$$

If (2.2.1) does not hold for any $i \in M$ with $y_i^t < 0$, then the slack variable of the i-th constraint will remain negative for whatever solution augmented from J_t by assigning 1 to some variables in T^t. In other words, it is impossible for J_t to have a feasible completion which can be adopted to improve the current incumbent value and thereby J_t is fathomed.

If (2.2.1) holds for all $i \in M$ with $y_i^t < 0$, we could augment J_t on the right. A suitable criterion in selecting a free variable from T^t is to use the following formulation:

$$j^t = \arg\max_{j \in T^t} \sum_{i=1}^{m} \min\{y_i^t - a_{ij}, 0\}. \tag{2.2.2}$$

If j^t is chosen according to the above formulation, $J_{t+1} = J_t \cup \{j^t\}$ has the "least" degree of the violation of the constraints.

The back-track scheme can be used to clearly interpret the additive algorithm of Balas and has been adopted to simplify the additive algorithm of Balas such that not only the solution logic in the algorithm becomes much clearer, but also the memory requirement of computation is significantly reduced. Based on the back-track scheme, the additive algorithm of Balas can be explained via the following flow chart in Figure 2.3.

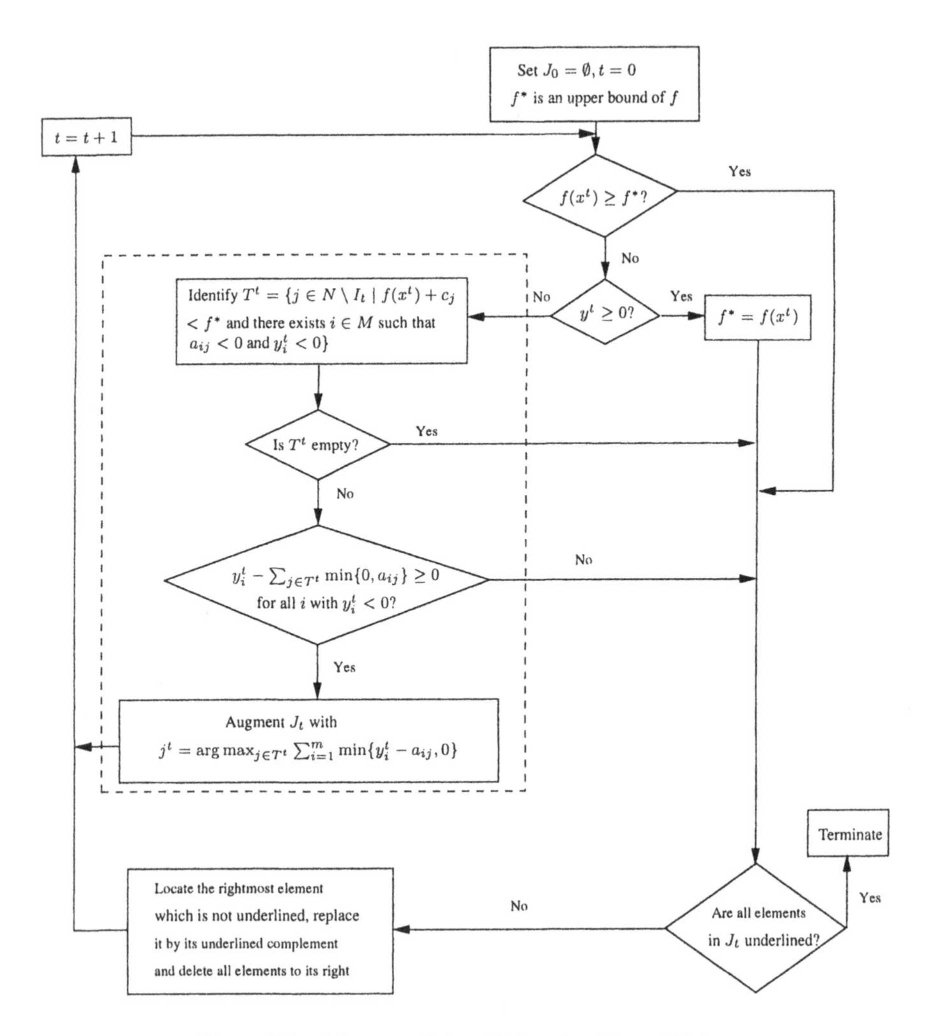

Figure 2.3. Diagram of the additive algorithm of Balas.

The following linear 0-1 programming problem serves as an example to illustrate the back-track scheme in the additive algorithm.

EXAMPLE 2.1

$$\begin{aligned}
\min\ & 5x_1 + 7x_2 + 10x_3 + 3x_4 + x_5 \\
\text{s.t.}\ & -x_1 + 3x_2 - 5x_3 - x_4 + 4x_5 \le -2, \\
& 2x_1 - 6x_2 + 3x_3 + 2x_4 - 2x_5 \le 0, \\
& x_2 - 2x_3 + x_4 + x_5 \le -1, \\
& x_1, x_2, x_3, x_4, x_5 \in \{0, 1\}.
\end{aligned}$$

Adding slack variables yields the following standard formulation,

$$\begin{aligned}
\min\ & 5x_1 + 7x_2 + 10x_3 + 3x_4 + x_5 \\
\text{s.t.}\ & -x_1 + 3x_2 - 5x_3 - x_4 + 4x_5 + y_1 = -2, \\
& 2x_1 - 6x_2 + 3x_3 + 2x_4 - 2x_5 + y_2 = 0, \\
& x_2 - 2x_3 + x_4 + x_5 + y_3 = -1, \\
& x_1, x_2, x_3, x_4, x_5 \in \{0, 1\},\ y_1, y_2, y_3 \ge 0.
\end{aligned}$$

Initial Iteration

Step 0. Set $J_0 = \emptyset$ and $f^* = \sum_{j=1}^{5} c_j = 26$.

Iteration 1 ($t = 0$)

Step 1. $x^0 = (0,0,0,0,0)^T$, $f(x^0) = 0 < f^* = 26$ and $y^0 = (-2, 0, -1)^T \not\geq 0 \Rightarrow$ Augmenting J_0.

Step 2. Notice that all x_1, x_2, x_3, x_4, x_5 are free variables and $T^0 = \{1, 3, 4\}$.

Step 3. For $i = 1$, $y_1^0 - \sum_{j \in T^0} \min\{0, a_{1j}\} = -2 - (-1 - 5 - 1) = 5 \ge 0$; For $i = 3$, $y_3^0 - \sum_{j \in T^0} \min\{0, a_{3j}\} = -1 - (-2) = 1 \ge 0$.

Step 4. $j^0 = \arg\max_{j \in T^0}\{\sum_{i=1}^{3} \min(y_i^0 - a_{ij}, 0)\} = \arg\max\{-1-2-1, -3, -1-2-2\} = 3 \Rightarrow J_1 = \{3\}$.

Iteration 2 ($t = 1$)

Step 1. $x^1 = (0,0,1,0,0)^T$, $f(x^1) = 10 < f^* = 26$ and $y^1 = (3, -3, 1)^T \not\geq 0 \Rightarrow$ Augmenting J_1.

Step 2. Notice x_1, x_2, x_4, x_5 are free variables and $T^1 = \{2, 5\}$.

Step 3. For $i = 2$, $y_2^1 - \sum_{j \in T^1} \min(0, a_{2j}) = -3 - (-6 - 2) = 5 \ge 0$.

Step 4. $j^1 = \arg\max_{j \in T^1}\{\sum_{i=1}^{3} \min(y_i^1 - a_{ij}, 0)\} = \arg\max\{0, -1-1\} = 2$. Thus $J_2 = \{3, 2\}$.

Iteration 3 ($t = 2$)

Step 1. $x^2 = (0, 1, 1, 0, 0)^T$, $f(x^2) = 17 < f^* = 26$ and $y^2 = (0, 3, 0)^T \geq 0$ $\Rightarrow$ Record $x^* = \{0, 1, 1, 0, 0\}$, set $f^* = 17$ and J_2 is fathomed.

Step 2. Back track and get $J_3 = \{3, \underline{-2}\}$.

Iteration 4 ($t = 3$)

Step 1. $x^3 = (0, 0, 1, 0, 0)^T$, $f(x^3) = 10 < f^* = 17$ and $y^3 = (3, -3, 1)^T \ngeq 0 \Rightarrow$ Augmenting J_3.

Step 2. Notice that x_1, x_4, x_5 are free variables and $T^3 = \{5\}$.

Step 3. For $i = 2$, $y_2^3 - \sum_{j \in T^3} \min(0, a_{2j}) = -3 - (-2) = -1 < 0 \Rightarrow$ J_3 is fathomed.

Step 4. Back track and get $J_4 = \{\underline{-3}\}$.

Iteration 5 ($t = 4$)

Step 1. $x^4 = (0, 0, 0, 0, 0)^T$, $f(x^4) = 0 < f^* = 17$ and $y^4 = (-2, 0, -1)^T \ngeq 0 \Rightarrow$ Augmenting J_4.

Step 2. Notice that x_1, x_2, x_4, x_5 are free variables and $T^4 = \{1, 4\}$.

Step 3. For $i = 3$, $y_3^4 - \sum_{j \in T^4} \min(0, a_{3j}) = -1 - (0) = -1 < 0 \Rightarrow J_4$ is fathomed.

Step 4. No element in J_4 is not underlined. $\Rightarrow$ The algorithm terminates with an optimal solution $x^* = \{0, 1, 1, 0, 0\}$ and $f^* = 17$.

2.3 Continuous Relaxation and Lagrangian Relaxation

Let $v(Q)$ denote the optimal value of problem (Q). A problem $(R(\xi))$ with a parameter ξ is called a relaxation of the primal problem (P) if $v(R(\xi)) \leq v(P)$ holds for all possible values of ξ. In other words, solving a relaxation problem offers a lower bound of the optimal value of the primal problem. The dual problem, (D), is formulated to search for an optimal parameter, ξ^*, such that the duality gap of $v(P) - v(R(\xi))$ is minimized at $\xi = \xi^*$. The quality of a relaxation should be thus judged by two measures. The first measure is how easier the relaxation problem can be solved when compared to the primal problem. The second measure is how tight the lower bound can be, in other words, how small the duality gap can be reduced to.

2.3.1 Continuous relaxation

The continuous relaxation of (P) can be expressed as follows:

$$
\begin{aligned}
(\overline{P}) \qquad & \min\ f(x) \\
& \text{s.t. } g_i(x) \leq b_i,\ i = 1, \ldots, m, \\
& \qquad h_k(x) = c_k,\ k = 1, \ldots, l, \\
& \qquad x \in conv(X),
\end{aligned}
$$

where $conv(X)$ is the convex hull of the integer set X. Problem $(\overline{P})$ is a general constrained nonlinear programming problem. Since $X \subset conv(X)$, it holds $v(\overline{P}) \leq f^*$. Generally speaking, a continuous relaxation problem is easier to solve than the primal nonlinear integer programming problem.

When all f and g_i's are convex and all h_k's are linear in $(\overline{P})$, the continuous relaxation problem is convex. For continuous convex minimization problems, many efficient solution methods have been developed over the last four decades. Below is a list of some of the well-known solution methods for convex constrained optimization (see e.g. [13][58][148]):

- Penalty Methods;
- Successive Quadratic Programming (SQP) methods;
- Feasible Direction Methods:
  - Wolfe's Reduced Gradient Method for linearly constrained problems;
  - The Generalized Reduced Gradient Method for nonlinearly constrained problems;
  - Rosen's Gradient Projection Methods.
- Trust Region Methods.

There does not exist a general-purpose solution method, however, for searching for a global solution for nonconvex constrained optimization problems. Nevertheless, there are several solution algorithms developed in *global optimization* for nonconvex problems with certain special structures, for example, outer approximation methods for concave minimization with linear constraints ([105] [174]) and convexification methods for monotone optimization problems ([136] [207]).

2.3.2 Lagrangian relaxation

Define the following Lagrangian function of (P) for $\lambda \in \mathbb{R}^m_+$ and $\mu \in \mathbb{R}^l$:

$$L(x, \lambda, \mu) = f(x) + \sum_{i=1}^{m} \lambda_i (g_i(x) - b_i) + \sum_{k=1}^{l} \mu_k (h_k(x) - c_k).$$

The Lagrangian relaxation problem of (P) is posted as follows:

$$(L_{\lambda,\mu}) \quad d(\lambda, \mu) = \min_{x \in X} L(x, \lambda, \mu). \tag{2.3.1}$$

Denote the feasible region of (P) by

$$S = \{x \in X \mid g_i(x) \leq b_i, i = 1, \ldots, m, \ h_k(x) = c_k, \ k = 1, \ldots, l\}.$$

The following *weak duality* relation will be derived in the next chapter:

$$d(\lambda, \mu) \leq f(x), \quad \forall \lambda \in \mathbb{R}^m_+, \ \mu \in \mathbb{R}^l, \ x \in S. \tag{2.3.2}$$

This ensures that solving $(L_{\lambda,\mu})$ gives a lower bound of f^*, the optimal value of (P). The dual problem of (P) is to search for the best lower bound provided by the Lagrangian relaxation:

$$(D) \quad \max_{\lambda \in \mathbb{R}^m_+, \mu \in \mathbb{R}^l} d(\lambda, \mu). \tag{2.3.3}$$

2.3.3 Continuous bound versus Lagrangian bound

We first establish a relationship between the continuous bound and the Lagrangian bound in convex cases of (P). We need the following assumption.

ASSUMPTION 2.1 *Functions f and g_i ($i = 1, \ldots, m$) are convex, functions h_k ($k = 1, \ldots, l$) are linear, and certain constraint qualification holds for $(\overline{P})$.*

One sufficient condition to ensure the satisfaction of the constraint qualification in Assumption 2.1 is that the gradients of the active inequality constraints and that of the equality constraints at the optimal solution to $(\overline{P})$ are linearly independent.

The following theorem shows that the Lagrangian bound for convex integer programming problem (P) is at least as good as the bound obtained by the continuous relaxation.

THEOREM 2.4 *Under Assumption 2.1, it holds $v(D) \geq v(\overline{P})$.*

Proof. Since $X \subseteq conv(X)$, we have

$$\begin{aligned} v(D) &= \max_{\lambda \in \mathbb{R}^m_+, \mu \in \mathbb{R}^l} \min_{x \in X} L(x, \lambda, \mu) \\ &\geq \max_{\lambda \in \mathbb{R}^m_+, \mu \in \mathbb{R}^l} \min_{x \in conv(X)} L(x, \lambda, \mu) \\ &= v(\overline{P}). \end{aligned}$$

The last equality is due to the strong duality theorem of convex programming under Assumption 2.1. □

The tightness of the Lagrangian bound has been also witnessed in many combinatorial optimization problems. In the case of nonlinear integer programming, to compute the Lagrangian bound $v(D)$, one has to solve the Lagrangian relaxation problem (2.3.1). When all functions f, g_i's and h_k's and set X are separable, the Lagrangian relaxation problem (2.3.1) can be solved efficiently via decomposition which we are going to discuss in Chapter 3. When some of the functions f, g_i's and h_k's are nonseparable, problem (2.3.1) is not easier to solve than the original problem (P). Nevertheless, the Lagrangian bound of a quadratic 0-1 programming problem can still be computed efficiently (see Chapter 11). Lagrangian bounds for linearly constrained convex integer programming problems can also be computed via certain decomposition schemes (see Chapter 3).

Next, we compare the continuous bound with the Lagrangian bound for a nonconvex case of (P), more specifically, the following linearly constrained concave integer programming problem:

$$
\begin{aligned}
(P_v) \qquad & \min\ f(x) \\
& \text{s.t. } Ax \le b, \\
& \quad\ Bx = d, \\
& \quad\ x \in X = \{x \in \mathbb{Z}^n \mid l_j \le x_j \le u_j,\ j = 1, \ldots, n\},
\end{aligned}
$$

where $f(x)$ is a concave function, A is an $m \times n$ matrix, B is an $l \times n$ matrix, $b \in \mathbb{R}^m$, $d \in \mathbb{R}^l$, l_j and u_j are integer lower bound and upper bound of x_j, respectively. Let $(\overline{P}_v)$ denote the continuous relaxation problem of (P_v).

The Lagrangian dual problem of (P_v) is:

$$(D_v) \qquad \max_{\lambda \in \mathbb{R}^m_+, \mu \in \mathbb{R}^l} d_v(\lambda, \mu),$$

where

$$d_v(\lambda, \mu) = \min_{x \in X} [f(x) + \lambda^T (Ax - b) + \mu^T (Bx - d)],$$

for $\lambda \in \mathbb{R}^m_+$ and $\mu \in \mathbb{R}^l$.

The following result shows that, on the contrary to the convex case of (P), the continuous relaxation of (P_v) always generates a lower bound of (P_v) at least as good as that by the Lagrangian dual.

THEOREM 2.5 *Assume that f is a concave function on X in (P_v). Then $v(D_v) \le v(\overline{P}_v)$.*

Proof. Let Ω denote the set of extreme points of $conv(X)$:

$$\Omega = \{x^i \mid i = 1, \ldots, K\},$$

where $K = 2^n$. Consider the following convex envelope of f over $conv(X)$:

$$\phi(x) = \min\{\sum_{i=1}^{K} \gamma_i f(x^i) \mid \sum_{i=1}^{K} \gamma_i x^i = x,\ \gamma \in \Lambda\}, \tag{2.3.4}$$

where $\Lambda = \{\gamma \in \mathbb{R}^K \mid \sum_{i=1}^K \gamma_i = 1,\ \gamma_i \geq 0,\ i = 1, \ldots, K\}$. It is clear that ϕ is a piecewise linear convex function on $conv(X)$. By the concavity of f, we have

$$f(x) \geq \phi(x), \quad \forall x \in conv(X) \tag{2.3.5}$$

and $f(x) = \phi(x)$ for all $x \in \Omega$. Recall that $f(x)$ and $\phi(x)$ have the same global optimal value over $conv(X)$ (see [182]). Notice that a concave function always achieves its minimum over a polyhedron at one of the extreme points. Also, the extreme points of $conv(X)$ are integer points. Thus, we have

$$\begin{aligned} v(D_v) &= \max_{\lambda \in \mathbb{R}_+^m, \mu \in \mathbb{R}^l} \min_{x \in X} [f(x) + \lambda^T(Ax - b) + \mu^T(Bx - d)] \\ &= \max_{\lambda \in \mathbb{R}_+^m, \mu \in \mathbb{R}^l} \min_{x \in conv(X)} [f(x) + \lambda^T(Ax - b) + \mu^T(Bx - d)] \\ &= \max_{\lambda \in \mathbb{R}_+^m, \mu \in \mathbb{R}^l} \min_{x \in conv(X)} [\phi(x) + \lambda^T(Ax - b) + \mu^T(Bx - d)] \\ &= \min_{x \in conv(X)} \max_{\lambda \in \mathbb{R}_+^m, \mu \in \mathbb{R}^l} [\phi(x) + \lambda^T(Ax - b) + \mu^T(Bx - d)] \\ &= \min_{x \in conv(X)} \{\phi(x) \mid Ax \leq b,\ Bx = d\} \\ &\leq \min_{x \in conv(X)} \{f(x) \mid Ax \leq b,\ Bx = d\} \\ &= v(\overline{P}_v). \end{aligned}$$

The fourth equation in the above derivation is due to the strong duality theorem for piecewise linear programming. □

Combining Theorems 2.4 and 2.5 gives rise to the well-known result in classical linear integer programming theory: The Lagrangian dual bound is identical to the continuous bound for linear integer programming.

COROLLARY 2.1 *If f is a linear function in (P_v), then $v(D_v) = v(\overline{P}_v)$.*

2.4 Proximity between Continuous Solution and Integer Solution

A natural and simple way to solve (P) is to relax the integrality of x and to solve the continuous version of (P) as a nonlinear programming problem. The optimal solution to the continuous relaxation is then rounded to its nearest integer point in X which sometimes happens to be a good sub-optimal feasible solution to (P). In many situations, however, the idea of rounding the continuous solution may result in an integer solution that is not only far away from the optimal solution of (P) but also infeasible. Thus, it is important to study the relationship between the integer and continuous solutions in mathematical programming problems.

2.4.1 Linear integer program

Consider a linear integer program

$$\begin{aligned} \min\ & c^T x \\ \text{s.t.}\ & Ax \leq b, \\ & x \in \mathbb{Z}^n, \end{aligned} \tag{2.4.1}$$

and its continuous relaxation:

$$\begin{aligned} \min\ & c^T x \\ \text{s.t.}\ & Ax \leq b, \\ & x \in \mathbb{R}^n, \end{aligned} \tag{2.4.2}$$

where A is an integer $m \times n$ matrix and $c \in \mathbb{R}^n$ and $b \in \mathbb{R}^m$. Denote by $\Delta(A)$ the maximum among the absolute values of all sub-determinants of matrix A.

THEOREM 2.6 *Assume that the optimal solutions of problems (2.4.1) and (2.4.2) both exist. Then:*

(i) *For each optimal solution $\bar{x}$ to (2.4.2), there exists an optimal solution z^* to (2.4.1) such that*

$$\|\bar{x} - z^*\|_\infty \leq n\Delta(A). \tag{2.4.3}$$

(ii) *For each optimal solution $\bar{z}$ to (2.4.1), there exists an optimal solution x^* to (2.4.2) such that*

$$\|x^* - \bar{z}\|_\infty \leq n\Delta(A). \tag{2.4.4}$$

Proof. Let $\bar{x}$ and $\bar{z}$ be optimal solutions to (2.4.2) and (2.4.1), respectively. Partition A into $A^T = [A_1^T, A_2^T]$, where $A_1\bar{x} \geq A_1\bar{z}$ and $A_2\bar{x} < A_2\bar{z}$, and partition b into b^1 and b^2 accordingly. Note that $A_2\bar{x} < A_2\bar{z} \leq b^2$. Let $\bar{\lambda}_1 \geq 0$ and $\bar{\lambda}_2 \geq 0$ be optimal dual variables corresponding to A_1 and A_2, respectively, for (2.4.2). By the complementary slackness condition, $\bar{\lambda}_2 = 0$ and thus we have $A_1^T\bar{\lambda}_1 = -c$. Consider the following cone:

$$C = \{x \mid A_1 x \geq 0,\ A_2 x \leq 0\}.$$

Obviously, $\bar{x} - \bar{z} \in C$. Furthermore $c^T x \leq 0$ for all $x \in C$, since $c^T x = -\bar{\lambda}_1^T A_1 x \leq 0$ for all $x \in C$. By Carathéodory's theorem, there exist t ($t \leq n$) integer vectors $d^i \in C$, $i = 1, \ldots, t$, and $\mu_i \geq 0$, $i = 1, \ldots, t$, such that

$$\bar{x} - \bar{z} = \mu_1 d^1 + \cdots + \mu_t d^t. \tag{2.4.5}$$

By Cramer's rule, we can assume that $\|d^i\|_\infty \leq \Delta(A)$, $i = 1, \ldots, t$.

Let

$$z^* = \bar{z} + \lfloor \mu_1 \rfloor d^1 + \cdots + \lfloor \mu_t \rfloor d^t. \tag{2.4.6}$$

where $\lfloor x \rfloor$ is the maximum integer number less than or equal to x. By (2.4.5), we have

$$z^* = \bar{x} + (\lfloor \mu_1 \rfloor - \mu_1)d^1 + \cdots + (\lfloor \mu_t \rfloor - \mu_t)d^t. \tag{2.4.7}$$

Thus,

$$\begin{aligned} A_1 z^* &= A_1\bar{x} + (\lfloor \mu_1 \rfloor - \mu_1)A_1 d^1 + \cdots + (\lfloor \mu_t \rfloor - \mu_t)A_1 d^t \leq A_1\bar{x} \leq b^1, \\ A_2 z^* &= A_2\bar{z} + \lfloor \mu_1 \rfloor A_2 d^1 + \cdots + \lfloor \mu_t \rfloor A_2 d^t \leq A_2\bar{z} \leq b^2. \end{aligned}$$

So $Az^* \leq b$. Moreover, since $c^T d^i = -\bar{\lambda}_1^T A_1 d^i \leq 0$ for all $i = 1, \ldots, t$, we imply from (2.4.6) that $c^T z^* \leq c^T \bar{z}$. Therefore, z^* is an optimal solution to (2.4.1) and by (2.4.7), we get

$$\|z^* - \bar{x}\|_\infty \leq \|d^1\|_\infty + \cdots + \|d^t\|_\infty \leq n\Delta(A),$$

which is (2.4.3). Moreover, combining $c^T z^* \leq c^T \bar{z}$ with the optimality of $\bar{z}$ and (2.4.6) leads to $c^T d^i = 0$ for i with $\mu_i \geq 1$.

Now, let

$$x^* = \bar{x} - \lfloor \mu_1 \rfloor d^1 - \cdots - \lfloor \mu_t \rfloor d^t. \tag{2.4.8}$$

Then,

$$A_1 x^* = A_1\bar{x} - \lfloor \mu_1 \rfloor A_1 d^1 - \cdots - \lfloor \mu_t \rfloor A_1 d^t \leq A_1\bar{x} \leq b^1. \tag{2.4.9}$$

Also, by (2.4.5), it holds

$$x^* = \bar{z} + (\mu_1 - \lfloor \mu_1 \rfloor)d^1 + \cdots + (\mu_t - \lfloor \mu_t \rfloor)d^t.$$

Thus, we obtain $\|x^* - \bar{z}\|_\infty \leq n\Delta(A)$ using the similar arguments as in part (i). Moreover, we have

$$A_2 x^* = A_2\bar{z} + (\mu_1 - \lfloor \mu_1 \rfloor)A_2 d^1 + \cdots + (\mu_t - \lfloor \mu_t \rfloor)A_2 d^t \leq A_2\bar{z} \leq b^2. \tag{2.4.10}$$

Combining (2.4.9) with (2.4.10) gives rise to $Ax^* \leq b$. Since $c^T d^i = 0$ for i with $\mu_i \geq 1$ and $\lfloor \mu_i \rfloor = 0$ for i with $0 \leq \mu_i < 1$, we obtain from (2.4.8) that $c^T x^* = c^T \bar{x}$. Thus, x^* is an optimal solution to (2.4.2). □

2.4.2 Linearly constrained separable convex integer program

The proximity results in the previous subsection can be extended to separable convex programming problems. Consider the following problems:

$$\begin{aligned} \min\ f(x) = \sum_{j=1}^{n} f_j(x_j) \\ \text{s.t. } Ax \le b, \\ x \in \mathbb{Z}^n, \end{aligned} \tag{2.4.11}$$

and its continuous relaxation:

$$\begin{aligned} \min\ f(x) = \sum_{j=1}^{n} f_j(x_j) \\ \text{s.t. } Ax \le b, \\ x \in \mathbb{R}^n, \end{aligned} \tag{2.4.12}$$

where $f_j(x_j)$, $j = 1, \ldots, n$, are all convex functions on $\mathbb{R}$, A is an integer $m \times n$ matrix and $b \in \mathbb{R}^m$. The following result generalizes Theorem 2.6.

THEOREM 2.7 *Assume that the optimal solutions of problems (2.4.11) and (2.4.12) both exist. Then:*

(i) *For each optimal solution $\bar{x}$ to (2.4.12), there exists an optimal solution z^* to (2.4.11) such that*

$$\|\bar{x} - z^*\|_\infty \le n\Delta(A). \tag{2.4.13}$$

(ii) *For each optimal solution $\bar{z}$ to (2.4.11), there exists an optimal solution x^* to (2.4.12) such that*

$$\|x^* - \bar{z}\|_\infty \le n\Delta(A). \tag{2.4.14}$$

Proof. Let $\bar{x}$ and $\bar{z}$ be optimal solutions to (2.4.12) and (2.4.11), respectively. Let S^* be the intersection of the feasible region of (2.4.12) with the minimal box that contains $\bar{x}$ and $\bar{z}$. Let

$$A^* = \begin{bmatrix} A \\ I_{n\times n} \\ -I_{n\times n} \end{bmatrix}, \quad b^* = \begin{bmatrix} b \\ \max(\bar{x}, \bar{z}) \\ -\min(\bar{x}, \bar{z}) \end{bmatrix}. \tag{2.4.15}$$

Then S^* can be expressed as $\{x \in \mathbb{R}^n \mid A^*x \le b^*\}$. Now, consider the linear over-estimation of $f_j(x_j)$. Let $c_j^* = (f_j(\bar{x}_j) - f_j(\bar{z}_j))/(\bar{x}_j - \bar{z}_j)$. Without loss of generality, we can assume that $\bar{z}_j = f_j(\bar{z}_j) = 0$. So $f_j(\bar{x}_j) = c_j^*\bar{x}_j$.

Moreover, by the convexity of f_j, we have $f_j(x_j) \leq c_j^* x_j$ for all $j = 1, \ldots, n$ and $x \in S^*$. Consider the following linear program:

$$\begin{aligned} \min\ & (c^*)^T x \\ \text{s.t. } & A^* x \leq b^*, \\ & x \in \mathbb{R}^n. \end{aligned} \tag{2.4.16}$$

Since $(c^*)^T \bar{x} = f(\bar{x}) \leq f(x) \leq (c^*)^T x$ for all $x \in S^*$, $\bar{x}$ is also an optimal solution to (2.4.16). Note that the upper bound of the absolute values of subdeterminants of A^* remains $\Delta(A)$.

By Theorem 2.6, there exists an integer $z^* \in S^*$ such that $\|\bar{x} - z^*\|_\infty \leq n\Delta(A)$ and $(c^*)^T z^* \leq (c^*)^T z$ for all integer $z \in S^*$. Note that $f(z^*) \leq (c^*)^T z^* \leq (c^*)^T \bar{z} = f(\bar{z})$. It follows that z^* is an optimal solution to (2.4.11). This proves part (i) of the theorem. Part (ii) can be proved similarly. □

2.4.3 Unconstrained convex integer program

In this subsection, we establish some proximity results for general unconstrained convex integer programs which are not necessarily separable. For a separable convex function the distance (in ∞-norm) between its integer and real minimizers is bounded by 1. This is simply because the distance between the integer and real minimizers of a univariate convex function is always dominated by 1. Thus, we first concentrate in this subsection on a proximity bound for nonseparable quadratic functions and then extend it to strictly convex functions. We further discuss an extension to mixed-integer cases.

Let Q be an $n \times n$ symmetric positive definite matrix. Define

$$q(x) = (x - x_0)^T Q (x - x_0).$$

Consider

$$\min\{q(x) \mid x \in \mathbb{R}^n\} \tag{2.4.17}$$

and

$$\min\{q(x) \mid x \in \mathbb{Z}^n\}. \tag{2.4.18}$$

Obviously, x_0 is the unique minimizer of (2.4.17). For any $n \times n$ real symmetric matrix P, denote by $\lambda_{\max}(P)$ and $\lambda_{\min}(P)$ its largest and smallest eigenvalues, respectively.

THEOREM 2.8 *For any optimal solution $\bar{x}$ to (2.4.18), it holds*

$$\|\bar{x} - x_0\|_2 \leq \frac{1}{2}\sqrt{n\kappa}, \tag{2.4.19}$$

where $\kappa = \lambda_{\max}(Q)/\lambda_{\min}(Q)$ *is the condition number of* Q.

Proof. Let

$$q(\bar{x}) = (\bar{x} - x_0)^T Q(\bar{x} - x_0) = r. \tag{2.4.20}$$

We assume without loss of generality that $\bar{x} \neq x_0$ and thus $r > 0$. By the optimality of $\bar{x}$, no integer point is contained in the interior of the following ellipsoid:

$$E = \{x \in \mathbb{R}^n \mid (x - x_0)^T Q(x - x_0) \leq r\}.$$

Since the diameter of the circumscribed sphere of a unit cube in $\mathbb{R}^n$ is $\sqrt{n}$, the interior of a ball in $\mathbb{R}^n$ with diameter greater than $\sqrt{n}$ must contain at least one integer point. It is clear that ellipsoid E contains the ball centered at x_0 with diameter $2\sqrt{r\lambda_{\min}(Q^{-1})}$. Hence, we have

$$2\sqrt{r\lambda_{\min}(Q^{-1})} \leq \sqrt{n}. \tag{2.4.21}$$

Notice also that ellipsoid E is enclosed in the ball centered at x_0 with diameter $2\sqrt{r\lambda_{\max}(Q^{-1})}$. We therefore find from (2.4.20) and (2.4.21) that

$$\begin{aligned} \|\bar{x} - x_0\|_2 &\leq \sqrt{r\lambda_{\max}(Q^{-1})} \\ &\leq \frac{1}{2}\sqrt{\frac{n\lambda_{\max}(Q^{-1})}{\lambda_{\min}(Q^{-1})}} \\ &= \frac{1}{2}\sqrt{\frac{n\lambda_{\max}(Q)}{\lambda_{\min}(Q)}} \\ &= \frac{1}{2}\sqrt{n\kappa}. \end{aligned}$$

□

Let $f : \mathbb{R}^n \to \mathbb{R}$ be a twice differentiable convex function satisfying the following strong convexity condition:

$$0 < m \leq \lambda_{\min}(\nabla^2 f(x)) \leq \lambda_{\max}(\nabla^2 f(x)) \leq M, \quad \forall x \in \mathbb{R}^n. \tag{2.4.22}$$

Consider

$$\min\{f(x) \mid x \in \mathbb{R}^n\} \tag{2.4.23}$$

and

$$\min\{f(x) \mid x \in \mathbb{Z}^n\}. \tag{2.4.24}$$

THEOREM 2.9 *Let x_0 be the unique optimal solution to (2.4.23). Then for any optimal solution $\bar{x}$ to (2.4.24), it holds*

$$\|\bar{x} - x_0\|_2 \le \frac{1}{2}\sqrt{n\kappa_1},$$

where $\kappa_1 = M/m$.

Proof. Note that the condition (2.4.22) and Taylor's Theorem imply

$$\frac{1}{2}m\|x - x_0\|_2^2 \le f(x) - f(x_0) \le \frac{1}{2}M\|x - x_0\|_2^2, \quad \forall x \in \mathbb{R}^n. \quad (2.4.25)$$

Let $r = f(\bar{x}) - f(x_0)$. By (2.4.25), the convex level set $\{x \in \mathbb{R}^n \mid f(x) - f(x_0) \le r\}$ contains a sphere with diameter $2\sqrt{2rM^{-1}}$ and is enclosed in a sphere with diameter $2\sqrt{2rm^{-1}}$. The theorem then follows by using the same arguments as in the proof of Theorem 2.8. □

Now we consider the mixed-integer convex program:

$$\min\{f(x) \mid x = (y, z)^T,\ y \in \mathbb{Z}^l,\ z \in \mathbb{R}^k\}, \quad (2.4.26)$$

where $l > 0$, $k > 0$, $l + k = n$ and $f(x)$ satisfies condition (2.4.22).

THEOREM 2.10 *Let x_0 be the unique optimal solution to (2.4.23). Then for any optimal solution $\bar{x}$ of (2.4.26), it holds*

$$\|\bar{x} - x_0\|_2 \le \frac{1}{2}\sqrt{n\kappa_1},$$

where $\kappa_1 = M/m$.

Proof. Note that every sphere in $\mathbb{R}^n$ with diameter $\sqrt{n}$ has a nonempty intersection with a k-dimensional hyperplane $\{x \in \mathbb{R}^n \mid x = (y, z)^T,\ y = a\}$ for some integer $a \in \mathbb{Z}^l$. The theorem can then be proved along the same line as in the proof of Theorem 2.8. □

One promising application of the above proximity results is their usage in reducing the set of feasible solutions in integer programming problems.

EXAMPLE 2.2 Consider the following unconstrained quadratic integer program:

$$\begin{aligned} &\min\ q(x) = 27x_1^2 - 18x_1x_2 + 4x_2^2 - 3x_2 \\ &\text{s.t. } x \in \mathbb{Z}^2. \end{aligned}$$

The optimal solution of the continuous relaxation of this example is $x_0 = (0.5, 1.5)^T$ with $q(x_0) = -2.25$. Theorem 2.8 can be used to reduce the feasible region by setting the bounds for the integer variables. It is easy to verify that $\kappa = 33.5627$. From (2.4.19) we have $\|\bar{x} - x_0\|_2 \le (1/2)\sqrt{2\kappa} = 4.0965$. We can thus attach a box constraint $-3 \le x_1 \le 4$, $-2 \le x_2 \le 5$ to Example 2.2. This significant reduction in the feasible region may help the solution process when a branch-and-bound algorithm is used as a solution scheme. Applying a branch-and-bound procedure to Example 2.2 with the box constraint yields an optimal solution $\bar{x} = (1,3)^T$ with $q(\bar{x}) = 0$. We note that $\bar{x}$ cannot be obtained by rounding the continuous optimal solution x_0 since $q((0,1)^T) = q((1,2)^T) = 1$, $q((1,1)^T) = q((0,2)^T) = 10$.

The following example shows that the bound in (2.4.19) can be achieved in some situations.

EXAMPLE 2.3 Consider the following problem:

$$\min\{\sum_{i=1}^{n}(x_i - \frac{1}{2})^2 \mid x \in \mathbb{Z}^n\}.$$

It is easy to see that all vertices of the unit cube $[0,1]^n$ are the optimal integer solutions of this problem. Since $x_0 = (1/2, 1/2, \ldots, 1/2)^T$, we get $\|\bar{x} - x_0\|_2 = \sqrt{n}/2$. On the other hand, since $Q = I$, we have $\kappa = 1$ and $\sqrt{n\kappa}/2 = \sqrt{n}/2$.

Now, we give another example in which the strict inequality in (2.4.19) holds while both $\|\bar{x} - x_0\|_2$ and κ tend to infinity simultaneously. As a by-product, we can get a method in constructing nonseparable quadratic test problems where the distance between the continuous and integer solutions can be predetermined.

Let $v_1 = (\cos\theta, \sin\theta)^T$, $v_2 = (-\sin\theta, \cos\theta)^T$. Then v_1 and v_2 are orthonormal and the angle between v_1 and x_1-axis is θ. For $\lambda_1 \ge \lambda_2 > 0$, let

$$\begin{aligned} P &= \lambda_1 v_1 v_1^T + \lambda_2 v_2 v_2^T \\ &= \begin{pmatrix} \lambda_1\cos^2\theta + \lambda_2\sin^2\theta & (\lambda_1 - \lambda_2)\sin\theta\cos\theta \\ (\lambda_1 - \lambda_2)\sin\theta\cos\theta & \lambda_1\sin^2\theta + \lambda_2\cos^2\theta \end{pmatrix}. \end{aligned} \quad (2.4.27)$$

It follows that P is a 2×2 symmetric positive definite matrix and it has eigenvalues λ_1 and λ_2 with corresponding eigenvectors v_1 and v_2, respectively.

EXAMPLE 2.4 Consider the following problem:

$$\min\{q(x) := (x - x_0)^T P^{-1}(x - x_0) \mid x \in \mathbb{Z}^2\}, \quad (2.4.28)$$

where P is defined by (2.4.27), $x_0 = (0, 1/2)^T$ and $\lambda_2 \in (0, 1/4)$.

For any positive integer $m > 0$ and $\lambda_2 \in (0, 1/4)$, we can determine the values of θ and λ_1 such that axis $x_2 = 0$ supports ellipsoid $E(x_0, P^{-1}) = \{x \in$

$\mathbb{R}^2 \mid (x - x_0)^T P^{-1}(x - x_0) \leq 1\}$ at $(-m, 0)$. For $t \in \mathbb{R}$, consider equation $q((t, 0)^T) = 1$. From (2.4.27) and (2.4.28), this equation is equivalent to

$$a_1 t^2 + a_2 t + a_3 = 0, \tag{2.4.29}$$

where

$$\begin{aligned} a_1 &= \lambda_1 \sin^2\theta + \lambda_2 \cos^2\theta, \\ a_2 &= (\lambda_1 - \lambda_2)\sin\theta\cos\theta, \\ a_3 &= \frac{1}{4}\lambda_1 \cos^2\theta + \frac{1}{4}\lambda_2 \sin^2\theta - \lambda_1\lambda_2. \end{aligned}$$

Note that

$$a_2^2 - 4a_1 a_3 = -\lambda_1\lambda_2 + 4\lambda_1\lambda_2(\lambda_1 \sin^2\theta + \lambda_2\cos^2\theta).$$

Therefore, equation (2.4.29) has a unique real root if and only if

$$\lambda_1 \sin^2\theta + \lambda_2 \cos^2\theta = \frac{1}{4}. \tag{2.4.30}$$

If condition (2.4.30) holds, $a_1 = \frac{1}{4}$ and the root of equation (2.4.29) is

$$t = -\frac{a_2}{2a_1} = -2(\lambda_1 - \lambda_2)\sin\theta\cos\theta. \tag{2.4.31}$$

Setting $t = -m$ in (2.4.31), we get

$$(\lambda_1 - \lambda_2)\sin(2\theta) = m. \tag{2.4.32}$$

Equations (2.4.30) and (2.4.32) uniquely determine the values of $\theta \in (0, \pi/2)$ and $\lambda_1 > 1/4$ for which $(-m, 0)^T$ is the unique intersection point of ellipsoid $E(x_0, P^{-1})$ and x_1-axis. By the symmetry of the ellipsoid, $(m, 1)^T$ is the unique intersection point of $E(x_0, P^{-1})$ and the line $x_2 = 1$. Since no integer point other than $(-m, 0)^T$ and $(m, 1)^T$ lies in $E(x_0, P^{-1})$, $\bar{x}_1(m) = (-m, 0)^T$ and $\bar{x}_2(m) = (m, 1)^T$ are the optimal solutions of (2.4.28).

Now, we set $\lambda_2 = 1/5$. For any positive integer m, let $\theta(m)$ and $\lambda_1(m)$ be determined from (2.4.30) and (2.4.32). Denote $\kappa(m) = \lambda_1(m)/\lambda_2 = 5\lambda_1(m)$.

By (2.4.30) and (2.4.32), we have

$$\begin{aligned} & \|\bar{x}_1(m) - x_0\|_2 = \|\bar{x}_2(m) - x_0\|_2 \\ = & \sqrt{m^2 + \frac{1}{4}} \\ = & \sqrt{4(\lambda_1(m) - 1/5)^2 \sin^2[\theta(m)](1 - \sin^2[\theta(m)]) + \frac{1}{4}} \\ = & \sqrt{4(\lambda_1(m) - 1/5)^2 \left(\frac{1/4 - 1/5}{\lambda_1(m) - 1/5}\right)\left(1 - \frac{1/4 - 1/5}{\lambda_1(m) - 1/5}\right) + \frac{1}{4}} \\ = & \sqrt{\frac{1}{5}\lambda_1(m) + \frac{1}{5}} \\ = & \sqrt{\frac{1}{25}\kappa(m) + \frac{1}{5}}. \end{aligned}$$

Thus, $\|\bar{x}_1(m) - x_0\|_2 = \|\bar{x}_2(m) - x_0\|_2 \to \infty$ and $\kappa(m) \to \infty$ when $m \to \infty$. Moreover, since $\kappa(m) > 1$, we have

$$\|\bar{x}_1(m) - x_0\|_2 = \sqrt{\frac{1}{25}\kappa(m) + \frac{1}{5}} < \frac{1}{2}\sqrt{2\kappa(m)}.$$

2.5 Penalty Function Approach

Generally speaking, an unconstrained integer programming problem is easier to solve than a constrained one. We discuss in this section how to convert a general constrained integer programming problem into an unconstrained one by using an exact penalty method. Consider the following problem:

$$\begin{aligned} (P) \qquad & \min\ f(x) \\ & \text{s.t. } g_i(x) \le 0,\ i = 1, \ldots, m, \\ & \qquad h_j(x) = 0,\ j = 1, \ldots, l, \\ & \qquad x \in X, \end{aligned}$$

where f, $g_i(x)$ $(i = 1, \ldots, m)$ and $h_j(x)$ $(j = 1, \ldots, l)$ are continuous functions, and X is a finite set in $\mathbb{Z}^n$. Let

$$S = \{x \in X \mid g_i(x) \le 0,\ i = 1, \ldots, m,\ h_j(x) = 0,\ j = 1, \ldots, l\}.$$

Define a penalty function $P(x)$ such that: $P(x) = 0$ for $x \in S$ and $P(x) \ge \epsilon > 0$ for $x \notin S$. A typical penalty function for (P) is

$$P(x) = \sum_{i=1}^{m} \max(g_i(x), 0) + \sum_{j=1}^{l} h_j^2(x). \qquad (2.5.1)$$

Define the penalty problem of (P) as follows:

$$(PEN) \quad \min_{x \in X} T(x, \mu) = f(x) + \mu P(x), \quad \mu > 0.$$

Since $T(x, \mu) = f(x)$ for $x \in S$ and $S \subseteq X$, we have $v(P) \geq v(PEN)$.

THEOREM 2.11 *Let $\underline{f}$ be a lower bound of $\min_{x \in X} f(x)$ and $\gamma > 0$ be a lower bound of $\min_{x \in X \setminus S} P(x)$. Suppose that $X \setminus S \neq \emptyset$. Then, there exists a μ_0 such that for any $\mu > \mu_0$, any solution x^* that solves (PEN) also solves (P) and $v(PEN) = v(P)$.*

Proof. Let

$$\mu_0 = \frac{v(P) - \underline{f}}{\gamma}. \tag{2.5.2}$$

For any $x \in X \setminus S$ and any $\mu > \mu_0$,

$$\begin{aligned} T(x, \mu) &= f(x) + \mu P(x) \\ &> f(x) + \mu_0 P(x) \\ &\geq f(x) + (v(P) - \underline{f}) \\ &\geq v(P). \end{aligned}$$

Therefore, the minimum of $T(x, \mu)$ over X must be achieved in S. Since $T(x, \mu) = f(x)$ for any $x \in S$, we conclude that x^* solves (P) and $v(PEN) = T(x^*, \mu) = f(x^*) = v(P)$. □

COROLLARY 2.2 *Let $\overline{f}$ be an upper bound of $v(P)$. If $m = 0$ and h_j ($j = 1, \ldots, l$) are integer-valued functions on X, then for any $\mu \geq \mu_0 = \overline{f} - \underline{f}$, any solution x^* solves (PEN) also solves (P), where $P(x) = \sum_{j=1}^{l} h_j^2(x)$ in problem (PEN).*

Proof. Since h_j is integer-valued, we deduce that $P(x) \geq 1$ for any $x \in X \setminus S$ and hence γ can be taken as 1. Moreover, $v(P) \geq \underline{f}$, thus, by (2.5.2), $\mu_0 \leq \overline{f} - \underline{f}$. The conclusion then follows from Theorem 2.11. □

If $h_j(x) \geq 0$ for any $x \in X$, $j = 1, \ldots, l$, then $P(x)$ in Corollary 2.2 can be taken as $P(x) = \sum_{j=1}^{l} h_j(x)$.

2.6 Optimality Conditions for Unconstrained Binary Quadratic Problems

2.6.1 General case

We consider the following unconstrained binary quadratic optimization problem:

$$(BQ) \quad \min_{x \in \{-1,1\}^n} q(x) = \frac{1}{2} x^T Q x + b^T x,$$

where Q is a symmetric matrix in $\mathbb{R}^{n\times n}$ and $b \in \mathbb{R}^n$. Notice that any binary quadratic problem with $y_i \in \{l_i, u_i\}$, $i = 1, \ldots, n$, can be transformed into the form of (BQ) by the linear transformation: $y_i = l_i + (u_i - l_i)(x_i + 1)/2$, $i = 1, \ldots, n$. It is clear that (BQ) is equivalent to the following continuous quadratic problem:

$$(CQ) \qquad \min\ q(x) = \frac{1}{2}x^T Qx + b^T x, \quad \text{s.t. } x_i^2 = 1,\ i = 1, \ldots, n.$$

Problem (CQ) is essentially a nonconvex continuous optimization problem even if matrix Q is positive semidefinite. Thus, problem (CQ) is the same as hard as the primal problem (BQ).

To motivate the derivation of the global optimality conditions, let's consider the relationship between the solutions of the following two scalar optimization problems with $a > 0$:

$$(SQ) \quad \min\ \{\frac{1}{2}ax^2 + bx \mid x \in \{-1, 1\}\}$$

and

$$(\overline{SQ}) \quad \min\ \{\frac{1}{2}ax^2 + bx \mid -1 \le x \le 1\}.$$

We are interested in conditions under which $v(SQ) = v(\overline{SQ})$, and furthermore (SQ) and $(\overline{SQ})$ have the same optimal solution. Note that we can rewrite $\frac{1}{2}ax^2$ + bx as $\frac{1}{2}a(x + \frac{b}{a})^2 - \frac{b^2}{2a}$. It can be verified that when $a \le |b|$ and $b > 0$, x^* = -1 solves both (SQ) and $(\overline{SQ})$ and when $a \le |b|$ and $b < 0$, $x^* = 1$ solves both (SQ) and $(\overline{SQ})$. In summary, $a \le |b|$ is both a necessary and sufficient condition for generating an optimal solution of the integer optimization problem (SQ) by its continuous optimization problem $(\overline{SQ})$.

Consider the following Lagrangian relaxation of problem (CQ):

$$\min_{x\in\mathbb{R}^n} L(x, y) = q(x) + \sum_{i=1}^{n} y_i(x_i^2 - 1),$$

where $y_i \in \mathbb{R}$ is the Lagrangian multiplier for constraint $x_i^2 = 1$, $i = 1, \ldots, n$. Define two $n \times n$ diagonal matrices $X = diag(x)$ and $Y = diag(y)$. The Lagrangian relaxation problem of (CQ) can be expressed as

$$(LCQ) \quad h(y) = \min_{x\in\mathbb{R}^n} [\frac{1}{2}x^T(Q + 2Y)x + b^T x - e^T y],$$

where e is an n dimensional vector with all components equal to 1. The dual problem of (CQ) is then given as

$$(DQ) \qquad \max_{y\in\text{dom } h} h(y),$$

where

$$\mathrm{dom}\, h = \{y \in \mathbb{R}^n \mid h(y) > -\infty\}.$$

Note that the necessary and sufficient conditions for $h(y) > -\infty$ are:

(i) There exists an x such that $(Q + 2Y)x + b = 0$;

(ii) The matrix $Q + 2Y$ is positive semidefinite.

Although problem (CQ) is nonconvex, if we are lucky enough to find out an $\bar{x}$ that is feasible in (CQ) and $\bar{y} \in \mathrm{dom}\, h$ such that $q(\bar{x}) = h(\bar{y})$, then $\bar{x}$ must be a global optimal solution to (CQ).

THEOREM 2.12 *Let $\bar{x} = \bar{X}e$ be feasible in (CQ). If*

$$\bar{X}Q\bar{X}e + \bar{X}b \leq \lambda_{min}(Q)e, \tag{2.6.1}$$

where $\lambda_{min}(Q)$ is the minimum eigenvalue of matrix Q, then $\bar{x}$ is a global optimal solution of (CQ) or (BQ).

Proof. Let

$$\bar{y} = -\frac{1}{2}(\bar{X}Q\bar{X}e + \bar{X}b). \tag{2.6.2}$$

Let $\bar{Y} = diag(\bar{y})$. Then

$$\begin{aligned} (Q + 2\bar{Y})\bar{x} + b &= Q\bar{X}e + 2\bar{Y}\bar{X}e + b \\ &= Q\bar{X}e + 2\bar{X}\bar{y} + b \\ &= Q\bar{X}e - \bar{X}^2Q\bar{X}e - \bar{X}^2b + b \\ &= 0, \end{aligned}$$

where the last equality is due to $\bar{X}^2 = I$ when $\bar{x}$ is feasible to (CQ). This implies that $\bar{x}$ is a solution to (LCQ) with $y = \bar{y}$ when $Q + 2\bar{Y}$ is positive semidefinite.

From (2.6.1) and (2.6.2), we have

$$\lambda_{min}(2\bar{Y}) = \min_{1 \leq i \leq n} (-\bar{X}Q\bar{X}e - \bar{X}b)_i \geq -\lambda_{min}(Q).$$

Thus,

$$\lambda_{min}(Q + 2\bar{Y}) \geq \lambda_{min}(Q) + \lambda_{min}(2\bar{Y}) \geq 0.$$

We can conclude that matrix $Q + 2\bar{Y}$ is positive semidefinite. Thus $\bar{y}$ defined in (2.6.2) belongs to $\mathrm{dom}\, h$. The remaining task in deriving the sufficient global optimality condition is to prove that the dual value $h(\bar{y})$ attains the objective value of the feasible solution $\bar{x}$,

$$
\begin{aligned}
h(\bar{y}) &= \min_{x\in\mathbb{R}^n}\{\frac{1}{2}x^T(Q+2\bar{Y})x+b^Tx-e^T\bar{y}\} \\
&= -\frac{1}{2}\bar{x}^T(Q+2\bar{Y})\bar{x}-e^T\bar{y} \\
&= -\frac{1}{2}e^T\bar{X}(Q+2\bar{Y})\bar{X}e-e^T\bar{y} \\
&= -\frac{1}{2}e^T\bar{X}Q\bar{X}e-2e^T\bar{y} \\
&= \frac{1}{2}e^T\bar{X}Q\bar{X}e+b^T\bar{X}e \\
&= q(\bar{x}),
\end{aligned}
$$

where the fact of $\bar{X}\bar{Y}\bar{X}e=\bar{X}^2\bar{Y}e=\bar{Y}e$ is used in the fourth equality and (2.6.2) is applied in the fifth equality. □

The next theorem gives a necessary global optimality condition for (CQ) or (BQ).

THEOREM 2.13 *If $x^* = X^*e$ is a global optimal solution to (CQ), then*

$$X^*QX^*e+X^*b\le diag(Q)e, \tag{2.6.3}$$

where $diag(Q)$ is a diagonal matrix formed from matrix Q by setting all its nondiagonal elements at zero.

Proof. Let e_i be the i-th unit vector in $\mathbb{R}^n$. If x^* is optimal to (CQ), then $q(x^*) \le q(z)$ for every feasible z to (CQ). Especially, setting $z = x^* - 2x_i^*e_i$ in the above relation yields

$$x_i^*e_i^TQx^*+x_i^*b^Te_i\le q_{ii},\quad i=1,\dots,n,$$

where q_{ii} is the i-th diagonal element of Q. □

The above derived sufficient and necessary global optimality conditions for the unconstrained binary quadratic problem (BQ) can be rewritten in the following form where the two bear a resemblance,

$$
\begin{aligned}
&\text{Sufficient Condition for } (BQ): && X(Q-\lambda_{min}(Q)I)Xe\le -Xb,\\
&\text{Necessary Condition for } (BQ): && X(Q-diag(Q)I)Xe\le -Xb.
\end{aligned}
$$

Note that $q_{ii} \ge \lambda_{min}(Q)$ for all $i = 1, \dots, n$. Thus, $diag(Q)e \ge \lambda_{min}(Q)e$. Obviously, sufficient condition (2.6.1) implies necessary condition (2.6.3).

2.6.2 Convex case

We now consider a special case of (BQ) where matrix Q is positive semidefinite. Consider the following relaxation of (BQ):

$$(\overline{BQ}) \qquad \min\ q(x) = \frac{1}{2}x^T Qx + b^T x,$$
$$\text{s.t. } x_i^2 \le 1,\ i = 1, \dots, n.$$

It is clear that $(\overline{BQ})$ is a continuous convex minimization problem when q is convex. It is also obvious that if $x \in \{-1,1\}^n$ is optimal to $(\overline{BQ})$, then x is also optimal to problem (BQ). On the other hand, if $x^* \in \{-1,1\}^n$ is optimal to problem (BQ), then $v(\overline{BQ}) \le q(x^*)$.

THEOREM 2.14 *Assume that Q is positive semidefinite. Then $x^* \in \{-1,1\}^n$ is an optimal solution to both (BQ) and $(\overline{BQ})$ if and only if*

$$X^*QX^*e + X^*b \le 0, \tag{2.6.4}$$

where $X^* = diag(x^*)$ and $e = (1, \dots, 1)^T$.

Proof. Assume that x^* satisfies (2.6.4). For any $y \in \mathbb{R}^n_+$, consider the Lagrangian relaxation of problem $(\overline{BQ})$:

$$h(y) = \min_{x \in \mathbb{R}^n} L(x, y) = q(x) + \sum_{i=1}^n y_i(x_i^2 - 1). \tag{2.6.5}$$

Let

$$y^* = -\frac{1}{2}(X^*QX^*e + X^*b),$$

which is nonnegative according to the assumption in (2.6.4). Furthermore, matrix $(Q + 2Y^*)$ is positive semidefinite, where $Y^* = diag(y^*)$. As the same as in proving Theorem 2.12, we can prove that x^* solves problem (2.6.5) and $h(y^*) = q(x^*)$. Thus, $x^* \in \{-1,1\}^n$ is optimal to $(\overline{BQ})$, thus an optimal solution to (BQ).

To prove the converse, assume that $x^* \in \{-1,1\}^n$ solves both $(\overline{BQ})$ and (BQ). Then from the KKT conditions for $(\overline{BQ})$, there exists a $\bar{y} \in \mathbb{R}^n_+$ such that $(Q + 2\bar{Y})x^* + b = 0$, where $\bar{Y} = diag(\bar{y})$. Thus,

$$\begin{aligned} X^*QX^*e + X^*b &= X^*(Qx^* + b) \\ &= -2X^*\bar{Y}x^* \\ &= -2\bar{Y}e \le 0. \end{aligned}$$

□

Notice that problem $(\overline{BQ})$ is a box constrained convex quadratic programming problem and hence is much easier to solve than (BQ). Solving $(\overline{BQ})$, however, in general only yields a real solution. The next result gives a sufficient condition for getting a nearby integer optimal solution to (BQ) based on a real optimal solution to $(\overline{BQ})$.

THEOREM 2.15 *Assume that Q is a real positive semidefinite matrix and x^* is an optimal solution to $(\overline{BQ})$. If $z^* \in \{-1,1\}^n$ satisfies the following conditions:*

(i) $z_i^* = x_i^*$ *for* $x_i^* \in \{-1,1\}$, *and*

(ii) $Z^*Q(z^* - x^*) \leq \lambda_{min}(Q)e$, *where* $Z^* = diag(z^*)$ *and* $\lambda_{min}(Q)$ *is the minimum eigenvalue of Q,*

then z^* is an optimal solution to (BQ).

Proof. There exists Lagrangian multiplier vector $y \in \mathbb{R}^n_+$ such that x^* satisfies the following KKT conditions for $(\overline{BQ})$:

$$\begin{aligned}
&(Q + 2Y)x^* + b = 0,\\
&y_i[(x_i^*)^2 - 1] = 0,\ i = 1, \ldots, n,
\end{aligned}$$

where $Y = diag(y)$. Let $\delta = z^* - x^*$ and $\Delta = diag(\delta)$. It can be verified that $y_i\delta_i = 0$, $i = 1, \ldots, n$. Thus $\Delta Y = 0$. We have

$$\begin{aligned}
Z^*QZ^*e + Z^*b &= Z^*Qz^* + Z^*b\\
&= Z^*[Q(x^* + \delta) + b]\\
&= Z^*(-2Yx^* + Q\delta)\\
&= (X^* + \Delta)(-2Yx^* + Q\delta)\\
&= -2y + Z^*Q\delta - 2\Delta Y x^*\\
&= -2y + Z^*Q(z^* - x^*)\\
&\leq Z^*Q(z^* - x^*).
\end{aligned}$$

Thus $Z^*Q(z^* - x^*) \leq \lambda_{min}(Q)e$ implies $Z^*QZ^*e + Z^*b \leq \lambda_{min}(Q)e$. Applying Theorem 2.12 concludes that z^* is optimal to (BQ). □

The above theorem can be used to check the global optimality of an integer solution by rounding off a continuous solution.

EXAMPLE 2.5 Consider problem (BQ) with

$$Q = \begin{pmatrix} 4 & 2 & 0 & 2\\ 2 & 4 & 0 & 2\\ 0 & 0 & 4 & 2\\ 2 & 2 & 2 & 4 \end{pmatrix},\quad b = (4,4,3,3)^T.$$

For this problem, we have $\lambda_{min}(Q) = 1.0376$ and the optimal solution to $(\overline{BQ})$ is $x^* = (-0.875, -0.875, -1, 0.625)^T$. Rounding x^* to its nearest integer point in $\{-1, 1\}^n$, we obtain $z^* = (-1, -1, -1, 1)^T$. It can be verified that $Z^*Q(z^* - x^*) = (0, 0, -0.75, 1)^T \leq 1.0376 \times e = \lambda_{min}(Q)e$ is satisfied. Thus, by Theorem 2.15, z^* is an optimal solution to (BQ).

2.7 Notes

The concept of relaxation in integer programming was first formally presented in [76]. The framework of the branch-and-bound method for integer programming was first presented in [124]. More about implicit enumeration techniques can be found in [176].

In 1965, Glover first introduced the back-track scheme in his algorithm for solving linear 0-1 programming problems [77]. Based on Glover's previous work, Geoffrion [73] proposed a framework for implicit enumeration using the concept of the back-tracking scheme which was used later to simplify the well-known additive algorithm of Balas [7] for linear 0-1 programming problems. Both Glover [77] and Geoffrion [73] proved Theorem 2.3 separately using induction.

The relationship between the integer and continuous solutions in mathematical programming problems has been an interesting and challenging topic discussed in the literature. Proximity results were first established in [43] (see also [28][191]) for linear integer programming and then extended to linearly constrained convex separable integer programming problems in [102][225] (see also [11]). The proximity results for nonseparable convex function were obtained in [204].

There is almost no optimality condition derived in the literature for nonlinear integer programming problems. The binary quadratic optimization problem may be the only exception for which optimality conditions were investigated (see e.g., [15][179]).

Chapter 3

LAGRANGIAN DUALITY THEORY

The concept of the duality plays an important role in continuous and discrete optimization. The duality theory is one of the fundamental tools for the development of efficient algorithms for general nonlinear integer programming problems. Without doubt, the Lagrangian dual formulation is one of the most widely used dual formulations in integer optimization, largely due to the associated rich duality theory and its solution elegance in dealing with separable integer optimization problems.

3.1 Lagrangian Relaxation and Dual Formulation

The general bounded integer programming problem can be formulated as follows:

$$\begin{aligned} (P) \qquad & \min\ f(x) \\ & \text{s.t. } g_i(x) \le b_i,\ i = 1, 2, \ldots, m, \\ & \quad\ \ x \in X \subseteq \mathbb{Z}^n, \end{aligned}$$

where X is a finite integer set and $\mathbb{Z}^n$ is the set of all integer numbers in $\mathbb{R}^n$. Problem (P) is called the primal problem. The constraints of $g_i(x) \le b_i,\ i = 1, \ldots, m$, are termed Lagrangian constraints. Let $g(x) = (g_1(x), \ldots, g_m(x))^T$ and $b = (b_1, \ldots, b_m)^T$. The feasible region of problem (P) is defined to be $S = \{x \in X \mid g(x) \le b\}$. Let $f^* = \min_{x \in S} f(x)$.

Incorporating the Lagrangian constraints into the objective function yields a Lagrangian relaxation. Mathematically, a Lagrangian function is constructed by attaching the Lagrangian constraints to the objective with an introduction of a nonnegative Lagrangian multiplier vector, $\lambda = (\lambda_1, \lambda_2, \ldots, \lambda_m)^T \in \mathbb{R}^m_+$,

$$L(x, \lambda) = f(x) + \lambda^T (g(x) - b).$$

The Lagrangian relaxation of problem (P) is then formed by minimizing the Lagrangian function for a given λ:

$$(L_\lambda) \qquad d(\lambda) = \min_{x \in X} L(x, \lambda) = f(x) + \lambda^T(g(x) - b), \quad (3.1.1)$$

where $d(\lambda)$ is called a dual function. The Lagrangian dual is a maximization problem over the dual function with respect to λ, namely,

$$(D) \qquad \max_{\lambda \in \mathbb{R}^m_+} d(\lambda).$$

Let $v(Q)$ be the optimal value of problem (Q). The following theorem shows that for any $\lambda \in \mathbb{R}^m_+$, problem (L_λ) is a relaxation of the primal problem (P), since the minimum value of (L_λ) never exceeds the minimum value of (P).

THEOREM 3.1 (WEAK LAGRANGIAN DUALITY) *For all* $\lambda \in \mathbb{R}^m_+$,

$$v(L_\lambda) \le v(P). \quad (3.1.2)$$

Furthermore,

$$v(D) \le v(P). \quad (3.1.3)$$

Proof. The following is evident for any $\lambda \in \mathbb{R}^m_+$,

$$\begin{aligned} v(P) &= \min\{f(x) \mid g(x) \le b, \ x \in X\} \\ &\ge \min\{f(x) \mid \lambda^T(g(x) - b) \le 0, \ x \in X\} \\ &\ge \min\{f(x) + \lambda^T(g(x) - b) \mid \lambda^T(g(x) - b) \le 0, \ x \in X\} \\ &\ge \min\{f(x) + \lambda^T(g(x) - b) \mid x \in X\} \\ &= v(L_\lambda). \end{aligned}$$

This yields (3.1.2). Since $v(L_\lambda) \le v(P)$ holds for all $\lambda \in \mathbb{R}^m_+$, we imply that (3.1.3) holds true. □

It is clear that the optimal Lagrangian dual value $v(D)$ always provides a lower bound for $v(P)$.

THEOREM 3.2 (STRONG LAGRANGIAN DUALITY) *If* $x^* \in X$ *solves* (L_{λ^*}) *with* $\lambda^* \in \mathbb{R}^m_+$, *and, in addition, the following conditions are satisfied:*

$$g_i(x^*) \le b_i, \ \ i = 1, \ldots, m, \quad (3.1.4)$$

$$\lambda_i^*(g_i(x^*) - b_i) = 0, \ \ i = 1, \ldots, m, \quad (3.1.5)$$

then x^* *solves problem* (P) *and* $v(D) = v(P)$.

Proof. It is clear that x^* is feasible in (P) and thus $f(x^*) \geq v(P)$. From the weak Lagrangian duality and the assumption, we have

$$v(P) \geq v(D) \geq v(L_{\lambda^*}) = f(x^*) + (\lambda^*)^T(g(x^*) - b) = f(x^*) \geq v(P).$$

Thus, $v(D) = v(P) = f(x^*)$ and x^* solves problem (P). □

Unfortunately, it is rare that the strong Lagrangian duality is satisfied in integer programming. More specifically, the strong duality conditions rarely hold true for an optimal solution of integer programming problem since the constraint $g_i(x) \leq b_i$ is often inactive at x^* for index i with $\lambda_i^* > 0$. The difference $v(P) - v(D)$ is called *duality gap* between problems (P) and (D). For any feasible solution $x \in S$, the difference $f(x) - v(D)$ is called a *duality bound.*

The following theorem reveals that performing dual search separately on individual sub-domains is never worse than performing dual search on the entire domain as a whole.

THEOREM 3.3 *Suppose that the domain X in (P) can be decomposed into a union of sub-domains, $X = \cup_{k=1}^{K} X^k$. Let*

$$\begin{aligned} S &= \{x \in X \mid g(x) \leq b\}, \\ S_k &= \{x \in X^k \mid g(x) \leq b\}, \\ d(\lambda) &= \min_{x \in X} L(x, \lambda), \\ d_k(\lambda) &= \min_{x \in X^k} L(x, \lambda),\ k = 1, \ldots, K. \end{aligned}$$

Furthermore, let λ^* be the solution to the dual problem (D) and λ_k^* be the solution to the dual problem on X^k, $\max_{\lambda \in \mathbb{R}_+^m} d_k(\lambda)$, $k = 1, \ldots, K$. Then,

$$d(\lambda^*) \leq \min_{1 \leq k \leq K} d_k(\lambda_k^*) \leq \min_{x \in S} f(x) = v(P) \tag{3.1.6}$$

Proof. Since $X^k \subseteq X$, $d(\lambda) \leq d_k(\lambda)$ for all $\lambda \in \mathbb{R}_+^m$ and $k = 1, \ldots, K$. We thus have $d(\lambda^*) \leq d_k(\lambda_k^*)$ for $k = 1, \ldots, K$. This further leads to the first inequality in (3.1.6). On the other hand, from the weak duality, we have $d_k(\lambda_k^*) \leq \min_{x \in S_k} f(x)$. Thus

$$\min_{1 \leq k \leq K} d_k(\lambda_k^*) \leq \min_{1 \leq k \leq K} \min_{x \in S_k} f(x) = \min_{x \in S} f(x),$$

which is the second inequality in (3.1.6). □

A basic property of the dual function is summarized in the following theorem.

THEOREM 3.4 *The dual function $d(\lambda)$ is a piecewise linear concave function on $\mathbb{R}^m_+$.*

Proof. By definition, for any $\lambda \in \mathbb{R}^n_+$,

$$d(\lambda) = \min_{x \in X}[f(x) + \lambda^T(g(x) - b)].$$

Since X is finite, d is the minimum of a finite number of linear functions of λ. Thus, $d(\lambda)$ is a piecewise linear concave function. □

Recall that $\xi \in \mathbb{R}^m$ is a subgradient of d at λ if

$$d(\mu) \leq d(\lambda) + \xi^T(\mu - \lambda), \quad \forall \mu \in \mathbb{R}^m_+.$$

THEOREM 3.5 *Let x_λ be an optimal solution to the Lagrangian relaxation problem (L_λ), then $\xi = g(x_\lambda) - b$ is a subgradient of $d(\lambda)$ at λ.*

Proof. Since x_λ is an optimal solution to (L_λ), it holds

$$d(\lambda) = f(x_\lambda) + \lambda^T(g(x_\lambda) - b).$$

For any $\mu \in \mathbb{R}^m_+$, we have

$$\begin{aligned} d(\mu) &= \min_{x \in X}[f(x) + \mu^T(g(x) - b)] \\ &\leq f(x_\lambda) + \mu^T(g(x_\lambda) - b) \\ &= f(x_\lambda) + \lambda^T(g(x_\lambda) - b) + (g(x_\lambda) - b)^T(\mu - \lambda) \\ &= d(\lambda) + \xi^T(\mu - \lambda). \end{aligned}$$

Thus, $\xi = g(x_\lambda) - b$ is a subgradient of the dual function. □

Define the subdifferential $\partial d(\lambda)$ to be the set of all subgradients of d at λ, namely,

$$\partial d(\lambda) = \{\xi \mid d(\mu) \leq d(\lambda) + \xi^T(\mu - \lambda), \forall \mu \in \mathbb{R}^m_+\}.$$

It is easy to see that any vector in the convex hull of all the subgradients in the form of $\xi = g(x_\lambda) - b$ is also a subgradient of d at λ. In fact, $\partial d(\lambda)$ can be totally characterized by the subgradients in the form of $g(x_\lambda) - b$ (see [100]):

$$\partial d(\lambda) = conv\{g(x) - b \mid d(\lambda) = f(x) + \lambda^T(g(x) - b), \ x \in X\}.$$

Consider a special case of the primal problem where f, g and X are of separable structures:

$$\min\ f(x) = \sum_{j=1}^{n} f_j(x_j) \tag{3.1.7}$$

$$\text{s.t. } g_i(x) = \sum_{j=1}^{n} g_{ij}(x_j) \leq b_i, i = 1, \ldots, m,$$

$$x \in X_1 \times X_2 \times \ldots \times X_n \subseteq \mathbb{Z}^n.$$

The Lagrangian function of (3.1.7) can be expressed as a summation of n univariate functions,

$$L(x, \lambda) = \sum_{j=1}^{n} L_j(x_j, \lambda) - \lambda^T b,$$

where

$$L_j(x_j, \lambda) = f_j(x_j) + \sum_{i=1}^{m} \lambda_i g_{ij}(x_j).$$

Then the Lagrangian relaxation problem (L_λ) can be decomposed into n one-dimensional subproblems,

$$\begin{aligned} d(\lambda) &= \min_{x \in X} L(x, \lambda) \\ &= \min_{x \in X} \sum_{j=1}^{n} L_j(x_j, \lambda) - \lambda^T b \\ &= \sum_{j=1}^{n} [\min_{x_j \in X_j} L_j(x_j, \lambda)] - \lambda^T b. \end{aligned} \tag{3.1.8}$$

In a worst case scenario, the total enumeration scheme for computing $d(\lambda)$ requires $O(\sum_{j=1}^{n}(u_j - l_j + 1))$ evaluations of L_j's and comparisons. Comparing with (L_λ), an integer programming problem over integer set X with $|X| = \prod_{j=1}^{n}(u_j - l_j + 1)$, the above one-dimensional integer optimization is much easier to solve. Therefore, Lagrangian dual method is very powerful when dealing with separable integer optimization problems, due to the decomposition scheme.

Since the Lagrangian relaxation problem (L_λ) has to be solved many times for different λ in a dual search procedure, it is desirable to derive methods more efficient than the total enumeration for evaluating $d(\lambda)$. Consider the linearly constrained case of (3.1.7) where $g_i(x) = \sum_{j=1}^{n} a_{ij}x_j, i = 1, \ldots, m$. The j-th one-dimensional subproblem in (3.1.8) becomes:

$$\min_{x_j \in X_j} L_j(x_j, \lambda) = \min_{x_j \in X_j} [f_j(x_j) + \sum_{i=1}^{m} \lambda_i a_{ij} x_j].$$

Let $y_j = \sum_{i=1}^{m} \lambda_i a_{ij}$. Denote by x_{jt}, $t = 1, \ldots, T_j$, the integer values in X_j. Let $Q_j = \{1, \ldots, T_j\}$ and $f_{jt} = f_j(x_{jt})$, $t \in Q_j$. Then the j-th one-dimensional subproblem in (3.1.8) can be expressed as

$$(SP)_j \quad \min_{t \in Q_j} [f_{jt} + y_j x_{jt}].$$

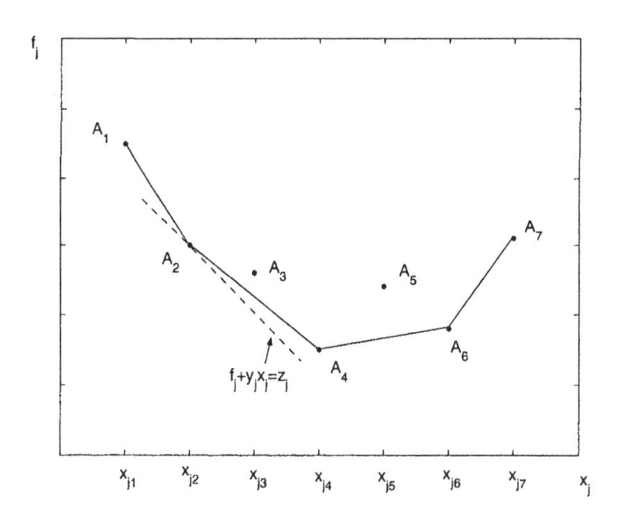

Figure 3.1. Illustration of the solution scheme for (SP_j).

As illustrated in Figure 3.1, the process of minimizing $f_{jt}+y_j x_{jt}$ over $t \in Q_j$ corresponds to moving the line $z_j = f_j + y_j x_j$ along the direction $(-y_j, -1)$ until the line last touches the lower convex envelope of points $(x_{jt}, f_{jt}), t \in Q_j$. It is clear that the minimum value of $(SP)_j$ is achieved at one of the extreme points of the lower convex envelope. Furthermore, we have the following observations from Figure 3.1: (a) points A_3 and A_5 are not on the lower convex envelope and thus cannot be touched by the line corresponding to the optimal solution to $(SP)_j$; (b) the slope $-y_j$ is bounded from below by the slope of the line connecting A_1 and A_2, and bounded from above by the slope of the line connecting A_2 and A_4, when $f_j + y_j x_j$ achieves the minimum value at A_2. In general, we have the following propositions ([55]).

PROPOSITION 3.1 *Let p, q, $r \in Q_j$ be such that*

$$x_{jp} < x_{jq} < x_{jr}, \tag{3.1.9}$$
$$\frac{f_{jr} - f_{jp}}{x_{jr} - x_{jp}} \geq \frac{f_{jr} - f_{jq}}{x_{jr} - x_{jq}}. \tag{3.1.10}$$

Then, there is an optimal solution x_j^* to $(SP)_j$ such that $x_j^* \neq x_{jq}$.

Proof. By (3.1.9), x_{jq} can be expressed as a convex combination of x_{jp} and x_{jr}. More specifically, $x_{jq} = \alpha x_{jp} + (1-\alpha)x_{jr}$ with $\alpha = (x_{jr} - x_{jq})/(x_{jr} - x_{jp})$. The inequality (3.1.10) becomes

$$f_{jq} \geq \alpha f_{jp} + (1-\alpha) f_{jr}. \tag{3.1.11}$$

Suppose, on the contrary, x_{jq} solves $(SP)_j$ uniquely. Then $f_{jq} + y_j x_{jq} < f_{jp} + y_j x_{jp}$ and $f_{jq} + y_j x_{jq} < f_{jr} + y_j x_{jr}$. This yields

$$f_{jq} + y_j x_{jq} < \alpha f_{jp} + (1-\alpha) f_{jr} + y_j(\alpha x_{jp} + (1-\alpha) x_{jr}).$$

Since $x_{jq} = \alpha x_{jp} + (1-\alpha)x_{jr}$, it follows from the above inequality that $f_{jq} < \alpha f_{jp} + (1-\alpha) f_{jr}$. A contradiction to (3.1.11). □

Proposition 3.1 implies that if a point $x_{jq} \in X_j$ satisfies conditions (3.1.9) and (3.1.10), then it cannot be an extreme point of the lower convex envelope of points (x_{jt}, f_{jt}). Let $\widetilde{Q}_j$ be the subset of Q_j after removing those q's with x_{jq} satisfying (3.1.9) and (3.1.10). Obviously, $\widetilde{Q}_j$ is the index set of the extreme points of the lower convex envelope of points (x_{jt}, f_{jt}), $t \in Q_j$. The index set $\widetilde{Q}_j$ can be efficiently determined by an $O(T_j)$ search scheme ([55]).

Denote $\widetilde{Q}_j = \{1, 2, \ldots, R_j\}$ after relabeling $\widetilde{Q}_j$. Define

$$\beta_{jd} = \frac{f_{jd} - f_{j,d-1}}{x_{jd} - x_{j,d-1}}, \quad d = 2, \ldots, R_j, \ \beta_{j1} = -\infty, \ \beta_{j,R_j+1} = +\infty.$$

Then, $\{\beta_{jd}\}$ is a nondecreasing sequence for $d = 1, \ldots, R_j + 1$.

PROPOSITION 3.2 *Let p be such that $\beta_{jp} \leq -y_j \leq \beta_{j,p+1}$. Then, x_{jp} is an optimal solution to $(SP)_j$.*

Proof. For any $d = 2, \ldots, R_j$, by the definition of β_{jd}, we have

$$(f_{jd} + y_j x_{jd}) - (f_{j,d-1} + y_j x_{j,d-1}) = (x_{jd} - x_{j,d-1})(y_j + \beta_{jd}). \tag{3.1.12}$$

Let $L_d = f_{jd} + y_j x_{jd}$. Since $\beta_{j1} \leq \ldots \leq \beta_{j,p-1} \leq \beta_{jp} \leq -y_j \leq \beta_{j,p+1} \leq \beta_{j,p+2} \leq \ldots \leq \beta_{j,R_j+1}$, it follows from (3.1.12) that $L_d \leq L_{d-1}$ for $d \leq p$ and $L_d \geq L_{d-1}$ for $d \geq p+1$. Thus, $\{L_d\}$ is nonincreasing for $d \leq p$ and is nondecreasing for $d \geq p+1$. Therefore, L_p is the minimum of all L_d's and hence x_{jp} solves $(SP)_j$. □

Notice that the set $\widetilde{Q}_j$ and the sequence $\{\beta_{jd}\}$, $d = 2, \ldots, R_j$, are independent of y_j and λ. Thus, they are only needed to be computed once and can be stored and used in the process of a dual search procedure where (L_λ) has to be solved many times for different λ. Proposition 3.1 suggests that if $\widetilde{Q}_j$ and β_{jd} ($d = 2, \ldots, R_j$) are available, the optimal solution x_j^* to $(SP)_j$ can be determined by $O(R_j) = O(T_j)$ comparisons (using bisection). The Lagrangian relaxation (L_λ), therefore, can be solved by $O(\sum_{j=1}^n T_j)$ comparisons. Clearly, this may result in a significant saving in computation when the problem (L_λ) is solved repeatedly in the dual search procedure.

3.2 Dual Search Methods

In this section, we study solution methods for solving the dual problem,

$$(D) \quad \max_{\lambda \geq 0} d(\lambda).$$

As discussed in Section 3.1, the dual function is a piecewise linear concave function on $\mathbb{R}^m_+$ and one of its subgradients at λ is readily obtained after solving the corresponding Lagrangian relaxation problem (L_λ). These properties facilitate the search of an optimal solution to (D).

3.2.1 Subgradient method

Unlike the gradient of a smooth function, the direction of a subgradient of a nonsmooth concave function is not necessarily an ascent direction of the function. Nevertheless, it can be shown that the subgradient is indeed a descent direction of the Euclidean distance to the set of the maximum points of the dual function (see [198]). This property leads to the well-known *subgradient method* in nonsmooth optimization.

The basic subgradient method for solving (D) iterates as follows:

$$\lambda^{k+1} = P^+(\lambda^k + s_k \xi^k / \|\xi^k\|), \tag{3.2.1}$$

where ξ^k is a subgradient of d at λ^k, s_k is the stepsize, and P^+ is the projection operator that projects $\mathbb{R}^m$ into $\mathbb{R}^m_+$, i.e.,

$$P^+(\lambda) = \max(0, \lambda) = (\max(0, \lambda_1), \ldots, \max(0, \lambda_m))^T.$$

PROCEDURE 3.1 (BASIC SUBGRADIENT METHOD FOR (D))

Step 0. Choose any $\lambda^1 \geq 0$. Set $v^1 = -\infty$, $k = 1$.

Step 1. Solve the Lagrangian problem

$$(L_{\lambda^k}) \qquad d(\lambda^k) = \min_{x \in X} L(x, \lambda^k)$$

and obtain an optimal solution x^k. Set $\xi^k = g(x^k) - b$ and $v^{k+1} = \max(v^k, d(\lambda^k))$. If $\xi^k = 0$, stop and λ^k is the optimal solution to (D) due to the strong duality.

Step 2. Compute

$$\lambda^{k+1} = P^+(\lambda^k + s_k \xi^k / \|\xi^k\|),$$

where $s_k > 0$ is the stepsize.

Step 3. Set $k := k + 1$, go to Step 1.

We have the following basic lemma for the above procedure.

LEMMA 3.1 *Let $\lambda^* \geq 0$ be an optimal solution to (D). Then, for any k, we have*

$$d(\lambda^*) - v^k \leq \frac{\|\lambda^1 - \lambda^*\|^2 + \sum_{i=1}^k s_i^2}{2\sum_{i=1}^k (s_i/\|\xi^i\|)}. \tag{3.2.2}$$

Proof. Since ξ^i is a subgradient of d at λ^i, we have

$$d(\lambda^*) \leq d(\lambda^i) + (\xi^i)^T(\lambda^* - \lambda^i).$$

Thus,

$$\begin{aligned} \|\lambda^{i+1} - \lambda^*\|^2 &= \|P^+(\lambda^i + s_i\xi^i/\|\xi^i\|) - P^+(\lambda^*)\|^2 \\ &\leq \|\lambda^i + s_i\xi^i/\|\xi^i\| - \lambda^*\|^2 \\ &= \|\lambda^i - \lambda^*\|^2 + (2s_i/\|\xi^i\|)(\xi^i)^T(\lambda^i - \lambda^*) + s_i^2 \\ &\leq \|\lambda^i - \lambda^*\|^2 + (2s_i/\|\xi^i\|)[d(\lambda^i) - d(\lambda^*)] + s_i^2. \end{aligned} \tag{3.2.3}$$

Summing up (3.2.3) for $i = 1, \ldots, k$, we obtain

$$0 \leq \|\lambda^{k+1} - \lambda^*\|^2 \leq \|\lambda^1 - \lambda^*\|^2 + 2\sum_{i=1}^k (s_i/\|\xi^i\|)[d(\lambda^i) - d(\lambda^*)] + \sum_{i=1}^k s_i^2.$$

Therefore,

$$\begin{aligned} d(\lambda^*) - v^k &= d(\lambda^*) - \max_{i=1,\ldots,k} d(\lambda^i) \\ &\leq \frac{\sum_{i=1}^k (s_i/\|\xi^i\|)[d(\lambda^*) - d(\lambda^i)]}{\sum_{i=1}^k (s_i/\|\xi^i\|)} \\ &\leq \frac{\|\lambda^1 - \lambda^*\|^2 + \sum_{i=1}^k s_i^2}{2\sum_{i=1}^k (s_i/\|\xi^i\|)}. \end{aligned}$$

□

Various stepsize rules for choosing s_k have been proposed ([118][119][180]). In the following, we discuss three basic stepsize rules.

(i) Rule 1 for stepsize (constant):

$$s_k = \epsilon, \tag{3.2.4}$$

where $\epsilon > 0$ is a constant.

(ii) Rule 2 for stepsize:

$$\sum_{k=1}^{+\infty} s_k^2 < +\infty \text{ and } \sum_{k=1}^{+\infty} s_k = +\infty. \tag{3.2.5}$$

(iii) Rule 3 for stepsize:

$$s_k \to 0,\ k \to +\infty, \text{ and } \sum_{k=1}^{+\infty} s_k = +\infty. \tag{3.2.6}$$

Notice that there exists $M > 0$ such that $\|\xi^k\| = \|g(x^k) - b\| \le M$ for any k since $x^k \in X$ and X is a finite integer set.

THEOREM 3.6 (i) *If Rule 1 for stepsize is used in Procedure 3.1, then*

$$\liminf_{k\to\infty} v^k \ge d(\lambda^*) - (1/2)\epsilon M. \tag{3.2.7}$$

(ii) *If Rule 2 or Rule 3 for stepsize is used in Procedure 3.1, then*

$$\lim_{k\to+\infty} v^k = d(\lambda^*). \tag{3.2.8}$$

Proof. (i) Note that $\|\xi^i\| \le M$ for any i. By Lemma 3.1, we have

$$d(\lambda^*) - v^k \le \frac{\|\lambda^1 - \lambda^*\|^2 + \epsilon^2 k}{2\epsilon k/M} \to (1/2)\epsilon M \quad (k \to \infty).$$

This is (3.2.7).

(ii) If Stepsize 2 is used, then, by Lemma 3.1, we have

$$0 \le d(\lambda^*) - v^k \le \frac{\|\lambda^1 - \lambda^*\|^2 + \sum_{i=1}^k s_i^2}{2\sum_{i=1}^k (s_i/M)} \to 0, \quad (k \to \infty). \tag{3.2.9}$$

Thus, (3.2.8) holds true. Suppose now Stepsize 3 is used. We claim that the right-hand side of (3.2.9) converges to 0. Otherwise, there must exist $\eta > 0$ such that

$$\frac{\|\lambda^1 - \lambda^*\|^2 + \sum_{i=1}^k s_i^2}{2\sum_{i=1}^k (s_i/M)} \geq \eta, \ \forall k,$$

or

$$\sum_{i=1}^k s_i^2 - 2(\eta/M)\sum_{i=1}^k s_i \geq -\|\lambda^1 - \lambda^*\|^2, \ \forall k. \tag{3.2.10}$$

Since $s_i \to 0$ $(i \to \infty)$, there exists N_1 such that $s_i \leq \eta/M$ when $i > N_1$. Thus,

$$\begin{aligned}
&\sum_{i=1}^k s_i^2 - 2(\eta/M)\sum_{i=1}^k s_i \\
&= (\sum_{i=1}^{N_1} s_i^2 + \sum_{i=N_1+1}^{k} s_i^2) - (\eta/M)(\sum_{i=1}^k s_i + \sum_{i=1}^k s_i) \\
&\leq \sum_{i=1}^{N_1} s_i^2 + (\eta/M)\sum_{i=N_1+1}^{k} s_i - (\eta/M)(\sum_{i=1}^k s_i + \sum_{i=N_1+1}^{k} s_i) \\
&= \sum_{i=1}^{N_1} s_i^2 - (\eta/M)\sum_{i=1}^k s_i \to -\infty \quad (k \to \infty).
\end{aligned}$$

This contradicts (3.2.10). □

Now, consider a more sophisticated stepsize rule:

$$s_k = \rho\frac{w_k - d(\lambda^k)}{\|\xi^k\|}, \quad 0 < \rho < 2, \tag{3.2.11}$$

where w_k is an approximation of the optimal value $v(D)$, $w_k \geq d(\lambda^k)$ and $\xi^k \neq 0$.

THEOREM 3.7 *Let $\{\lambda^k\}$ be the sequence generated by Procedure 3.1 where s_k is defined by (3.2.11). If $\{w_k\}$ is monotonically increasing and $\lim_{k\to\infty} w_k = w \leq v(D)$, then*

$$\lim_{k\to+\infty} d(\lambda^k) = w, \text{ and } \lim_{k\to\infty} \lambda^k = \lambda^*, \text{ with } d(\lambda^*) = w.$$

Proof. For any $\lambda \in \Lambda(w) = \{\lambda \in \mathbb{R}^m_+ \mid d(\lambda) \geq w\}$, we have

$$\begin{aligned}
\|\lambda - \lambda^{k+1}\|^2 &= \|P^+(\lambda) - P^+(\lambda^k + s_k\xi^k/\|\xi^k\|)\|^2 \\
&\leq \|\lambda - \lambda^k - s_k(\xi^k/\|\xi^k\|)\|^2 \\
&= \|\lambda - \lambda^k\|^2 + s_k^2 - 2(s_k/\|\xi^k\|)(\xi^k)^T(\lambda - \lambda^k) \\
&\leq \|\lambda - \lambda^k\|^2 + s_k^2 - 2(s_k/\|\xi^k\|)(d(\lambda) - d(\lambda^k)) \\
&\leq \|\lambda - \lambda^k\|^2 + s_k^2 - 2s_k(w_k - d(\lambda^k))/\|\xi^k\| \\
&= \|\lambda - \lambda^k\|^2 - \rho(2-\rho)(w_k - d(\lambda^k))^2/\|\xi^k\|^2.
\end{aligned} \tag{3.2.12}$$

Thus, $\{\|\lambda - \lambda^k\|\}$ is a monotonically decreasing sequence and hence converges. Taking limits on the both sides of the above inequality and noting that $\{\|\xi^k\|\}$ is a bounded sequence, we deduce that

$$\lim_{k \to +\infty} d(\lambda^k) = w.$$

Let λ^* be a limit point of the bounded sequence $\{\lambda^k\}$. Then $d(\lambda^*) = w$ by the continuity of d and hence $\lambda^* \in \Lambda(w)$. Since $\{\|\lambda^* - \lambda^k\|\}$ is monotonically decreasing, we conclude that $\lim_{k\to\infty} \lambda^k = \lambda^*$. □

Notice that if we know the exact value of $v(D)$, then choosing $w^k = w = v(D)$ in (3.2.11) leads to a convergent subgradient method. Let's consider an example to illustrate the computational effects of using different types of stepsizes.

EXAMPLE 3.1 Consider the following quadratic integer programming problem:

$$\begin{aligned}
\min\ & f(x) = \sum_{j=1}^{n} (\alpha_j x_j^2 + \beta_j x_j) \\
\text{s.t. } & Ax \leq b, \\
& x \in X = \{x \in \mathbb{Z}^n \mid l \leq x \leq u\},
\end{aligned}$$

where A is an $m \times n$ matrix.

We use subgradient methods to solve the dual problem of the example with the three different rules of stepsize. We take $n = 20$, $m = 10$, $l = (1, \ldots, 1)^T$ and $u = (5, \ldots, 5)^T$. The data α_j, β_j, $A = (a_{ij})$ and b_i are taken from uniform distributions with $\alpha_j \in (0, 10]$, $\beta_j \in [-120, -100]$, $a_{ij} \in [1, 50]$ and $b_i = 0.5 \times \sum_{j=1}^{n} a_{ij}(l_j + u_j)$.

1. For Rule 1 of stepsize, we take $s_k = \epsilon$. Figures 3.2 and 3.3 depict the error bounds $d(\lambda^*) - d(\lambda^k)$ and $d(\lambda^*) - v^k$ for $\epsilon = 0.02, 0.05$ ($k = 1, \ldots, 500$).

2. For Rule 2 of stepsize, we take $s_k = \epsilon/k$. Figures 3.4 and 3.5 depict the error bounds $d(\lambda^*) - d(\lambda^k)$ and $d(\lambda^*) - v^k$ for $\epsilon = 0.3, 0.5$ ($k = 1, \ldots, 500$).

3. For Rule 3 of stepsize, we take $s_k = \epsilon/\sqrt{k}$. Figures 3.6 and 3.7 depict the error bounds $d(\lambda^*) - d(\lambda^k)$ and $d(\lambda^*) - v^k$ for $\epsilon = 2, 3$ ($k = 1, \ldots, 500$).

From Figures 3.2 and 3.3, we can see that for the subgradient method with the constant stepsize, a smaller ϵ results in a slower convergence and a larger ϵ causes a wider variance of $d(\lambda^k)$ values, thus leading to a slow convergence in later stages of the iterations. Similar phenomenon can be observed from Figures 3.4–3.7 for the subgradient methods with the stepsize $s_k = \epsilon/k$ and $s_k = \epsilon/\sqrt{k}$. Hybrid strategies of using different rules of stepsize can be adopted to achieve the best trade-off. In practice, a suitable parameter ϵ can be obtained empirically.

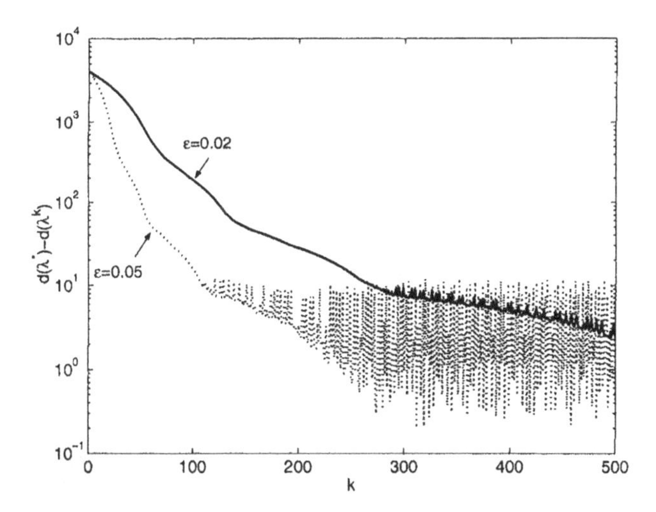

Figure 3.2. Error bound $d(\lambda^*) - d(\lambda^k)$ in the subgradient method with Rule 1 for stepsize for Example 3.1.

3.2.2 Outer Lagrangian linearization method

The dual problem (D) can be rewritten as a linear programming problem:

$$(LD) \qquad \max \mu$$
$$\text{s.t. } \mu \le f(x) + \lambda^T(g(x) - b), \quad \text{for all } x \in X,$$
$$\lambda \ge 0.$$

Apparently, the number of constraints in (LD) is equal to the cardinality of X. Thus, it is difficult to solve this linear programming problem directly due to its

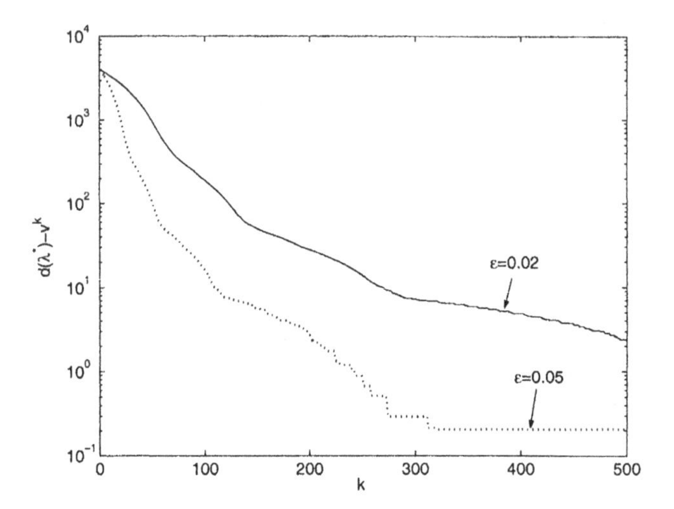

Figure 3.3. Error bound $d(\lambda^*) - v^k$ in the subgradient method with Rule 1 for stepsize for Example 3.1.

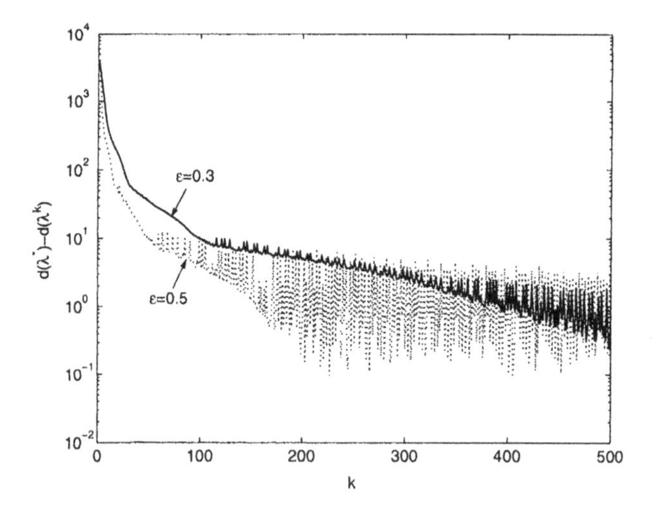

Figure 3.4. Error bound $d(\lambda^*) - d(\lambda^k)$ in the subgradient method with Rule 2 for stepsize for Example 3.1.

huge number of constraints. Nevertheless, we can successively approximate the dual function by adding linear constraints (cutting plane). Geometrically, we construct a cutting-plane approximation to the surface of dual function d near the optimal solution λ^*.

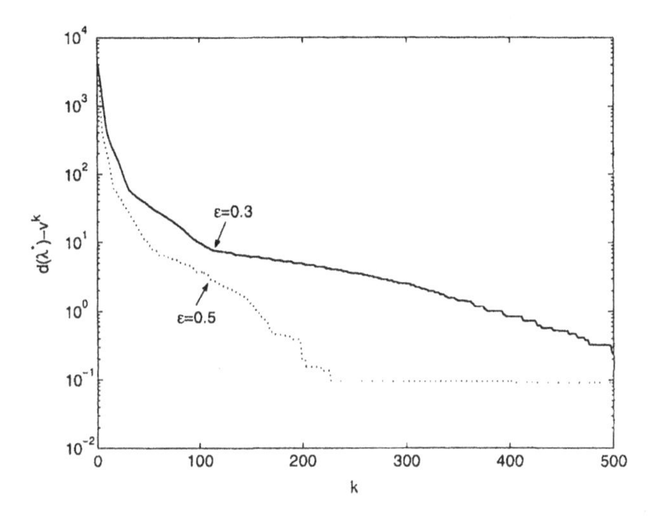

Figure 3.5. Error bound $d(\lambda^*) - v^k$ in the subgradient method with Rule 2 for stepsize for Example 3.1.

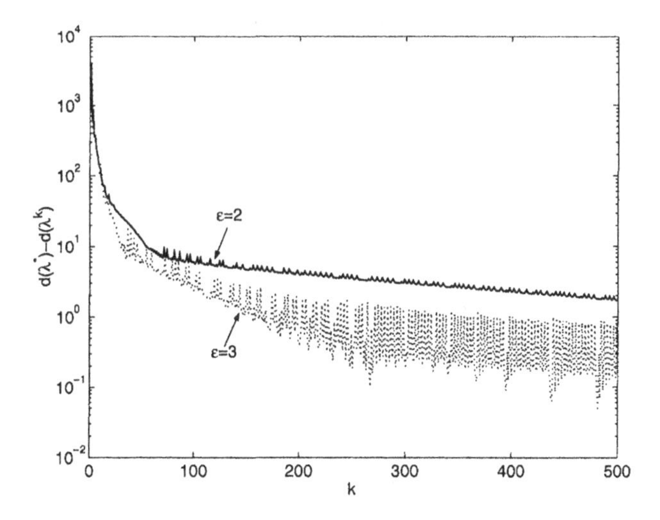

Figure 3.6. Error bound $d(\lambda^*) - d(\lambda^k)$ in the subgradient method with Rule 3 for stepsize for Example 3.1.

PROCEDURE 3.2 (OUTER LAGRANGIAN LINEARIZATION METHOD FOR (D))

Step 0. Choose subset T^1 of X such that T^1 contains at least one feasible solution x^0. Set $k = 1$.

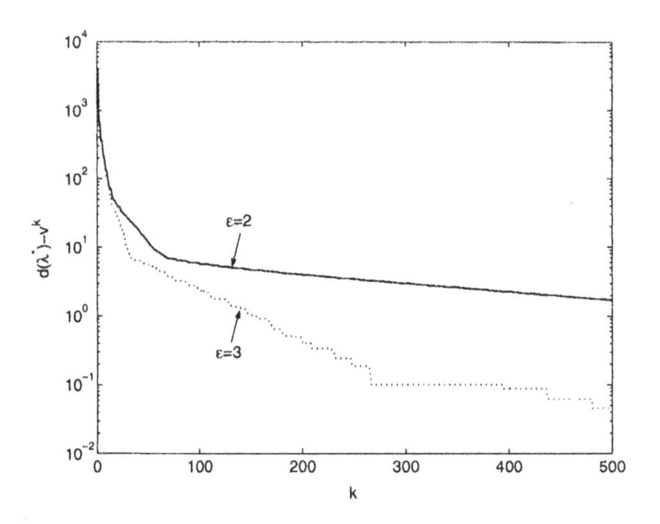

Figure 3.7. Error bound $d(\lambda^*) - v^k$ in the subgradient method with Rule 3 for stepsize for Example 3.1.

Step 1. Solve the linear programming problem

$$(LD^k) \qquad \max \ \mu$$
$$\text{s.t.} \ \ \mu \le f(x^j) + \lambda^T(g(x^j) - b), \quad \text{for all } x^j \in T^k,$$
$$\lambda \ge 0.$$

Let (μ^k, λ^k) be an optimal solution to (LD^k).

Step 2. Solve the Lagrangian relaxation problem (L_{λ^k}) and obtain the dual value $d(\lambda^k)$ and an optimal solution $x^k \in X$.

Step 3. If

$$(\lambda^k)^T[g(x^k) - b] = 0, \ g(x^k) \le b, \tag{3.2.13}$$

then stop, x^k is the optimal solution to (P) and λ^k is the optimal solution to (D) with $v(P) = v(D)$. If

$$\mu^k \le d(\lambda^k), \tag{3.2.14}$$

stop and λ^k is the optimal solution to (D) with $\mu^k = v(D)$.

Step 4. Update T^k by adding x^k,

$$T^{k+1} = T^k \cup \{x^k\}.$$

Set $k := k + 1$, go to Step 1.

REMARK 3.1 Notice that in Step 1 if $g(x^0) \leq b$, then $\mu \leq f(x^0)+\lambda^T(g(x^0)-b) \leq f(x^0)$. Therefore, an initial feasible solution x^0 is needed to guarantee that linear programming problem (LD^k) has a finite optimal value. Otherwise, (LD^k) may be unbounded and Step 1 of the algorithm is not well-defined.

THEOREM 3.8 *Procedure 3.2 stops at an optimal solution to (D) in a finite number of iterations.*

Proof. We first notice that $(\mu, \lambda) = (\min_{x^j \in T^k} f(x^j), 0)$ is always feasible to (LD^k). Since T^k contains at least one feasible x^0 to (P), problem (LD^k) has a finite solution for each k. If the procedure stops at Step 3 with (3.2.13) satisfied, then the strong duality conditions (3.1.4)–(3.1.5) hold. Thus, x^k is the optimal solution to (P) and λ^k is the optimal solution to (D) with $v(P) = v(D)$. If the procedure stops at Step 3 with (3.2.14) satisfied, then we have

$$v(D) \geq d(\lambda^k) = f(x^k) + (\lambda^k)^T(g(x^k) - b) \geq \mu^k.$$

On the other hand, since the feasible region of (LD) is a subset of that of (LD^k), we have $\mu^k \geq v(D)$. Therefore, $\mu^k = v(D)$ and λ^k is an optimal solution to (D).

If the procedure does not stop at Step 3, then (3.2.14) is not satisfied and hence $x^k \notin T^k$. Therefore, a new point x^k is included in T^{k+1}. Since X is finite, the procedure will terminate in a finite number of iterations. □

Note that linear programming problem (LD^{k+1}) is formed by adding one constraint to (LD^k). The optimal solution to (LD^k) $(k = 1, 2, \ldots)$ can be efficiently computed if the dual simplex method is used.

Since X is a finite set, we can express X as $\{x^t\}_{t=1}^T$. The dual problem of (LD) is

$$(DLD) \qquad \min \sum_{t=1}^T f(x^t)\mu_t$$
$$\text{s.t.} \sum_{t=1}^T g(x^t)\mu_t \leq b,$$
$$\sum_{t=1}^T \mu_t = 1, \ \mu_t \geq 0, \ t = 1, \ldots, T.$$

Dantzig-Wolfe decomposition ([227]) can be applied to the formulation (DLD) and has recently drawn much attention in solving large-scale linear integer programming problems.

To illustrate Procedure 3.2, let us consider the following example.

EXAMPLE 3.2

$$\begin{aligned} \min\ & f(x) = 3x_1^2 + 2x_2^2 \\ \text{s.t.}\ & g_1(x) = 10 - 5x_1 - 2x_2 \leq 7, \\ & g_2(x) = 15 - 2x_1 - 5x_2 \leq 12, \\ & x \in X = \left\{ \begin{array}{c} \text{integer} \\ 0 \leq x_1 \leq 1, 0 \leq x_2 \leq 2 \\ 8x_1 + 8x_2 \geq 1 \end{array} \right\}. \end{aligned}$$

The explicit expression of set X is $X = \{(0,1)^T, (0,2)^T, (1,0)^T, (1,1)^T, (1,2)^T\}$. It is easy to check that the feasible solutions are $(0,2)^T$, $(1,1)^T$ and $(1,2)^T$. The optimal solution is $x^* = (1,1)^T$ with $f(x^*) = 5$.

The iteration process of Procedure 3.2 for this example is described as follows.

Step 0. Choose $x^0 = (1,1)^T$, $T^1 = \{x^0\}$. Set $k = 1$.

Iteration 1

Step 1. Solve the linear programming problem

$$(LD^1) \qquad \begin{aligned} \max\ & \mu \\ \text{s.t.}\ & \mu \leq 5 - 4\lambda_1 - 4\lambda_2, \\ & \lambda_1 \geq 0,\ \lambda_2 \geq 0. \end{aligned}$$

We obtain $\mu^1 = 5$ and $\lambda^1 = (0,0)^T$.

Step 2. Solving Lagrangian relaxation problem (L_{λ^1}) gives $d(\lambda^1) = 2$ and $x^1 = (0,1)^T$.

Step 3. $\mu^1 = 5 > 2 = d(\lambda^1)$.

Step 4. Set $T^1 = \{x^0, x^1\}$.

Iteration 2

Step 1. Solve the linear programming problem

$$(LD^2) \qquad \begin{aligned} \max\ & \mu \\ \text{s.t.}\ & \mu \leq 5 - 4\lambda_1 - 4\lambda_2, \\ & \mu \leq 2 + \lambda_1 - 2\lambda_2, \\ & \lambda_1 \geq 0,\ \lambda_2 \geq 0. \end{aligned}$$

We obtain $\mu^2 = 2.6$ and $\lambda^2 = (0.6,0)^T$.

Step 2. Solving Lagrangian relaxation problem (L_{λ^2}) gives $d(\lambda^2) = 1.8$ and $x^2 = (1,0)^T$.

Step 3. $\mu^2 = 2.6 > 1.8 = d(\lambda^2)$.

Step 4. Set $T^2 = \{x^0, x^1, x^2\}$.

Iteration 3

Step 1. Solve the linear programming problem

$$
\begin{aligned}
(LD^3) \qquad & \max\ \mu \\
& \text{s.t. } \mu \le 5 - 4\lambda_1 - 4\lambda_2, \\
& \quad\ \ \mu \le 2 + \lambda_1 - 2\lambda_2, \\
& \quad\ \ \mu \le 3 - 2\lambda_1 + \lambda_2, \\
& \quad\ \ \lambda_1 \ge 0,\ \lambda_2 \ge 0.
\end{aligned}
$$

We obtain $\mu^3 = 2\frac{1}{3}$ and $\lambda^3 = (\frac{1}{3}, 0)^T$.

Step 2. Solving Lagrangian relaxation problem (L_{λ^3}) gives $d(\lambda^3) = 2\frac{1}{3}$ and $x^3 = (0, 1)^T$.

Step 3. Since $\mu^3 = 2\frac{1}{3} = d(\lambda^3)$, stop and the optimal dual value is $2\frac{1}{3}$.

There are two disadvantages of the above outer Lagrangian linearization procedure. First, it is sometimes difficult to find an initial feasible solution to (P) to start with. Second, all the past cutting-planes have to be stored which may cause numerical problems in solving large-scale problems. To overcome these disadvantages, stabilization techniques were proposed to ensure the solvability of the subproblems and certain strategies to drop some previous constraints in (LD^k) were suggested (see [121]).

Next, we consider the singly constrained case of (P):

$$
\begin{aligned}
(P_s) \qquad & \min\ f(x) \\
& \text{s.t. } g(x) \le b \\
& \quad\ \ x \in X.
\end{aligned}
$$

By taking advantage of the property in singly constrained situations, a specific outer approximation dual search scheme can be derived. Consider the following example.

EXAMPLE 3.3

$$
\begin{aligned}
& \min\ f(x) = x_1 x_2 - x_1 + 4x_2 + x_3 \\
& \text{s.t. } g(x) = x_1 - 2x_2 + x_3 \le -0.5, \\
& \quad\ \ x \in X = \{0, 1\}^3.
\end{aligned}
$$

Figure 3.8 illustrates the dual function of the example. Geometrically, the dual function is obtained by taking the minimum of all lines: $y = f(x) + \lambda^T(g(x) - b)$, $x \in X$, for each value of λ. The line with the least slope $g(x) - b$ is the most right segment of $d(\lambda)$ and the line with the minimum value of $f(x)$ is the most left segment of $d(\lambda)$. In this example, l_1: $y = -1 + 1.5\lambda$ is the most left segment of the dual function, and l_2: $y = 4 - 1.5\lambda$ is the most right segment.

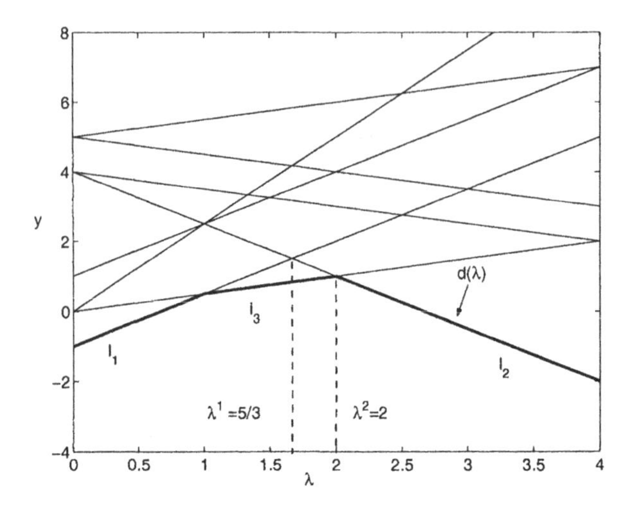

Figure 3.8. Dual function of Example 3.3.

The intersection point of l_1 and l_2 is $\lambda^1 = 5/3$. The lowest line at $\lambda = 5/3$ is l_3: $y = 0.5\lambda$. The intersection point of l_2 and l_3 is $\lambda^2 = 2$ which is the maximum point of $d(\lambda)$, the optimal solution to the dual problem (D).

This motivates a dual search procedure that starts with the intersection point λ^1 of the line of the most left segment and the line of the most right segment of the dual function $d(\lambda)$. At the k-th iteration, λ^{k+1} is calculated by intersecting the line of the segment of $d(\lambda)$ that intersects the line $\lambda = \lambda^k$ with one of the previous two lines. The procedure terminates when λ^{k+1} is a breaking point of $d(\lambda)$ itself.

PROCEDURE 3.3 (DUAL SEARCH PROCEDURE A FOR (P_s))

Step 1. Calculate

$$x^0 = \arg\min_{x \in X} g(x), \; y^0 = \arg\min_{x \in X} f(x).$$

(i) If $g(x^0) > b$, stop and problem (P_s) is infeasible.

(ii) If $g(y^0) \le b$, y^0 is an optimal solution to (P_s) and $\lambda^* = 0$ is the optimal solution to (D).

(iii) If $g(x^0) \le b < g(y^0)$, set $f_0^- = f(x^0)$, $g_0^- = g(x^0)$, $f_0^+ = f(y^0)$, $g_0^+ = g(y^0)$. Set $k = 1$.

Step 2. Compute

$$\lambda^k = -\frac{f_{k-1}^+ - f_{k-1}^-}{g_{k-1}^+ - g_{k-1}^-}. \qquad (3.2.15)$$

Step 3. Solve (L_{λ^k}). Let x^k_{min} and x^k_{max} be the optimal solutions to (L_{λ^k}) with minimum and maximum values of g, respectively.

(i) If $g(x^k_{max}) \le b$, then, set

$$x^k = x^k_{max},\ y^k = y^{k-1},$$
$$f^-_k = f(x^k),\ f^+_k = f^+_{k-1},$$
$$g^-_k = g(x^k),\ g^+_k = g^+_{k-1}.$$

Set $k := k+1$. Return to Step 2.

(ii) If $g(x^k_{min}) > b$, then, set

$$x^k = x^{k-1},\ y^k = x^k_{min},$$
$$f^-_k = f^-_{k-1},\ f^+_k = f(y^k),$$
$$g^-_k = g^-_{k-1},\ g^+_k = g(y^k).$$

Set $k := k+1$. Return to Step 2.

(iii) If $g(x^k_{min}) \le b < g(x^k_{max})$, set $\lambda^* = \lambda^k$, $x^* = x^k_{min}$ and $y^* = x^k_{max}$, stop and λ^* is the optimal solution to the dual problem (D).

THEOREM 3.9 *Procedure 3.3 stops at an optimal solution to (D) within a finite number of iterations.*

Proof. Suppose that (P_s) is feasible. It is obvious that if the algorithm stops at Step 1 (ii), then $\lambda^* = 0$ is the optimal solution to (D). We now suppose that the algorithm stops at Step 3 (iii) after k iterations. Then both x^k_{min} and x^k_{max} solve (L_{λ^k}) and $g(x^k_{min}) \le b < g(x^k_{max})$. Notice from Steps 1 and 3 that $g^-_i \le b < g^+_i$ and $f^+_i \le f^* \le f^-_i$ for $i = 0, 1, \ldots, k-1$. Thus, by (3.2.15), $\lambda^i \ge 0$ for $i = 1, \ldots, k$. Now, for any $\lambda \ge 0$, if $\lambda < \lambda^k$, we have

$$\begin{aligned} d(\lambda) &\le L(x^k_{max}, \lambda) \\ &= f(x^k_{max}) + \lambda(g(x^k_{max}) - b) \\ &< f(x^k_{max}) + \lambda^k(g(x^k_{max}) - b) \\ &= L(x^k_{max}, \lambda^k) = d(\lambda^k). \end{aligned} \tag{3.2.16}$$

If $\lambda > \lambda^k$, then

$$\begin{aligned} d(\lambda) &\le L(x^k_{min}, \lambda) \\ &= f(x^k_{min}) + \lambda(g(x^k_{min}) - b) \\ &\le f(x^k_{min}) + \lambda^k(g(x^k_{min}) - b) \\ &= L(x^k_{min}, \lambda^k) = d(\lambda^k). \end{aligned} \tag{3.2.17}$$

Combining (3.2.16) and (3.2.17) implies that λ^k is an optimal solution to (D).

We now prove the finite termination of the algorithm. Suppose that the algorithm iterates infinitely. Then, either x^k_{max} is feasible or x^k_{min} is infeasible at Step 3 of each iteration. Let $k \geq 1$. Suppose x^k_{max} is feasible. Then $x^k = x^k_{max}$ and $y^k = y^{k-1}$. It follows from (3.2.15) that

$$L(x^{k-1}, \lambda^k) = L(y^{k-1}, \lambda^k).$$

We must have

$$L(x^k, \lambda^k) < L(y^{k-1}, \lambda^k), \tag{3.2.18}$$

otherwise, both x^{k-1} and y^{k-1} solve (L_{λ^k}) and the algorithm will stop at Step 3 (iii). Since $y^k = y^{k-1}$, (3.2.18) yields

$$\lambda^k > -\frac{f(y^k) - f(x^k)}{g(y^k) - g(x^k)} = \lambda^{k+1}.$$

Similarly, if x^k_{min} is infeasible, it holds $\lambda^k < \lambda^{k+1}$. One of the following four cases occurs: (i) x^k_{max} and x^{k+1}_{max} are feasible; (ii) x^k_{min} and x^{k+1}_{min} are infeasible; (iii) x^k_{max} is feasible and x^{k+1}_{min} is infeasible; (iv) x^k_{min} is infeasible and x^{k+1}_{max} is feasible. From the above discussion, we know that $\lambda^k > \lambda^{k+1} > \lambda^{k+2}$ if case (i) occurs and $\lambda^k < \lambda^{k+1} < \lambda^{k+2}$ if case (ii) occurs. Suppose now case (iii) occurs. We claim that $\lambda^{k+1} < \lambda^{k+2} < \lambda^k$. In fact, we have $\lambda^{k+1} < \lambda^k$ and $\lambda^{k+1} < \lambda^{k+2}$. If $\lambda^{k+2} \geq \lambda^k$, then by (3.2.15), we have

$$-\frac{f(y^{k+1}) - f(x^{k+1})}{g(y^{k+1}) - g(x^{k+1})} = \lambda^{k+2} \geq \lambda^k. \tag{3.2.19}$$

Since x^{k+1}_{min} is infeasible, by Step 3 (ii), $x^{k+1} = x^k$ and $y^{k+1} = x^{k+1}_{min}$. Hence, by (3.2.19),

$$\begin{aligned} L(x^{k+1}, \lambda^k) &= f(x^{k+1}) + \lambda^k(g(x^{k+1}) - b) \\ &\geq f(y^{k+1}) + \lambda^k(g(y^{k+1}) - b) \\ &= L(y^{k+1}, \lambda^k). \end{aligned}$$

Since x^k_{max} is feasible, by Step 3 (i), $x^k = x^k_{max}$ is an optimal solution to (L_{λ^k}). The above inequality implies that y^{k+1} also solves (L_{λ^k}). Since y^{k+1} is infeasible to (P_s), the algorithm must have stopped at Step 3 (iii) of the k-th iteration, a contradiction. Thus, $\lambda^{k+1} < \lambda^{k+2} < \lambda^k$. Using similar arguments, we can prove that $\lambda^k < \lambda^{k+2} < \lambda^{k+1}$ if case (iv) occurs. In summary, the sequence $\{\lambda^k\}$ does not repeat with each other. Since X is a finite integer set, there exists only a finite number of different sequences of λ^k's computed by (3.2.15). Therefore, the algorithm must stop in finite iterations. □

If problem (P_s) is feasible and Procedure 3.3 does not stop at Step 1 (ii), then Procedure 3.3 produces an optimal dual solution λ^* to (P_s) together with two solutions of (L_{λ^*}), x^* and y^*, where x^* is feasible and y^* is infeasible.

At Step 3 of Procedure 3.3, the optimal solutions x^k_{min} and x^k_{max} to (L_{λ^k}) could be identical if (L_{λ^k}) has a unique optimal solution. The solutions x^k_{min} and x^k_{max} can be easily computed for separable integer programming problems where $f(x)$ and $g(x)$ are summations of univariate functions.

For nonseparable integer programming problems, for example, quadratic 0-1 programming problems, computing two solutions x^k_{min} and x^k_{max} to (L_{λ^k}) can be very expensive. In this case, a revised dual search procedure for (P_s) can be devised as follows.

PROCEDURE 3.4 (DUAL SEARCH PROCEDURE B FOR (P_s))

Step 1. Calculate

$$x^0 = \arg\min_{x\in X} g(x),\ y^0 = \arg\min_{x\in X} f(x).$$

(i) If $g(x^0) > b$, stop and problem (P_s) is infeasible.

(ii) If $g(y^0) \le b$, y^0 is an optimal solution to (P_s) and $\lambda^* = 0$ is the optimal solution to (D).

(iii) If $g(x^0) \le b < g(y^0)$, set $f_0^- = f(x^0)$, $g_0^- = g(x^0)$, $f_0^+ = f(y^0)$, $g_0^+ = g(y^0)$. Set $z^0 = x^0$, $\lambda^0 = 0$ and $k = 1$.

Step 2. Compute

$$\lambda^k = -\frac{f^+_{k-1} - f^-_{k-1}}{g^+_{k-1} - g^-_{k-1}}. \tag{3.2.20}$$

If $\lambda^k = \lambda^{k-1}$, set $\lambda^* = \lambda^k$, $x^* = z^{k-1}$, stop and λ^* is the optimal solution to the dual problem (D).

Step 3. Solve (L_{λ^k}) to obtain an optimal solution z^k.

(i) If $g(z^k) \le b$, then, set

$$\begin{aligned}x^k &= z^k,\ y^k = y^{k-1},\\ f_k^- &= f(x^k),\ f_k^+ = f^+_{k-1},\\ g_k^- &= g(x^k),\ g_k^+ = g^+_{k-1}.\end{aligned}$$

(ii) If $g(z^k) > b$, then, set

$$\begin{aligned}x^k &= x^{k-1},\ y^k = z^k,\\ f_k^- &= f^-_{k-1},\ f_k^+ = f(y^k),\\ g_k^- &= g^-_{k-1},\ g_k^+ = g(y^k).\end{aligned}$$

Set $k := k + 1$. Return to Step 2.

The output of the above procedure is an optimal dual solution λ^* and an optimal solution x^* to (L_{λ^*}). Notice that x^* could be either feasible or infeasible to (P_s). The optimality of λ^* and the finite termination of Procedure 3.4 can be proved using similar arguments as in the proof of Theorem 3.9.

3.2.3 Bundle method

The subgradient method discussed in Section 3.2.1 does not guarantee a strict increase of the dual function at each iteration since the direction along the subgradient of the dual function is not necessarily an ascent direction. Information more than a single subgradient is needed to construct an ascent direction. Suppose that we can compute the set of subgradients $\partial d(\lambda)$. Let $\eta \in \mathbb{R}^m$. Let $d'(\lambda, \eta)$ denote the directional derivative of d at λ along the direction η. From convex analysis (see, e.g, [182]), $d'(\lambda, \eta)$ can be expressed as

$$d'(\lambda, \eta) = \min_{\xi \in \partial d(\lambda)} \xi^T \eta.$$

For a given λ, as in the smooth optimization, we can then find the steepest ascent direction by maximizing $d'(\lambda, \eta)$ over all the possible directions, η, in a unit ball. The resulting problem is

$$\max_{\|\eta\| \le 1} \min_{\xi \in \partial d(\lambda)} \xi^T \eta,$$

where $\|\cdot\|$ is the 2-norm in $\mathbb{R}^m$. Since the unit ball and $\partial d(\lambda)$ are compact convex sets, we can exchange the order of the max and min in the above expression, which gives rise to

$$\min_{\xi \in \partial d(\lambda)} \max_{\|\eta\| \le 1} \xi^T \eta = \min_{\xi \in \partial d(\lambda)} \|\xi\|. \tag{3.2.21}$$

Thus, finding an ascent direction is equivalent to finding the minimum norm of the subdifferential of d at λ. If $0 \notin \partial d(\lambda)$, then problem (3.2.21) gives an ascent direction. We notice, however, that the direction found by (3.2.21) reduces to the steepest ascent direction in smooth optimization. Thus, this method will suffer from the same problem of a slow convergence as in the steepest ascent method. To overcome this drawback, the ϵ-subdifferential can be introduced to replace the subdifferential in (3.2.21). Define the ϵ-subdifferential of d as

$$\partial_\epsilon d(\lambda) = \{\xi \mid d(\mu) \le d(\lambda) + \xi^T(\mu - \lambda) + \epsilon, \forall \mu \in \mathbb{R}^m_+\},$$

where $\epsilon > 0$. Define the ϵ-directional derivative as

$$d'_\epsilon(\lambda, \eta) = \max_{s>0} \frac{d(\lambda + s\eta) - d(\lambda) - \epsilon}{s}.$$

It can be proved ([100]) that

$$d'_\epsilon(\lambda, \eta) = \min_{\xi \in \partial_\epsilon d(\lambda)} \xi^T \eta.$$

Similar to (3.2.21), let's consider the following problem to find a search direction:

$$\max_{\|\eta\| \leq 1} d'_\epsilon(\lambda, \eta) = \max_{\|\eta\| \leq 1} \min_{\xi \in \partial_\epsilon d(\lambda)} \xi^T \eta = \min_{\xi \in \partial_\epsilon d(\lambda)} \|\xi\|. \tag{3.2.22}$$

Similar analysis shows that if $0 \notin \partial_\epsilon d(\lambda)$, then the minimum norm of the ϵ-subdifferential provides an ascent direction of $d(\lambda)$ along which the dual function can be increased by at least ϵ. If $0 \in \partial_\epsilon d(\lambda)$, then λ is an ϵ-optimal solution that satisfies $d(\lambda) \geq d(\lambda^*) - \epsilon$.

Notice that the full knowledge of the subdifferential or the ϵ-subdifferential of the dual function is difficult to obtain since, in most situations, only one optimal solution can be found from solving the Lagrangian relaxation problem. The key idea of the bundle method is to construct an inner approximation of the ϵ-subdifferential by accumulating subgradients at the previous iteration points up to the current iteration. Let $\xi^j = g(x^j) - b$, $j = 1, \ldots, k$, where x^j is an optimal solution to the Lagrangian relaxation problem (L_{λ^j}). Let

$$p_j = d(\lambda^j) + (\xi^j)^T(\lambda^k - \lambda^j) - d(\lambda^k), \quad j = 1, \ldots, k. \tag{3.2.23}$$

Since ξ^j is a subgradient of d at λ^j, it holds $p_j \geq 0$, for all $j = 1, \ldots, k$. Moveover, for any $\mu \in \mathbb{R}^m_+$, we have

$$d(\mu) \leq d(\lambda^j) + (\xi^j)^T(\mu - \lambda^j), \quad j = 1, \ldots, k. \tag{3.2.24}$$

Therefore, for any $\theta_j \geq 0$, $j = 1, \ldots, k$, such that $\sum_{j=1}^k \theta_j = 1$, it follows from (3.2.23) and (3.2.24) that

$$d(\mu) \leq d(\lambda^k) + (\sum_{j=1}^k \theta_j \xi^j)^T(\mu - \lambda^k) + \sum_{j=1}^k \theta_j p_j, \quad \forall \mu \in \mathbb{R}^+_m. \tag{3.2.25}$$

Define the following set,

$$P^k_\epsilon = \{\sum_{j=1}^k \theta_j \xi^j \mid \theta_j \geq 0, \sum_{j=1}^k \theta_j = 1, \sum_{j=1}^k \theta_j p_j \leq \epsilon\}.$$

It follows from (3.2.25) that $P_\epsilon^k \subseteq \partial_\epsilon d(\lambda^k)$. Therefore, we can approximate problem (3.2.22) by the following quadratic program:

$$\min \frac{1}{2}\|\sum_{j=1}^{k}\theta_j\xi^j\|^2 \tag{3.2.26}$$

$$\text{s.t.} \sum_{j=1}^{k}\theta_j p_j \le \epsilon,$$

$$\sum_{j=1}^{k}\theta_j = 1,\ \theta_j \ge 0, j = 1,\ldots,k.$$

A basic bundle method for the dual problem (D) consists of two main steps: To find an ascent direction and to perform a line search. In the first step, a search direction is obtained by solving the quadratic program (3.2.26). In the second step, a line search procedure is employed, resulting either a serious step when a new point along the search direction gives a sufficient increase of *d*, or a null step otherwise. The convergence analysis of the basic bundle method can be found in [100][127]. Note that a quadratic programming problem (3.2.26) has to be solved in finding an ascent direction at every iteration of the bundle method. Moreover, the line search may require additional computational efforts. Therefore, the bundle method may be time-consuming in order to guarantee an increase of the dual value at each consecutive step.

3.3 Perturbation Function

The perturbation function has served as a key in investigating the duality theory for general integer programming. Especially, the perturbation function offers insights into prominent features of integer programming problems when the Lagrangian relaxation method is adopted.

We make the following assumption on problem (P):

ASSUMPTION 3.1 *$S \neq \emptyset$ and there is at least one $x \in X \setminus S$ such that $f(x) < f^*$.*

Assumption 3.1 ensures that the problem (P) is feasible and cannot be trivially reduced to an unconstrained integer programming problem. For any vectors x and $y \in \mathbb{R}^m$, $x \le y$ iff $x_i \le y_i$, $i = 1,\ldots,m$. A function $h(x)$ defined on $\mathbb{R}^m$ is said to be nonincreasing if for any x and $y \in \mathbb{R}^m$, $x \le y$ implies $h(x) \ge h(y)$.

Let $b = (b_1,\ldots,b_m)^T$. The perturbation function associated with (P) is defined as

$$w(y) = \min\{f(x) \mid g(x) \le y,\ x \in X\}, \quad y \in \mathbb{R}^m. \tag{3.3.1}$$

The domain of w is

$$Y = \{y \in \mathbb{R}^m \mid \text{there exists } x \in X \text{ such that } g(x) \leq y\}. \tag{3.3.2}$$

Note that Y is not always a convex set. The perturbation function w can be extended to the convex hull of Y by defining $w(y) = +\infty$ for $y \in conv(Y)\backslash Y$. Furthermore, w is a nonincreasing and piecewise constant $(+\infty)$ function of y on $conv(Y)$. By definition (3.3.1), $w(g(x)) \leq f(x)$ for any $x \in X$ and $w(b) = f^*$. In a process of increasing y, if there is a new point $\tilde{x} \in X$ such that $f(\tilde{x}) < w(y)$ for any $y \in \{z \in Y \mid z \leq g(\tilde{x}),\ z \neq g(\tilde{x})\}$, the perturbation function w has a downward jump at $y = g(\tilde{x})$. The point $g(\tilde{x})$ corresponding to this new point $\tilde{x}$ is called a *corner point* of the perturbation function w in the y space. Since f and g_i's are continuous functions and X is a finite integer set, there is only a finite number of corner points, say K corner points, c_1, c_2, ..., c_K. Let $f_i = w(c_i)$, $i = 1, \ldots, K$. Define the sets of corner points in the y space and the $\{y, w(y)\}$ space, respectively, as follows,

$$C = \{c_i = (c_{i1}, c_{i2}, \ldots, c_{im}) \mid i = 1, \ldots, K\},$$
$$\Phi_c = \{(c_i, f_i) \mid i = 1, \ldots, K\}.$$

It is clear that $(y, w(y)) \in \Phi_c$ iff for any $z \in Y$ satisfying $z \leq y$ and $z \neq y$, it holds $w(z) > w(y)$.

By the definition of w, if $y \in Y$ then $\prod_{i=1}^m [y_i, +\infty) \subseteq Y$. Let e^i denote the i-th unit vector in $\mathbb{R}^m$. Then, e^i's are the extreme directions of $conv(Y)$. Also, the set of extreme points of $conv(Y)$ is a subset of C. Denote

$$\Lambda = \{\mu \in \mathbb{R}^K \mid \sum_{i=1}^K \mu_i = 1,\ \mu_i \geq 0,\ i = 1, \ldots, K\}.$$

The convex hull of Y can be expressed as

$$\begin{aligned} conv(Y) &= \{\sum_{i=1}^K \mu_i c_i + \sum_{i=1}^m \alpha_i e^i \mid \mu \in \Lambda,\ \alpha_i \geq 0,\ i = 1, \ldots, m\} \\ &= \{y \mid y \geq \sum_{i=1}^K \mu_i c_i,\ \mu \in \Lambda\}. \end{aligned} \tag{3.3.3}$$

From the definition of the corner point, the domain Y can be decomposed into K subsets with each c_i as the lower end of each subset Y_i. More specifically, we have $Y = \cup_{i=1}^K Y_i$ with $c_{ij} = \min\{y_j \mid y \in Y_i\}$, $j = 1, \ldots, m$, and w takes a constant f_i over Y_i:

$$w(y) = f_i, \quad \forall y \in Y_i,\ i = 1, \ldots, K. \tag{3.3.4}$$

Define

$$\Phi = \{(y, w(y)) \mid y \in Y\}.$$

By the definition of Y_i, $c_i \in Y_i$ and $w(c_i) = f_i$ for each i. Thus $\Phi_c \subset \Phi$.

Consider the following example.

EXAMPLE 3.4

$$\begin{aligned} \min\ & f(x) = 4 + x_1x_2x_3x_4 - x_1 + 3x_2 + x_3 - 2x_4 \\ \text{s.t.}\ & g_1(x) = x_1 - 2x_2 + x_3 + 3 \leq 2.5, \\ & x \in X = \{0,1\}^4. \end{aligned}$$

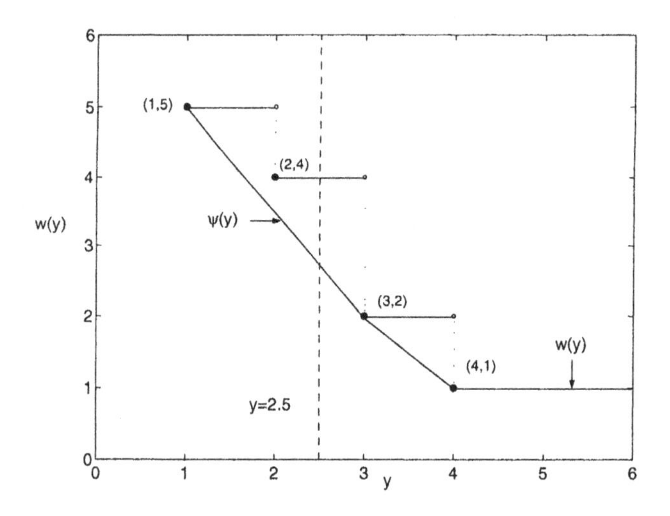

Figure 3.9. Perturbation function of Example 3.4.

Figure 3.9 illustrates the perturbation function of Example 3.4. All the points on the line $y = 2.5$ and to its left are feasible points, while all the points to the right of $y = 2.5$ are infeasible. We can see from Figure 3.9 that point $(y, w(y))$ = $(2, 4)$, the map of $x = (1, 1, 0, 1)^T$, has the lowest value of $f(x)$ among the points located on the left of line $y = 2.5$. Thus the optimal solution of this example is $x^* = (1, 1, 0, 1)^T$ with $w(2.5) = f(x^*) = 4$. There are four corner points in Φ_c, $(1, 5)$, $(2, 4)$, $(3, 2)$, and $(4, 1)$. It is evident from the figure that at least one corner point in Φ_c is optimal to the primal problem.

Note that for multiply constrained problems some Y_i's may not be a single rectangular strip and there may exist different Y_i's on which $w(y)$ takes the same value.

EXAMPLE 3.5

$$\begin{aligned} \min\ & f(x) = 3 - x_1 - x_2 - x_1x_2 \\ \text{s.t. } & g_1(x) = x_1 \le 1, \\ & g_2(x) = x_2 \le 0.5, \\ & x \in X = \{(0,0)^T, (0,1)^T, (1,1)^T, (0,2)^T, (2,0)^T, (2,2)^T\}. \end{aligned}$$

By definition, we have

$$\begin{aligned} & Y_1 = [0,2) \times [0,1),\ c_1 = (0,0)^T,\ f_1 = 3, \\ & Y_2 = [0,1) \times [1,2),\ c_2 = (0,1)^T,\ f_2 = 2, \\ & Y_3 = [0,1) \times [2,+\infty),\ c_3 = (0,2)^T,\ f_3 = 1, \\ & Y_4 = [2,+\infty) \times [0,1),\ c_4 = (2,0)^T,\ f_4 = 1, \\ & Y_5 = [1,2) \times [1,+\infty) \cup [1,+\infty) \times [1,2),\ c_5 = (1,1)^T,\ f_5 = 0, \\ & Y_6 = [2,+\infty) \times [2,+\infty),\ c_6 = (2,2)^T,\ f_6 = -5. \end{aligned}$$

We see that Y_5 is not a single rectangular strip and $w(y)$ takes the same value 1 over Y_3 and Y_4. Figure 3.10 illustrates the perturbation function of this example.

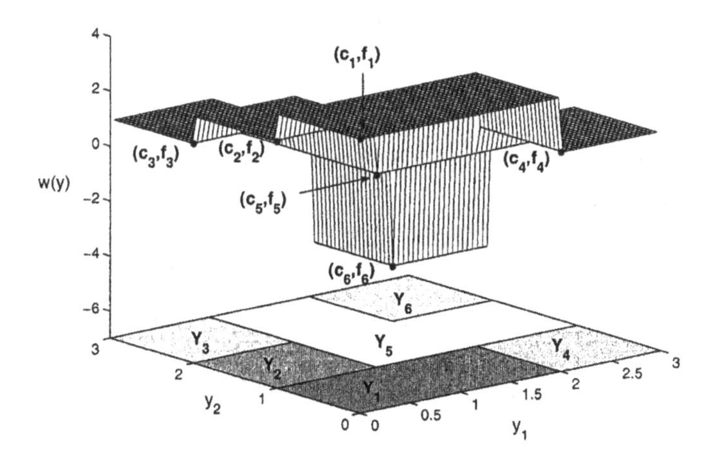

Figure 3.10. Illustration of perturbation function w and decomposition of Y.

A point $x \in X$ is said to be *noninferior* if there is no $\bar{x} \in X$ with $w(g(\bar{x}))$ $= w(g(x))$ such that $g(\bar{x}) \le g(x)$ and $g(\bar{x}) \ne g(x)$. The following lemma shows some useful properties of the perturbation function. Most importantly, the lemma proves that any noninferior optimal solution of (P) is corresponding to a corner point.

LEMMA 3.2 (i) *For any $y \in Y$, if x_y solves the perturbation problem*

$$w(y) = \min\{f(x) \mid g(x) \le y,\ x \in X\},$$

then $(g(x_y), f(x_y)) \in \Phi$.

(ii) *For any $c_i \in C$, there exists $\bar{x} \in X$ such that $(c_i, f_i) = (g(\bar{x}), f(\bar{x})) \in \Phi_c$.*

(iii) *For any noninferior optimal solution x^* to (P), $(g(x^*), f(x^*)) \in \Phi_c$.*

(iv) *If $b \in Y_k$ for some $k \in \{1, \ldots, K\}$, then $f^* = f_k$ and any optimal solution to the perturbation problem*

$$w(c_k) = \min\{f(x) \mid g(x) \le c_k,\ x \in X\}$$

is a noninferior optimal solution to (P).

(v) *For any $\lambda \ge 0$, there exists $x_\lambda \in X$ that solves (L_λ) and satisfies $(g(x_\lambda), f(x_\lambda)) \in \Phi_c$.*

Proof. (i) Since $g(x_y) \le y$ and w is a nonincreasing function, we have $f(x_y) = w(y) \le w(g(x_y))$. On the other hand, since x_y is feasible in the perturbation problem

$$w(g(x_y)) = \min\{f(x) \mid g(x) \le g(x_y),\ x \in X\},$$

we have $w(g(x_y)) \le f(x_y)$. Thus, $w(g(x_y)) = f(x_y)$, i.e., $(g(x_y), f(x_y)) \in \Phi$.

(ii) Suppose that $\bar{x}$ solves the perturbation problem $w(c_i) = \min\{f(x) \mid g(x) \le c_i,\ x \in X\}$, then $f(\bar{x}) = w(c_i) = f_i$ and $g(\bar{x}) \le c_i$. By part (i), we have $(g(\bar{x}), f(\bar{x})) \in \Phi$. It then follows from the definition of c_i that $g(\bar{x}) = c_i$ and so $(g(\bar{x}), f(\bar{x})) = (c_i, f_i)$.

(iii) By part (i), we have $(g(x^*), f(x^*)) \in \Phi$. Let $z \in Y$ be such that $z \le g(x^*)$ and $z \ne g(x^*)$. Suppose that $\bar{x}$ solves the perturbation problem

$$w(z) = \min\{f(x) \mid g(x) \le z,\ x \in X\}.$$

Then $w(z) = f(\bar{x})$ and $g(\bar{x}) \le z \le g(x^*)$ with $g(\bar{x}) \ne g(x^*)$. Since x^* is a noninferior optimal solution, we must have $w(z) = f(\bar{x}) > f(x^*) = w(g(x^*))$. Thus $(g(x^*), f(x^*)) \in \Phi_c$.

(iv) Suppose that x^* solves the problem $w(c_k) = \min\{f(x) \mid g(x) \le c_k,\ x \in X\}$. Then, by (3.3.4), $f^* = w(b) = w(c_k) = f_k = f(x^*)$. So x^* is an optimal solution to (P). If there exists another optimal solution $\bar{x}$ to (P) such that $g(\bar{x}) \le g(x^*)$ and $g(\bar{x}) \ne g(x^*)$, then $g(\bar{x}) \le g(x^*) \le c_k \le b$ and $g(\bar{x}) \ne c_k$. Since (c_k, f_k) is a corner point and $w(y)$ is a nonincreasing function, we have

$$f(\bar{x}) \ge w(g(\bar{x})) > w(c_k) = f_k = f(x^*),$$

which contradicts the optimality of $\bar{x}$. Therefore, x^* is a noninferior optimal solution of (P).

(v) Let $\bar{x} \in X$ be an optimal solution to (L_λ). We claim that $f(\bar{x}) = w(g(\bar{x}))$. Otherwise, $f(\bar{x}) > w(g(\bar{x}))$. Let $\tilde{x} \in X$ solve $\min\{f(x) \mid g(x) \leq g(\bar{x}),\ x \in X\}$. Then $g(\tilde{x}) \leq g(\bar{x})$ and $f(\bar{x}) > w(g(\bar{x})) = f(\tilde{x})$. We have

$$L(\tilde{x}, \lambda) = f(\tilde{x}) + \lambda^T(g(\tilde{x}) - b) < f(\bar{x}) + \lambda^T(g(\bar{x}) - b) = L(\bar{x}, \lambda),$$

which contradicts the optimality of $\bar{x}$ to (L_λ). Now, let $g(\bar{x}) \in Y_k$ for some $k \in \{1, \ldots, K\}$. Then $c_k \leq g(\bar{x})$. By part (ii), there exists $x_\lambda \in X$ such that $(c_k, f_k) = (g(x_\lambda), f(x_\lambda))$. By (3.3.4), $f(x_\lambda) = f_k = w(c_k) = w(g(\bar{x})) = f(\bar{x})$. We have $L(x_\lambda, \lambda) \leq L(\bar{x}, \lambda)$ and hence x_λ is also an optimal solution to (L_λ) and $(g(x_\lambda), f(x_\lambda)) \in \Phi_c$. □

Let E denote the epigraph of w:

$$E := epi(w) = \{(y, z) \mid z \geq w(y),\ y \in conv(Y)\}. \tag{3.3.5}$$

Define the *convex envelope function* of w on $conv(Y)$:

$$\psi(y) \quad = \quad \min\{z \mid (y, z) \in conv(E)\}. \tag{3.3.6}$$

By definitions (3.3.5) and (3.3.6), it holds

$$w(y) \geq \psi(y), \quad y \in conv(Y). \tag{3.3.7}$$

Note that $f_i = w(c_i)$, $i = 1, \ldots, K$. By (3.3.3) and (3.3.5), $conv(E)$ can be expressed as

$$conv(E) = \{(y, z) \mid (y, z) \geq (\sum_{i=1}^{K} \mu_i c_i, \sum_{i=1}^{K} \mu_i f_i), \mu \in \Lambda\}.$$

Therefore, (3.3.6) is in turn equivalent to

$$\begin{aligned} \psi(y) \quad = \quad & \min \sum_{i=1}^{K} \mu_i f_i \\ & \text{s.t.} \ \sum_{i=1}^{K} \mu_i c_i \leq y,\ \mu \in \Lambda. \end{aligned} \tag{3.3.8}$$

The dual problem of (3.3.8) is

$$\begin{aligned} \psi(y) \quad = \quad & \max \ -\lambda^T y + r \\ & \text{s.t.} \ -\lambda^T c_i + r \leq f_i,\ i = 1, \ldots, K, \\ & \quad \lambda \in \mathbb{R}^m_+,\ r \in \mathbb{R}. \end{aligned} \tag{3.3.9}$$

We see from (3.3.9) that ψ is a nonincreasing piecewise linear convex function on $conv(Y)$.

THEOREM 3.10 *Let μ^* and $(-\lambda^*, r^*)$ be optimal solutions to (3.3.8) and (3.3.9) with $y = b$, respectively. Then*

(i) λ^* *is an optimal solution to the dual problem* (D) *and*

$$\psi(b) = \max_{\lambda \in \mathbb{R}^m_+} d(\lambda) = d(\lambda^*).$$

(ii) *For each* i *with* $\mu_i^* > 0$, *any* $\bar{x} \in X$ *satisfying* $(g(\bar{x}), f(\bar{x})) = (c_i, f_i)$ *is an optimal solution to the Lagrangian problem* (L_{λ^*}).

Proof. (i) For any $\lambda \in \mathbb{R}^m_+$, by Lemma 3.2 (v), there exists $j \in \{1, \ldots, K\}$ such that

$$\min_{x \in X} L(x, \lambda) = \min\{f_i + \lambda^T(c_i - b) \mid i = 1, \ldots, K\} = f_j + \lambda^T(c_j - b).$$

Let $r_\lambda = f_j + \lambda^T c_j$, then $f_i + \lambda^T c_i \geq r_\lambda$, $i = 1, \ldots, K$. Thus

$$\begin{aligned} \max_{\lambda \in \mathbb{R}^m_+} d(\lambda) &= \max_{\lambda \in \mathbb{R}^m_+} \min_{x \in X} L(x, \lambda) \\ &= \max_{\lambda \in \mathbb{R}^m_+} (-\lambda^T b + r_\lambda) \\ &= \max_{\lambda \in \mathbb{R}^m_+,\, r \in \mathbb{R}} \{-\lambda^T b + r \mid f_i + \lambda^T c_i \geq r,\ i = 1, \ldots, K\} \\ &= \max_{\lambda \in \mathbb{R}^m_+,\, r \in \mathbb{R}} \{-\lambda^T b + r \mid -\lambda^T c_i + r \leq f_i,\ i = 1, \ldots, K\}. \end{aligned} \tag{3.3.10}$$

On the other hand, by (3.3.9), we have

$$\begin{aligned} \psi(b) = \ & \max -\lambda^T b + r \\ & \text{s.t.} \ -\lambda^T c_i + r \leq f_i,\ i = 1, \ldots, K, \\ & \qquad \lambda \in \mathbb{R}^m_+,\ r \in \mathbb{R}. \end{aligned} \tag{3.3.11}$$

Comparing (3.3.10) with (3.3.11) leads to

$$\psi(b) = \max_{\lambda \in \mathbb{R}^m_+} d(\lambda) = d(\lambda^*).$$

Thus λ^* is a dual optimal solution.

(ii) By the complementary slackness condition of linear program (3.3.11), we have

$$\mu_i^*[(-\lambda^*)^T c_i + r^* - f_i] = 0, \quad i = 1, \ldots, K.$$

So for each $\mu_i^* > 0$, it holds $r^* = f_i + (\lambda^*)^T c_i$. Hence

$$d(\lambda^*) = \psi(b) = (-\lambda^*)^T b + r^* = f_i + (\lambda^*)^T (c_i - b). \qquad (3.3.12)$$

By Lemma 3.2 there exists $\bar{x} \in X$ such that $(g(\bar{x}), f(\bar{x})) = (c_i, f_i)$. It then follows from (3.3.12) that $d(\lambda^*) = L(\bar{x}, \lambda^*)$, which means $\bar{x}$ is an optimal solution to (L_{λ^*}). □

3.4 Optimal Generating Multiplier and Optimal Primal-Dual Pair

Consider the general integer programming problem (P) in Section 3.1. If an optimal solution x^* to (P) is also optimal to (L_λ) with $\lambda = \lambda^*$, then we say that λ^* is an *optimal generating multiplier* of (P) for x^*. If the dual optimal solution λ^* is an optimal generating multiplier for an optimal solution x^* to (P), then (x^*, λ^*) is said to be an *optimal primal-dual pair* of (P).

While the Lagrangian method is a powerful constructive dual search method, it often fails to identify an optimal solution of the primal integer programming problem. Two critical situations could be present that prevent the Lagrangian method from succeeding in the dual search. Firstly, the optimal solution of (P) may not even be generated by solving (L_λ) for any $\lambda \in \mathbb{R}^m_+$. Secondly, the optimal solution to (L_{λ^*}), with λ^* being an optimal solution to the dual problem (D), is not necessarily an optimal solution to (P), or even not feasible. The first situation mentioned above is associated with the existence of an optimal generating Lagrangian multiplier vector. The second situation is related to the existence of an optimal primal-dual pair. Example 3.4 can be used to serve the purpose to illustrate the above two situations. As seen in Figure 3.9 for Example 3.4, there does not exist a Lagrangian multiplier that enables an identification of the optimal point $(2, 4)$ via solving a Lagrangian relaxation problem. Thus, there does not exist an optimal generating multiplier for Example 3.4. Furthermore, the optimal dual solution is $\lambda^* = 1.5$ and the optimal solutions to $(L_{1.5})$ are $(0, 0, 0, 1)^T$ and $(0, 1, 0, 1)^T$, none of which is an optimal solution to Example 3.4. Thus, there does not exist an optimal primal-dual pair for Example 3.4.

A vector $-\lambda$ with $\lambda \in \mathbb{R}^m$ is said to be a subgradient of $w(\cdot)$ at $y = \hat{y}$ if

$$w(y) \geq w(\hat{y}) - \lambda^T (y - \hat{y}), \quad \forall\, y \in Y.$$

LEMMA 3.3 *Let $\hat{x} \in X$ and $\hat{\lambda} \in \mathbb{R}^m_+$.*

(i) *If $\hat{x}$ solves problem $(L_{\hat{\lambda}})$, then $-\hat{\lambda}$ is a subgradient of $w(\cdot)$ at $y = \hat{y} = g(\hat{x})$.*

(ii) *If $-\hat{\lambda}$ is a subgradient of $w(\cdot)$ at $y = \hat{y} = g(\hat{x})$, and if $w(\hat{y}) = f(\hat{x})$, then $\hat{x}$ solves problem $(L_{\hat{\lambda}})$.*

Proof. (i) To show that $-\hat{\lambda}$ is a subgradient of $w(\cdot)$ at $y = \hat{y}$, we must show that

$$w(y) \geq w(\hat{y}) + \hat{\lambda}^T(\hat{y} - y), \quad \forall\, y \in Y.$$

Suppose on contrary that there exists some $\tilde{y} \in Y$ such that:

$$w(\tilde{y}) < w(\hat{y}) + \hat{\lambda}^T(g(\hat{x}) - \tilde{y}).$$

Then, noting that $w(\hat{y}) \leq f(\hat{x})$, we have:

$$w(\tilde{y}) + \hat{\lambda}^T(\tilde{y} - b) < w(\hat{y}) + \hat{\lambda}^T(g(\hat{x}) - b) \leq f(\hat{x}) + \hat{\lambda}^T(g(\hat{x}) - b). \quad (3.4.1)$$

Suppose that $w(\tilde{y})$ is realized by $\tilde{x}$. We have $g(\tilde{x}) \leq \tilde{y}$. Since $f(\tilde{x}) = w(\tilde{y})$ and $\hat{\lambda} \in \mathbb{R}^m_+$, (3.4.1) implies the following:

$$f(\tilde{x}) + \hat{\lambda}^T(g(\tilde{x}) - b) \ < f(\hat{x}) + \hat{\lambda}^T(g(\hat{x}) - b).$$

This is a contradiction to the assumption that $\hat{x}$ solves problem $(L_{\hat{\lambda}})$.

(ii) Since $-\hat{\lambda}$ is a subgradient of $w(\cdot)$ at $y = \hat{y} = g(\hat{x})$ and $w(\hat{y}) = f(\hat{x})$, we have:

$$w(y) \geq f(\hat{x}) + \hat{\lambda}^T(g(\hat{x}) - y), \quad \forall\, y \in Y. \quad (3.4.2)$$

Let $(y, y_0) \in E$, where E is defined in (3.3.5). The following is satisfied:

$$y_0 \geq f(\hat{x}) + \hat{\lambda}^T(g(\hat{x}) - y).$$

Consider the set $E_1 = \{(g(x), f(x)) \mid x \in X\}$. Choose $\tilde{x} \in X$ and form the vector $(g(\tilde{x}), f(\tilde{x})) \in E_1$. If we set $\tilde{y} = g(\tilde{x})$, then $\tilde{y} \in Y$ and $f(\tilde{x}) \geq w(\tilde{y})$. Thus $(g(\tilde{x}), f(\tilde{x})) \in E$. Since $\tilde{x}$ is an arbitrary element of X, we have $E_1 \subseteq E$. Therefore, using (3.4.2), we have:

$$f(x) \geq f(\hat{x}) + \hat{\lambda}^T(g(\hat{x}) - g(x)), \quad \forall\, x \in X.$$

Finally, we have

$$f(x) + \hat{\lambda}^T(g(x) - b) \geq f(\hat{x}) + \hat{\lambda}^T(g(\hat{x}) - b), \quad \forall\, x \in X.$$

So $\hat{x}$ solves problem $(L_{\hat{\lambda}})$. □

The following theorem concerning the existence of an optimal generating Lagrangian multiplier vector is evident from Lemma 3.3.

THEOREM 3.11 *Let x^* solve the primal problem (P). Then x^* is an optimal solution to problem (L_{λ^*}) for some $\lambda^* \in \mathbb{R}^m_+$ iff $-\lambda^*$ is a subgradient of $w(\cdot)$ at $y = g(x^*)$.*

Proof. Notice that the optimality of x^* in (P) implies $w(g(x^*)) = f(x^*)$. The theorem then follows from Lemma 3.3. □

The above theorem can be further enhanced by showing that the existence of an optimal generating multiplier for an optimal solution x^* to (P) is equivalent to the coincidence of the perturbation function and its convex envelope at $g(x^*)$.

THEOREM 3.12 *Let x^* be an optimal solution to (P). Then there exists an optimal generating multiplier for x^* if and only if $w(g(x^*)) = \psi(g(x^*))$.*

Proof. Let $-\lambda^* \leq 0$ be a subgradient of ψ at $g(x^*) \in Y$. We have

$$\psi(y) \geq \psi(g(x^*)) + (-\lambda^*)^T(y - g(x^*)), \quad \forall y \in Y. \tag{3.4.3}$$

For any $x \in X$, setting $y = g(x) \in Y$ in (3.4.3) and using (3.3.7), we get

$$f(x) \geq w(g(x)) \geq \psi(g(x)) \geq \psi(g(x^*)) + (-\lambda^*)^T(g(x) - g(x^*)). \tag{3.4.4}$$

Since x^* is an optimal solution to (P), from Lemma 3.2 (i), we have $f(x^*) = w(g(x^*))$. If the condition $w(g(x^*)) = \psi(g(x^*))$ holds, then we deduce from (3.4.4) that

$$f(x) + (\lambda^*)^T(g(x) - b) \geq f(x^*) + (\lambda^*)^T(g(x^*) - b), \quad \forall x \in X, \tag{3.4.5}$$

which means x^* is an optimal solution to (L_{λ^*}) and hence λ^* is an optimal generating multiplier for x^*.

Conversely, if there exists an optimal generating multiplier $\lambda^* \geq 0$ for x^*, then (3.4.5) holds. For any $y \in Y$, there exists $x \in X$ satisfying $f(x) = w(y)$ and $g(x) \leq y$. From (3.4.5), we have

$$\begin{aligned} w(y) &= f(x) \\ &\geq f(x^*) - (\lambda^*)^T(g(x) - g(x^*)) \\ &\geq w(g(x^*)) - (\lambda^*)^T(y - g(x^*)) \end{aligned} \tag{3.4.6}$$

for all $y \in Y$. Recall that ψ is the greatest convex function majorized by w. We therefore deduce from (3.4.6) that

$$\psi(y) \geq w(g(x^*)) - (\lambda^*)^T(y - g(x^*)), \quad \forall y \in Y.$$

Letting $y = g(x^*)$ in the above inequality yields $\psi(g(x^*)) \geq w(g(x^*))$. Together with (3.3.7), this implies $w(g(x^*)) = \psi(g(x^*))$. □

COROLLARY 3.1 *Let x^* be a noninferior optimal solution to (P). If*

$$\psi(c_i) = f_i, \quad i = 1, \ldots, K, \tag{3.4.7}$$

then there exists an optimal generating multiplier vector for x^*.

Proof. From Lemma 3.2 (iii), $(g(x^*), f(x^*)) \in \Phi_c$. By the assumption, $\psi(g(x^*)) = f(x^*) = w(g(x^*))$. The conclusion then follows from Theorem 3.12. □

We can conclude from Theorems 3.11 and 3.12 that in order to generate x^*, an optimal solution of problem (P), using Lagrangian relaxation, the existence of a subgradient of $w(\cdot)$ at $y = g(x^*)$ plays a central role. Figure 3.11 depicts a situation where there does not exist a subgradient of $w(\cdot)$ at $y = g(x^*)$ with x^* being an optimal solution of problem (P). Figure 3.12 depicts a situation where a subgradient of $w(\cdot)$ exists at $y = g(x^*)$ with x^* being an optimal solution of problem (P). It is clear from Theorems 3.11 and 3.12 that only in situations such as in Figure 3.12 can the optimal solutions of problem (P) be generated via problem (L_λ) for some $\lambda \in \mathbb{R}^m_+$.

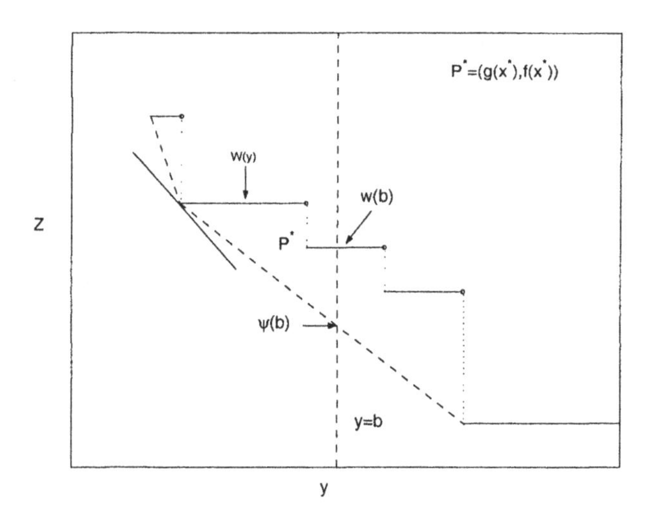

Figure 3.11. Perturbation function where there exists no subgradient of w at $y = g(x^*)$.

If $-\hat{\lambda}$ is a subgradient of w at $\hat{y} = g(\hat{x})$ and $w(\hat{y}) = f(\hat{x})$, then

$$\begin{aligned} y_0 &= w(\hat{y}) - \hat{\lambda}^T(y - \hat{y}) \\ &= f(\hat{x}) + \hat{\lambda}^T(g(\hat{x}) - y) \end{aligned}$$

is a supporting hyperplane of the set E at $(y, y_0) = (\hat{y}, w(\hat{y}))$. The intercept of this supporting plane with axis of $y = b$ is $f(\hat{x}) + \hat{\lambda}^T(g(\hat{x}) - b)$. Thus, the geometric interpretation of the dual problem (D) is to maximize the intercept

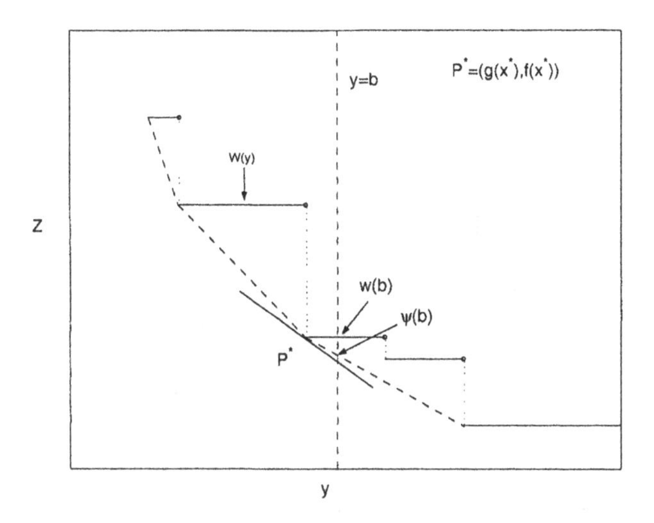

Figure 3.12. Perturbation function where there exists a subgradient of w at $y = g(x^*)$.

of the supporting planes with axis $y = b$. Graphically, the maximum intercept on the axis of $y = b$ is achieved at $\psi(b)$.

It is clear that the duality gap is given by

$$v(P) - v(D) = w(b) - \psi(b).$$

A condition can be now given for the Lagrangian relaxation method to be successful in identifying an optimal solution of problem (P) via the maximization of $v(L_\lambda)$ with respect to $\lambda \in \mathbb{R}^m_+$.

THEOREM 3.13 *Let x^* solve problem (P), and $\lambda^* \in \mathbb{R}^m_+$. Then, (x^*, λ^*) is an optimal primal-dual pair iff the hyperplane given by $y_0 = f(x^*) + (\lambda^*)^T(g(x^*) - y)$ is a supporting hyperplane of E at $(g(x^*), f(x^*))$ and contains the point $(b, \psi(b))$.*

Proof. The proof follows from Theorems 3.10 and 3.11. □

Notice that in integer programming the duality gap is often nonzero even when there exists an optimal primal-dual pair.

When condition (3.4.7) is satisfied, the existence of an optimal generating multiplier can be ensured. The following example, however, shows that condition (3.4.7) is not enough to guarantee the existence of an optimal primal-dual pair of (P).

EXAMPLE 3.6

$$\begin{aligned} \min\ & -3\sqrt{x_1} - 2x_2 \\ \text{s.t. } & x_1 \le 5, \\ & x_2 \le 5, \\ & x \in X = \{(1,4)^T, (2,2)^T, (5,7)^T, (8,8)^T, (9,7)^T\}. \end{aligned}$$

The optimal solution of this problem is $x^* = (1,4)^T$ with $f(x^*) = -11$. The corner points are: $c_1 = (1,4)^T$, $f_1 = -11$, $c_2 = (2,2)^T$, $f_2 = -8.2426$, $c_3 = (5,7)^T$, $f_3 = -20.7082$, $c_4 = (8,8)^T$, $f_4 = -24.4853$, $c_5 = (9,7)^T$, $f_5 = -23$. The optimal solution to (D) is $\lambda^* = (0.57287, 2.14946)^T$ with $d(\lambda^*) = -16.4095$. There are three optimal solutions to the Lagrangian problem (L_{λ^*}): $(2,2)^T$, $(5,7)^T$ and $(9,7)^T$, among which only $(2,2)^T$ is feasible. However, $(2,2)^T$ with $f((2,2)^T) = -8.2426$ is not an optimal solution to the primal problem. Hence there is no optimal primal-dual pair in this problem. We can verify, however, condition (3.4.7) is satisfied and $\lambda = (1.01311, 1.88524)^T$ is an optimal generating multiplier vector for $x^* = (1,4)^T$. This example also shows that an optimal generating multiplier vector is not necessarily an optimal solution to the dual problem (D).

The condition (3.4.7) is, however, sufficient to guarantee the existence of an optimal primal-dual pair of (P) in singly constrained situations. Notice that the corner point set $\Phi_c = \{(c_i, f_i) \mid i = 1, \ldots, K\}$ in singly constrained situations is a set in $\mathbb{R}^2$ and by the monotonicity of w we can assume without loss of generality that $c_1 < c_2 < \cdots < c_K$ and $f_1 > f_2 > \cdots > f_K$. The domain of w is $Y = [c_1, +\infty)$.

Define the envelope function of w in singly constrained cases as

$$\phi(y) = \begin{cases} f_1 + \xi_1(y - c_1), & c_1 \le y < c_2 \\ f_2 + \xi_2(y - c_2), & c_2 \le y < c_3 \\ \cdots & \cdots \\ f_{K-1} + \xi_{K-1}(y - c_{K-1}), & c_{K-1} \le y < c_K \\ f_K, & c_K \le y < \infty \end{cases} \tag{3.4.8}$$

where

$$\xi_i = \frac{f_{i+1} - f_i}{c_{i+1} - c_i} < 0, \quad 1 \le i \le K - 1. \tag{3.4.9}$$

It is clear that ϕ is a convex function if and only if $\xi_1 \le \xi_2 \le \cdots \le \xi_{K-1}$. We have the following theorem.

THEOREM 3.14 *Suppose that $m = 1$ and ϕ is convex on $Y = [c_1, \infty)$. If x^* is a noninferior optimal solution to (P), then there exists $\lambda^* \ge 0$ such that (x^*, λ^*) is an optimal primal-dual pair of (P).*

Proof. By Assumption 3.1 and Lemma 3.2 (iii), there exists $k \in \{1, \ldots, K-1\}$ satisfying $b \in [c_k, c_{k+1})$ and $(g(x^*), f(x^*)) = (c_k, f_k) \in \Phi_c$. Let $\lambda^* = -\xi_k$. We first prove that x^* solves problem (L_{λ^*}). Since ξ_k is a subgradient of ϕ at $y = g(x^*) = c_k$, we have

$$w(y) \geq \phi(y) \geq \phi(g(x^*)) + \xi_k(y - g(x^*)) = f(x^*) + \xi_k(y - g(x^*)), \quad \forall y \in Y. \tag{3.4.10}$$

For any $x \in X$, let $y = g(x)$. It follows from (3.4.10) that

$$\begin{aligned} f(x) \geq w(g(x)) &= w(y) \geq f(x^*) + \xi_k(y - g(x^*)) \\ &= f(x^*) + \xi_k(g(x) - g(x^*)), \end{aligned}$$

which in turn yields

$$\begin{aligned} L(x, \lambda^*) &= f(x) + \lambda^*(g(x) - b) \\ &\geq f(x^*) + \lambda^*(g(x^*) - b) = L(x^*, \lambda^*). \end{aligned} \tag{3.4.11}$$

Thus x^* solves (L_{λ^*}). Next, we prove that λ^* solves the dual problem (D). For any fixed $\lambda \in \mathbb{R}^m_+$, suppose that x_λ solves (L_λ). Then, we have from Lemma 3.3 (i)

$$f_i \geq f(x_\lambda) - \lambda(c_i - g(x_\lambda)), \quad i = k, k+1. \tag{3.4.12}$$

Also, since $b \in [c_k, c_{k+1})$, there exists a $\mu \in (0, 1]$ such that $b = \mu c_k + (1 - \mu)c_{k+1}$. We thus obtain from (3.4.8), (3.4.9) and (3.4.12) that

$$\begin{aligned} d(\lambda^*) &= \min_{x \in X} L(x, \lambda^*) \\ &= f(x^*) - \xi_k(g(x^*) - b) \\ &= f_k - \frac{f_{k+1} - f_k}{c_{k+1} - c_k}[c_k - (\mu c_k + (1 - \mu)c_{k+1})] \\ &= \mu f_k + (1 - \mu) f_{k+1} \\ &\geq \mu[f(x_\lambda) - \lambda(c_k - g(x_\lambda))] + (1 - \mu)[f(x_\lambda) - \lambda(c_{k+1} - g(x_\lambda))] \\ &= f(x_\lambda) + \lambda(g(x_\lambda) - b) \\ &= \min_{x \in X} L(x, \lambda) = d(\lambda). \end{aligned}$$

Hence λ^* solves (D). Therefore, (x^*, λ^*) is an optimal primal-dual pair of (P). □

3.5 Solution Properties of the Dual Problem

In this section, we focus on the solution properties of Lagrangian relaxation problem (L_{λ^*}):

$$d(\lambda^*) = \min_{x \in X} L(x, \lambda^*), \tag{3.5.1}$$

where λ^* is an optimal solution to the dual problem (D).

A key question arises from the problem (L_{λ^*}): Is there always an optimal solution to (L_{λ^*}) which is feasible in the primal problem? The answer is negative in general situations as shown in the following example.

EXAMPLE 3.7

$$\begin{aligned} \min\ & f(x) = 3x_1 + 2x_2 - 1.5x_1^2 \\ \text{s.t. } & g_1(x) = 15 - 7x_1 + 2x_2 \le 12, \\ & g_2(x) = 15 + 2x_1^2 - 7x_2 \le 12, \\ & x \in X = \{(0,1)^T, (0,2)^T, (1,0)^T, (1,1)^T, (2,0)^T\}. \end{aligned}$$

The optimal solution to the example is $x^* = (1,1)^T$ with $f(x^*) = 3.5$. The optimal solution to the dual problem (D) is $\lambda^* = (0.1951, 0.3415)^T$ with $d(\lambda^*) = 1.6095$. The Lagrangian relaxation problem (L_λ) with $\lambda = \lambda^*$ has three optimal solutions: $(0,1)^T$, $(0,2)^T$ and $(2,0)^T$, none of which is feasible.

Nevertheless, we will show that the answer is positive in single-constraint cases.

THEOREM 3.15 *If $m = 1$, then there exists at least one optimal solution to the Lagrangian problem (L_{λ^*}) which is feasible in the primal problem.*

Proof. Suppose on the contrary there is no feasible optimal solution to (L_{λ^*}). Then

$$L(x, \lambda^*) > L(x^*, \lambda^*), \quad \forall x \in S, \tag{3.5.2}$$

where $x^* \in X \setminus S$ is an optimal solution to (L_{λ^*}) which is infeasible in the primal problem. Let

$$\bar{\lambda} = \min_{x \in S} \frac{f(x) - f(x^*)}{g(x^*) - g(x)}. \tag{3.5.3}$$

Then by (3.5.2), we have $\bar{\lambda} > \lambda^*$. Let $\bar{x} \in S$ be such that

$$L(\bar{x}, \bar{\lambda}) = \min_{x \in S} L(x, \bar{\lambda}). \tag{3.5.4}$$

Now for any $x \in X \setminus S$, since $g(x) - b > 0$, we have

$$\begin{aligned} L(x, \bar{\lambda}) &= f(x) + \bar{\lambda}(g(x) - b) > f(x) + \lambda^*(g(x) - b) \\ &= L(x, \lambda^*) \ge L(x^*, \lambda^*). \end{aligned} \tag{3.5.5}$$

On the other hand, for any $x \in S$, by (3.5.3) and (3.5.4), we have

$$\begin{aligned} L(x,\bar{\lambda}) &\geq L(\bar{x},\bar{\lambda}) \\ &= f(\bar{x}) + \bar{\lambda}(g(\bar{x}) - b) \\ &= f(\bar{x}) + \bar{\lambda}(g(\bar{x}) - g(x^*)) + \bar{\lambda}(g(x^*) - b) \\ &\geq f(\bar{x}) + \frac{f(\bar{x}) - f(x^*)}{g(x^*) - g(\bar{x})}(g(\bar{x}) - g(x^*)) + \bar{\lambda}(g(x^*) - b) \\ &> f(x^*) + \lambda^*(g(x^*) - b) \\ &= L(x^*, \lambda^*). \end{aligned} \tag{3.5.6}$$

Combining (3.5.5) with (3.5.6), we infer that

$$d(\bar{\lambda}) = \min_{x \in X} L(x,\bar{\lambda}) > L(x^*,\lambda^*) = d(\lambda^*),$$

which contradicts the optimality of λ^*. □

Interestingly, the following theorem and corollary reveal that the primal infeasibility is assured for at least one optimal solution to (L_{λ^*}) in general situations, including both singly-constrained and multiply-constrained cases, where there exists a nonzero duality gap.

THEOREM 3.16 *Assume that $\psi(y) < w(y)$ for some $y \in conv(Y)$. Let μ^* be an optimal solution to (3.3.8). Then there is at least an $i \in \{1,\ldots,K\}$ such that $\mu_i^* > 0$ and $c_i \in C$ with $c_i \not\leq y$.*

Proof. For any $y \in Y$, by (3.3.8), there exists $\mu^* \in \Lambda$ that solves the following problem:

$$\psi(y) = \min \sum_{i=1}^{K} \mu_i f_i, \tag{3.5.7}$$

$$\text{s.t. } \sum_{i=1}^{K} \mu_i c_i \leq y, \ \mu \in \Lambda.$$

Let $I = \{i \mid \mu_i^* > 0\}$. Suppose that $c_i \leq y$ for all $i \in I$. We claim that $f_k = f_l$ for any $k, l \in I$. Otherwise, suppose that $f_k > f_l$ for some $k, l \in I$. Define $\tilde{\mu} = (\tilde{\mu}_1, \ldots, \tilde{\mu}_K)$ as follows: $\tilde{\mu}_i = \mu_i^*$, if $i \neq k$ and $i \neq l$; and $\tilde{\mu}_k = \mu_k^* - \epsilon$, $\tilde{\mu}_l = \mu_l^* + \epsilon$, with $\epsilon > 0$ being small enough such that $\tilde{\mu}_k > 0$.

Note that $\tilde{\mu} \in \Lambda$ and $\tilde{\mu}_j = 0$ iff $\mu_j^* = 0$. Since by assumption $c_i \leq y$ for all $i \in I$, it follows that $\sum_{i=1}^K \tilde{\mu}_i c_i = \sum_{i \in I} \tilde{\mu}_i c_i \leq \sum_{i \in I} \tilde{\mu}_i y = y$. Thus, $\tilde{\mu}$ is feasible to problem (3.5.7). Moreover,

$$\sum_{i=1}^{K} (\tilde{\mu}_i f_i - \mu_i^* f_i) = \epsilon(f_l - f_k) < 0,$$

which contradicts that μ^* is an optimal solution to (3.5.7). Therefore, $f_k = f_l$ for any $k, l \in I$. It then follows that $\psi(y) = f_i$ for any $i \in I$. Since $c_i \leq y$ for all $i \in I$, $w(c_i) \geq w(y)$. Thus, $\psi(y) = f_i = w(c_i) \geq w(y)$, contradicting the assumption that $\psi(y) < w(y)$. □

COROLLARY 3.2 *Assume that the duality gap between (P) and (D) is nonzero, i.e., $d(\lambda^*) < f^*$. Then there is at least one optimal solution to the Lagrangian problem (L_{λ^*}) which is infeasible in the primal problem.*

Proof. Notice from Theorem 3.10 (i) that $\psi(b) = d(\lambda^*)$. Thus, $\psi(b) < f^* = w(b)$. Applying Theorem 3.16 with $y = b$, we conclude that there exists an $i \in I$ such that $c_i \nleq b$. Let $\bar{x}$ be such that $(g(\bar{x}), f(\bar{x})) = (c_i, f_i)$. Then $\bar{x}$ is infeasible and by Theorem 3.10 (ii), $\bar{x}$ solves (L_{λ^*}). □

It is important to investigate the solution properties associated with the situations where the optimal solution to the dual problem, λ^*, is equal to zero. It is interesting to notice from the following discussion that there is a substantial difference between the singly constrained and multiply constrained situations.

THEOREM 3.17 *If the dual optimal solution $\lambda^* = 0$, then any feasible solution to (L_{λ^*}) is an optimal solution to (P) and $f^* = d(\lambda^*)$. Conversely, if there is a feasible solution x^* in the optimal solution set of (L_λ) with $\lambda = 0$, then $\lambda = 0$ is an optimal solution to (D) and x^* is an optimal solution to (P).*

Proof. Let x^* be a feasible solution to (L_{λ^*}) with $\lambda^* = 0$. Since

$$f(x^*) = \min_{x \in X} L(x, 0) = \min_{x \in X} f(x) \leq \min_{x \in S} f(x) = f^*, \tag{3.5.8}$$

we imply that x^* is optimal to (P) and $f(x^*) = f^* = d(\lambda^*)$. Conversely, by (3.5.8), if a solution to $(L_{\lambda=0})$, x^*, is feasible, then x^* must be optimal to (P). Moreover, by weak duality, we have $d(\lambda) \leq f(x^*) = d(0)$ for all $\lambda \in \mathbb{R}^m_+$. Thus $\lambda = 0$ is the dual optimal solution. □

Theorems 3.15 and 3.17 imply that if a zero dual optimal solution is found for a singly constrained integer programming problem, then $\lambda = 0$ is the optimal generating multiplier, there is no duality gap for this problem, and there must be a solution of $(L_{\lambda=0})$ that is feasible to (P). For multiply constrained situations, however, there could exist cases where none of the solutions to $(L_{\lambda=0})$ is feasible to (P) when zero is the optimal dual solution. The following example illustrates this situation.

EXAMPLE 3.8

$$\begin{aligned} \min\ & 2x_1 \\ \text{s.t. }\ & 1.7x_1 + 2x_2 + 2x_4 \le 1.9, \\ & 1.7x_1 + 2x_3 + 2x_4 \le 1.9, \\ & x \in X = \{e_1, e_2, e_3, e_4\}, \end{aligned}$$

where e_i is the i-th unit vector in $\mathbb{R}^4$, $i = 1, 2, 3, 4$.

The problem has a unique feasible solution e_1. The dual problem $\max_{\lambda \in \mathbb{R}^m_+} d(\lambda)$ can be written explicitly as

$$\begin{aligned} \max\ & \min_{x \in X}\ 2x_1 + \lambda_1(1.7x_1 + 2x_2 + 2x_4 - 1.9) \\ & \qquad + \lambda_2(1.7x_1 + 2x_3 + 2x_4 - 1.9) \\ \text{s.t. }\ & \lambda_1 \ge 0, \lambda_2 \ge 0, \end{aligned}$$

or equivalently

$$\begin{aligned} \max\ & \mu \\ \text{s.t. }\ & 2 + \lambda_1(1.7 - 1.9) + \lambda_2(1.7 - 1.9) \ge \mu, \\ & 0 + \lambda_1(2 - 1.9) + \lambda_2(0 - 1.9) \ge \mu, \\ & 0 + \lambda_1(0 - 1.9) + \lambda_2(2 - 1.9) \ge \mu, \\ & 0 + \lambda_1(2 - 1.9) + \lambda_2(2 - 1.9) \ge \mu, \\ & \lambda_1 \ge 0, \lambda_2 \ge 0. \end{aligned}$$

Notice that $(\lambda_1, \lambda_2, \mu) = (0, 0, 0)$ is feasible to the above problem. Since adding the second constraint to the third constraint yields $-0.9\lambda_1 - 0.9\lambda_2 \ge \mu$, any feasible solution $(\lambda_1, \lambda_2, \mu)$ with $\lambda_1 > 0$ or $\lambda_2 > 0$ will lead to a negative μ. Thus $\lambda = (0, 0)^T$ is the optimal solution to the dual problem and only the three infeasible solutions, e_1, e_2 and e_3, solve $(L_{\lambda=0})$.

Geometrically, the above example shows that there exist multiply constrained cases where more than one points $(g(x), f(x))$ with $g(x) \not\le b$ surround the axis $y = b$ and span a horizontal plane (corresponding to $\lambda = 0$) with $f(x)$ being the lowest objective value over X. Algorithmically, the dual search method will fail in this situation to raise the dual value higher than the lowest objective value.

3.6 Lagrangian Decomposition via Copying Constraints

In this section, we focus on the following linearly constrained nonlinear integer programming problem:

$$
\begin{array}{lll}
(P_l) & \min & f(x) \\
& \text{s.t.} & Ax \leq b, \\
& & Bx = d, \\
& & x \in X \subseteq \mathbb{Z}^n,
\end{array}
$$

where f is a continuous nonlinear (possibly nonseparable) function on X, A is an $m \times n$ matrix, B is a $q \times n$ matrix, $b \in \mathbb{R}^m$, $d \in \mathbb{R}^q$, and X is a finite integer set.

3.6.1 General Lagrangian decomposition schemes

Since the objective function $f(x)$ could be nonseparable, a direct adoption of the Lagrangian dual formulation in Section 3.1 results in a nonseparable Lagrangian relaxation problem (L_λ), which is difficult to solve in most situations. The motivation of the Lagrangian decomposition via copying constraints is to separate the nonlinearity and nonseparability from the integrality and thus to reduce the extent of difficulty of the Lagrangian relaxation problem. It is clear that (P_l) is equivalent to the following problem

$$
\begin{array}{lll}
(\widetilde{P}_l) & \min & f(y) \\
& \text{s.t.} & Ay \leq b, \\
& & By = d, \\
& & y \in conv(X), \\
& & y = x, \\
& & Ax \leq b, \\
& & Bx = d, \\
& & x \in X.
\end{array}
$$

Define

$$
\begin{aligned}
X_I &= \{x \in X \mid Ax \leq b, Bx = d\}, \\
X_{IR} &= \{x \in conv(X) \mid Ax \leq b, Bx = d\}.
\end{aligned}
$$

Let $\mu \in \mathbb{R}^n$ be the Lagrangian multiplier vector for the link constraint $y = x$ in $(\widetilde{P}_l)$. Then the Lagrangian relaxation problem of $(\widetilde{P}_l)$ is

$$
\begin{aligned}
\ell(\mu) \quad = \quad & \min \ [f(y) + \mu^T(x - y)], \\
& \text{s.t.} \ \ y \in X_{IR}, \ x \in X_I.
\end{aligned}
\tag{3.6.1}
$$

It is easy to see that problem (3.6.1) can be decomposed into a continuous nonlinear optimization problem and a linear integer programming problem:

$$\ell(\mu) = \min_{y \in X_{IR}} [f(y) - \mu^T y] + \min_{x \in X_I} \mu^T x. \tag{3.6.2}$$

Define

$$(L^{\ell}_{\mu y}) \qquad \min_{y \in X_{IR}} f(y) - \mu^T y,$$

and

$$(L^{\ell}_{\mu x}) \qquad \min_{x \in X_I} \mu^T x.$$

Let $v(\cdot)$ denote the optimal value of problem $(\cdot)$. Then we have $\ell(\mu) = v(L^{\ell}_{\mu y}) + v(L^{\ell}_{\mu x})$. If f is a convex function, then problem $(L^{\ell}_{\mu y})$ is a linearly constrained convex programming problem which can be solved by many existing efficient solution methods. On the other hand, linear integer programming has been extensively studied and efficient algorithms such as branch-and-bound methods have been developed for solving $(L^{\ell}_{\mu x})$. Furthermore, more efficient methods for $(L^{\ell}_{\mu x})$ can be adopted when the discrete polyhedron X_I assumes some special structure.

The following weak duality inequality holds for any $\mu \in \mathbb{R}^n$,

$$\ell(\mu) \leq v(\widetilde{P}_l) = v(P_l).$$

The dual problem of $(\widetilde{P}_l)$ is

$$(D_{\ell}) \quad \max_{\mu \in \mathbb{R}^n} \ell(\mu).$$

It is easy to see that $\ell(\mu)$ is a concave function of μ. Let x_μ solve $(L^{\ell}_{\mu x})$ and y_μ solve $(L^{\ell}_{\mu y})$. Then, $x_\mu - y_\mu$ is a subgradient of ℓ at μ.

Let $(\overline{P}_l)$ denote the continuous relaxation of (P_l). Then we have

$$v(\overline{P}_l) \leq v(D_{\ell}) \leq v(P_l). \tag{3.6.3}$$

The first inequality is due to the fact $v(\overline{P}_l) = \ell(0) \leq v(D_{\ell})$. Inequality (3.6.3) implies that the lower bound derived from the Lagrangian decomposition is at least as good as that of the continuous relaxation of (P_l). However, we realize that more computational effort is needed to obtain the Lagrangian bound $v(D_{\ell})$.

To understand more about the dual problem, let us consider the following continuous problem by replacing the constraints in (P_l) by its convex hull:

$$(P_l^*) \qquad \begin{aligned} &\min\ f(x) \\ &\text{s.t. } x \in conv(X_I). \end{aligned}$$

Since $conv(X_I) \subseteq X_{IR}$, we obtain the following equivalent problem of (P_l^*) by placing a link equality $x = y$,

$$(\widetilde{P}_l^*) \qquad \min\ f(y) \\ \text{s.t. } y \in X_{IR},\ y = x,\ x \in conv(X_I).$$

We have the following result.

THEOREM 3.18 *If f is convex, then*

$$v(\overline{P}_l) \le v(P_l^*) = v(D_\ell) \le v(P_l). \tag{3.6.4}$$

Proof. Dualizing $x = y$ in $(\widetilde{P}_l^*)$ and using the convex duality theory give rise to

$$\begin{aligned} v(P_l^*) &= v(\widetilde{P}_l^*) \\ &= \max_{\mu \in \mathbb{R}^n}\{\min_{y \in X_{IR}} [f(y) - \mu^T y] + \min_{x \in conv(X_I)} \mu^T x\} \\ &= \max_{\mu \in \mathbb{R}^n}\{\min_{y \in X_{IR}} [f(y) - \mu^T y] + \min_{x \in X_I} \mu^T x\} \\ &= \max_{\mu \in \mathbb{R}^n} \ell(\mu) \\ &= v(D_\ell). \end{aligned}$$

The third equality is due to the fact that linear program achieves its optimum at one of its extreme points while all the extreme points of $conv(X_I)$ are integral. Therefore, the dual value $v(D_\ell)$ is nothing but the optimal value obtained by solving the convexified problem (P_l^*). The inequality (3.6.4) then follows from (3.6.3). □

Next, we consider an alternative way of Lagrangian decomposition. Let Y be such that $X_I \subseteq Y \subseteq conv(X)$. Problem (P_l) is equivalent to the following problem:

$$(P_l^\circ) \qquad \min\ f(y) \\ \text{s.t. } y \in Y, \\ y = x, \\ x \in X_I.$$

Dualizing constraint $y = x$ yields the following decomposition:

$$l(\mu) = \min_{y \in Y}(f(y) - \mu^T y) + \min_{x \in X_I} \mu^T x. \tag{3.6.5}$$

Let $(L_{\mu y}^l)$ and $(L_{\mu x}^l)$ denote the first problem and the second problem in $l(\mu)$, respectively. Again, l is a concave function on $\mathbb{R}^n$ and $x_\mu - y_\mu$ is a subgradient of l at μ, where y_μ and x_μ are the optimal solutions to $(L^l_{\mu y})$ and $(L^l_{\mu x})$, respectively.

The dual problem corresponding to l is

$$(D_l) \quad \max_{\mu \in \mathbb{R}^n} l(\mu).$$

Let $Y = conv(X)$. Consider the following problem:

$$\begin{array}{ll}(\widetilde{P}_l^\circ) & \min\ f(y) \\ & \text{s.t. } y \in conv(X),\ y = x,\ x \in conv(X_I).\end{array}$$

We have the following results:

THEOREM 3.19 *If f is convex and $Y = conv(X)$, then*

$$v(D_\ell) = v(D_l) = v(\widetilde{P}_l^\circ) \le v(P_l^\circ).$$

Proof. Since $conv(X_I) \subseteq X_{IR} \subseteq conv(X)$, problem $(\widetilde{P}_l^*)$ is equivalent to problem $(\widetilde{P}_l^\circ)$. Thus, by Theorem 3.18, we have

$$v(\widetilde{P}_l^\circ) = v(\widetilde{P}_l^*) = v(P_l^*) = v(D_\ell)$$

Using similar arguments as in the proof of Theorem 3.18, we can prove that $v(\widetilde{P}_l^\circ) = v(D_l)$. This proves the theorem. □

Thus, if $Y = conv(X)$, the decomposition formulations (D_ℓ) and (D_l) produce the same lower bounds. Note that the first part $(L^l_{\mu y})$ in l is a nonlinear continuous optimization problem without constraint $Ax \le b$ and $Bx = d$, and hence is easier to solve than $(L^\ell_{\mu y})$ in ℓ.

Comparing to the classical Lagrangian dual function $d(\lambda)$ defined in (3.1.1), which is piecewise linear concave, the dual function ℓ and l are not necessarily piecewise linear (see [161]). Therefore, the subgradient method seems to be the only suitable dual search procedure for solving the dual problems (D_ℓ) or (D_l).

3.6.2 0-1 quadratic case

Now, we consider the 0-1 quadratic case of (P_l) in the following form:

$$\begin{array}{ll}(0\text{-}1QP) & \min\ f(x) = x^T Q x + c^T x \\ & \text{s.t. } Ax \le b, \\ & \qquad x \in X = \{0,1\}^n.\end{array}$$

where Q is an $n \times n$ symmetric matrix, A is an $m \times n$ matrix and $b \in \mathbb{R}^m$.

Two ways of choosing set Y in the Lagrangian decomposition dual (D_l) will be considered: $Y = [0,1]^n$ and $Y = \{0,1\}^n$. We have the following two Lagrangian decomposition dual problems:

$$(DQ_1) \quad \max_{\mu \in \mathbb{R}^n} l^1(\mu),$$

where

$$l^1(\mu) = \min_{y \in [0,1]^n} [y^T Q y + y^T (c - \mu)] + \min_{x \in \{0,1\}^n} \{\mu^T x \mid Ax \leq b\}, \quad (3.6.6)$$

and

$$(DQ_2) \quad \max_{\mu \in \mathbb{R}^n} l^2(\mu),$$

where

$$l^2(\mu) = \min_{y \in \{0,1\}^n} [y^T Q y + y^T (c - \mu)] + \min_{x \in \{0,1\}^n} \{\mu^T x \mid Ax \leq b\}. \quad (3.6.7)$$

Also, the classical Lagrangian relaxation dual problem of (0-1QP) is:

$$(DQ) \quad \max_{\lambda \in \mathbb{R}^m_+} d(\lambda),$$

where

$$d(\lambda) = \min_{x \in \{0,1\}^n} [x^T Q x + c^T x + \lambda^T (Ax - b)]. \quad (3.6.8)$$

We are going to study the relationship between these three dual bounds. Define the following two problems:

$$(\overline{0\text{-}1QP}) \qquad \min x^T Q x + c^T x \quad \text{s.t. } x \in \{x \in [0,1]^n \mid Ax \leq b\}.$$

and

$$(0\text{-}1QP^*) \qquad \min x^T Q x + c^T x \quad \text{s.t. } x \in conv\{x \in \{0,1\}^n \mid Ax \leq b\}.$$

From Theorem 3.19, we have

$$v(\overline{0\text{-}1QP}) \leq v(DQ_1) = v(0\text{-}1QP^*) \leq v(0\text{-}1QP). \quad (3.6.9)$$

Now, we discuss a simplification of the dual problem (DQ_1). Let

$$U = \{\mu = 2Qx + c \mid x \in [0,1]^n\}.$$

LEMMA 3.4 *If Q is positive definite, then the dual function l^1 is strongly concave on U and for any $\mu \in U$,*

$$l^1(\mu) = -\frac{1}{4}(c-\mu)^T Q^{-1}(c-\mu) + \min_{x \in \{0,1\}^n} \{\mu^T x \mid Ax \le b\}. \quad (3.6.10)$$

Proof. For any $\mu \in U$, there exists $z \in [0,1]^n$ such that $\mu = 2Qz + c$. Thus, the KKT conditions for the convex quadratic problem $\min_{y \in [0,1]^n} y^T Q y + y^T(c-\mu)$ holds at z and $z = \frac{1}{2}Q^{-1}(\mu - c)$ is its unique optimal solution. The expression (3.6.10) then follows immediately. □

The dual problem (DQ_1) can be simplified by using the following lemma.

LEMMA 3.5 ([162]) *If Q is positive definite, then there exists at least an optimal solution of (DQ_1) in U.*

THEOREM 3.20 *If Q is positive definite, then (DQ_1) is equivalent to*

$$\max_{\gamma \in [0,1]^n} \{-\gamma^T Q \gamma + \min_{x \in \{0,1\}^n} \{(2Q\gamma + c)^T x \mid Ax \le b\}\}. \quad (3.6.11)$$

Proof. From Lemmas 3.4 and 3.5, (DQ_1) is equivalent to

$$\max_{\mu \in U}\{-\frac{1}{4}(c-\mu)^T Q^{-1}(c-\mu) + \min_{x \in \{0,1\}^n} \{\mu^T x \mid Ax \le b\}\},$$

which is in turn equivalent to (3.6.11) by letting $\gamma = \frac{1}{2}Q^{-1}(\mu - c)$. □

Next, we turn to study the dual problem (DQ_2) and its relation with (DQ_1) and (DQ). Rewrite the quadratic function $f(x) = x^T Q x + c^T x$ as

$$f(x) = \sum_{1 \le i < j \le n} q_{ij} x_i x_j + \sum_{i=1}^n q_i x_i.$$

Note that the first subproblem in $l^2(\mu)$ and the Lagrangian relaxation problem $d(\lambda)$ in (DQ) are 0-1 unconstrained quadratic optimization problems. We only consider the case when $q_{ij} \le 0$ for $1 \le i < j \le n$. Under this condition, it has been shown [178] that the first subproblem in l^2 and the Lagrangian relaxation problem $d(\lambda)$ are polynomially solvable (see Chapter 10). Since $l^1(\mu) \le l^2(\mu)$ for any μ, (DQ_2) produces better lower bound than (DQ_1), i.e.,

$$v(DQ_1) \le v(DQ_2).$$

In order to compare the bounds $v(DQ)$ and $v(DQ_2)$, we define two problems:

$$(0\text{-}1QP_1) \qquad \min\ \bar{f}(x) = \sum_{1\le i<j\le n} q_{ij}\min(x_i,x_j) + \sum_{i=1}^{n} q_i x_i$$
$$\text{s.t. } Ax \le b,$$
$$x \in [0,1]^n,$$

and

$$(0\text{-}1QP_2) \qquad \min\ \bar{f}(x) = \sum_{1\le i<j\le n} q_{ij}\min(x_i,x_j) + \sum_{i=1}^{n} q_i x_i$$
$$\text{s.t. } x \in conv\{x \in \{0,1\}^n \mid Ax \le b\}.$$

By assumption, $q_{ij} \le 0$ for $1 \le i < j \le n$, thus $\bar{f}(x)$ is convex. Note also that $f(x) = \bar{f}(x)$ for all $x \in \{0,1\}^n$. Hence problems $(0\text{-}1QP_1)$ and $(0\text{-}1QP_2)$ are convex continuous relaxations of $(0\text{-}1QP)$ and

$$v(0\text{-}1QP_1) \le v(0\text{-}1QP_2) \le v(0\text{-}1QP).$$

We need the following lemma.

LEMMA 3.6 ([178]) *For any $\mu \in \mathbb{R}^n$, it holds*

$$\min_{x\in[0,1]^n} (\bar{f}(x) - \mu^T x) = \min_{x\in\{0,1\}^n} (\bar{f}(x) - \mu^T x).$$

By the strong duality theorem for convex optimization and Lemma 3.6, we have

$$\begin{aligned} v(0\text{-}1QP_1) &= \min\{\bar{f}(x) \mid Ax \le b,\ x \in [0,1]^n\} \\ &= \max_{\lambda\in\mathbb{R}^m_+} \min_{x\in[0,1]^n} \{\bar{f}(x) + \lambda^T(Ax-b)\} \\ &= \max_{\lambda\in\mathbb{R}^m_+} \min_{x\in\{0,1\}^n} \{\bar{f}(x) + \lambda^T(Ax-b)\} \\ &= v(DQ). \end{aligned}$$

Also, we have

$$\begin{aligned} & v(0\text{-}1QP_2) \\ &= \min\{\bar{f}(x) \mid x \in conv\{x \in \{0,1\}^n \mid Ax \le b\}\} \\ &= \min\{\bar{f}(y) \mid x \in conv\{x \in \{0,1\}^n \mid Ax \le b\},\ x = y,\ y \in [0,1]^n\} \\ &= \max_{\mu\in\mathbb{R}^n}\{\min_{y\in[0,1]^n} (\bar{f}(y) - \mu^T y) + \min\{\mu^T x \mid x \in conv\{x \in \{0,1\}^n \mid Ax \le b\}\}\} \\ &= \max_{\mu\in\mathbb{R}^n}\{\min_{y\in\{0,1\}^n} (f(y) - \mu^T y) + \min_{x\in\{0,1\}^n} \{\mu^T x \mid Ax \le b\}\} \\ &= \max_{\mu\in\mathbb{R}^n} l^2(\mu) \\ &= v(DQ_2). \end{aligned}$$

Therefore, we obtain the following theorem.

THEOREM 3.21 *If $q_{ij} \leq 0$ for $1 \leq i < j \leq n$, then*

$$v(DQ) = v(0\text{-}1QP_1) \leq v(0\text{-}1QP_2) = v(DQ_2) \leq v(0\text{-}1QP). \quad (3.6.12)$$

The above theorem shows that for 0-1 quadratic problem, the Lagrangian decomposition dual (DQ_2) can produce better lower bound than the conventional Lagrangian dual (DQ). The following result indicates that in some special cases, the conventional dual problem (DQ) is better than the Lagrangian decomposition dual (DQ_1).

COROLLARY 3.3 *If $q_{ij} \leq 0$ for $1 \leq i < j \leq n$ and every extreme point of $\{x \in [0,1]^n \mid Ax \leq b\}$ is integer, then*

$$v(DQ_1) \leq v(DQ_2) = v(DQ). \quad (3.6.13)$$

Proof. Under the assumption of the corollary, it holds

$$conv\{x \in \{0,1\}^n \mid Ax \leq b\} = \{x \in [0,1]^n \mid Ax \leq b\}.$$

Thus, $v(0\text{-}1QP_1) = v(0\text{-}1QP_2)$. Inequality (3.6.13) then follows from Theorem 3.21. □

3.7 Notes

The basic properties of Lagrangian duality theory for integer programming were first presented in [107]. Lagrangian methods for linear integer programming were extensively studied in the literature (see for example [17][56][57][75] [168]). A survey of the use of Lagrangian techniques in integer programming can be found in [192]. The properties of the Lagrangian relaxation in Section 3.1 for linearly constrained convex integer programming problems were analyzed and exploited in [55].

The use of the subgradient method in solving integer programming was first proposed in [97]. Subgradient methods for general nonsmooth convex minimization were summarized in [198]. The outer Lagrangian linearization method for the dual search in linear integer programming was discussed in [176][192]. Procedure 3.3 for singly constrained problems was presented in [134]. Extensive discussions about bundle methods for nonsmooth convex optimization can be found in [100][127]. Bundle-type methods for Lagrangian dual search were also proposed in [167][235].

The relationship between the perturbation function and the dual function was established in [128][134][143]. Many new properties associated with the Lagrangian dual were presented in [135] based on the perturbation analysis.

Lagrangian decomposition method via copying constraints was first proposed in [86] for linear integer programming and was later extended for convex integer programming in [161] and 0-1 quadratic programming [162].

Chapter 4

SURROGATE DUALITY THEORY

Along with the Lagrangian duality theory, the surrogate duality theory has been widely used in solving integer programming problems. While the Lagrangian dual formulation generates a relaxation by incorporating the constraints into the objective function, the surrogate dual generates a relaxation by aggregating multiple constraints into a single surrogate constraint.

4.1 Conventional Surrogate Dual Method

Consider the following general integer programming problem with multiple inequality constraints:

$$
\begin{aligned}
(P) \qquad & \min f(x) \\
& \text{s.t. } g_i(x) \le b_i, \quad i = 1, 2, \ldots, m, \\
& \qquad x \in X \subseteq \mathbb{Z}^n,
\end{aligned}
$$

where $m \ge 2$, X is a finite set and $\mathbb{Z}^n$ is the set of all integer points in $\mathbb{R}^n$. Constraints $g_i(x) \le b_i$, $i = 1, 2, \ldots, m$, are called major constraints. Define S to be the feasible region of decision vectors in (P),

$$S = \{x \in X \mid g_i(x) \le b_i,\ i = 1, 2, \ldots, m\}.$$

4.1.1 Surrogate dual and its properties

Let $g(x) = (g_1(x), \ldots, g_m(x))^T$ and $b = (b_1, \ldots, b_m)^T$. Aggregating the multiple major constraints of (P) into a single surrogate constraint generates a surrogate relaxation,

$$
\begin{aligned}
(P_\mu) \qquad & \min\ f(x) \\
& \text{s.t. } \mu^T(g(x) - b) \le 0, \\
& \qquad x \in X,
\end{aligned}
$$

where $\mu = (\mu_1, \ldots, \mu_m)^T \in \mathbb{R}^m_+$ is a vector of surrogate multipliers. Define $S(\mu)$ to be the feasible region of decision vectors in (P_μ),

$$S(\mu) = \{x \in X \mid \mu^T(g(x) - b) \leq 0\}. \tag{4.1.1}$$

Denote by $v(Q)$ the optimal value of an optimization problem (Q). The surrogate dual is an optimization problem in μ,

$$\begin{array}{lll} (D_S) & & \max\ v(P_\mu) \\ & & \text{s.t.}\ \ \mu \in \mathbb{R}^m_+. \end{array}$$

Since $S \subseteq S(\mu), \forall\, \mu \in \mathbb{R}^m_+$, (P_μ) is a relaxation of (P). The following weak surrogate duality is evident,

$$v(P_\mu) \leq v(P), \quad \forall\, \mu \in \mathbb{R}^m_+.$$

Consequently, the surrogate dual provides a lower bound for $v(P)$.

$$v(D_S) \leq v(P).$$

THEOREM 4.1 (STRONG SURROGATE DUALITY) *If an x^* solves (P_{μ^*}) for a $\mu^* \in \mathbb{R}^m_+$ and x^* is feasible in (P), then x^* solves (P) and $v(D_S) = v(P)$.*

Proof. Note that problems (P) and (P_μ) have the same objective function. Since $S \subseteq S(\mu), \forall\, \mu \in \mathbb{R}^m_+$, a minimizer, x^*, over $S(\mu^*)$ with $\mu^* \in \mathbb{R}^m_+$ and $x^* \in S$ must be also a minimizer over S. Thus, x^* solves (P). Furthermore, from the weak surrogate duality, we have $f(x^*) = v(P_{\mu^*}) \leq v(D_S) \leq v(P) = f(x^*)$. Therefore, $v(D_S) = v(P)$. □

It is clear that $v(P_\mu) = v(P_{\theta\mu})$ for any $\theta > 0$. Thus, the surrogate dual problem (D_S) can be normalized to an equivalent problem with a compact feasible region:

$$\begin{array}{lll} (D_S^n) & & \max\ v(P_\mu) \\ & & \text{s.t.}\ \ \mu \in \Lambda, \end{array}$$

where $\Lambda = \{\mu \in \mathbb{R}^m_+ \mid e^T\mu \leq 1\}$ and $e = (1, \ldots, 1)^T$.

Let (L_λ) be the Lagrangian relaxation of (P) with a given Lagrangian multiplier vector λ and (D) be the Lagrangian dual of (P). We have the following theorem to reveal the relationship between the Lagrangian dual and the surrogate dual.

THEOREM 4.2 *The surrogate dual generates a bound tighter than the Lagrangian dual, i.e., $v(D) \leq v(D_S)$. Furthermore, if $v(D) = v(D_S)$, then for any*

Lagrangian multiplier vector $\hat{\lambda} \in \mathbb{R}^m_+$ *that solves* (D), *there exists an* $\hat{x}$ *such that* $\hat{\lambda}^T(g(\hat{x}) - b) = 0$.

Proof. For any $\lambda \in \mathbb{R}^m_+$, we have

$$\begin{aligned} v(L_\lambda) &= \min\{f(x) + \lambda^T(g(x) - b) \mid x \in X\} \\ &\le \min\{f(x) + \lambda^T(g(x) - b) \mid \lambda^T(g(x) - b) \le 0,\ x \in X\} \\ &\le \min\{f(x) \mid \lambda^T(g(x) - b) \le 0,\ x \in X\} \\ &= v(P_\lambda). \end{aligned}$$

One immediate result of the above inequality is

$$v(D) = \max_{\lambda \ge 0} v(L_\lambda) \le \max_{\lambda \ge 0} v(P_\lambda) = v(D_S). \tag{4.1.2}$$

This completes the first part of the theorem. Now let $\hat{\lambda}$ solve (D) and let $\hat{x}$ solve the surrogate relaxation $(P_{\hat{\lambda}})$. Feasibility of $\hat{x}$ in $(P_{\hat{\lambda}})$ implies $\hat{\lambda}^T(g(\hat{x}) - b) \le 0$. Since $\hat{x} \in X$, $\hat{x}$ is also feasible in $(L_{\hat{\lambda}})$. Thus,

$$v(D) \le f(\hat{x}) + \hat{\lambda}^T(g(\hat{x}) - b) \le f(\hat{x}) \le v(D_S).$$

The assumption $v(D) = v(D_S)$ leads to the conclusion that $\hat{\lambda}^T(g(\hat{x}) - b) = 0$.
□

4.1.2 Surrogate dual search

A key issue in applying the surrogate dual method is how to solve the surrogate dual problem, more specifically, how to update the surrogate multipliers. Several surrogate dual search methods have been developed for linear integer programming and they can be also applied to nonlinear integer programming problems.

For $\alpha \in \mathbb{R}$, let $X(\alpha)$ denote the level set of $f(x)$, $X(\alpha) = \{x \in X \mid f(x) \le \alpha\}$. For given $\mu \in \Lambda$ and $\alpha \in \mathbb{R}$, $v(P_\mu) \le \alpha$ if and only if

$$S(\mu) \cap X(\alpha) \ne \emptyset, \tag{4.1.3}$$

where $S(\mu)$ is defined by (4.1.1). Consider the following problem

$$(P(\alpha, \mu)) \qquad \begin{aligned} &\min\ \mu^T(g(x) - b) \\ &\text{s.t. } x \in X(\alpha). \end{aligned}$$

We notice that (4.1.3) holds if and only if $v(P(\alpha, \mu)) \le 0$. Since $v(D^n_S) = \max\{v(P_\mu) \mid \mu \in \Lambda\}$, it follows that $v(D^n_S) \le \alpha$ if and only if $v(P(\alpha, \mu)) \le 0$

for all $\mu \in \Lambda$. Similar to the Lagrangian dual, we can define the following dual problem:

$$(D(\alpha)) \qquad \max\ v(P(\alpha, \mu)) \quad \text{s.t. } \mu \in \Lambda.$$

The above discussion leads to the following theorem.

THEOREM 4.3 *For given $\alpha \in \mathbb{R}$, $v(D_S^n) \le \alpha$ if and only if $v(D(\alpha)) \le 0$.*

An immediate corollary of Theorem 4.3 is as follows.

COROLLARY 4.1 *The optimal surrogate dual value $v(D_S^n)$ is the minimum $\alpha \in \mathbb{R}$ such that $v(D(\alpha)) \le 0$.*

The cutting plane method can be used to solve $(D(\alpha))$. Notice that $(D(\alpha))$ is equivalent to the following linear program:

$$\begin{aligned} &\max_{(\beta,\mu)} \ \beta \\ &\text{s.t. } \beta \le \mu^T(g(x) - b), \ \forall x \in X(\alpha), \\ &\qquad \mu \in \Lambda. \end{aligned}$$

For each $x \in X(\alpha)$, the first constraint forms a cutting plane. Similar to the outer Lagrangian linearization method for Lagrangian dual search, we can construct $T^k \subset X(\alpha)$ step by step, thus approximating $v(D(\alpha))$ successively by solving the following linear program:

$$(LP_k) \qquad \begin{aligned} &\max_{(\beta,\mu)} \ \beta \\ &\text{s.t. } \beta \le \mu^T(g(x) - b), \ \forall x \in T^k, \\ &\qquad \mu \in \Lambda. \end{aligned}$$

PROCEDURE 4.1 (CUTTING PLANE PROCEDURE FOR (D_S^n))

Step 0 (Initialization). Set $\alpha^0 = -\infty$, $T^0 = \emptyset$. Choose any $\mu^1 \in \Lambda$. Set $k = 1$.

Step 1 (Surrogate relaxation). Solve the surrogate relaxation problem (P_{μ^k}) and obtain an optimal solution x^k. If $g(x^k) \le b$, stop and x^k is an optimal solution to (P) and $v(D_S^n) = v(P)$.

Step 2 (Updating lower bound). If $f(x^k) > \alpha^{k-1}$, then set $\alpha^k = f(x^k)$. Otherwise, set $\alpha^k = \alpha^{k-1}$.

Step 3 (Updating multiplier). Set $T^k = T^{k-1} \cup \{x^k\}$. Solve the linear program (LP_k) and obtain an optimal solution (β^k, μ^k). If $\beta^k \le 0$, stop and $\alpha^k = v(D_S^n)$. Otherwise, set $\mu^{k+1} = \mu^k$ and $k := k + 1$, go to Step 1.

THEOREM 4.4 *Algorithm 4.1 finds an optimal value of* (D_S^n) *within a finite number of iterations.*

Proof. If the procedure stops at Step 1, then by Theorem 4.1, the strong duality holds for (P) and x^k solves (P) and μ^k solves (D_S^n) with $v(P) = v(D_S^n)$. Suppose now the procedure stops at Step 3 of the k-th iteration. By Step 2, for any $1 \le i \le k$, if $f(x^i) > \alpha^{i-1}$, then $\alpha^i = f(x^i) > \alpha^{i-1}$; if $f(x^i) \le \alpha^{i-1}$, then $\alpha^i = \alpha^{i-1} \ge f(x^i)$. Thus, $f(x^i) \le \alpha^k$ for $1 \le i \le k$ which implies that $x^i \in X(\alpha^k)$ for any $x^i \in T^k$. Therefore,

$$v(D(\alpha^k)) \le v(LP_k) = \beta^k \le 0. \tag{4.1.4}$$

It then follows from Theorem 4.3 that $v(D_S^n) \le \alpha^k$. On the other hand, by Step 2 and the weak duality of the surrogate dual, there exists an $i \le k$ such that $\alpha^k = f(x^i) = v(P_{\mu^i}) \le v(D_S^n)$. Thus, $v(D_S^n) = \alpha^k$.

To show the finite termination of the procedure, suppose that at the k-th iteration, the procedure does not stop at Step 1 or Step 3. Then

$$0 < \beta^k = \min_{x^i \in T^k} (\mu^k)^T (g(x^i) - b).$$

This implies that all x^i's $\in T^k$ are infeasible in (P_{μ^k}) and they will not be added again to T^k in later stages. Since for any optimal solution x of (P_μ), $f(x) \le v(D_S^n)$, it will eventually hold $T^k = X(v(D_S^n))$ if the procedure does not stop at Step 1 or Step 3. Thus problem (LP_k) is then equivalent to problem $(D(\alpha))$ with $\alpha = v(D_S^n)$ and $\beta^k = v(D(\alpha))$. By Theorem 4.3, this implies $\beta^k = v(D(\alpha)) \le 0$. Therefore, the procedure will finally stop at Step 3. □

To illustrate Procedure 4.1, consider again Example 3.2:

EXAMPLE 4.1

$$\begin{aligned} \min\ & f(x) = 3x_1^2 + 2x_2^2 \\ \text{s.t.}\ & g_1(x) = 10 - 5x_1 - 2x_2 \le 7, \\ & g_2(x) = 15 - 2x_1 - 5x_2 \le 12, \\ & x \in X = \left\{ \begin{array}{c} \text{integer} \\ 0 \le x_1 \le 1,\ 0 \le x_2 \le 2 \\ 8x_1 + 8x_2 \ge 1 \end{array} \right\}. \end{aligned} \tag{4.1.5}$$

The optimal solution is $x^* = (1,1)^T$ with $f(x^*) = 5$. As computed in Example 3.2, the Lagrangian dual value of Example 4.1 is $2\frac{1}{3}$.

The iteration process of Procedure 4.1 for this example is described as follows:

Step 0. Set $\alpha^0 = -\infty$, $T^0 = \emptyset$. Choose $\mu^1 = (0.5, 0.5)^T$. Set $k = 1$.

Iteration 1

Step 1. Solve the surrogate problem

$$(P_{\mu^1}) \qquad \begin{aligned} &\min\ 3x_1^2 + 2x_2^2 \\ &\text{s.t. } 0.5 \times (10 - 5x_1 - 2x_2) + 0.5 \times (15 - 2x_1 - 5x_2) \le 9.5, \\ &\qquad x \in X. \end{aligned}$$

We obtain $x^1 = (0, 1)^T$ with $g(x^1) = (8, 10)^T \not\le (7, 12)^T$.

Step 2. Since $f(x^1) = 2 > \alpha^0$, set $\alpha^1 = 2$.

Step 3. Set $T^1 = \{x^1\}$. Solve the linear program:

$$(LP_1) \qquad \begin{aligned} &\max_{(\beta,\mu)}\ \beta \\ &\text{s.t. } \beta \le \mu_1 - 2\mu_2, \\ &\qquad \mu_1 + \mu_2 \le 1, \\ &\qquad \mu_1 \ge 0,\ \mu_2 \ge 0. \end{aligned}$$

We obtain $\beta^1 = 1 > 0$ and $\mu^1 = (1, 0)^T$. Set $k = 2$ and $\mu^2 = \mu^1$.

Iteration 2

Step 1. Solve the surrogate problem

$$(P_{\mu^2}) \qquad \begin{aligned} &\min\ 3x_1^2 + 2x_2^2 \\ &\text{s.t. } 1 \times (10 - 5x_1 - 2x_2) + 0 \times (15 - 2x_1 - 5x_2) \le 7, \\ &\qquad x \in X. \end{aligned}$$

We obtain $x^2 = (1, 0)^T$ with $g(x^2) = (5, 13)^T \not\le (7, 12)^T$.

Step 2. Since $f(x^2) = 3 > \alpha^1$, set $\alpha^2 = 3$.

Step 3. Set $T^2 = \{x^1, x^2\}$. Solve the linear program:

$$(LP_2) \qquad \begin{aligned} &\max_{(\beta,\mu)}\ \beta \\ &\text{s.t. } \beta \le \mu_1 - 2\mu_2, \\ &\qquad \beta \le -2\mu_1 + \mu_2, \\ &\qquad \mu_1 + \mu_2 \le 1, \\ &\qquad \mu_1 \ge 0,\ \mu_2 \ge 0. \end{aligned}$$

We obtain $\beta^2 = 0$ and $\mu^2 = (0, 0)^T$. Stop and the optimal surrogate dual value is $v(D_S^n) = \alpha^2 = 3$. Note that the surrogate dual value, 3, is better than the Lagrangian dual value, $2\frac{1}{3}$.

Similar to the Lagrangian dual, the dual function $v(P(\alpha, \cdot))$ in the surrogate dual also possesses a concavity as seen in the following lemma.

LEMMA 4.1 *Function $v(P(\alpha,\cdot))$ is concave on Λ and $\xi(\mu) = g(x_\mu) - b$ is a subgradient of $v(P(\alpha,\cdot))$ at μ, where x_μ is an optimal solution to $(P(\alpha,\mu))$.*

Proof. Since x_μ solves $(P(\alpha,\mu))$, we have $v(P(\alpha,\mu)) = \mu^T(g(x_\mu) - b)$. For any $\gamma \in \Lambda$, since $x_\mu \in X(\alpha)$, it holds

$$v(P(\alpha,\gamma)) \leq \gamma^T(g(x_\mu) - b). \tag{4.1.6}$$

Thus

$$v(P(\alpha,\gamma)) \leq v(P(\alpha,\mu)) + \xi(\mu)^T(\gamma - \mu), \quad \forall \gamma \in \Lambda.$$

This implies that $v(P(\alpha,\cdot))$ is concave and $\xi(\mu)$ is a subgradient of $v(P(\alpha,\cdot))$ at μ. □

In view of the concavity of $v(P(\alpha,\mu))$ and the availability of the subgradient, it is also natural to use the subgradient method to search for the optimal solution of $(D(\alpha))$. Moreover, it is easy to see that $v(D(\alpha))$ is a monotonically decreasing function of α and is lower semicontinuous on $\mathbb{R}$. This motivates a surrogate dual search method based on the subgradient method.

PROCEDURE 4.2 (SUBGRADIENT PROCEDURE FOR (D_S^n))

Step 0 (Initialization). Choose parameter $\epsilon > 0$. Set $\alpha^0 = -\infty$, $T^0 = \emptyset$. Choose any $\mu^1 \in \Lambda$. Set $k = 1$.

Step 1 (Surrogate relaxation). Solve the surrogate relaxation problem (P_{μ^k}) and obtain an optimal solution x^k. If $g(x^k) \leq b$, stop and x^k is an optimal solution to (P) and $v(D_S^n) = v(P)$.

Step 2 (Updating lower bound). If $f(x^k) > \alpha^{k-1}$, then set $\alpha^k = f(x^k)$. Otherwise, set $\alpha^k = \alpha^{k-1}$.

Step 3 (Updating multiplier). Compute

$$t^k = (\epsilon - (\mu^k)^T\xi^k)/\|\xi^k\|^2,$$
$$\mu^{k+1} = Proj(\mu^k - t^k\xi^k),$$

where $\xi^k = g(x^k) - b$ is the subgradient of $v(P(\alpha^k,\cdot))$ at $\mu = \mu^k$, t^k is the stepsize, and $Proj$ is the projection on Λ. Set $k := k+1$, go to Step 1.

It can be proved that the lower bound $\{\alpha^k\}$ generated by Procedure 4.2 converges to $v(D_S^n)$ (see [115]).

From Theorem 4.2, we see that the surrogate dual bound is tighter than the Lagrangian dual bound. We note, however, that a surrogate problem (P_μ) has to be solved at each iteration of a surrogate dual search procedure which turns out to be much more difficult to solve than the Lagrangian relaxation problem (L_λ). Therefore, the surrogate dual search is more expensive in computation than the Lagrangian dual search.

4.2 Nonlinear Surrogate Dual Method

The surrogate constraint method does not always solve the primal problem, i.e., the surrogate dual does not guarantee generation of an optimal solution of the primal problem. When $v(D_S)$ is strictly less than $v(P)$, a duality gap exists between (D_S) and (P). We will analyze, in the following, the reason for the existence of the duality gap in the surrogate dual, and the discussion will lead naturally to the development of the p-norm surrogate constraint method with which the duality gap can be eliminated.

The surrogate relaxation, (P_μ), differs from the primal problem, (P), only in the feasible region. In general, the feasible region of the surrogate relaxation enlarges the feasible region of the primal problem. If this enlarged feasible region contains a point that is infeasible with respect to the major constraints of the primal problem and has a smaller objective value than $v(P)$, then the surrogate relaxation, (P_μ), will fail to identify an optimal solution of the primal problem, (P), while searching for the minimum in this enlarged feasible region.

To illustrate this argument further, we consider Example 4.1 again. Applying the conventional surrogate constraint method to solve (4.1.5) yields,

$$\begin{aligned} &\min\ 3x_1^2 + 2x_2^2 \qquad\qquad (4.2.1)\\ &\text{s.t. } \mu_1(10 - 5x_1 - 2x_2) + \mu_2(15 - 2x_1 - 5x_2) \le 7\mu_1 + 12\mu_2,\\ &\quad x \in X = \left\{ \begin{array}{c} \text{integer} \\ 0 \le x_1 \le 1,\ 0 \le x_2 \le 2 \\ 8x_1 + 8x_2 \ge 1 \end{array} \right\}. \end{aligned}$$

It can be seen from Figure 4.1 that the surrogate constraint defines a closed half space in the $\{g_1, g_2\}$ space. For whatever value of μ chosen, the resulting closed half space always includes an infeasible solution of the primal problem. Both infeasible solutions, $(0,1)^T$ and $(1,0)^T$, in this example have objective values smaller than $v(P)$. As a result, the conventional surrogate constraint method fails to generate the optimal solution of the primal problem in this example. The resulting maximum dual value is $v(P_\mu) = 3$ with $\mu = (1,0)^T$ as been computed in Example 4.1, and a duality gap exists.

It becomes clear now that a sufficient requirement to eliminate the duality gap in the surrogate constraint method is to make the feasible region in the constraint space, defined by a single surrogate constraint, the same as the feasible region in the primal problem. This goal can be achieved by some nonlinear surrogate constraint methods.

We discuss first a p-norm surrogate constraint method for integer programming. Without loss of generality, $g_i(x)$, $i = 1, 2, \ldots, m$, are assumed to be strictly positive for all $x \in X$.

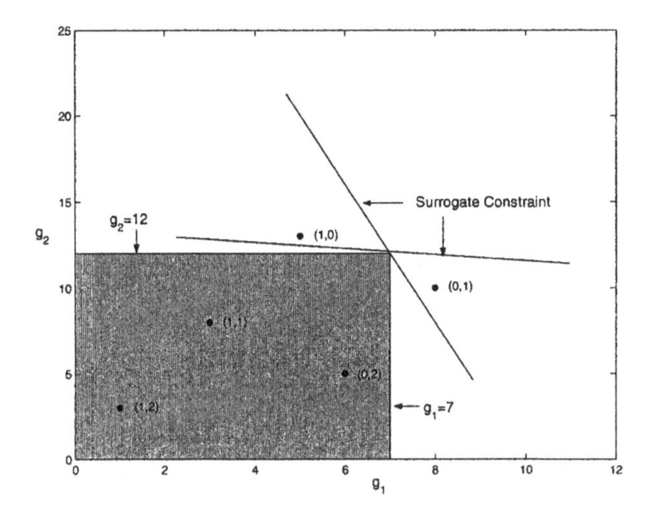

Figure 4.1. Surrogate constraints in the constraint space of Example 4.1.

Let $M = diag(\mu_1, \ldots, \mu_m)$. We define the following weighted p-norms, for a real number p with $1 \le p < \infty$, as:

$$\|Mg(x)\|_p = \Big\{ \sum_{i=1}^{m} [\mu_i g_i(x)]^p \Big\}^{1/p},$$

$$\|Mb\|_p = \Big\{ \sum_{i=1}^{m} [\mu_i b_i]^p \Big\}^{1/p},$$

and the weighted ∞-norm as

$$\|Mg(x)\|_\infty = \max \big\{ \mu_1 g_1(x), \mu_2 g_2(x), \ldots, \mu_m g_m(x) \big\},$$
$$\|Mb\|_\infty = \max \big\{ \mu_1 b_1, \mu_2 b_2, \ldots, \mu_m b_m \big\}.$$

The following are well known,

$$\lim_{p \to \infty} \|Mg(x)\|_p = \|Mg(x)\|_\infty, \quad \forall\, x \in \mathbb{R}^n,$$
$$\lim_{p \to \infty} \|Mb\|_p = \|Mb\|_\infty.$$

The p-norm surrogate constraint formulation of (P) is now formed as follows for $1 \le p \le \infty$:

$$(P_\mu^p) \qquad \begin{aligned} &\min\ f(x) \\ &\text{s.t. } \|Mg(x)\|_p \le \|Mb\|_p, \\ &\qquad x \in X, \end{aligned}$$

where μ satisfies the following,

$$\mu_1 b_1 = \mu_2 b_2 = \ldots = \mu_m b_m. \tag{4.2.2}$$

Let B be a positive real number that is defined as follows for a μ that satisfies (4.2.2),

$$B = \mu_i b_i, \quad i = 1, 2, \ldots, m.$$

Define $S^p(\mu)$ to be the feasible region of decision vector in (P_μ^p),

$$S^p(\mu) = \{x \in X \mid \|Mg(x)\|_p \le \|Mb\|_p\}.$$

When x satisfies $g_i(x) \le b_i$, $i = 1, 2, \ldots, m$, x also satisfies $\|Mg(x)\|_p \le \|Mb\|_p$ for $1 \le p \le \infty$. Thus, $S \subseteq S^p(\mu)$ when $1 \le p \le \infty$. If $x \in S^\infty(\mu)$ with μ satisfying (4.2.2), $\max\{\mu_1 g_1(x), \mu_2 g_2(x), \ldots, \mu_m g_m(x)\} \le \max\{\mu_1 b_1, \mu_2 b_2, \ldots, \mu_m b_m\} = B$ implies that all $g_i(x) \le b_i$, $i = 1, 2, \ldots, m$. Then, x belongs to S. Thus, $S = S^\infty(\mu)$ when μ satisfies (4.2.2). The ∞-norm formulation is an equivalent formulation of the original problem, (P), when μ satisfies (4.2.2), and we have

$$v(P_\mu^\infty) = v(P)$$

with μ satisfying (4.2.2). When $1 \le p < \infty$, problem (P_μ^p) is a relaxation of problem (P) and we have

$$v(P_\mu^p) \le v(P), \quad \forall\, 1 \le p < \infty.$$

Note that (P_μ^p) still constitutes a relaxation of problem (P) even when μ does not satisfy (4.2.2). However, the p-norm surrogate constraint method confines itself to use only those μ's that satisfy (4.2.2), due to several important properties associated with μ satisfying (4.2.2).

For $\mu \in \mathbb{R}_+^m$ satisfying (4.2.2), let $G^p(\mu)$ denote the feasible region formed by the p-norm surrogate constraint in the g space,

$$G^p(\mu) = \{g \in \mathbb{R}_+^m \mid \|Mg\|_p \le \|Mb\|_p\}.$$

Figure 4.2 graphically demonstrates the feasible regions in the $\{g_1, g_2\}$ space defined by the p-norm surrogate constraint for different values of p. A nice property of inclusion can be seen for $G^p(\mu)$ from Figure 4.2. Mathematically, we have the following theorem.

THEOREM 4.5 *For* $\infty > p > q$,

$$G^p(\mu) \subseteq G^q(\mu),$$

where μ *satisfies (4.2.2).*

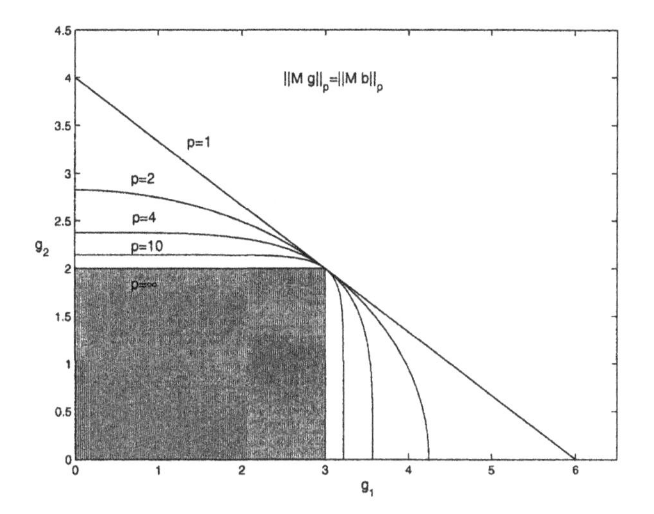

Figure 4.2. p-norm surrogate constraints in the constraint space with $\mu = (0.4, 0.6)^T$ and $b = (3, 2)^T$.

Proof. If $g(\tilde{x}) \in G^p(\mu)$, we have

$$\{\sum_{i=1}^{m}[\mu_i g_i(\tilde{x})]^p\}^{1/p} \leq \{\sum_{i=1}^{m}[\mu_i b_i]^p\}^{1/p} = Bm^{1/p}. \tag{4.2.3}$$

Equation (4.2.3) can be rewritten as

$$\{\sum_{i=1}^{m}\frac{1}{m}[\mu_i g_i(\tilde{x})]^p\}^{1/p} \leq B.$$

An inequality for the mean of order t ([16], pp. 17) states that $\{\sum_{i=1}^{m}\frac{1}{m}z_i^t\}^{1/t}$ is a nondecreasing function of t. Thus, we have the following for $p > q$,

$$\{\sum_{i=1}^{m}\frac{1}{m}[\mu_i g_i(\tilde{x})]^q\}^{1/q} \leq \{\sum_{i=1}^{m}\frac{1}{m}[\mu_i g_i(\tilde{x})]^p\}^{1/p} \leq B.$$

Thus $g(\tilde{x}) \in G^q(\mu)$, and $G^p(\mu)$ is proven to be a subset of $G^q(\mu)$ for $p > q$. □

Note that $G^p(\mu) \subseteq G^q(\mu)$ for $p > q$ is not generally true if μ does not satisfy $\mu_1 b_1 = \mu_2 b_2 = \ldots = \mu_m b_m$. Since $\|Mg(x)\|_\infty \leq \|Mb\|_\infty = B$ implies $g_i \leq b_i$ for all $i = 1, 2, \ldots, m$, $G^p(\mu)$ converges to the feasible region of the primal problem in the g space when p approaches infinity. This convergence property of the feasible region in the constraint space is good. We need, however, a stronger result for a finite value of p with which the equivalence between (P_μ^p) and (P) can be established with respect to the feasible region of decision vectors.

THEOREM 4.6 *If μ satisfies (4.2.2), then there exists a finite q such that*

$$S = S^p(\mu)$$

for all $p > q$.

Proof. We know that $S \subseteq S^p(\mu)$ when $1 \leq p \leq \infty$. Let $B = \mu_i b_i$, $i = 1, \ldots, m$. If $\hat{x} \in S^p(\mu)$, we have

$$\{\sum_{i=1}^{m}[\mu_i g_i(\hat{x})]^p\}^{1/p} \leq \{\sum_{i=1}^{m}[\mu_i b_i]^p\}^{1/p} = Bm^{1/p}. \tag{4.2.4}$$

Let

$$\mu_{i^*} g_{i^*}(\hat{x}) = \max_{1 \leq i \leq m} \{\mu_i g_i(\hat{x})\}.$$

Then (4.2.4) can be rewritten as

$$\mu_{i^*} g_{i^*}(\hat{x})\{\sum_{i=1}^{m} \frac{[\mu_i g_i(\hat{x})]^p}{[\mu_{i^*} g_{i^*}(\hat{x})]^p}\}^{1/p} \leq Bm^{1/p}. \tag{4.2.5}$$

Since $1 \leq \sum_{i=1}^{m}[\mu_i g_i(\hat{x})]^p/[\mu_{i^*} g_{i^*}(\hat{x})]^p$, we have

$$\frac{\mu_{i^*} g_{i^*}(\hat{x})}{B} \leq m^{1/p}. \tag{4.2.6}$$

The result in (4.2.6) leads to

$$\frac{g_i(\hat{x})}{b_i} \leq m^{1/p}, \quad i = 1, 2, \ldots, m. \tag{4.2.7}$$

Since $\lim_{p \to \infty} m^{1/p} = 1$, no infeasible $\tilde{x}$ ($\in X$) with one or more $g_i(\tilde{x}) > b_i$ will satisfy (4.2.7) when p is sufficiently large. Notice that X is finite. We can define

$$U_i = \min\{\frac{g_i(x)}{b_i} \mid x \in X, \ g_i(x) > b_i\}.$$

Define further

$$U = \min_{1 \leq i \leq m} U_i.$$

Let

$$q = \frac{\ln(m)}{\ln(U)}. \tag{4.2.8}$$

When $p > q$, no infeasible $\tilde{x}$ ($\in X$) with one or more $g_i(\tilde{x}) > b_i$ can satisfy (4.2.7). Thus, for $p > q$, $\hat{x} \in S^p(\mu)$ implies $\hat{x} \in S$, i.e., $S^p(\mu) \subseteq S$. Finally, we have $S^p(\mu) = S$ for $p > q$. □

In general, obtaining U could be at least as difficult as solving (P) itself. For an important general class of integer programming problems, however, a lower bound of p can be easily calculated.

COROLLARY 4.2 *Suppose that all g_i, $i = 1, 2, \ldots, m$, are integer-valued functions, e.g., polynomial functions with integer coefficients. Then for μ satisfying (4.2.2),*

$$S = S^p(\mu).$$

when $p > \ln(m)/\ln[\min_{1 \le i \le m}(b_i + 1)/b_i]$.

Proof. Notice that

$$\begin{aligned} U_i &= \min\left\{\frac{g_i(x)}{b_i} \mid x \in X,\ g_i(x) > b_i\right\} \\ &\ge \frac{b_i + 1}{b_i}. \end{aligned} \tag{4.2.9}$$

Then the proof follows from Theorem 4.6. □

In other situations when implementing the p-norm surrogate constraint method, the selection of p may need to be carried out by trial and error, since the value of q defined in (4.2.8) is unknown. Using Theorem 4.5, we have for $p > q$,

$$S^p(\mu) \subseteq S^q(\mu) \tag{4.2.10}$$

where μ satisfies (4.2.2). Thus, we further have for $p > q$,

$$v(S^p_\mu) \ge v(S^q_\mu) \tag{4.2.11}$$

where μ satisfies (4.2.2). This monotonicity will guarantee the success of the p-norm surrogate constraint method when increasing the value of p to a certain level.

Theorem 4.6 and Corollary 4.2 provide interesting results in separation. By selecting a sufficiently large p, all infeasible solutions of the primal problem will be excluded from $S^p(\mu)$. In other words, the feasible set defined by the p-norm surrogate constraint, $S^p(\mu)$, will exactly match the feasible set of the primal problem, S, when $p > q$. In summary, an appropriately selected single surrogate constraint can be always constructed by aggregating multiple major constraints of the primal problem such that a surrogate relaxation and the primal problem are exactly equivalent. This result offers a basis in achieving zero duality gap in integer programming when adopting the p-norm surrogate constraint method.

THEOREM 4.7 *If μ satisfies (4.2.2), then*

$$v(P^p_\mu) = v(P) \tag{4.2.12}$$

for $p > q$, where q is defined in (4.2.8).

Proof. From Theorem 4.6, we have $S^p(\mu) = S$ when $p > q$ defined in (4.2.8). Thus, problems (P^p_μ) and (P) are exactly the same. So are their optimal objective values. □

The above results confirm the existence of a kind of saddle point in integer programming. Define

$$K_p(x, \mu) = \begin{cases} f(x), & \text{if } \|Mg(x)\|_p \leq \|Mb\|_p \\ \infty, & \text{if } \|Mg(x)\|_p > \|Mb\|_p \end{cases}$$

where $M = diag(\mu_1, \ldots, \mu_m)$ and $x \in X$.

THEOREM 4.8 *The solution x^* solves (P) if and only if (x^*, μ^*) is a saddle point of $K_p(x, \mu)$ on $X \times \mathbb{R}^m_+$, i.e.,*

$$K_p(x^*, \mu) \leq K_p(x^*, \mu^*) \leq K_p(x, \mu^*),$$

for $p > q$ where q is defined in (4.2.8) and μ^* $(\in \mathbb{R}^m_+)$ satisfies (4.2.2).

Proof. Necessity: Let M^* denote diag$(\mu_1^*, \mu_2^*, \ldots, \mu_m^*)$. For every $p > q$ defined in (4.2.8) and μ^* that satisfies (4.2.2), $S^p(\mu^*)$ is equal to S. Now, $x \in X$ is feasible in (P) if and only if it satisfies $\|M^*g(x)\|_p \leq \|M^*b\|_p$ for $p > q$. Thus, by the optimality of x^* and the definition of $K_p(x, \mu)$, $K_p(x^*, \mu^*) = f(x^*) \leq K_p(x, \mu^*) = f(x)$ for all $x \in X$ satisfying $\|M^*g(x)\|_p \leq \|M^*b\|_p$. Notice that $K_p(x, \mu^*) = \infty$ for all infeasible $x \in X$ when $p > q$. Since for any $\mu \in \mathbb{R}^m_+$, x^* is feasible in (P^p_μ), we have $K_p(x^*, \mu) = K_p(x^*, \mu^*) = f(x^*)$. In summary, (x^*, μ^*) with μ^* satisfying (4.2.2) is a saddle point of $K_p(x, \mu)$ for every $p > q$ defined in (4.2.8).

Sufficiency: For every $p > q$ defined in (4.2.8), $S^p(\mu^*)$ is equal to S with μ^* satisfying (4.2.2). Thus, a finite value of $K_p(x^*, \mu^*)$ implies that x^* belongs to $S^p(\mu^*)$, and hence, $x^* \in S$. $K_p(x^*, \mu) \leq K_p(x^*, \mu^*)$, with $K_p(x^*, \mu^*)$ being finite, means that x^* is feasible in every (P^p_μ) with $\mu \in \mathbb{R}^m_+$. Note that $x \in S$ implies $x \in S^p(\mu)$ for any $\mu \in \mathbb{R}^m_+$. Thus, $K_p(x^*, \mu^*) \leq K_p(x, \mu^*)$ for all $x \in S^p(\mu^*)$ implies $f(x^*) \leq f(x)$ for all $x \in S$, i.e., x^* solves the primal problem (P). □

A point to be emphasized is that the p-norm surrogate constraint method does not require a search for an optimal μ vector. The value of the μ vector can be simply assigned by solving (4.2.2).

Now we come back to Example 4.1 which the conventional surrogate constraint method fails to solve as we discussed before. Applying the p-norm surrogate constraint method yields the following formulation,

$$\begin{aligned}
&\min\ 3x_1^2 + 2x_2^2 \\
&\text{s.t. } \{\mu_1^p[10 - 5x_1 - 2x_2]^p + \mu_2^p[15 - 2x_1 - 5x_2]^p\}^{1/p} \\
&\qquad \le \{\mu_1^p \times 7^p + \mu_2^p \times 12^p\}^{1/p} \\
&\qquad x \in X.
\end{aligned}$$

One normalized solution for $\mu_1 \times 7 = \mu_2 \times 12$ is $(\hat{\mu}_1, \hat{\mu}_2) = (0.6316, 0.3684)$. The value of B is equal to $\hat{\mu}_1 \times 7 = \hat{\mu}_2 \times 12 = 4.4211$. Figure 4.3 shows $S^p(\hat{\mu})$ for $p = 1, 2, 6, 9$. It can be clearly observed that when $p = 9$, $S^p(\hat{\mu}) = S$ and the p-norm surrogate method successfully identifies the optimal solution $x^* = (0, 2)^T$ with zero duality gap.

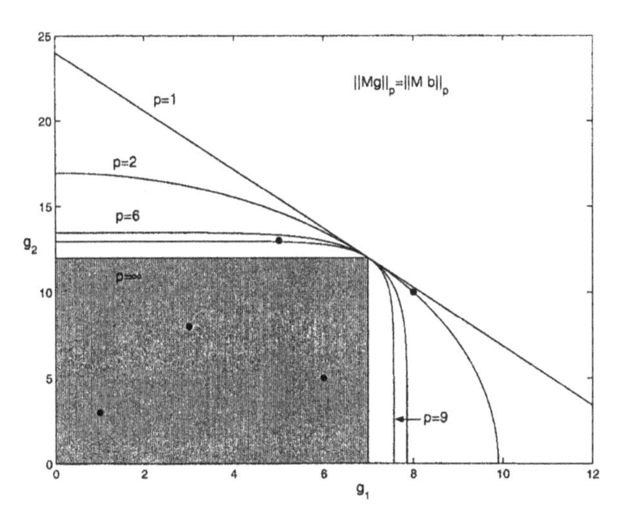

Figure 4.3. p-norm surrogate constraints in the constraint space of Example 4.1 with $\mu = (0.6316, 0.3684)^T$.

In continuous optimization, Luenberger [147] has shown that for quasi-convex programming problems, there is no duality gap between the surrogate dual and the primal problem. The zero duality results presented in this section for general integer programming problems via using the nonlinear surrogate constraint method are even stronger in the sense that there is no assumption of any convexity.

While the p-norm surrogate constraint method greatly simplifies the dual search at the upper level, i.e., there is no need to search for the optimal multiplier vector, the resulting surrogate relaxation problem at the lower level, in general, becomes more difficult to solve, when comparing with the conventional surrogate constraint method. For example, when the original problem is of a linear form, the p-norm surrogate constraint method will make the problem highly nonlinear. There exists, however, an exception. Notice that any power of a zero-one variable is itself. Polynomial zero-one programming problem thus is an area where the p-norm surrogate constraint method could show its computational promise in problem-solving practice, as we will witness in Chapter 12 of this book.

Note that many other different nonlinear surrogate constraint formulations can also achieve the same as the p-norm surrogate constraint method does. For example, let us consider another surrogate constraint formulation of (P),

$$
\begin{array}{lll}
(E_t) & & \min\ f(x) \\
& & \text{s.t. } g_t(x) = \frac{1}{t}\ln\sum_{i=1}^{m}\exp(t\frac{g_i(x)}{b_i}) \leq 1 + \frac{\ln m}{t} \\
& & x \in X \subseteq \mathbb{Z}^n.
\end{array}
$$

Essentially, no surrogate multipliers are needed in this nonlinear surrogate formulation, except for a parameter t. Denote by $\tilde{S}^t$ the feasible region of (E_t),

$$\tilde{S}^t = \{x \in X \mid \frac{1}{t}\ln\sum_{i=1}^{m}\exp(t\frac{g_i(x)}{b_i}) \leq 1 + \frac{\ln m}{t}\}.$$

It is clear that $S \subseteq \tilde{S}^t$ for all positive t and (E_t) is a relaxation of (P). We can prove further that S and $\tilde{S}^t$ exactly match when t is sufficiently large.

THEOREM 4.9 *It holds $S = \tilde{S}^t$ for all $t > t_0$, where*

$$
\begin{aligned}
t_0 &= \frac{\ln m}{U-1}, \\
U &= \min_{1\leq i\leq m} U_i, \\
U_i &= \min\{\frac{g_i(x)}{b_i} \mid x \in X, g_i(x) > b_i\}, \quad i = 1, 2, \ldots, m,
\end{aligned}
$$

where U_i is defined to be ∞ if there is no $g_i(x)$ greater than b_i for all $x \in X$.

4.3 Notes

The surrogate dual was first investigated in [84] and [147] for continuous optimization problems. The surrogate dual was then applied to linear integer programming in [69][114][115]. Several surrogate dual search methods were developed for linear integer programming in [54][114][115][189]. Variants of Procedure 4.2 were proposed in [120][189]. In particular, a finite convergence surrogate dual search was proposed in [120] by using a more sophisticated stepsize rule of the subgradient method.

The development of nonlinear surrogate constraint methods started with the p-norm surrogate method presented in [133]. Nonlinear surrogate constraint methods were also discussed in [134] and [143].

Chapter 5

NONLINEAR LAGRANGIAN AND STRONG DUALITY

Although the conventional Lagrangian duality theory is a powerful solution methodology to find out a lower bound for integer programming problems, being efficient especially for separable integer programming problems, the conventional Lagrangian dual search, in general, does not converge to an exact solution of the primal problem as discussed in Chapter 3. This chapter discusses how to extend the conventional Lagrangian duality theory to nonlinear Lagrangian theory in order to achieve strong duality.

5.1 Convexification and Nonlinear Support: p-th power Nonlinear Lagrangian Formulation

Consider the general bounded integer programming problem:

$$(P) \qquad \min\ f(x)$$
$$\text{s.t. } g_i(x) \le b_i,\ i = 1, 2, \ldots, m,$$
$$x \in X \subseteq \mathbb{Z}^n.$$

As discussed in Chapter 3, there are situations where no optimal generating multiplier λ^* exists such that an optimal solution x^* to the primal problem (P) is also an optimal solution to the Lagrangian relaxation problem (L_{λ^*}). As seen from Corollary 3.1, (3.4.7) is a sufficient condition for ensuring the existence of an optimal generating multiplier vector.

To motivate the development of the nonlinear Lagrangian theory described in this chapter, let us start with the locus of the unit circle in the first quadrant in a two-dimensional space. It is obvious from Figure 5.1 that x_2 $(= \sqrt{1 - x_1^2})$ is not a convex function of x_1 and the set $S = \{x \in \mathbb{R}^2 \mid x_2 \ge \sqrt{1 - x_1^2}, 0 \le x_1 \le 1\}$ is non-convex. If we change the coordinates from $\{x_1, x_2\}$ to $\{x_1^p, x_2^p\}$ with $p \ge 2$, then it can be verified that curve x_2^p $(= [1 - (x_1^p)^{2/p}]^{p/2})$ becomes

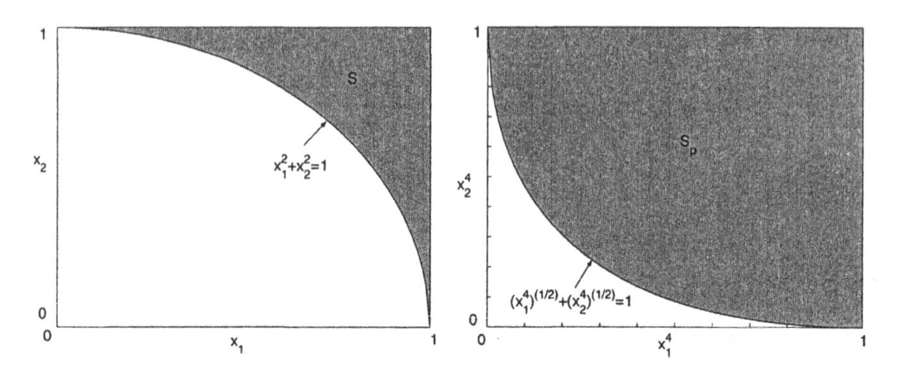

Figure 5.1. Set S in $\{x_1, x_2\}$ space. *Figure 5.2.* Set S_p in $\{x_1^p, x_2^p\}$ space.

a convex function of x_1^p in the $\{x_1^p, x_2^p\}$ space and the set $S_p = \{(x_1^p, x_2^p)^T \in \mathbb{R}^2 \mid x_2^p \geq [1 - (x_1^p)^{2/p}]^{p/2}, 0 \leq x_1^p \leq 1\}$ becomes convex too. Figure 5.2 illustrates the set S_p with $p = 4$.

A key observation from this illustrative example is that a monotone nonconvex function can be convexified by a *nonlinear* transformation. Recall that the values of the perturbation function at the corner points are decreasing. This motivates the use of p-th power convexification scheme to convexify the corner points, thus guaranteeing the sufficient condition (3.4.7) for the existence of an optimal generating multiplier vector.

In addition to Assumption 3.1 for (P), we make the following assumption for (P).

ASSUMPTION 5.1 *Function f and all constraint functions g_i (i= 1, ..., m) in (P) are nonnegative on X.*

Assumption 5.1 can be always satisfied via some suitable equivalent transformations on (P). Let Y be defined in (3.3.2). For $y \in Y$ and $p > 0$, denote $y^p = [(y_1)^p, (y_2)^p, \ldots, (y_m)^p]^T$. Consider the following equivalent form of problem (P):

$$(P_p) \qquad \begin{aligned} &\min\ [f(x)]^p \\ &\text{s.t. } [g_i(x)]^p \leq (b_i)^p, \quad i = 1, \ldots, m, \\ &\qquad x \in X \subseteq \mathbb{Z}^n. \end{aligned}$$

The perturbation function of (P_p) is

$$w_p(y) = \min\ \{[f(x)]^p \mid [g_i(x)]^p \leq y_i, i = 1, \ldots, m, x \in X\}. \qquad (5.1.1)$$

It is easy to see that $w_p(y) = [w(y^{1/p})]^p$ for any y, where w is the perturbation function defined in (3.3.1). The set of corner points of (P_p) is

$$\Phi_c^p = \{(c_i^p, f_i^p) \mid i = 1, \ldots, K\},$$

where $\Phi_c = \{(c_i, f_i) \mid i = 1, \ldots, K\}$ is the set of corner points of (P). Let E_p be the epigraph of w_p. Then

$$\begin{aligned} E_p &= \{(y, z) \mid z \geq w_p(y),\ y \in conv(Y)\} \\ &= \{(y, z) \mid z \geq [w(y^{1/p})]^p,\ y \in conv(Y)\} \\ &= \{(y, z) \mid (y^{1/p}, z^{1/p}) \in epi(w)\}. \end{aligned}$$

Note that $\Phi_c^p \subset E_p$ and the extreme points of $conv(E_p)$ are in Φ_c^p. Moreover, e^i $(i = 1, \ldots, m)$ are the extreme directions of $conv(E_p)$. By definition, the convex envelope function of w_p can then be expressed as

$$\begin{aligned} \psi_p(y) &= \min\{z \mid (y, z) \in conv(E_p)\} \\ &= \min\{z \mid (y, z) \geq \sum_{i=1}^{K} \mu_i (c_i^p, f_i^p),\ \mu \in \Lambda\} \\ &= \min\{\sum_{i=1}^{K} \mu_i f_i^p \mid y \geq \sum_{i=1}^{K} \mu_i c_i^p,\ \mu \in \Lambda\}, \end{aligned} \tag{5.1.2}$$

where $\Lambda = \{\mu \in \mathbb{R}_+^K \mid \sum_{i=1}^{K} \mu_i = 1\}$.

THEOREM 5.1 *There exists $p_0 > 0$ such that*

$$\psi_p(c_i^p) = f_i^p, \quad i = 1, \ldots, K, \tag{5.1.3}$$

when $p \geq p_0$.

Proof. Note that $f_i^p = w_p(c_i^p) \geq \psi_p(c_i^p)$ for each i. Assume that (5.1.3) does not hold. Then there exists $l \in \{1, \ldots, K\}$ and $\{p_k\}$ such that $f_l^{p_k} > \psi_{p_k}(c_l^{p_k})$ and $p_k \to +\infty$. By (5.1.2), there exists $\mu^k \in \Lambda$ such that

$$\sum_{i \in I_k} \mu_i^k f_i^{p_k} < f_l^{p_k}, \tag{5.1.4}$$

$$\sum_{i \in I_k} \mu_i^k c_i^{p_k} \leq c_l^{p_k}, \tag{5.1.5}$$

where $I_k = \{i \mid \mu_i^k > 0\}$. It is clear that $c_l \neq c_i$ for any $i \in I_k$, otherwise, $\psi_p(c_l^p) = f_l^p$. Since $\mu^k \in \Lambda$, there exists $i_k \in I_k$ such that $\mu_{i_k}^k \geq 1/|I_k| \geq 1/K$, where $|I_k|$ is the cardinality of I_k. Thus, by (5.1.4), (5.1.5) and Assumption 5.1, we have

$$(1/K) f_{i_k}^{p_k} \leq \mu_{i_k}^k f_{i_k}^{p_k} < f_l^{p_k}, \tag{5.1.6}$$

$$(1/K) c_{i_k}^{p_k} \leq \mu_{i_k}^k c_{i_k}^{p_k} \leq c_l^{p_k}. \tag{5.1.7}$$

Since $p_k \to +\infty$, (5.1.6) implies that $f_{i_k} \le f_l$ for sufficient large k. Since the corner points are noninferior and $c_{i_k} \le c_l$ implies $f_{i_k} = w(c_{i_k}) > w(c_l) = f_l$, we must have $c_{i_k} \not\le c_l$. Let $j_k \in \{1, \ldots, m\}$ be such that $c_{i_k, j_k} > c_{l, j_k}$. Let

$$\delta = \min\{c_{ij}/c_{lj} \mid c_{ij} > c_{lj},\ i = 1, \ldots, K,\ j = 1, \ldots, m\} > 1.$$

It follows from (5.1.7) that

$$\delta^{p_k} \le (c_{i_k, j_k}/c_{l, j_k})^{p_k} \le K.$$

This contradicts $\delta > 1$ and $p_k \to +\infty$. □

Using Corollary 3.1, Theorem 5.1 leads immediately to the following corollary and the development of the p-th power Lagrangian method.

COROLLARY 5.1 *If $p \ge p_0$, then every noninferior optimal solution $\hat{x}$ of the primal problem (P) is guaranteed to be generated by a Lagrangian relaxation of (P_p), i.e., there exists an optimal generating multiplier vector for $\hat{x}$.*

For any $\lambda \in \mathbb{R}^m_+$, define

$$d_p(\lambda) = \min_{x \in X} L_p(x, \lambda) = [f(x)]^p + \sum_{i=1}^{m} \lambda_i[(g_i(x))^p - (b_i)^p]. \tag{5.1.8}$$

The Lagrangian dual problem of (P_p) is:

$$(D_p) \quad \max_{\lambda \in \mathbb{R}^m_+} d_p(\lambda). \tag{5.1.9}$$

The above derived p-th power Lagrangian method simply involves a two-phase procedure. The first phase is to perform a p-th power transformation on both the objective function and all constraints of problem (P). The second phase is to apply the conventional Lagrangian method on problem (P_p) resultant from phase 1. We can conclude from Corollary 5.1 that for each noninferior optimal solution of problem (P), the existence of an associated optimal generating Lagrangian multiplier vector is guaranteed when applying the Lagrangian method on problem (P_p) with a sufficiently large p.

EXAMPLE 5.1 Consider the p-th power transformation of Example 3.4:

$$\begin{aligned} &\min\ f^p(x) = (4 + x_1x_2x_3x_4 - x_1 + 3x_2 + x_3 - 2x_4)^p \\ &\text{s.t. } (g_1(x))^p = (x_1 - 2x_2 + x_3 + 3)^p \le 2.5^p, \\ &\qquad x \in X = \{0, 1\}^4. \end{aligned}$$

The corner points are $(c_1, f_1) = (1, 5)$, $(c_2, f_2) = (2, 4)$, $(c_3, f_3) = (3, 2)$, $(c_4, f_4) = (4, 1)$. Notice that $\psi(c_2) = 3.5 < 4 = f_2$. Figure 5.3 depicts the

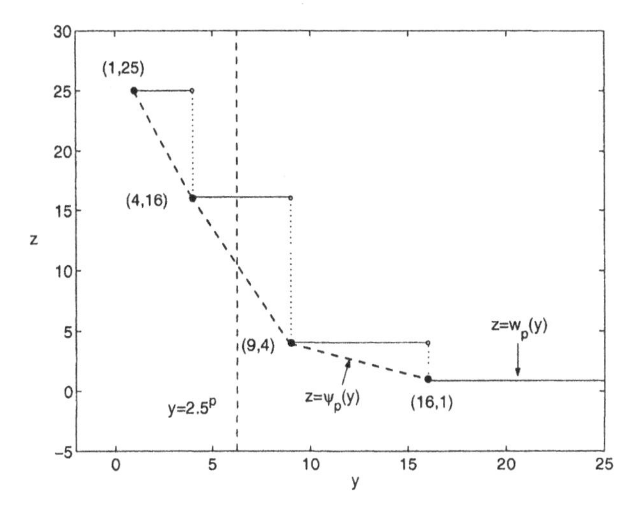

Figure 5.3. Illustration of $w_p(y)$ and $\psi_p(y)$ for Example 3.4 with $p = 2$.

functions w_p and ψ_p for $p = 2$. We can see from the figure that $\psi_2(c_i^2) = f_i^2$ for $i = 1, 2, 3, 4$, thus guaranteeing the existence of an optimal generating multiplier. Let $\lambda^1 = -(2^2 - 4^2)/(3^2 - 2^2) = 2.4$ and $\lambda^2 = -(4^2 - 5^2)/(2^2 - 1^2) = 3$. It can be verified that any $\lambda \in [\lambda^1, \lambda^2]$ is an optimal generating multiplier vector for $x^* = (1, 1, 0, 1)^T$ and λ^1 is an optimal solution to the dual problem (5.1.9).

A prominent feature which the p-th power Lagrangian formulation offers is its ability to convexify the envelope function of the perturbation function. Let's now examine the p-th power Lagrangian formulation from another angle. Recall in Figure 3.9 that there does not exist a linear support at $(2, 4)^T$, which corresponds to the optimal solution $x^* = (1, 1, 0, 1)^T$. Notice that the conventional Lagrangian is a linear function of the objective function f and the constraint functions g_i, $i = 1, 2, \ldots, m$, i.e., $L(x, \lambda) = f(x) + \lambda^T[g(x) - b]$. The p-th power Lagrangian function, on the other hand, is a nonlinear Lagrangian in terms of the objective function and the constraint functions, $L_p(x, \lambda) = f^p(x) + \lambda^T[g^p(x) - b^p]$. Figure 5.4 demonstrates how a p-th power Lagrangian function serves as a nonlinear support at the optimal point for Example 5.1. It is clear that the larger the value of p, the sharper the nonlinear support becomes. By selecting a large enough value of p, the contour of this nonlinear Lagrangian forms a nonlinear support at the optimal point, thus offering an optimal generating multiplier. In summary, while a linear support associated with the conventional Lagrangian may not exist, a nonlinear support corresponding to a suitable nonlinear Lagrangian always exists.

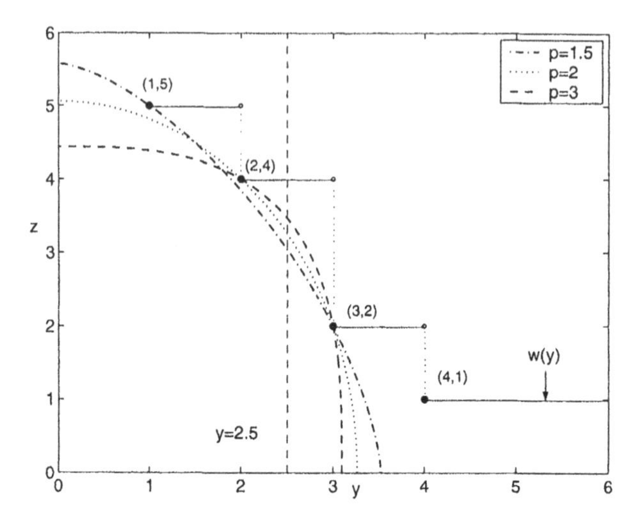

Figure 5.4. Contours of $z^p + \lambda^T(y^p - b^p) = r$ with different values of p.

5.2 Nonlinear Lagrangian Theory Using Equivalent Reformulation

The key concept in introducing nonlinear Lagrangian formulations is the construction of a nonlinear support of the perturbation function at the optimal point. There could be many different forms of nonlinear supports. The natural question of what are the common characteristics of various nonlinear Lagrangian formulations arises. More specifically, what is a general form of nonlinear functions of the objective function and the constraint functions that can serve as a nonlinear Lagrangian function? To be qualified as a nonlinear Lagrangian function, it is required that the corresponding nonlinear Lagrangian formulation guarantees the identification of an optimal solution of the primal problem via a dual search, i.e., an insurance of the existence of an optimal primal-dual pair.

We consider now the following transformation of problem (P),

$$
\begin{aligned}
(P_{qp}) \qquad & \min\ t_0(f(x), q) \\
& \text{s.t. } t_i(g_i(x), p) \le t_i(b_i, p), \quad i = 1, \ldots, m, \\
& \qquad x \in X \subseteq \mathbb{Z}^n,
\end{aligned}
$$

where t_i $(i = 0, 1, \ldots, m)$ are continuous functions defined on $\mathbb{R}^1_+ \times \mathbb{R}^1_+$, p and q are parameters. We assume the following conditions for t_i.

ASSUMPTION 5.2 (i) *For any given $r > 0$, $t_i(\cdot, r)$, $i = 0, 1, \ldots, m$, are strictly increasing functions on $\mathbb{R}_+$.*

(ii) *For any given* $0 \le y_1 < y_2$,

$$\lim_{p\to\infty} \frac{t_i(y_1,p)}{t_i(y_2,p)} = 0, \quad i = 1,\dots,m.$$

Examples of t_i satisfying Assumption 5.2 include $t_i(y,r) = y^r$ and $t_i(y,r) = \exp(ry)$ for $i = 0,1,\dots,m$, and $t_0(y,r) = y$. It is clear from the above assumption that the transformed problem (P_{qp}) is equivalent to the original problem (P), i.e., x^* is an optimal solution to (P) if and only if it is an optimal solution to (P_{qp}) with optimal value $t_0(f(x^*),q) = t_0(f^*,q)$. It is evident that (P_p) is a special case of (P_{qp}) when $t_i(y,p) = y^p$ for $i = 0,1,\dots,m$.

The Lagrangian relaxation of (P_{qp}) is

$$d_{qp}(\lambda) = \min_{x\in X} L_{qp}(x,\lambda), \tag{5.2.1}$$

where $\lambda \in \mathbb{R}^m_+$ and

$$L_{qp}(x,\lambda) = t_0(f(x),q) + \sum_{i=1}^{m} \lambda_i[t_i(g_i(x),p) - t_i(b_i,p)]. \tag{5.2.2}$$

The Lagrangian dual problem of (P_{qp}) is

$$(D_{qp}) \quad \theta_{qp} = d_{qp}(\lambda^*_{qp}) = \max_{\lambda\in\mathbb{R}^m_+} d_{qp}(\lambda), \tag{5.2.3}$$

where λ^*_{qp} is the optimal dual solution to (P_{qp}). Denote

$$t(y,p) = (t_1(y_1,p),\dots,t_m(y_m,p))^T$$

for any $y \in \mathbb{R}^m_+$. The perturbation function of problem (P_{qp}) is

$$w_{qp}(y) = \min_{x\in X}\{t_0(f(x),q) \mid t(g(x),p) \le y\}. \tag{5.2.4}$$

It is clear that the domain of w_{qp} is

$$Y_p = \{(t(y,p)) \mid y \in Y\},$$

where Y is the domain of the perturbation function $w(\cdot)$ of the original problem (P). The corner points of w_{qp} are

$$\Phi^{qp}_c = \{(t(c_i,p), t_0(f_i,q)) \mid i = 1,\dots,K\}.$$

Similar to (5.1.2), the convex envelope function of perturbation function w_{qp} can be written as

$$\begin{aligned} \psi_{qp}(y) \quad = \quad & \min \sum_{i=1}^{K} \mu_i t_0(f_i,q) \\ & \text{s.t. } \sum_{i=1}^{K} \mu_i t(c_i,p) \le y,\ \mu \in \Lambda. \end{aligned} \tag{5.2.5}$$

We have

$$w_{qp}(y) \geq \psi_{qp}(y), \quad \forall y \in Y_p. \tag{5.2.6}$$

THEOREM 5.2 *There exists $p_0 \geq 0$ such that*

$$\psi_{qp}(t(c_i, p)) = t_0(f_i, q), \quad i = 1, \ldots, K, \tag{5.2.7}$$

when $p \geq p_0$.

Proof. From (5.2.6), we have

$$t_0(f_i, q) = w_{qp}(t(c_i, p)) \geq \psi_{qp}(t(c_i, p)).$$

We prove the theorem by contradiction. Suppose that the conclusion of the theorem does not hold. Then there exists $l \in \{1, \ldots, K\}$ and a sequence $\{p_k\}$ with $p_k \to \infty$ such that

$$t_0(f_l, q) > \psi_{qp_k}(t(c_l, p_k)), \quad \forall k. \tag{5.2.8}$$

Let μ^k be an optimal solution to (5.2.5) with $y = t(c_l, p_k)$. Then

$$\psi_{qp_k}(t(c_l, p_k)) = \sum_{i=1}^{K} \mu_i^k t_0(f_i, q), \tag{5.2.9}$$

$$\sum_{i=1}^{K} \mu_i^k t(c_i, p_k) \leq t(c_l, p_k). \tag{5.2.10}$$

Define

$$I^k = \{i \in \{1, \ldots, K\} \mid \mu_i^k > 0\}.$$

We claim that $\mu_l^k = 0$ for any k, i.e., $l \notin I^k$ for any k. We note first that $\mu_l^k \neq 1$, since by (5.2.9) $\mu_l^k = 1$ implies $\psi_{qp_k}(t(c_l, p_k)) = t_0(f_l, q)$, contradicting (5.2.8). If $0 < \mu_l^k < 1$, then we can rewrite (5.2.10) as $\sum_{i=1}^{K} \hat{\mu}_i^k t(c_i, p_k) \leq t(c_l, p_k)$, where $\hat{\mu}_i^k = \mu_i^k/(1 - \mu_l^k)$ for $i \neq l$ and $\hat{\mu}_l^k = 0$. Thus, $\hat{\mu}^k \in \Lambda$ and $\hat{\mu}^k$ is feasible to (5.2.5) with $y = t(c_l, p_k)$. Moreover, we have

$$\begin{aligned} \sum_{i=1}^{K} \hat{\mu}_i^k t_0(f_i, q) &= \frac{\psi_{qp_k}(t(c_l, p_k)) - \mu_l^k t_0(f_l, p_k)}{1 - \mu_l^k} \\ &= \psi_{qp_k}(t(c_l, p_k)) + \frac{\mu_l^k}{1 - \mu_l^k}(\psi_{qp_k}(t(c_l, p_k)) - t_0(f_l, q)). \end{aligned} \tag{5.2.11}$$

Since, by (5.2.8), $t_0(f_l, q) > \psi_{qp_k}(t(c_l, p_k))$, (5.2.11) implies $\sum_{i=1}^{K} \hat{\mu}_i^k t_0(f_i, q) < \psi_{qp_k}(t(c_l, p_k))$, contradicting the optimality of μ^k. Let

$$I_1^k = \{i \in \{1, \ldots, K\} \mid i \in I^k, \text{ and } c_i \leq c_l\},$$
$$I_2^k = I^k \setminus I_1^k.$$

Note that $w_{qp_k}(t(c_l, p_k)) = t_0(f_l, q) > \psi_{qp_k}(t(c_l, p_k))$. Applying Theorem 3.16 to problem (P_{qp_k}), we deduce that there exists i such that $t(c_i, p_k) \not\leq t(c_l, p_k)$, which in turn implies that $I_2^k \neq \emptyset$ for all k. Next, we prove that

$$\lim_{k \to \infty} \sum_{i \in I_2^k} \mu_i^k = 0. \tag{5.2.12}$$

Suppose on the contrary that (5.2.12) does not hold. Then there exists a subsequence $\mathcal{K}$ of $\{1, 2, \ldots\}$ such that $\sum_{i \in I_2^k} \mu_i^k \geq \epsilon > 0$ for all $k \in \mathcal{K}$. Since $|I_2^k| \leq K$, there must exist a $j \in I_2^k$ and $\mathcal{K}' \subseteq \mathcal{K}$, such that for each $k \in \mathcal{K}'$, $\mu_j^k \geq \epsilon/K$ holds. Moreover, since $c_j \not\leq c_l$, there exists $s \in \{1, 2, \ldots, m\}$ such that $c_{js} > c_{ls}$. Thus, by (5.2.10), we have

$$\begin{aligned} t_s(c_{ls}, p_k) &\geq \sum_{i \in I^k} \mu_i^k t_s(c_{is}, p_k) \geq \sum_{i \in I_2^k} \mu_i^k t_s(c_{is}, p_k) \geq \mu_j^k t_s(c_{js}, p_k) \\ &\geq (\epsilon/K) t_s(c_{js}, p_k), \quad \forall k \in \mathcal{K}'. \end{aligned} \tag{5.2.13}$$

On the other hand, by Assumption 5.2 (ii), there exists $k' \in \mathcal{K}$ such that

$$\frac{t_s(c_{ls}, p_k)}{t_s(c_{js}, p_k)} < \epsilon/K, \quad k \geq k', \ k \in \mathcal{K}'.$$

This contradicts (5.2.13). Therefore (5.2.12) holds. Since $\mu^k \in \Lambda$, (5.2.12) implies

$$\lim_{k \to \infty} \sum_{i \in I_1^k} \mu_i^k = 1, \tag{5.2.14}$$

which in turn implies $I_1^k \neq \emptyset$ for sufficiently large k.

Now, let

$$\delta = \min\{t_0(f_i, q) - t_0(f_l, q) \mid c_i \leq c_l, \ i \neq l, \ i \in \{1, \ldots, K\}\}.$$

Since any corner point (c_i, f_i) is noninferior and t_0 is strictly increasing, we have $\delta > 0$. Since $l \notin I_1^k$ for all k, we have

$$t_0(f_i, q) \geq t_0(f_l, q) + \delta, \quad \forall i \in I_1^k, \ \forall k.$$

Thus, by (5.2.12) and (5.2.14), there exists k_0 such that when $k \geq k_0$, $I_1^k \neq \emptyset$ holds and

$$\sum_{i \in I_1^k} \mu_i^k t_0(f_i, q) \geq \sum_{i \in I_1^k} \mu_i^k (t_0(f_l, q) + \delta) \geq t_0(f_l, q) + \frac{1}{2}\delta, \quad (5.2.15)$$

$$\sum_{i \in I_2^k} \mu_i^k t_0(f_i, q) \geq -\frac{1}{4}\delta. \quad (5.2.16)$$

Combining (5.2.9) with (5.2.15) and (5.2.16) yields

$$\begin{aligned} \psi_{p_k}(t(c_l, p_k)) &= \sum_{i \in I_1^k} \mu_i^k t_0(f_i, q) + \sum_{i \in I_2^k} \mu_i^k t_0(f_i, q) \\ &\geq t_0(f_l, q) + \frac{1}{2}\delta - \frac{1}{4}\delta \\ &= t_0(f_l, q) + \frac{1}{4}\delta \end{aligned} \quad (5.2.17)$$

for $k \geq k_0$. Inequality (5.2.17) contradicts (5.2.8). The proof is completed. □

We point out that Theorem 5.2 is a generalization of Theorem 5.1. In fact, when $t_i(y, p) = y^p$ for $i = 0, 1, \ldots, m$, we obtain Theorem 5.1 from Theorem 5.2.

The following theorem further shows that primal feasibility of (5.2.1) with $\lambda = \lambda_{qp}^*$ and the existence of an optimal primal-dual pair of (P_{qp}) can be also ensured when p is larger than a threshold value. Moreover, problem (P_{qp}) possesses an asymptotic strong duality.

THEOREM 5.3 (i) *There exists $p_1 \geq 1$ such that there exists at least an optimal solution to (5.2.1) with $\lambda = \lambda_{qp}^*$ that is feasible to (P) when $p \geq p_1$, where λ_{qp}^* is an optimal solution to (D_{qp}).*

(ii) $\lim_{p \to \infty} \theta_{qp} = t_0(f^*, q)$, *where f^* is the optimal objective value of (P).*

(iii) *There exists $p_2 \geq p_1$ such that (x^*, λ_{qp}^*) is an optimal primal-dual pair of (P_{qp}) when $p \geq p_2$, where x^* is a noninferior solution of (P).*

Proof. We first notice from Theorem 3.10 (i) that $\theta_{qp} = d_{qp}(\lambda_{qp}^*) = \psi_{qp}(t(b, p))$. Moreover, by (5.2.5), there exist $\mu(p) \in \Lambda$ such that

$$\psi_{qp}(t(b, p)) = \sum_{i=1}^{K} \mu_i(p) t_0(f_i, q), \quad (5.2.18)$$

$$t(b, p) \geq \sum_{i=1}^{K} \mu_i(p) t(c_i, p). \quad (5.2.19)$$

Let

$$\begin{aligned} I(p) &= \{i \in \{1, \ldots, K\} \mid \mu_i(p) > 0\}, \\ I_1(p) &= \{i \in \{1, \ldots, K\} \mid \mu_i(p) > 0 \text{ and } c_i \leq b\}, \\ I_2(p) &= I(p) \setminus I_1(p). \end{aligned}$$

(i) Note that if $I_1(p) \neq \emptyset$, then for any $i \in I_1(p)$, by Lemma 3.2 (ii), there is $\bar{x} \in X$ satisfying $t(g(\bar{x}), p) = t(c_i, p) \leq t(b, p)$. Moreover, by Theorem 3.10 (ii), $\bar{x}$ is an optimal solution to (5.2.1) with $\lambda = \lambda^*_{qp}$. Thus, it suffices to show that there exists $p_1 > 0$ such that $I_1(p) \neq \emptyset$ when $p \geq p_1$. Since $I_2(p) = \emptyset$ implies $I_1(p) = I(p) \neq \emptyset$, we assume in the following $I_2(p) \neq \emptyset$. Similar to (5.2.12) in the proof of Theorem 5.2, we have

$$\lim_{p \to \infty} \sum_{i \in I_2(p)} \mu_i(p) = 0 \tag{5.2.20}$$

and consequently

$$\lim_{p \to \infty} \sum_{i \in I_1(p)} \mu_i(p) = 1. \tag{5.2.21}$$

Therefore, $I_1(p) \neq \emptyset$ for sufficiently large p.

(ii) By part (i), $I_1(p) \neq \emptyset$ for $p \geq p_1$. For any $i \in I_1(p)$, $t_0(f_i, q) = w_{qp}(t(c_i, p)) \geq w_{qp}(t(b, p)) = t_0(f^*, q)$. We obtain from (5.2.18) and (5.2.21) that

$$\begin{aligned} \psi_{qp}(t(b,p)) &= \sum_{i \in I_1(p)} \mu_i(p) t_0(f_i, q) + \sum_{i \in I_2(p)} \mu_i(p) t_0(f_i, q) \\ &\geq \sum_{i \in I_1(p)} \mu_i(p) t_0(f^*, q) \\ &\to t_0(f^*, q), \quad p \to \infty. \end{aligned} \tag{5.2.22}$$

On the other hand, the weak duality relation and Theorem 3.10 (i) give

$$t_0(f^*, q) \geq d_{qp}(\lambda^*_{qp}) = \psi_{qp}(t(b, p)). \tag{5.2.23}$$

Combining (5.2.22) with (5.2.23) yields part (ii).

(iii) Notice first that if $f_i = f^*$ for some $i \in I_1(p)$, then, by Lemma 3.2 (ii), there exists $x^* \in S$ such that $(t(g(x^*), p), t_0(f(x^*), q)) = (t(c_i, p), t_0(f_i, q)) = (t(c_i, p), t_0(f^*, q))$. Hence (x^*, λ^*_{qp}) is an optimal primal-dual pair of (P_{qp}). We now prove that there exists $i \in I_1(p)$ satisfying $f_i = f^*$ when p ($\geq p_1$) is sufficiently large. Suppose on the contrary there exists a sequence $\{p_k\}$ with $p_k \to \infty$ and for each k, $f_i > f^*$ for all $i \in I_1(p_k)$. Let

$$\delta^* = \min\{t_0(f_i, q) - t_0(f^*, q) \mid c_i \leq b,\ f_i \neq f^*,\ i \in \{1, \ldots, K\}\} > 0.$$

Using the similar arguments as in the proof of (5.2.17), we can deduce from (5.2.20) and (5.2.21) that

$$\theta_{qp_k} = \psi_{qp_k}(t(b, p_k)) > t_0(f^*, q) + \frac{1}{4}\delta^*$$

when k is sufficiently large. This, however, contradicts part (ii). □

In the above discussion, we assume that parameter q is fixed. As it is clear from Assumption 5.2, the requirement on function t_0 is much weaker than the requirement on function t. Essentially, it is evident from the proof of Theorem 5.3 that the strong duality can be achieved only via reformulation of the constraint functions. The reason to also introduce transformation on the objective function will be explained later.

Now let us consider a partial p-th power formulation of (P):

$$(P_{1p}) \qquad \begin{array}{ll} \min f(x) \\ \text{s.t. } [g(x)]^p \leq b^p, \\ \quad\ x \in X. \end{array}$$

Notice that (P_{1p}) is a special form of (P_{qp}) by taking $t_0(y, q) = y$ and $t_i(y, p) = y^p$ for $i = 1, \ldots, m$.

Let w_{1p} and ψ_{1p} denote the perturbation function and convex envelope function of (P_{1p}), respectively. Apply the partial p-th power reformulation to Example 3.4. Figure 5.5 depicts the functions w_{1p} and ψ_{1p} for $p = 3$. We can see from the figure that condition (5.2.7) is satisfied when $p = 3$. It can be verified that $\lambda_{13}^* = -(4-2)/(2^3 - 3^3) = 2/19$ is an optimal dual solution in this example and $\{x^* = (1, 1, 0, 1)^T, \lambda_{13}^* = 2/19\}$ is the optimal primal-dual pair.

Next, we study the relationship among the parameters p_0, p_1 and p_2 in Theorem 5.2 and Theorem 5.3 via the partial p-th power Lagrangian formulation. By the definition of the optimal primal-dual pair, it always holds $p_1 \leq p_2$. When $m = 1$, we also know from Theorem 3.14 and Theorem 3.15 that $p_1 = 1$ and $p_2 \leq p_0$. Thus, for singly constrained problems, we have

$$1 = p_1 \leq p_2 \leq p_0. \tag{5.2.24}$$

The strict inequality $p_2 < p_0$ in (5.2.24) could hold when condition (5.2.7) is satisfied for c_i's around $y = b$ and thus there may exist an optimal primal-dual pair, while condition (5.2.7) is not satisfied for c_i's far away from $y = b$. Consider Example 3.4 with $b = 3.5$. The perturbation function $w(y)$ and the convex envelope function $\psi(y)$ of this problem are illustrated in Figure 5.6. It can be verified that the optimal solution of this problem is $x^* = (0, 0, 0, 1)^T$

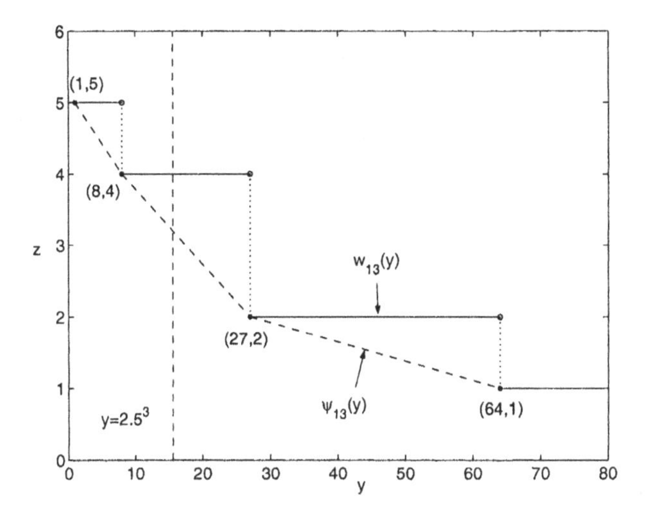

Figure 5.5. w_{1p} and ψ_{1p} of Example 3.4 with $p = 3$.

which corresponds to point $(3, 2)^T$ in Figure 5.6. Also, $\lambda^* = 1$ is the optimal solution to (D) and (x^*, λ^*) is an optimal primal-dual pair. However, as shown in Example 5.1, $\psi(c_2) = \psi(2) = 3.5 < 4$. Hence (5.2.7) is not satisfied and $1 = p_1 = p_2 < p_0$. It is noticed from Figure 5.5 that $p_0 \leq 3$.

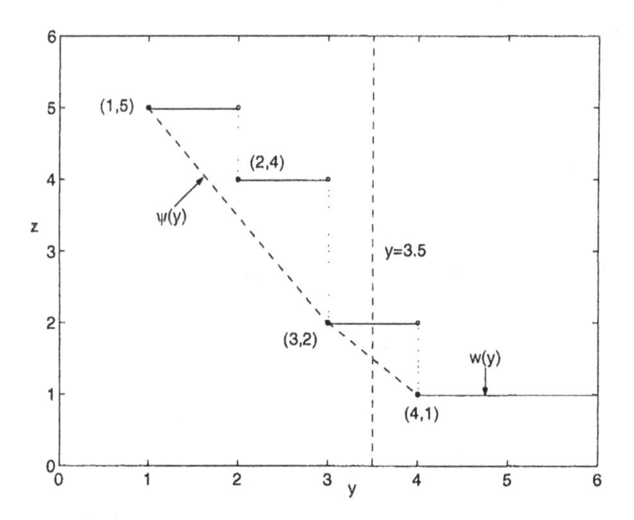

Figure 5.6. Illustration of $w(y)$ and $\psi(y)$ for Example 3.4 with $b = 3.5$.

For multiply constrained cases, the following two cases may happen:

$$1 \leq p_1 \leq p_2 \leq p_0, \tag{5.2.25}$$
$$1 \leq p_0 \leq p_1 \leq p_2. \tag{5.2.26}$$

EXAMPLE 5.2 Consider the following example:

$$\begin{aligned}
&\min f(x) = 3x_1 + 2x_2 - 1.5x_1^2 \\
&\text{s.t. } g_1(x) = \sqrt{15 - 7x_1 + 2x_2} \leq 2\sqrt{3}, \\
&\quad g_2(x) = \sqrt{15 + 2x_1^2 - 7x_2} \leq 2\sqrt{3}, \\
&\quad x \in X = \{(0,1)^T, (0,2)^T, (1,0)^T, (1,1)^T, (2,0)^T, (2,2)^T\}.
\end{aligned}$$

The optimal solution of the problem is $x^* = (1,1)^T$ with $f(x^*) = 3.5$. The optimal solution to the dual problem (D) is $\lambda^* = (0, 1.0166)^T$ with $d(\lambda^*) = 1.3538$. The Lagrangian relaxation problem (L_λ) with $\lambda = \lambda^*$ has two optimal solutions: $(0,1)^T$, $(2,0)^T$, none of which is feasible. Notice that the problem has only two feasible solutions $(1,1)^T$ and $(2,2)^T$.

The corner points of the example are: $c_1 = (4.1231, 2.8284)^T$, $f_1 = 2$, $c_2 = (4.3589, 1)^T$, $f_2 = 4$, $c_3 = (2.8284, 4.1231)^T$, $f_3 = 1.5$, $c_4 = (3.1623, 3.1623)^T$, $f_4 = 3.5$, $c_5 = (1, 4.7958)^T$, $f_5 = 0$, $c_6 = (2.2361, 3)^T$, $f_6 = 4$.

Applying the partial p-th power reformulation to the above example, it can be verified that primal feasibility of the p-th power Lagrangian relaxation problem (5.2.1) can be achieved when $p \geq 2$. However, this is not enough to guarantee the existence of the optimal primal-dual pair. For instance, take $p = 2$, we have $\lambda_{1p}^* = (0.1951, 0.3414)^T$ and the optimal solutions to $d_{1p}(\lambda_{1p}^*)$ are $(0,1)^T$, $(2,0)^T$, $(0,2)^T$ and $(2,2)^T$. Thus, (x^*, λ_{1p}^*) is not an optimal primal-dual pair when $p = 2$. For $p = 2$, we can verify that

$$\begin{aligned}
&\psi_{1p}(c_1^p) = 2 = f_1,\ \psi_{1p}(c_2^p) = 4 = f_2,\ \psi_{1p}(c_3^p) = 0.8 < 1.5 = f_3, \\
&\psi_{1p}(c_4^p) = 2.6829 < 3.5 = f_4,\ \psi_{1p}(c_5^p) = 0 = f_5,\ \psi_{1p}(c_6^p) = 4 = f_6.
\end{aligned}$$

So, condition (5.2.7) is not satisfied. We can also verify that there is no optimal generating multiplier vector for x^* when $p = 2$. We can further increase the value of p. When $p \geq 6.3$, condition (5.2.7) is satisfied and (x^*, λ_{1p}^*) becomes an optimal primal-dual pair. For instance, take $p = 6.3$, we have $\lambda_{1p}^* = (0.2874 \times 10^{-3}, 0.3609 \times 10^{-3})^T$ and the optimal solution to (5.2.1) are $(0,1)^T$, $(1,0)^T$, $(1,1)^T$ and $(2,2)^T$. Therefore, we have $1 < p_1 < p_2 = p_0$ and hence (5.2.25) holds in this example.

To show that (5.2.26) may happen, let us consider Example 3.6. Although condition (3.4.7) is satisfied, there does not exist an optimal primal-dual pair in the original problem setting. Applying the partial p-th power reformulation to Example 3.6 with $p = 3$, we make the generation of an optimal primal-dual pair with $\lambda_{1p}^* = (0.0038, 0.0331)^T$ and $x^* = (1, 4)^T$. Thus, $1 = p_0 = p_1 < p_2$.

Although the partial p-th power reformulation can guarantee the identification of the optimal primal-dual pair, it is sometimes beneficial computationally to adopt a transformation on the objective function at the same time. This can be clearly seen from Example 5.1 for which the optimal primal-dual pair of the p-th power reformulation can be guaranteed to exist when $(q, p) = (2, 2)$ in problem (P_{qp}) (see Figure 5.3). Yet $p = 2$ does not guarantee the existence of an optimal primal-dual pair of the partial p-th power reformulation (corresponding to $q = 1$) as seen from Figure 5.7. The impact of parameter q can be further seen from the following data set of some combinations of (q, p) that guarantee the existence of an optimal primal-dual pair in the reformulation (P_{qp}) of Example 5.1: (q, p)=(1,3), (2,2), (3,1.5). In general, the larger the q value in problem (P_{qp}), the smaller value of p we need to ensure the existence of an optimal primal-dual pair.

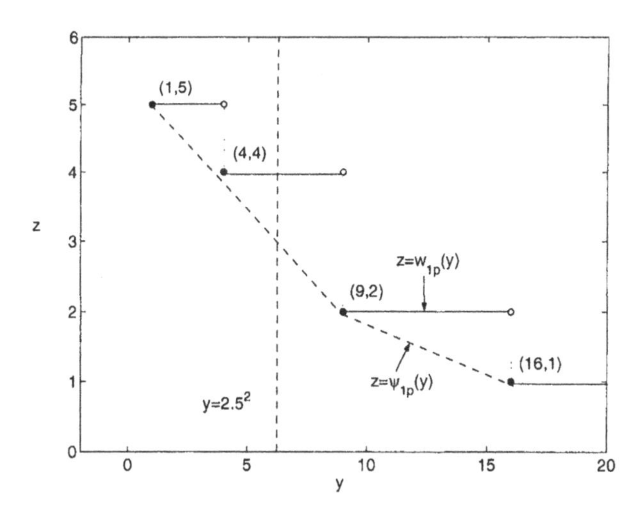

Figure 5.7. Illustration of $w_{1p}(y)$ and $\psi_{1p}(y)$ for Example 3.4 with $p = 2$.

5.3 Nonlinear Lagrangian Theory Using Logarithmic-Exponential Dual Formulation

The nonlinear Lagrangian theory developed in the previous sections investigates promising nonlinear transformations on both objective function and constraints such that the conventional Lagrangian theory can be successfully applied to identify an optimal solution of the primal problem via a dual search. While the resulting nonlinear Lagrangian formulations developed in the previous sections are nonlinear with respect to constraints and (sometimes) to the objective function, they are still linear with respect to the Lagrangian multiplier. In this and the next sections, we are going to explore more general forms of nonlinear Lagrangian formulations with a success guarantee of the dual search.

We consider in this section the following modified version of (P) by setting b as a zero vector:

$$(P_0) \qquad \begin{aligned} &\min\ f(x) \\ &\text{s.t.}\ g_i(x) \le 0,\ i = 1, \ldots, m, \\ &\qquad x \in X. \end{aligned}$$

Denote by S_0 the feasible region of (P_0):

$$S_0 = \{x \in X \mid g_i(x) \le 0, i = 1, \ldots, m\}.$$

Without loss of generality, we make the following assumptions in (P_0):

ASSUMPTION 5.3 *$S_0 \neq \emptyset$ and $f(x) > 0$ for all $x \in X$.*

We will pursue new insights for dual search by studying single-constraint cases of (P_0) first. The results gained from single-constraint cases will motivate a formal investigation of nonlinear Lagrangian dual theory using a logarithmic-exponential formulation.

Let us consider the following example.

EXAMPLE 5.3

$$\begin{aligned} &\min\ f(x) = 0.2(x_1 - 3)^2 + 0.1(x_2 - 5)^2 + 0.1 \\ &\text{s.t.}\ g(x) = x_1^2 - 2.5x_1 + 1.2x_2 - 1 \le 0, \\ &\qquad x \in X = [0, 3]^2 \cap \mathbb{Z}^2. \end{aligned}$$

The image of X in the (z_1, z_2) plane under the mapping $(g(x), f(x))$ is shown in Figure 5.8. It is clear from Figure 5.8 that $P^* = (-0.1, 1.8)^T$ is the image of the primal optimum point, $x^* = (1, 2)^T$, with $f(x^*) = 1.8$. Since the optimal Lagrangian multiplier is $\lambda^* = 7/12$ with $d(\lambda^*) = 43/30 \approx 1.433 < 1.8$, a duality gap exists for the Lagrangian dual formulation. Moreover, the feasible optimal solution of the problem $\min\{L(x, \lambda^*) \mid x \in X\}$ is $(2, 1)^T$ and the corresponding image in the (z_1, z_2) plane is $(-0.8, 1.9)^T$. Thus, the linear Lagrangian dual search fails to find the primal optimum. A key observation from Figure 5.8 is that there is no supporting plane at the point P^*, or more specifically, for whatever value of $\lambda \ge 0$, it is impossible in this case for the linear contour of the Lagrangian function $L(z, \lambda) = z_2 + \lambda z_1$ with a minimum contour level to pass through the point P^*, which corresponds to the primal optimum x^* of (P_0). It is therefore natural to consider some classes of nonlinear supports, or more specifically, some classes of nonlinear functions whose nonlinear contours can pass through the point P^* in any situation.

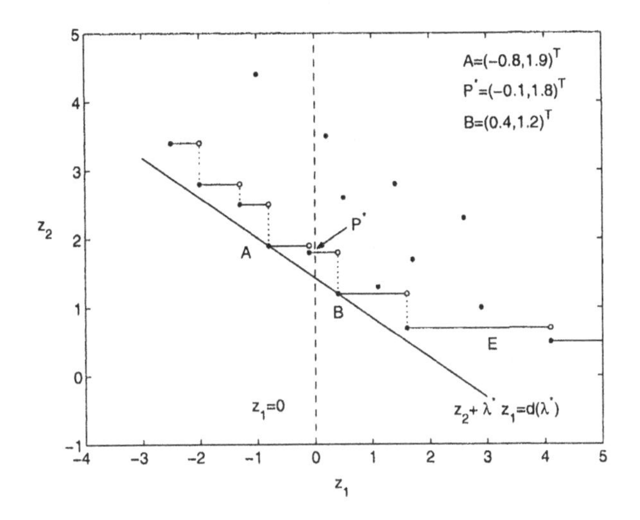

Figure 5.8. Illustration of the Lagrangian dual search for Example 5.3.

Let w be the perturbation function associated with the singly constrained problem of (P_0):

$$w(z_1) = \min\{f(x) \mid g(x) \le z_1,\ x \in X\}. \tag{5.3.1}$$

Define a set in $\mathbb{R}^2$:

$$E = \{(z_1, z_2) \mid z_1 \in Y,\ z_2 = w(z_1)\}, \tag{5.3.2}$$

where Y is the domain of w. Geometrically, E is the lower envelope of the image of X in the (z_1, z_2) plane under the mapping $(g(x), f(x))$ (see also Figure 5.8). In order to identify the point P^*, the image of the primal optimum point, the contour of a desired nonlinear function should be able to support E at the point P^*. It is also desirable to maintain the weak duality property in the new dual formulation. We are thus searching for a nonlinear function $C(z, \lambda)$ defined on the (z_1, z_2) plane with parameter λ that satisfies the following conditions:

(a) there exists a $\lambda \ge 0$ such that contour $C(z, \lambda) = \alpha$ supports E at a unique point P^* and lies completely below E, where $\alpha = C(P^*, \lambda)$;

(b) it holds that $C(z, \lambda) \le z_2$ whenever $z_1 \le 0$, $z_2 > 0$ and $\lambda \ge 0$.

One evident candidate for a nonlinear support is the polygonal line in the (z_1, z_2) plane with a positive $\lambda > 0$:

$$\begin{aligned} \Gamma_\lambda &= \{(z_1, z_2) \in \mathbb{R}^2 \mid \lambda z_1 \le z_2,\ z_2 = \alpha\} \\ &\quad \cup \{(z_1, z_2) \in \mathbb{R}^2 \mid \lambda z_1 = \alpha,\ \lambda z_1 > z_2\}, \end{aligned}$$

where $\alpha = w(0) > 0$. The polygonal line Γ_λ is exactly the contour of the function $C_M(z, \lambda) = \max\{z_2, \lambda z_1\}$ with a contour level α (see Figure 5.9, where $\alpha = 1.8$). Obviously, $C_M(x, \lambda)$ satisfies condition (b). For any $\alpha > 0$, there always exists a $\lambda > 0$ such that the line $z_1 = \alpha/\lambda$ falls between the point P^* and the point with the minimum value of z_1 in the half plane $\{(z_1, z_2) \mid z_1 > 0\}$ (point $(0.4, 1.2)^T$ in Figure 5.9). The polygonal line Γ_λ with a suitable $\lambda > 0$ is thus able to support E by a line segment including P^* and hence the function $C_M(z, \lambda)$ satisfies conditions (a) and (b) except for the unique supporting property. To achieve uniqueness, a logarithmic-exponential function that approximates C_M is constructed:

$$C_p(z, \lambda) = \frac{1}{p} \ln[\frac{1}{2}(\exp(pz_2) + \exp(p\lambda z_1))], \ \lambda \geq 0, \ p > 0. \tag{5.3.3}$$

Note that

$$-\frac{\ln(2)}{p} + C_M(z, \lambda) \leq C_p(z, \lambda) \leq C_M(z, \lambda), \ \ \lambda \geq 0, \ p > 0.$$

Thus, for any $\lambda \geq 0$, we have

$$C_p(z, \lambda) \leq z_2, \ \text{ if } z_1 \leq 0, \ z_2 > 0, \tag{5.3.4}$$

$$C_p(z, \lambda) \to C_M(z, \lambda), \ \ p \to \infty. \tag{5.3.5}$$

The inequality (5.3.4) is exactly the weak duality in condition (b) and (5.3.5) ensures condition (a) as shown below.

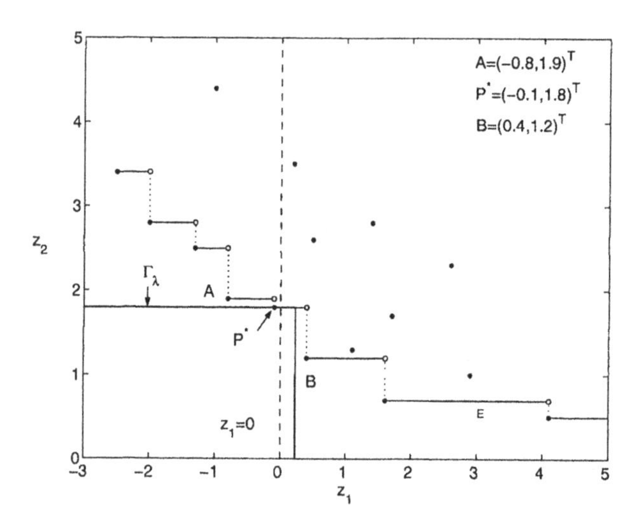

Figure 5.9. Illustration of the contour $C_M(z, \lambda) = \alpha$ for Example 5.3 ($\lambda = 8, \alpha = 1.8$).

Note that the contour $C_p(z, \lambda) = \alpha$ can be expressed as

$$\exp(pz_2) + \exp(p\lambda z_1) = 2\exp(p\alpha). \tag{5.3.6}$$

For any point $(z_1, z_2)^T$ on the contour, we have

$$\frac{dz_2}{dz_1} = -\frac{\lambda}{2\exp(p(\alpha - \lambda z_1)) - 1}. \qquad (5.3.7)$$

For $\lambda > 0$, the z_1 domain of the contour $C_p(z, \lambda) = \alpha$ is $(-\infty, (\frac{\ln(2)}{p} + \alpha)/\lambda)$. It follows from (5.3.7) that $dz_2/dz_1 < 0$ for any point $(z_1, z_2)^T$ on the contour when $\lambda > 0$. This implies that z_2 is a strictly decreasing function of z_1 when $\lambda > 0$. The following are evident from (5.3.7),

$$\frac{dz_2}{dz_1} \to 0, \quad \lambda \to \infty, \quad \text{for } z_1 \in (-\infty, 0), \qquad (5.3.8)$$

$$\frac{dz_2}{dz_1} \to -\infty, \quad \lambda \to \infty, \quad \text{for } z_1 \in (0, (\frac{\ln(2)}{p} + \alpha)/\lambda), \qquad (5.3.9)$$

$$\frac{dz_2}{dz_1} \to 0, \quad p \to \infty, \quad \text{for } z_1 \in (-\infty, \alpha/\lambda). \qquad (5.3.10)$$

It can be seen from (5.3.8) and (5.3.9) that if λ is chosen sufficiently large, then the value of z_2 on the contour $C_p(z, \lambda) = \alpha$ decreases very slowly when z_1 is negative while it decreases almost vertically when z_1 is positive, thus enforcing the contour to lie entirely below E. Figure 5.10 illustrates the behavior of the contour (5.3.6) for various values of λ. Moreover, (5.3.10) shows that the parameter p controls the slope of the contour on the interval $(-\infty, \alpha/\lambda)$ $(\lambda > 0)$, thus making the supporting contour touch the "hidden" point P^* (see Figure 5.11). Since z_2 is a strictly decreasing function of z_1 on any contour of $C_p(z, \lambda)$ with $\lambda > 0$, the supporting point of the contour to E must be unique. Therefore, the function $C_p(z, \lambda)$ satisfies condition (a).

We now illustrate the logarithmic-exponential function associated with Example (5.3). Take $\lambda = 4$ and $p = 1.2$. The contour $C_{1.2}(z, 4) = \alpha$ with contour level $\alpha = 1.2798$ passes through P^* and is located below E. See Figure 5.12. By solving an unconstrained integer optimization problem

$$\min\{C_{1.2}([g(x), f(x)], 4) \mid x \in X\},$$

we get exactly the primal optimal solution $x^* = (1, 2)^T$.

From the above discussion, we observe that the logarithmic-exponential function $C_p(z, \lambda)$ can serve well as a candidate function in carrying out a dual search. It makes use of the prominent features of the discrete structure in integer programming. Furthermore, if λ is viewed as a dual variable, then a new dual formulation can be established to exploit the zero duality gap and to guarantee a success for dual search in integer programming which are often not achievable by the conventional linear Lagrangian dual formulation.

The logarithmic-exponential Lagrangian function and its corresponding dual function are now formally described for problem (P_0). For any $\lambda \in \mathbb{R}^m_+$ and

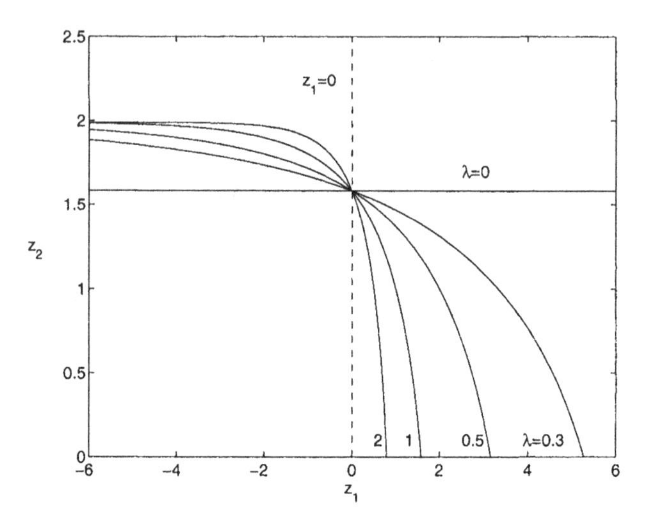

Figure 5.10. Behavior of $C_p(z, \lambda) = \alpha$ $(p = 0.7, \alpha = 1)$ for various λ.

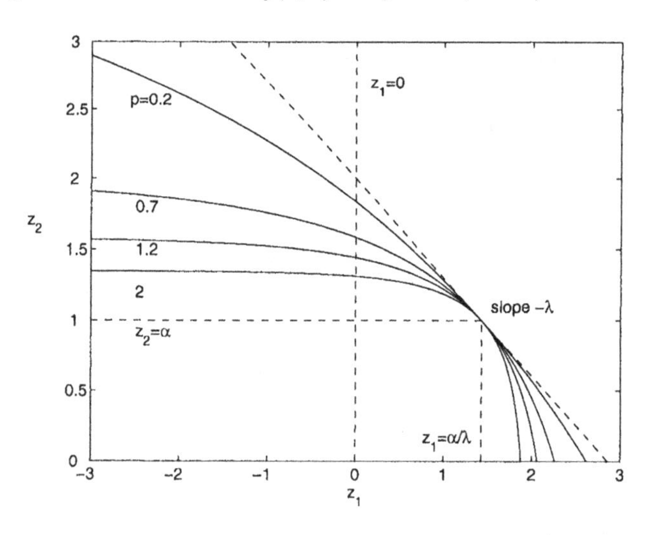

Figure 5.11. Behavior of $C_p(z, \lambda) = \alpha$ $(\lambda = 0.7, \alpha = 1)$ for various p.

$p > 0$, a logarithmic-exponential Lagrangian function is defined as follows:

$$Q_p(x, \lambda) = \frac{1}{p} \ln[\frac{1}{m+1}(\exp(pf(x)) + \sum_{i=1}^{m} \exp(p\lambda_i g_i(x)))]. \qquad (5.3.11)$$

The dual function associated with (P_0) is defined by

$$d_p^{LE}(\lambda) = \min_{x \in X} Q_p(x, \lambda). \qquad (5.3.12)$$

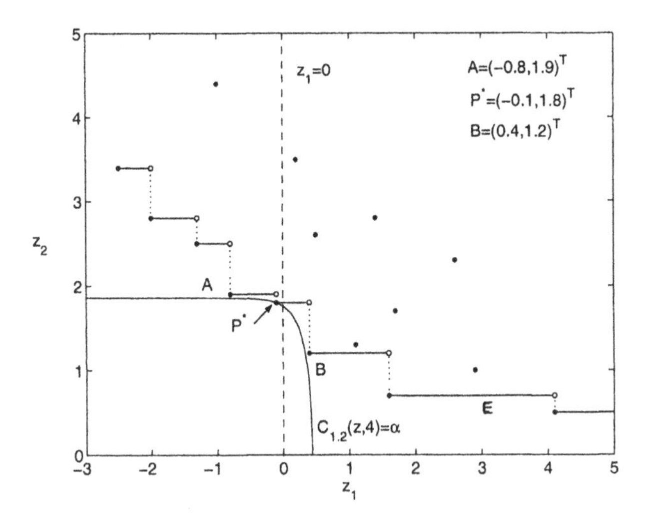

Figure 5.12. Contour $C_p(z, 4) = \alpha$ for Example 5.3 ($p = 1.2$, $\alpha = C_p(P^*, 4) = 1.2798$).

Similar to the classical Lagrangian dual formulation, the logarithmic-exponential dual problem of (P_0) is then formulated as

$$\theta_p^{LE} = \max_{\lambda \in \mathbb{R}_+^m} d_p^{LE}(\lambda). \tag{5.3.13}$$

The basic properties of the logarithmic-exponential Lagrangian function are investigated in the following.

LEMMA 5.1 (i) *For any $x \in X$, $\lambda \in \mathbb{R}_+^m$ and $p > 0$,*

$$\begin{aligned} Q_p(x, \lambda) &> -\frac{1}{p}\ln(m+1) + \max\{f(x), \lambda_1 g_1(x), \ldots, \lambda_m g_m(x)\}, \\ Q_p(x, \lambda) &\leq \max\{f(x), \lambda_1 g_1(x), \ldots, \lambda_m g_m(x)\}. \end{aligned}$$

(ii) *For any $x \in S_0$, $\lambda \in \mathbb{R}_+^m$ and $p > 0$,*

$$-\frac{1}{p}\ln(m+1) + f(x) < Q_p(x, \lambda) \leq f(x).$$

Proof. It can be verified that

$$\begin{aligned} &\exp(pf(x)) + \sum_{i=1}^{m} \exp(p\lambda_i g_i(x)) \\ &\quad > \exp(p \max\{f(x), \lambda_1 g_1(x), \ldots, \lambda_m g_m(x)\}), \\ &\exp(pf(x)) + \sum_{i=1}^{m} \exp(p\lambda_i g_i(x)) \\ &\quad \leq (m+1)\exp(p \max\{f(x), \lambda_1 g_1(x), \ldots, \lambda_m g_m(x)\}). \end{aligned}$$

Performing certain transformations on both sides of the above two inequalities and using (5.3.11) yield the results of (i). Part (ii) follows from part (i) and the assumption of $x \in S_0$. □

Part (ii) of Lemma 5.1 immediately leads to the following *weak duality* relation:

$$d_p^{LE}(\lambda) \leq f(x), \quad \text{for any } x \in S_0 \text{ and } \lambda \in \mathbb{R}_+^m. \tag{5.3.14}$$

LEMMA 5.2 (i) *For any $x \in X$ and $p > 0$, $Q_p(x, \lambda)$ is a convex function of λ.*
(ii) *For any $p > 0$, the dual function $d_p^{LE}(\lambda)$ is a continuous piecewise convex function of λ.*
(iii) *Suppose that f and the g_i's are convex functions. Then, for any $\lambda \in \mathbb{R}_+^m$ and any $p > 0$, $Q_p(x, \lambda)$ is also a convex function of x.*

Proof. The claim in part (i) can be easily checked. Part (ii) follows directly from part (i) and the finiteness of X. Let

$$W(x, \xi) = \ln \left[\sum_{i=1}^{k} \exp(\xi_i h_i(x)) \right], \quad \xi \in \mathbb{R}_+^k. \tag{5.3.15}$$

To prove (iii), it suffices to show that $W(x, \xi)$ is a convex function of x for any fixed $\xi \in \mathbb{R}_+^k$ whenever $h_i(x), i = 1, \ldots, k$, are convex functions. We need to prove that for any $x_1, x_2 \in \mathbb{R}^n$ and $\mu \in (0, 1)$, the following holds

$$W(\mu x_1 + (1 - \mu)x_2, \xi) \leq \mu W(x_1, \xi) + (1 - \mu)W(x_2, \xi). \tag{5.3.16}$$

From (5.3.15) and the convexity of the functions h_i, we have

$$\begin{aligned} & W(\mu x_1 + (1 - \mu)x_2, \xi) \\ = \ & \ln \left\{ \sum_{i=1}^{k} \exp[\xi_i h_i(\mu x_1 + (1 - \mu)x_2)] \right\} \\ \leq \ & \ln \left\{ \sum_{i=1}^{k} \exp[\mu \xi_i h_i(x_1) + (1 - \mu)\xi_i h_i(x_2)] \right\} \\ = \ & \ln \left\{ \sum_{i=1}^{k} [\exp(\xi_i h_i(x_1))]^{\mu} [\exp(\xi_i h_i(x_2))]^{1-\mu} \right\}. \end{aligned} \tag{5.3.17}$$

Let $a_i = \exp(\xi_i h_i(x_1))$, $b_i = \exp(\xi_i h_i(x_2))$. For any $\mu \in (0,1)$, by the Hölder inequality, we have

$$\begin{aligned}
&\sum_{i=1}^{k} [\exp(\xi_i h_i(x_1))]^{\mu} [\exp((\xi_i h_i(x_2))]^{1-\mu} \\
= &\sum_{i=1}^{k} a_i^{\mu} b_i^{1-\mu} \le \left(\sum_{i=1}^{k} a_i\right)^{\mu} \left(\sum_{i=1}^{k} b_i\right)^{1-\mu}. \qquad (5.3.18)
\end{aligned}$$

Combining (5.3.17) with (5.3.18) yields

$$\begin{aligned}
W(\mu x_1 + (1-\mu)x_2, \xi) &\le \mu \ln\left(\sum_{i=1}^{k} a_i\right) + (1-\mu)\ln\left(\sum_{i=1}^{k} b_i\right) \\
&= \mu W(x_1, \xi) + (1-\mu) W(x_2, \xi).
\end{aligned}$$

Therefore, the inequality (5.3.16) holds for all $\mu \in (0,1)$ and hence $W(x, \xi)$ is a convex function of x. □

The property of asymptotic strong duality will be now proven for the logarithmic-exponential dual formulation (5.3.11)–(5.3.13). The relationship between the solutions of $Q_p(x, \lambda)$ and (P_0) will be examined next. Denote

$$\begin{aligned}
&f^* = \min_{x \in S_0} f(x), \\
&S^* = \{x \in S_0 \mid f(x) = f^*\}, \\
&\delta = \min\{f(x) \mid x \in S_0 \setminus S^*\} - f^*.
\end{aligned}$$

THEOREM 5.4 *(Asymptotic strong duality)* $\lim_{p \to \infty} \theta_p^{LE} = f^*$.

Proof. If $S_0 = X$, then $\lim_{p \to \infty} \theta_p^{LE} = f^*$ holds trivially by (5.3.12), (5.3.13) and part (ii) of Lemma 5.1. We assume in the following that $X \setminus S_0 \ne \emptyset$. For any fixed $p > 0$, again from part (ii) of Lemma 5.1, we have

$$\theta_p^{LE} = \max_{\lambda \in \mathbb{R}_+^m} \min_{x \in X} Q_p(x, \lambda) \le \min_{x \in S_0} f(x) = f^*. \qquad (5.3.19)$$

For any $x \in X \setminus S_0$, there exists at least an i such that $g_i(x) > 0$. Since $X \setminus S_0$ is a finite set, we have

$$\mu = \min_{x \in X \setminus S_0} \max\{g_1(x), \dots, g_m(x)\} > 0. \qquad (5.3.20)$$

We claim that for any fixed $p > 0$, there must exist some $\lambda \in \mathbb{R}_+^m$ satisfying

$$\min_{x \in X \setminus S_0} Q_p(x, \lambda) \ge \min_{x \in S_0} Q_p(x, \lambda). \qquad (5.3.21)$$

Suppose that, on the contrary, there exists no $\lambda \in \mathbb{R}^m_+$ such that (5.3.21) holds. Then, for any $\lambda \in \mathbb{R}^m_+$, we have the following from part (i) of Lemma 5.1,

$$
\begin{aligned}
\theta_p^{LE} &\geq d_p^{LE}(\lambda) \\
&= \min_{x \in X} Q_p(x, \lambda) \\
&= \min\{\min_{x \in S_0} Q_p(x, \lambda), \min_{x \in X \setminus S_0} Q_p(x, \lambda)\} \\
&= \min_{x \in X \setminus S_0} Q_p(x, \lambda) \\
&> \min_{x \in X \setminus S_0} \max\{f(x), \lambda_1 g_1(x), \ldots, \lambda_m g_m(x)\} - \frac{1}{p} \ln(m+1) \\
&\geq \min_{x \in X \setminus S_0} \max\{\lambda_1 g_1(x), \ldots, \lambda_m g_m(x)\} - \frac{1}{p} \ln(m+1). \quad (5.3.22)
\end{aligned}
$$

Setting $\lambda_i = \gamma$, $i = 1, \ldots, m$, in (5.3.22), we get the following from (5.3.20),

$$
\theta_p^{LE} > \mu\gamma - \frac{1}{p} \ln(m+1).
$$

When γ is larger than $[\frac{1}{p} \ln(m+1) + f^*]/\mu$ in the above inequality, we get a contradiction to (5.3.19). Therefore, there must exist a $\bar{\lambda} \in \mathbb{R}^m_+$ such that (5.3.21) holds. We thus have the following from part (ii) of Lemma 5.1,

$$
\begin{aligned}
\theta_p^{LE} &\geq d_p^{LE}(\bar{\lambda}) \\
&= \min\{\min_{x \in S_0} Q_p(x, \bar{\lambda}), \min_{x \in X \setminus S_0} Q_p(x, \bar{\lambda})\} \\
&= \min_{x \in S_0} Q_p(x, \bar{\lambda}) \\
&> \min_{x \in S_0} f(x) - \frac{1}{p} \ln(m+1) \\
&= f^* - \frac{1}{p} \ln(m+1). \quad (5.3.23)
\end{aligned}
$$

Combining (5.3.19) with (5.3.23) yields the following

$$
-\frac{1}{p} \ln(m+1) + f^* < \theta_p^{LE} \leq f^*, \quad \text{for any } p > 0. \quad (5.3.24)
$$

The proof of the theorem follows from (5.3.24) by taking $p \to \infty$. □

THEOREM 5.5 *If $p > \frac{\ln(m+1)}{\delta}$, then any optimal solution x^* of (5.3.12) satisfying $x^* \in S_0$ is an optimal solution of (P_0).*

Proof. From part (ii) of Lemma 5.1, we have

$$\begin{aligned} f(x^*) &< Q_p(x^*, \lambda) + \frac{1}{p}\ln(m+1) \\ &= \min_{x\in X} Q_p(x, \lambda) + \frac{1}{p}\ln(m+1) \\ &\le \min_{x\in S} f(x) + \frac{1}{p}\ln(m+1) \\ &= f^* + \frac{1}{p}\ln(m+1). \end{aligned}$$

If $p > \frac{\ln(m+1)}{\delta}$, then

$$f(x^*) - f^* < \delta.$$

Since $x^* \in S_0$, we have $x^* \in S^*$ from the definition of δ. □

We can conclude from Theorem 5.4 that the logarithmic-exponential dual formulation possesses an asymptotic strong duality property. Furthermore, we can conclude from Theorem 5.5 that a successful dual search can be achieved for a sufficiently large p provided that primal feasibility holds. How to guarantee primal feasibility will be discussed in the next section.

5.4 Generalized Nonlinear Lagrangian Theory for Singly-Constrained Nonlinear Integer Programming Problems

We consider in this section the singly-constrained case of problem (P_0) where $m = 1$. Note that an integer programming problem with multiple constraints can be always converted into an equivalent singly-constrained problem by some nonlinear surrogate constraint methods discussed in Chapter 4.

From the analysis in the last section, a generalized Lagrangian function (GLF) should satisfy the followings: i) For any $x \in X \setminus S_0$, GLF tends to infinity as λ tends to infinity; and ii) for any $x \in S_0$, GLF does not depend on $g(x)$ when parameter p is sufficiently large. If we let GLF converge to $f(x)$ as parameter p becomes sufficiently large, then the GLF will not depend on $g(x)$. Now we introduce the definition of GLF.

DEFINITION 5.1 *A continuous function $L_p(g(x), f(x), \lambda)$ with parameters $p > 0$ and $\lambda > 0$ is called a generalized Lagrangian function (GLF) of problem (P_0) if it satisfies the following two conditions:*

(i) *For any $x \in S_0$, $L_p(g(x), f(x), \lambda) \to f(x)$ as $p \to \infty$.*

(ii) *For any $x \in X \setminus S_0$, $L_p(g(x), f(x), \lambda) \to +\infty$ as $\lambda \to \infty$.*

The following are examples of GLF that satisfy two conditions in Definition 5.1:

$$L_p(g(x), f(x), \lambda) = \frac{1}{p} \ln \left[\frac{1}{2} \left(\exp(pf(x)) + \exp(p\lambda g(x)) \right) \right], \ \lambda \geq 0 \tag{5.4.1}$$

$$L_p(g(x), f(x), \lambda) = f(x) + \frac{1}{\lambda} \exp(\lambda g(x)), \ \lambda \geq p > 0, \tag{5.4.2}$$

$$L_p(g(x), f(x), \lambda) = f(x) + \frac{\lambda}{p} \ln \left[1 + \exp(\lambda g(x)) \right], \tag{5.4.3}$$

$$L_p(g(x), f(x), \lambda) = \left[f(x)^p + \exp(p\lambda g(x)) \right]^{\frac{1}{p}}. \tag{5.4.4}$$

The conventional linear Lagrangian function $L(x, \lambda) = f(x) + \lambda g(x)$ is not a GLF, since the condition (i) of Definition 5.1 is unsatisfied, i.e., $L(x, \lambda) \not\to f(x)$ for any $x \in S_0$. We present some properties of a GLF in the following lemma without proof, since they are clear from the definition of the generalized Lagrangian function.

LEMMA 5.3 (i) *For a given $x \in S_0$ and any $\varepsilon > 0$, there exists a $p(x, \varepsilon) > 0$ such that for $p > p(x, \varepsilon)$,*

$$f(x) - \varepsilon \leq L_p(g(x), f(x), \lambda) \leq f(x) + \varepsilon. \tag{5.4.5}$$

(ii) *For a given $x \in X \setminus S_0$ and any $M > 0$, there exists a $\lambda(x, M) > 0$ such that for $\lambda > \lambda(x, M)$,*

$$L_p(g(x), f(x), \lambda) \geq M. \tag{5.4.6}$$

The GLF-based Lagrangian relaxation problem associated with (P_0) is defined as

$$d_p^{GLF}(\lambda) = \min_{x \in X} L_p(g(x), f(x), \lambda). \tag{5.4.7}$$

Furthermore, the GLF-based Lagrangian dual problem associated with (P_0) is defined as

$$\theta_p^{GLF} = \max_{\lambda > 0} d_p^{GLF}(\lambda). \tag{5.4.8}$$

Now we prove the asymptotic strong duality property of the generalized Lagrangian formulation given in (5.4.7) and (5.4.8). For simplicity, denote

$$f^* = \min_{x \in S_0} f(x).$$

From Assumption 5.3, we have $f^* > 0$.

THEOREM 5.6 *(Asymptotic Strong Duality) Suppose that $L_p(g(x), f(x), \lambda)$ is a GLF and θ_p^{GLF} is defined by (5.4.7) and (5.4.8). Then*

$$\lim_{p\to\infty} \theta_p^{GLF} = f^*.$$

Proof. If $S_0 = X$, then $\lim_{p\to\infty} \theta_p^{GLF} = \min_{x\in S_0} f(x)$ holds trivially by (5.4.7), (5.4.8) and part (i) of Lemma 5.3. Now suppose $X \setminus S_0 \neq \emptyset$. Again from part (i) of Lemma 5.3, for any $\varepsilon > 0$ and sufficiently large p, we have

$$\begin{aligned} \theta_p^{GLF} &= \max_{\lambda>0} \min_{x\in X} L_p(g(x), f(x), \lambda) \\ &\leq \max_{\lambda>0} \min_{x\in S_0} L_p(g(x), f(x), \lambda) \\ &\leq \max_{\lambda>0} \min_{x\in S_0} (f(x) + \varepsilon) \\ &= f^* + \varepsilon. \end{aligned} \tag{5.4.9}$$

Now we assert that for any sufficiently large $p > 0$, there exists a $\lambda > 0$ such that

$$\min_{x\in X\setminus S_0} L_p(g(x), f(x), \lambda) \geq \min_{x\in S_0} L_p(g(x), f(x), \lambda). \tag{5.4.10}$$

Suppose that, on the contrary, there exists no $\lambda > 0$ such that (5.4.10) holds. Then, for any $\lambda > 0$, we have

$$\begin{aligned} \theta_p^{GLF} &\geq d_p^{GLF}(\lambda) \\ &= \min\{\min_{x\in X\setminus S_0} L_p(g(x), f(x), \lambda), \min_{x\in S_0} L_p(g(x), f(x), \lambda)\} \\ &= \min_{x\in X\setminus S_0} L_p(g(x), f(x), \lambda). \end{aligned} \tag{5.4.11}$$

Let $M = f^* + 2\varepsilon$. From part (ii) of Lemma 5.3, $\forall x \in X\setminus S_0$, there exists a $\hat{\lambda} > 0$ such that $L_p(g(x), f(x), \hat{\lambda}) \geq f^* + 2\varepsilon$. Setting $\lambda = \hat{\lambda}$, we get from (5.4.11) that

$$\theta_p^{GLF} \geq \min_{x\in X\setminus S_0} L_p(g(x), f(x), \hat{\lambda}) \geq f^* + 2\varepsilon. \tag{5.4.12}$$

Equation (5.4.12) shows a contradiction to (5.4.9). Therefore, there must exist a $\tilde{\lambda} > 0$ such that (5.4.10) holds. In views of part (i) of Lemma 5.3 and (5.4.10), we have

$$\begin{aligned} \theta_p^{GLF} &\geq d_p^{GLF}(\tilde{\lambda}) \\ &= \min\{\min_{x\in X\setminus S_0} L_p(g(x), f(x), \tilde{\lambda}), \min_{x\in S_0} L_p(g(x), f(x), \tilde{\lambda})\} \\ &= \min_{x\in S_0} L_p(g(x), f(x), \tilde{\lambda}) \\ &\geq f^* - \varepsilon. \end{aligned} \tag{5.4.13}$$

Combining (5.4.9) with(5.4.13) yields that for any $\varepsilon > 0$ and sufficiently large $p > 0$, we have

$$f^* - \varepsilon \le \theta_p^{GLF} \le f^* + \varepsilon.$$

This comes to the conclusion. □

Theorem 5.6 reveals that the optimal value of a generalized nonlinear Lagrangian dual problem attains the optimal value of primal problem (P_0) when p approaches infinity. In implementation, we are more interested in achieving the primal optimality with a finite p. Once the parameter p exceeds a threshold, an optimal solution of primal problem (P_0) can be identified by the generalized nonlinear Lagrangian formulation. For convenience, the following two notations are introduced,

$$S_0^* = \{x \in S_0 \mid f(x) = f^*\},$$
$$\delta = \min\{f(x) \mid x \in S_0 \backslash S_0^*\} - f^*.$$

LEMMA 5.4 *There exists a $p^* > 0$ such that for any $p > p^*$, any optimum solution x^* of (5.4.7) satisfying $x^* \in S_0$ is an optimal solution of problem (P_0).*

Proof. In view of part (i) of Lemma 5.3 and Theorem 5.6, given $\varepsilon = \frac{\delta}{4}$, there exists p^* such that for any $p > p^*$,

$$f(x^*) - \varepsilon \le L_p(g(x^*), f(x^*), \lambda) = \min_{x \in X} L_p(g(x), f(x), \lambda) \le f^* + \varepsilon.$$

Hence

$$f(x^*) - f^* \le 2\varepsilon = \frac{\delta}{2}.$$

This implies $x^* \in S_0^*$ by the definition of δ. □

Notice that the dual function in the traditional linear Lagrangian formulation is concave, thus possessing the unimodality. As witnessed in Lemma 5.2, the dual function in the logarithmic-exponential dual formulation is continuous piecewise convex. Thus, the dual function in nonlinear Lagrangian formulations in general is not concave. We will show now the unimodality of the dual function for generalized Lagrangian functions. The property of the unimodality is important since it guarantees that the local maximum of the dual function is also a global maximum, thus facilitating the dual search.

It is clear that the monotonically increasing property of a nonlinear Lagrangian function with respect to both $f(\cdot)$ and $g(\cdot)$ is another desirable feature of nonlinear Lagrangian functions. Attaching this property to the definition of a GLF leads to the definition of a regular GLF.

DEFINITION 5.2 *A GLF is called regular if it satisfies the following additional three conditions:*

(i) *For any $x \in X \setminus S_0$, $L_p(g(x), f(x), \lambda)$ is strictly increasing with respect to λ;*

(ii) *For given $\lambda > 0$, $L_p(g(x), f(x), \lambda)$ is strictly increasing with respect to both $g(x)$ and $f(x)$;*

(iii) *For any $x \in S_0$, $L_p(g(x), f(x), \lambda)$ is decreasing with respect to λ.*

It is easy to verify that the nonlinear Lagrangian functions in (5.4.1), (5.4.3) and (5.4.4) are all regular, while the nonlinear Lagrangian function in (5.4.2) is regular when parameter p exceeds a certain threshold.

Let w be the perturbation function of (P_0). Let $\Phi_c = \{(c_i, f_i) \mid i = 1, \ldots, K\}$ be the set of corner points of w. Without loss of generality, we can assume that

$$c_1 < c_2 < \ldots < c_{K_0} \leq 0 < c_{K_0+1} < \ldots < c_K. \tag{5.4.14}$$

By Assumption 5.3, we have $K_0 \geq 1$. By the monotonicity of w and Assumption 5.3, we have

$$f_1 > f_2 > \ldots > f_{K_0} > f_{K_0+1} > \ldots > f_K > 0. \tag{5.4.15}$$

Note that the point c_i is associated with a feasible solution of problem (P_0) when $1 \leq i \leq K_0$ and with an infeasible solution of problem (P_0) when $K_0 + 1 \leq i \leq K$. The following lemma follows directly from Lemma 3.2 and its proof is omitted.

LEMMA 5.5 (i) *For any $p > 0$, if x^* is an optimum solution of (5.4.7) for a given $\lambda > 0$, then $(g(x^*), f(x^*)) \in \Phi_c$.*
(ii) *For a noninferior solution $x^* \in S_0^*$, there is a corresponding point $(g(x^*), f(x^*)) \in \Phi_c$.*

Let

$$l_p^i(\lambda) = L_p(c_i, f_i, \lambda), \ i = 1, \ldots, K. \tag{5.4.16}$$

Then by Lemma 5.5, we have

$$d_p^{GLF}(\lambda) = \min_{x \in X} L_p(g(x), f(x), \lambda) = \min_{1 \leq i \leq K} l_p^i(\lambda). \tag{5.4.17}$$

The theorem below reveals that the dual function defined by (5.4.7) is a unimodal function. Denote $I = I_1 \cup I_2$ where $I_1 = \{i \mid 1 \leq i \leq K_0\}$ and $I_2 = \{i \mid K_0 + 1 \leq i \leq K + 1\}$.

THEOREM 5.7 *Suppose that $L_p(g(x), f(x), \lambda)$ is a regular GLF. Then, for any $p > 0$, there exists a $\lambda^*(p) > 0$ such that the dual function $d_p^{GLF}(\lambda)$ is monotonically increasing in $[0, \lambda^*(p)]$ and monotonically decreasing in $[\lambda^*(p), \infty)$.*

Proof. First, we prove that for any $p > 0$, if there exists $\lambda_1 > 0$ such that $d_p^{GLF}(\lambda_1) = l_p^{i_1}(\lambda_1)$ where $i_1 \in I_1$, then for any $\lambda_2 > \lambda_1$, we must have $d_p^{GLF}(\lambda_2) = l_p^{i_2}(\lambda_2)$ satisfying $i_2 \in I_1$. Suppose, on the contrary, that there exists a $\lambda_2 > \lambda_1$ such that

$$\begin{aligned} d_p^{GLF}(\lambda_2) &= \min_{i \in I} l_p^i(\lambda_2) \\ &= \min\{\min_{i \in I_1} l_p^i(\lambda_2), \min_{i \in I_2} l_p^i(\lambda_2)\} \\ &= \min_{i \in I_2} l_p^i(\lambda_2) \\ &= l_p^{i_2}(\lambda_2),\ i_2 \in I_2. \end{aligned}$$

Then for any $i \in I_1$,

$$d_p^{GLF}(\lambda_2) = l_p^{i_2}(\lambda_2) \le l_p^i(\lambda_2). \tag{5.4.18}$$

From (iii) of Definition 5.2, $L_p(g(x), f(x), \lambda)$ is decreasing about λ when $x \in S_0$. Hence, for given $i_1 \in I_1$, we have

$$l_p^{i_1}(\lambda_2) \le l_p^{i_1}(\lambda_1). \tag{5.4.19}$$

Since $d_p^{GLF}(\lambda_1) = l_p^{i_1}(\lambda_1)$ where $i_1 \in I_1$, then for given $i_2 \in I_2$ in (5.4.18) we have

$$l_p^{i_1}(\lambda_1) \le l_p^{i_2}(\lambda_1). \tag{5.4.20}$$

Combining (5.4.18)–(5.4.20), we obtain

$$l_p^{i_1}(\lambda_1) \le l_p^{i_2}(\lambda_1) < l_p^{i_2}(\lambda_2) \le l_p^{i_1}(\lambda_1), \tag{5.4.21}$$

where the second inequality holds from (i) of Definition 5.2 and $\lambda_1 < \lambda_2$. This is a contradiction.

In the same way, we can also assert that for given $p > 0$, if there exists $\lambda_1 > 0$ such that $d_p^{GLF}(\lambda_1) = l_p^{i_1}(\lambda_1), i_1 \in I_2$, then for any $0 < \lambda_2 < \lambda_1$, we must have $d_p^{GLF}(\lambda_2) = l_p^{i_2}(\lambda_2), i_2 \in I_2$.

The conclusions above imply that there exists $\lambda^*(p) > 0$ such that for any $\lambda \in [0, \lambda^*(p)]$, $d_p^{GLF}(\lambda) = l_p^i(\lambda)$ with $i \in I_2$ and for any $\lambda \in [\lambda^*(p), \infty)$, $d_p^{GLF}(\lambda) = l_p^i(\lambda)$ with $i \in I_1$. Since the function $l_p^i(\lambda)$ corresponding to $i \in I_2$ is increasing by (i) of Definition 5.2 and that corresponding to $i \in I_1$ is decreasing by (iii) in Definition 5.2, the theorem is true. □

From Theorem 5.7, we can immediately obtain a corollary as follows.

COROLLARY 5.2 *For any $p > 0$, the dual problem (5.4.8) has a unique finite solution $\lambda^*(p)$.*

We now focus on how to obtain a primal optimum solution of problem (P_0) by solving a Lagrangian relaxation problem. The theorem below reveals that no actual dual search is needed when p is large enough.

LEMMA 5.6 *Suppose that $L_p(g(x), f(x), \lambda)$ is a regular GLF. Then, for any $p > 0$ and the corresponding $\lambda^*(p)$, there exists at least one optimal solution x^* of problem $\min_{x \in X} L_p(g(x), f(x), \lambda^*(p))$ such that x^* is a primal feasible solution of problem (P_0).*

Proof. Suppose that the optimal solutions of (5.4.7) corresponding to $\lambda^*(p)$ are all primal infeasible. Then, we have

$$\eta = \min_{x \in S_0} L_p(g(x), f(x), \lambda^*(p)) - \min_{x \in X \setminus S_0} L_p(g(x), f(x), \lambda^*(p)) > 0.$$

For any $x \in S_0$, by the continuity of $L_p(g(x), f(x), \lambda)$, there exists an $\varepsilon_1 > 0$ such that for any $0 \leq \varepsilon \leq \varepsilon_1$,

$$L_p(g(x), f(x), \lambda^*(p) + \varepsilon) > L_p(g(x), f(x), \lambda^*(p)) - \frac{\eta}{2},$$

which implies

$$\min_{x \in S_0} L_p(g(x), f(x), \lambda^*(p) + \varepsilon) > \min_{x \in S_0} L_p(g(x), f(x), \lambda^*(p)) - \frac{\eta}{2}. \quad (5.4.22)$$

Similarly, there exists an $\varepsilon_2 > 0$ such that for any $0 \leq \varepsilon \leq \varepsilon_2$,

$$\min_{x \in X \setminus S_0} L_p(g(x), f(x), \lambda^*(p) + \varepsilon) < \min_{x \in X \setminus S_0} L_p(g(x), f(x), \lambda^*(p)) + \frac{\eta}{2}. \quad (5.4.23)$$

Notice that

$$\min_{x \in S_0} L_p(g(x), f(x), \lambda^*(p)) - \frac{\eta}{2} = \min_{x \in X \setminus S_0} L_p(g(x), f(x), \lambda^*(p)) + \frac{\eta}{2}. \quad (5.4.24)$$

Choose an ε satisfying $0 < \varepsilon < \min\{\varepsilon_1, \varepsilon_2\}$. Then we have from (5.4.22), (5.4.23) and (5.4.24) that

$$\min_{x \in S_0} L_p(g(x), f(x), \lambda^*(p) + \varepsilon) > \min_{x \in X \setminus S_0} L_p(g(x), f(x), \lambda^*(p) + \varepsilon).$$

Since $L_p(g(x), f(x), \lambda)$ is regular, for $x \in X \setminus S_0$, we have

$$L_p(g(x), f(x), \lambda^*(p) + \varepsilon) > L_p(g(x), f(x), \lambda^*(p)).$$

Thus,

$$\begin{aligned}
d_p^{GLF}(\lambda^*(p)+\varepsilon) &= \min_{x\in X} L_p(g(x), f(x), \lambda^*(p)+\varepsilon) \\
&= \min_{x\in X\setminus S_0} L_p(g(x), f(x), \lambda^*(p)+\varepsilon) \\
&> \min_{x\in X\setminus S_0} L_p(g(x), f(x), \lambda^*(p)) \\
&= \min_{x\in X} L_p(g(x), f(x), \lambda^*(p)) \\
&= d_p^{GLF}(\lambda^*(p)).
\end{aligned}$$

This is a contradiction to the optimality of $\lambda^*(p)$ in problem (5.4.8). □

LEMMA 5.7 *Suppose that $L_p(g(x), f(x), \lambda)$ is a regular GLF. For any $p > 0$, any optimal solution of (5.4.7) with $\lambda > \lambda^*(p)$ is primal feasible for problem (P_0).*

Proof. From Lemma 5.6, there must exist an optimal solution x^* of (5.4.7) with $\lambda = \lambda^*(p)$ that is primal feasible. Since $g(x^*) \le 0$, for $\lambda > \lambda^*(p)$, we have

$$L_p(g(x^*), f(x^*), \lambda) \le L_p(g(x^*), f(x^*), \lambda^*(p)). \tag{5.4.25}$$

Since L_p is regular, for any $x \in X\setminus S_0$ and $\lambda > \lambda^*(p)$, we have

$$\begin{aligned}
L_p(g(x^*), f(x^*), \lambda^*(p)) &\le L_p(g(x), f(x), \lambda^*(p)) \\
&< L_p(g(x), f(x), \lambda).
\end{aligned} \tag{5.4.26}$$

Combining (5.4.25) with(5.4.26), we obtain

$$L_p(g(x^*), f(x^*), \lambda) < L_p(g(x), f(x), \lambda), \ \forall x \in X\setminus S_0.$$

Thus, any optimal solution of (5.4.7) must be primal feasible when $\lambda > \lambda^*(p)$. □

THEOREM 5.8 *Suppose that $L_p(g(x), f(x), \lambda)$ is regular. For sufficiently large p and $\lambda > \lambda^*(p)$, any optimal solution of problem (5.4.7) is a primal optimal solution of problem (P_0).*

Proof. This conclusion can be obtained directly from Lemmas 5.4 and 5.7. □

Let us now concentrate again on the logarithmic-exponential Lagrangian function discussed in the previous section.

COROLLARY 5.3 *If $p > \frac{\ln(2)}{\delta}$ and $\lambda > \lambda^*(p)$, then any solution that minimizes the logarithmic-exponential Lagrangian function given in (5.3.12) is also a primal optimal solution to (P_0).*

Proof. Note $m = 1$ in singly constrained cases. From the assumptions of $p > \frac{\ln(2)}{\delta}$ and $\lambda > \lambda^*(p)$, this corollary is proven by combining Theorems 5.5 and 5.8. □

An upper bound of $\lambda^*(p)$ can be estimated for the logarithmic-exponential Lagrangian formulation as follows. Let

$$\begin{aligned} \bar{f} &= \max_{x \in X} f(x), \\ \underline{g} &= \min_{x \in X \setminus S_0} g(x). \end{aligned}$$

It can be verified that any optimal solution of (5.3.12) is primal feasible when $\lambda \geq \bar{f}/\underline{g}$.

THEOREM 5.9 *If $\lambda \geq \bar{f}/\underline{g}$, then any optimal solution of (5.3.12) is primal feasible.*

Proof. Suppose that, by contradiction, an optimal solution, $\bar{x}$, to (5.3.12) is infeasible for $\lambda \geq \bar{f}/\underline{g}$. Then, we have $g(\bar{x}) > 0$. For any fixed $x \in S_0$, since $g(x) \leq 0$ and $f(\bar{x}) > 0$, we get

$$\begin{aligned} Q_p(\bar{x}, \lambda) &= 1/p \ln[1/2(\exp(pf(\bar{x})) + \exp(p\lambda g(\bar{x})))] \\ &\geq 1/p \ln[1/2(\exp(pf(\bar{x})) + \exp(p\bar{f}g(\bar{x})/\underline{g}))] \\ &\geq 1/p \ln[1/2(\exp(pf(\bar{x})) + \exp(p\bar{f}))] \\ &> 1/p \ln[1/2(\exp(p\lambda g(x)) + \exp(pf(x)))] \\ &= Q_p(x, \lambda). \end{aligned}$$

This is a contradiction to the optimality of $\bar{x}$ in (5.3.12). □

Thus, if $p > \frac{\ln(2)}{\delta}$ and $\lambda \geq \bar{f}/\underline{g}$, then any optimal solution of (5.3.12) is also an optimal solution of (P_0).

It can be concluded that in single-constraint cases, no actual dual search is necessary in the generalized nonlinear Lagrangian formulation if the values of p and λ are chosen sufficiently large.

EXAMPLE 5.4 Consider the following nonlinearly constrained convex integer programming problems

$$\begin{aligned} &\min \ f(x) \\ &\text{s.t. } g_i(x) \leq 0, \ i = 1, \ldots, m, \\ &\quad x \in X = \{x \in \mathbb{Z}^n \mid Ax \leq b\}, \end{aligned} \tag{5.4.27}$$

where $f, g_i, i = 1, \ldots, m$, are nonlinear convex functions, A is an $l \times n$ matrix and $b \in \mathbb{R}^l$. Let $X = \{x \in \mathbb{Z}^n \mid Ax \le b\}$. Assume that X is a finite set and $f(x) > 0$ for $x \in X$. Dualizing the nonlinear constraints $g_i(x) \le 0$ in problem (5.4.27), $i = 1, \ldots, m$, the logarithmic-exponential dual function is formed as

$$\begin{aligned} d_p^{LE}(\lambda) &= \min_{x \in X} Q_p(x, \lambda) \qquad (5.4.28) \\ &= \min_{x \in X} \frac{1}{p} \ln[\frac{1}{m+1}(\exp(pf(x)) + \sum_{i=1}^{m} \exp(p\lambda_i g_i(x))]. \end{aligned}$$

By Lemma 5.2 (iii) and the definition of X, (5.4.28) is a linearly constrained convex integer programming problem, for which various algorithms have been developed by exploiting the linear structure of the constraint set X (see [81] [87]).

Consider an instance of (5.4.27),

$$\begin{aligned} \min\ & f(x) = (x_1 - 2)^2 + (x_2 - 3)^2 + 1 \qquad (5.4.29) \\ \text{s.t. }\ & g_1(x) = x_1^2 + x_2^4 - 25 \le 0, \\ & g_2(x) = -x_1 + 2x_2 \le 4, \\ & g_3(x) = 2x_1 - x_2 \le 4, \\ & x \in X = [0, 4]^2 \cap \mathbb{Z}^2. \end{aligned}$$

It can be easily seen that $\delta \ge 1$, $\underline{g} \ge 1$ and $\bar{f} \le 14$. So $\frac{\ln(2)}{\delta} \le \ln(2) \approx 0.6931$ and $\bar{f}/\underline{g} \le 14$. If we take $p = 0.7$ and $\lambda = 15$ in problem (5.4.28) associated with (5.4.29), then solving (5.4.28) generates an optimal solution $x^* = (2, 2)^T$ with $f(x^*) = 2$. Since $p > \frac{\ln(2)}{\delta}$ and $\lambda \ge \bar{f}/\underline{g}$, $x^* = (2, 2)^T$ is also an optimal solution to problem (5.4.29).

EXAMPLE 5.5 Consider the following example:

$$\begin{aligned} \min\ & f(x) = x_1^2 - 8x_1 + x_2 + 11 \\ \text{s.t. }\ & g_1(x) = x_1^2 + x_2^2 - 3x_1 - 4x_2 + 7 \le 18, \\ & g_2(x) = x_1^2 + x_2^2 + 6x_1 - 6x_2 + 10 \le 20, \\ & x \in X = [0, 10]^2 \cap \mathbb{Z}^2. \end{aligned}$$

The above example can be converted into the following equivalent singly-constrained problem using the p-norm surrogate constraint method with $p = 15$,

$$\begin{aligned} \min\ & f(x) = x_1^2 - 8x_1 + x_2 + 11 \qquad (5.4.30) \\ \text{s.t. }\ & g(x) = (\mu_1^{15} g_1(x)^{15} + \mu_2^{15} g_2(x)^{15})^{1/15} - 9.928 \le 0, \\ & x \in X = [0, 10]^2 \cap \mathbb{Z}^2. \end{aligned}$$

where $\mu = (0.5263, 0.4737)^T$.

We now construct a regular GLF for (5.4.30) as follows.

$$L_p(g(x), f(x), \lambda) = t_p[F(f(x)) + G(g(x))]. \qquad (5.4.31)$$

Taking $t_p(y) = y^{\frac{1}{p}}$, $F(f(x)) = f(x)^p$ and $G(g(x)) = \exp(p\lambda g(x))$ in (5.4.31) yields the following nonlinear Lagrangian formulation,

$$L_p(g(x), f(x), \lambda) = [f(x)^p + \exp(p\lambda g(x))]^{\frac{1}{p}}. \qquad (5.4.32)$$

Applying formulation (5.4.32) to (5.4.30), we obtain the following relaxation problem,

$$\begin{aligned} &\min \ \{(x_1^2 - 8x_1 + x_2 + 11)^p + \exp[p\lambda((\mu_1^{15} g_1(x)^{15} + \mu_2^{15} g_2(x)^{15})^{1/15} \\ &\quad - 9.928)]\}^{1/p} \qquad (5.4.33) \\ &\text{s.t. } x \in X = [0, 10]^2 \cap \mathbb{Z}^2. \end{aligned}$$

For $p \geq 2$, we can solve the example problem by solving (5.4.33) for any $\lambda \geq 2$ by a branch-and-bound procedure. The algorithm identifies the optimal solution $x^* = (2, 2)^T$ with $f(x^*) = 1$.

Taking $t_p(x) = \frac{1}{p}\ln(x)$, $F(f(x)) = \exp(pf(x))$ and $G(g(x)) = \exp(p\lambda g(x))$ in (5.4.31) yields the following nonlinear Lagrangian formulation,

$$L_p(g(x), f(x), \lambda) = \frac{1}{p}\ln[\exp(pf(x)) + \exp(p\lambda g(x))]. \qquad (5.4.34)$$

Applying (5.4.34) to (5.4.30), we have the following relaxation problem,

$$\begin{aligned} &\min \ \frac{1}{p}\ln\{\exp[p(x_1^2 - 8x_1 + x_2 + 11)] \\ &\quad + \exp[p\lambda((\mu_1^{15} g_1(x)^{15} + \mu_2^{15} g_2(x)^{15})^{1/15} - 9.928)]\} \qquad (5.4.35) \\ &\text{s.t. } x \in X = [0, 10]^2 \cap \mathbb{Z}^2. \end{aligned}$$

For any $p \geq 2$, we can solve the example problem by solving (5.4.35) for any $\lambda \geq 1$ by a branch-and-bound procedure. The algorithm identifies the optimal solution $x^* = (2, 2)^T$ with $f(x^*) = 1$.

EXAMPLE 5.6 Consider a redundancy optimization problem in a network system consisting of n subsystems. The reliability of the i-th subsystem is $R_i(x_i) = 1 - (1 - r_i)^{x_i}$, where x_i is the number of the same components in parallel in the i-th subsystem and $r_i \in (0, 1)$ is the given reliability of the component in the i-th subsystem. Also, denote by C(x) the total resource consumed when adopting decision x. Consider an instance of this reliability optimization problem with five elements and a single constraint,

$$\begin{aligned} \min\ Q(x) &= 1 - R_1R_2 - (1-R_2)R_3R_4 - (1-R_1)R_2R_3R_4 \\ &\quad - R_1(1-R_2)(1-R_3)R_4R_5 - (1-R_1)R_2R_3(1-R_4)R_5 \end{aligned} \tag{5.4.36}$$

$$\begin{aligned} \text{s.t. } C(x) &= x_1x_2 + 3x_2x_3 + 3x_2x_4 + x_1x_5 \le 28, \\ x &\in X = [1,6]^5 \cap \mathbb{Z}^5, \end{aligned}$$

where $r_1 = 0.7, r_2 = 0.85, r_3 = 0.75, r_4 = 0.8, r_5 = 0.9$.

Applying formulation (5.4.1) to (5.4.36) with $p \ge 3$, we can solve Example 5.6 for any $\lambda \ge 3$. Applying formulation (5.4.2) to (5.4.36) with any $p \ge 6.5$, we can get the optimal solution for any $\lambda \ge p$. And applying formulation (5.4.4) to (5.4.36) with $p \ge 2$, we can solve this problem for any $\lambda \ge 8$. These three algorithms all identify the optimal solution $x^* = (2,1,4,4,1)^T$ with $Q(x^*) = 0.000656$.

As witnessed from above examples, adoption of the GLF transfers an integer programming problem with nonlinear constraints into an equivalent integer programming problem with box constraints that is easier to solve than the original problem. Note that for a p and a λ that are sufficiently large, no dual search is needed. Thus, only one resulting nonlinear Lagrangian problem needs to be solved by a branch-and-bound method.

5.5 Notes

The p-th power Lagrangian method was first proposed for achieving zero duality gap in nonconvex optimization in [132]. It was extended to integer programming in [134][143]. Strong duality properties of nonlinear reformulations of integer programming were further investigated in [135]. The logarithmic-exponential dual formulation was proposed in [202] (see also [203]). Generalized nonlinear Lagrangian theory for singly-constrained nonlinear integer programming was further discussed in [185][229][230].

Chapter 6

NONLINEAR KNAPSACK PROBLEMS

In this chapter, we investigate the solution methods for *nonlinear knapsack problems* of the following form:

$$(NKP) \qquad \max\ f(x) = \sum_{j=1}^{n} f_j(x_j)$$

$$\text{s.t. } g(x) = \sum_{j=1}^{n} g_j(x_j) \le b,$$

$$x \in X = \{x \in \mathbb{Z}^n \mid l_j \le x_j \le u_j,\ j = 1, \ldots, n\},$$

where $l_j < u_j$, l_j and u_j are integer numbers for $j = 1, \ldots, n$, and f_j and g_j, $j = 1, \ldots, n$, are continuous functions that satisfy the following monotonicity assumptions: f_j and g_j are *increasing* functions on $[l_j, u_j]$ for $j = 1, \ldots, n$. We first study the singly constrained nonlinear knapsack problem in (NKP). When there are multiple constraints in a nonlinear knapsack problem, the problem is called a *multi-dimensional nonlinear knapsack programming problem.*

Note that problems with all f_j's and all g_j's *decreasing* can be reduced to (NKP) by introducing a variable transformation $x_j = -y_j$, $j = 1, \ldots, n$.

Problem (NKP) and its multi-dimensional extension have a variety of applications, including production planning ([237]), marketing ([149]), reliability networks ([205][217]) and capital budgeting ([155]). Since monotonicity often is a natural property, either explicitly or implicitly, in optimal resource allocation problems, the solution methods developed for (NKP) can be used to solve generalized resource allocation problems ([33][34][36][106]).

This chapter is organized as follows. In Section 6.1, we will discuss branch-and-bound methods based on the continuous relaxation for convex (NKP). In Section 6.2, we will investigate 0-1 linearization methods for convex (NKP).

A convergent Lagrangian and domain cut method for general (NKP) and its multi-dimensional extension will be investigated in Section 6.3. In Section 6.4, a solution method for concave (NKP) will be studied. As an application of the branch-and-bound method, we will discuss in Section 6.5 a special class of multi-dimensional knapsack problems: The redundancy optimization problem in series-parallel reliability networks. Extensive computational results will be presented in Section 6.6.

6.1 Continuous-Relaxation-Based Branch-and-Bound Methods

The conventional branch-and-bound method can be applied to (NKP) as long as the continuous relaxation subproblems can be solved correctly and efficiently.

Consider the continuous relaxation subproblem of (NKP):

$$(\overline{NKP}) \qquad \max\ f(x) = \sum_{j=1}^{n} f_j(x_j)$$
$$\text{s.t. } g(x) = \sum_{j=1}^{n} g_j(x_j) \le b,$$
$$\alpha_j \le x_j \le \beta_j,\ j = 1, \ldots, n,$$

where $l_j \le \alpha_j \le \beta_j \le u_j$, $j = 1, \ldots, n$.

In order to enable the use of efficient continuous relaxation procedures and to guarantee the convergence of the branch-and-bound method for (NKP), we need the following additional assumption:

ASSUMPTION 6.1 (i) *f_j and g_j are differentiable functions.*

(ii) *f_j is concave and g_j is convex on $[l_j, u_j]$ for all $j = 1, \ldots, n$.*

(iii) *For any subproblem $(\overline{NKP})$, $\nabla g(x)$ is nonzero at the optimal solution to $(\overline{NKP})$.*

Part (iii) in the above assumption is equivalent to the linear independence constraint qualification for $(\overline{NKP})$ which ensures that the KKT conditions are sufficient and necessary optimality conditions for $(\overline{NKP})$ under Assumption 6.1.

Under Assumption 6.1, (NKP) is a convex knapsack problem. Solution methods developed for general constrained optimization are applicable to solve $(\overline{NKP})$. The optimal solution obtained in $(\overline{NKP})$ can then be used in the branch-and-bound method for (NKP). Furthermore, it is possible to design more efficient approaches to solve $(\overline{NKP})$ by exploiting the separability and the property of a single constraint of problem (NKP). Multiplier search method

[34] and pegging method [36] (see also [35]) are two specialized methods for solving $(\overline{NKP})$.

6.1.1 Multiplier search method

6.1.1.1 KKT conditions

The Karush-Kuhn-Tucker conditions for $(\overline{NKP})$ can be expressed as

$$f_j'(x_j) - \lambda g_j'(x_j) + v_j - w_j = 0, \quad j = 1, \ldots, n, \tag{6.1.1}$$

$$\lambda(\sum_{j=1}^{n} g_j(x_j) - b) = 0, \tag{6.1.2}$$

$$v_j(\alpha_j - x_j) = 0, \quad j = 1, \ldots, n, \tag{6.1.3}$$

$$w_j(x_j - \beta_j) = 0, \quad j = 1, \ldots, n, \tag{6.1.4}$$

$$v_j \geq 0, \quad j = 1, \ldots, n, \tag{6.1.5}$$

$$w_j \geq 0, \quad j = 1, \ldots, n, \tag{6.1.6}$$

$$\lambda \geq 0, \tag{6.1.7}$$

$$\sum_{j=1}^{n} g_j(x_j) \leq b, \tag{6.1.8}$$

$$\alpha_j \leq x_j \leq \beta_j, \quad j = 1, \ldots, n. \tag{6.1.9}$$

Under Assumption 6.1, the KKT system (6.1.1)–(6.1.9) is necessary and sufficient optimality conditions for $(\overline{NKP})$.

It is observed that if $\alpha_j < x_j < \beta_j$, $j = 1, \ldots, n$, then (6.1.3) and (6.1.4) imply that $v_j = w_j = 0$, $j = 1, \ldots, n$, and thus

$$f_j'(x_j) - \lambda g_j'(x_j) = 0, \quad j = 1, \ldots, n. \tag{6.1.10}$$

For any given $\lambda \geq 0$, suppose that a unique optimal solution $\bar{x}_j(\lambda)$ exists to (6.1.10). Then x_j, v_j and w_j can be expressed as a function of $\lambda \geq 0$ in terms of $\bar{x}_j(\lambda)$.

$$x_j(\lambda) = \begin{cases} \alpha_j, & \bar{x}_j(\lambda) < \alpha_j, \\ \bar{x}_j(\lambda), & \alpha_j \leq \bar{x}_j(\lambda) \leq \beta_j, \\ \beta_j, & \bar{x}_j(\lambda) > \beta_j, \end{cases} \tag{6.1.11}$$

$$v_j(\lambda) = \begin{cases} -f_j'(\alpha_j) + \lambda g_j'(\alpha_j), & \bar{x}_j(\lambda) \leq \alpha_j, \\ 0, & \bar{x}_j(\lambda) > \alpha_j, \end{cases} \tag{6.1.12}$$

$$w_j(\lambda) = \begin{cases} 0, & \bar{x}_j(\lambda) < \beta_j, \\ f_j'(\beta_j) - \lambda g_j'(\beta_j), & \bar{x}_j(\lambda) \geq \beta_j. \end{cases} \tag{6.1.13}$$

It can be verified (see [34]) that the above solutions $x_j(\lambda)$, $v_j(\lambda)$ and $w_j(\lambda)$ satisfy the KKT conditions of $(\overline{NKP})$ except (6.1.2) and (6.1.8).

6.1.1.2 Multiplier search procedure

The following procedure searches for an optimal λ^* such that all the conditions of (6.1.1)–(6.1.9) are satisfied. We point out that the multiplier search method for solving $(\overline{NKP})$ is applicable to general separable integer problems without the monotonicity for f_j and g_j.

PROCEDURE 6.1 (MULTIPLIER SEARCH PROCEDURE FOR $(\overline{NKP})$)

Step 1. For $j = 1, \ldots, n$, solve equation $f_j'(x_j) = 0$ and obtain a solution $\bar{x}_j(0)$. Calculate $x_j(0)$ by (6.1.11), $j = 1, \ldots, n$. If $x(0)$ satisfies (6.1.8), then stop and $\lambda^* = 0$. Otherwise, go to Step 2.

Step 2. Obtain the expression of $\bar{x}_j(\lambda)$ in terms of λ by solving nonlinear equation (6.1.10), $j = 1, \ldots, n$.

Step 3. Obtain $x_j(\lambda)$ by using (6.1.11), $j = 1, \ldots, n$. Use some iterative root finding procedure to solve equation

$$\sum_{j=1}^{n} g_j(x_j(\lambda)) = b \tag{6.1.14}$$

for λ and obtain an optimal multiplier $\lambda^* > 0$. Stop and the optimal solution to $(\overline{NKP})$ is $x_j(\lambda^*)$, $j = 1, \ldots, n$.

It can be verified (see [34]) that the solution $x_j(\lambda^*)$ obtained in Procedure 6.1 is an optimal solution to $(\overline{NKP})$.

An ability in carrying out Step 2 of the above procedure is essential for the multiplier search method. Fortunately, in some applications, the solution $\bar{x}_j(\lambda)$ of (6.1.10) has a closed form as a function of λ.

(1) *Quadratic knapsack problem* ([66][98][155][171]).

$$(QP) \qquad \max\ f(x) = \sum_{j=1}^{n}(a_j x_j - \frac{1}{2} d_j x_j^2)$$

$$\text{s.t. } g(x) = \sum_{j=1}^{n} b_j x_j \le b,$$

$$x \in X = \{x \in \mathbb{Z}^n \mid l_j \le x_j \le u_j,\ j = 1, \ldots, n\},$$

where $d_j > 0$ and $b_j > 0$ for $j = 1, \ldots, n$. We have

$$\bar{x}_j(\lambda) = (a_j - \lambda b_j)/d_j, \quad j = 1, \ldots, n. \tag{6.1.15}$$

Note that problem (QP) does not necessarily possess monotonicity.

(2) *Stratified sampling* ([33][42]).

$$(SAMP) \qquad \max\ f(x) = D - \sum_{j=1}^{n} d_j/x_j$$
$$\text{s.t. } g(x) = \sum_{j=1}^{n} b_j x_j \le b,$$
$$x \in X = \{x \in \mathbb{Z}^n \mid l_j \le x_j \le u_j,\ j = 1, \ldots, n\},$$

where $d_j > 0$ and $b_j > 0$ for $j = 1, \ldots, n$, and $D > 0$ is a constant. We have

$$\bar{x}_j(\lambda) = \sqrt{d_j/(\lambda b_j)}, \quad j = 1, \ldots, n. \qquad (6.1.16)$$

(3) *Manufacturing capacity planning* ([36]).

$$(MCP) \qquad \min\ f(x) = \sum_{j=1}^{n} c_j x_j$$
$$\text{s.t. } g(x) = \sum_{j=1}^{n} b_j \left(\frac{\gamma_j}{x_j - \gamma_j} \right) \le b,$$
$$x \in X = \{x \in \mathbb{Z}^n \mid l_j \le x_j \le u_j,\ j = 1, \ldots, n\},$$

where $c_j > 0$, $b_j > 0$ and $0 < \gamma_j < l_j$ for $j = 1, \ldots, n$. We have

$$\bar{x}_j(\lambda) = \gamma_j + \sqrt{(\lambda b_j \gamma_j)/c_j}, \quad j = 1, \ldots, n. \qquad (6.1.17)$$

Note that problem (MCP) does not necessarily possess monotonicity.

(4) *Linearly constrained redundancy optimization problem in reliability network* ([205][217][219]).

$$(LCROP) \qquad \max\ f(x) = \prod_{j=1}^{n} (1 - (1 - r_j)^{x_j})$$
$$\text{s.t. } g(x) = \sum_{j=1}^{n} b_j x_j \le b,$$
$$x \in X = \{x \in \mathbb{Z}^n \mid l_j \le x_j \le u_j,\ j = 1, \ldots, n\},$$

where $0 < r_j < 1$, $b_j > 0$ and $l_j \geq 1$ for $j = 1, \ldots, n$. The problem is equivalent to the following separable form:

$$\max\ \tilde{f}(x) = \sum_{j=1}^{n} \ln(1 - (1 - r_j)^{x_j})$$

$$\text{s.t. } g(x) = \sum_{j=1}^{n} b_j x_j \leq b,$$

$$x \in X = \{x \in \mathbb{Z}^n \mid l_j \leq x_j \leq u_j,\ j = 1, \ldots, n\}.$$

We have

$$\bar{x}_j(\lambda) = \frac{\ln(\lambda b_j) - \ln(\lambda b_j - \ln(q_j))}{\ln(q_j)}, \tag{6.1.18}$$

where $q_j = 1 - r_j$.

(5) *Linear cost minimization in reliability network* ([52][217][219]).

$$(LCOST) \qquad \min\ f(x) = \sum_{j=1}^{n} c_j x_j$$

$$\text{s.t. } g(x) = \prod_{j=1}^{n} (1 - (1 - r_j)^{x_j}) \geq R_0$$

$$x \in X = \{x \in \mathbb{Z}^n \mid l_j \leq x_j \leq u_j,\ j = 1, \ldots, n\},$$

where $0 < r_j < 1$, $c_j > 0$ and $l_j \geq 1$ for $j = 1, \ldots, n$, $0 < R_0 < 1$. The problem can be transformed into the form of (NKP) by letting $y_j = u_j - x_j$:

$$\max\ \tilde{f}(y) = \sum_{j=1}^{n} c_j y_j$$

$$\text{s.t. } \tilde{g}(y) = -\sum_{j=1}^{n} \ln(1 - (1 - r_j)^{u_j - y_j}) \leq -\ln(R_0)$$

$$y \in Y = \{y \in \mathbb{Z}^n \mid 0 \leq y_j \leq u_j - l_j,\ j = 1, \ldots, n\}.$$

For the above equivalent problem, we have

$$\bar{y}_j(\lambda) = u_j - \frac{\ln(c_j) - \ln(c_j - \lambda \ln(q_j))}{\ln(q_j)}. \tag{6.1.19}$$

In some other applications, such as capacity planning in manufacturing networks ([26][27]) and chemical production service facilities ([159]), there is no explicit expression for $\bar{x}(\lambda)$. In those cases, equations (6.1.10) and (6.1.14) need to be solved numerically.

6.1.1.3 Branch-and-bound method

Although the multiplier search method for solving $(\overline{NKP})$ is applicable to general separable integer problems without the monotonicity for f_j and g_j, the reoptimization procedure (see [34]) for an efficient implementation of the branch-and-bound method based on the multiplier search procedure does require certain monotonicity on f_j and g_j. Moreover, the pegging method which we will introduce in the next subsection also requires such assumptions. Therefore, we will focus on nonlinear knapsack problems where f_j and g_j are increasing functions. Notice that the continuous relaxation problem $(\overline{NKP})$ with g_j increasing and f_j concave but *not* necessarily increasing, such as problem (QP), can be reduced into an equivalent problem where f_j is increasing for $j = 1, \ldots, n$. Assume that $(\overline{NKP})$ is feasible. Let x_j^{max} denote the maximizer of f_j over $\mathbb{R}$. If $x_j^{max} \le \alpha_j$, then f_j is decreasing on $[\alpha_j, \beta_j]$ and $x_j = \alpha_j$ is the optimal solution to $(\overline{NKP})$; If $x_j^{max} \ge \beta_j$, then f_j is increasing on $[\alpha_j, \beta_j]$; If $\alpha_j \le x_j^{max} \le \beta_j$, then f_j is increasing on $[\alpha_j, x_j^{max}]$ and resetting $\beta_j = \lfloor x_j^{max} \rfloor$ does not change the optimal solution to $(\overline{NKP})$.

We first consider the situations where the following monotonicity condition of $\bar{x}_j(\lambda)$ is satisfied:

ASSUMPTION 6.2 *$\bar{x}_j(\lambda)$ is decreasing in λ for $j = 1, \ldots, n$.*

It can be verified that problems (QP), $(SAMP)$ and $(LCROP)$ in the Subsection 6.1.1.2 satisfy Assumption 6.2.

In the branch-and-bound process for solving (NKP), let x_k be the fractional variable in the optimal solution to the parent subproblem. Let $x_j^p(\lambda)$, α_j^p, β_j^p, λ_p^* denote the values of $x_j(\lambda)$, α_j, β_j and λ^* in the parent subproblem problem. Denote also by $x_j^L(\lambda)$, α_j^L, β_j^L and λ_L^* for the left subproblem, and $x_j^R(\lambda)$, α_j^R, β_j^R and λ_R^* for the right subproblem. It can be shown (see [34]) that the monotonicity of f_j and g_j and Assumption 6.2 imply that

$$\lambda_L^* \le \lambda_p^* \le \lambda_R^*, \tag{6.1.20}$$

$$x_k^L(\lambda_L^*) = \beta_k^L, x_k^R(\lambda_R^*) = \alpha_k^R, \tag{6.1.21}$$

$$x_j^p(\lambda_p^*) = \beta_j^p \Rightarrow x_j^L(\lambda_L^*) = \beta_j^p,\ j = 1, \ldots, n,\ j \ne k, \tag{6.1.22}$$

$$x_j^p(\lambda_p^*) = \alpha_j^p \Rightarrow x_j^R(\lambda_R^*) = \alpha_j^p,\ j = 1, \ldots, n,\ j \ne k. \tag{6.1.23}$$

The properties in (6.1.20)–(6.1.23) can be used to improve the performance of the branch-and-bound method for (NKP) in two aspects: reducing the range of λ when searching for the optimal multiplier λ^* and fixing certain variables before solving the subproblem $(\overline{NKP})$.

Similar to Assumption 6.2, there are other cases of the problem structure that may help to improve the efficiency of the branch-and-bound method.

ASSUMPTION 6.3 *Assume that one of the following conditions holds for* $(\overline{NKP})$:
(i) $g_j(x_j)$ *is decreasing in* x_j *and* $\bar{x}_j(\lambda)$ *is increasing in* λ *for* $j = 1, \ldots, n$;
(ii) $g_j(x_j)$ *is decreasing in* x_j *and* $\bar{x}_j(\lambda)$ *is decreasing in* λ *for* $j = 1, \ldots, n$;
(iii) $g_j(x_j)$ *is increasing in* x_j *and* $\bar{x}_j(\lambda)$ *is increasing in* λ *for* $j = 1, \ldots, n$.

Notice that (MCP) in Subsection 6.1.1.2 satisfies Assumption 6.3 (i).

The reoptimization procedure for Case (i) in Assumption 6.3 is similar to (6.1.20)–(6.1.23) while Case (ii) and Case (iii) lead to two special optimal solutions to $(\overline{NKP})$: $(\beta_1, \ldots, \beta_n)^T$ and $(\alpha_1, \ldots, \alpha_n)^T$, respectively.

The performance of the branch-and-bound method for (NKP) can also be improved by using heuristic search procedures for generating good initial integer feasible solutions and searching for a better feasible integer solution starting from an incumbent solution. The heuristic scheme is of a greedy type based on the monotonicity of the problem. Given a feasible point $x = (x_1, \ldots, x_n)^T$ with $g(x) < b$, the next trial point is the feasible point with maximum ratio along the axis:

$$x + k_0 e_{j_0} = \arg \max_{\substack{j=1,\ldots,n \\ k \in \mathbb{Z}^+}} \left\{ \frac{f_j(x_j + k) - f_j(x_j)}{g_j(x_j + k) - g_j(x_j)} \mid g(x + k e_j) \leq b \right\}, \tag{6.1.24}$$

where e_j denotes the j-th unit vector in $\mathbb{R}^n$ and $\mathbb{Z}^+$ the set of positive integers.

PROCEDURE 6.2 (GENERAL HEURISTIC FOR (NKP))

Step 1. If there exist $k_0 > 0$ and $j_0 \in \{1, \ldots, n\}$ such that (6.1.24) holds, then set $x := x + k_0 e_{j_0}$.

Step 2. Repeat Step 1 until there is no $j \in \{1, \ldots, n\}$ satisfying $g(x + e_j) \leq b$.

Notice that Procedure 6.2 does not require any convexity assumption for (NKP).

Consider convex knapsack problems where Assumption 6.1 is satisfied. It is easy to see that for any fixed j, the ratio in (6.1.24) is nonincreasing on k. Therefore, (6.1.24) can be replaced by

$$x + e_{j_0} = \arg \max_{j=1,\ldots,n} \left\{ \frac{f_j(x_j + 1) - f_j(x_j)}{g_j(x_j + 1) - g_j(x_j)} \mid g(x + e_j) \leq b \right\}. \tag{6.1.25}$$

PROCEDURE 6.3 (HEURISTIC FOR CONVEX KNAPSACK PROBLEMS)

Step 1. If there exists $j_0 \in \{1, \ldots, n\}$ such that (6.1.25) holds, then set $x := x + e_{j_0}$.

Step 2. Repeat Step 1 until there is no $j \in \{1, \ldots, n\}$ satisfying $g(x + e_j) \leq b$.

6.1.2 Pegging method

The basic idea of the pegging method or the variable relaxation method for solving the continuous subproblem $(\overline{NKP})$ is to omit the bound constraint $\alpha_j \leq x_j \leq \beta_j$, $j = 1, \ldots, n$, thus obtaining an easier subproblem. Using the monotonicity of f_j and g_j, the subproblem without the bound constraint becomes

$$\max \sum_{j=1}^{n} f_j(x_j) \tag{6.1.26}$$
$$\text{s.t.} \sum_{j=1}^{n} g_j(x_j) = b.$$

The KKT conditions for problem (6.1.26) can be expressed as

$$f_j'(x_j) - \lambda g_j'(x_j) = 0, \quad j = 1, \ldots, n, \tag{6.1.27}$$
$$\sum_{j=1}^{n} g_j(x_i) = b. \tag{6.1.28}$$

Notice that (6.1.27)–(6.1.28) can be solved more efficiently than the KKT system for $(\overline{NKP})$ (ref. (6.1.1)–(6.1.9)). The optimal solution may even have a closed form expression. If the optimal solution to problem (6.1.26) satisfies the bound constraint $\alpha_j \leq x_j \leq \beta_j$, $j = 1, \ldots, n$, then it is also an optimal solution to $(\overline{NKP})$. Otherwise, we can fix the variables that violate the bounds at the lower bound or the upper bound and solve the modified bound relaxation problem iteratively and eventually find the optimal solution to $(\overline{NKP})$. At the k-th iteration, the bound relaxation problem is

$$(RP_k) \qquad \max \sum_{j \in J^k} f_j(x_j) + \sum_{j \in L^k} f_j(\alpha_j) + \sum_{j \in U^k} f_j(\beta_j)$$
$$\text{s.t.} \sum_{j \in J^k} g_j(x_j) = b^k,$$

where $b^k = b - \sum_{j \in L^k} g_j(\alpha_j) - \sum_{j \in U^k} g_j(\beta_j)$, J^k is the index set of free variables, L^k the index set of variables fixed at lower bound α_j and U^k the index set of variables fixed at upper bound β_j.

PROCEDURE 6.4 (PEGGING METHOD FOR $(\overline{NKP})$)

Step 1. Set $\lambda = 0$ and let x^0 be the solution to $\nabla f(x) = 0$. If $g(x^0) \leq b$, stop and x^0 is the optimal solution to $(\overline{NKP})$. Otherwise, set $k = 1$, $J^1 = \{1, 2, \ldots, n\}$, $L^1 = \emptyset$, $U^1 = \emptyset$, $S_A^k = 0$, $S_B^k = 0$.

Step 2. Solve (RP_k) to obtain an optimal solution $x^k = (x_1^k, x_2^k, \ldots, x_n^k)$.

Step 3. Calculate

$$J_A^k = \{j \in J^k \mid x_j^k < \alpha_j\},$$
$$J_B^k = \{j \in J^k \mid x_j^k > \beta_j\},$$
$$S_A^k = \sum_{j \in J_A^k} (g_j(\alpha_j) - g_j(x_j^k)),$$
$$S_B^k = \sum_{j \in J_B^k} (g_j(x_j^k) - g_j(\beta_j)).$$

If $J_A^k = \emptyset$ and $J_B^k = \emptyset$, go to Step 5.

Step 4. If $S_A^k \geq S_B^k$, set $J^{k+1} = J^k \setminus J_A^k$, $L^{k+1} = L^k \cup J_A^k$, $U^{k+1} = U^k$. Otherwise, if $S_A^k < S_B^k$, set $J^{k+1} = J^k \setminus J_B^k$, $U^{k+1} = U^k \cup J_B^k$, $L^{k+1} = L^k$. Set $k = k + 1$ and go to Step 2.

Step 5. Stop and x^c defined by

$$x_j^c = \begin{cases} \alpha_j, & j \in L^k, \\ x_j^k, & j \in J^k, \\ \beta_j, & j \in U^k \end{cases}$$

is the optimal solution to $(\overline{NKP})$.

By Step 4 of Procedure 6.4, at least one variable can be fixed at the lower bound or the upper bound at each iteration. Thus, the method will terminate in a finite number of iterations. The optimality of the solution x^c in Step 5 can be proved using the monotonicity of f_j, g_j and $\bar{x}_j(\lambda)$ (see [36] for details).

THEOREM 6.1 *Under Assumptions 6.1 and 6.2, Procedure 6.4 terminates in a finite number of iterations at an optimal solution to* $(\overline{NKP})$.

With minor modifications, the pegging procedure can also be applied to problems that satisfy Assumption 6.3 (i).

As in the case of multiplier search method, the efficiency of the branch-and-bound method using Procedure 6.4 may largely rely on how fast the subproblem (RP_k) can be solved. The following lists three cases where the optimal solution x^k to the subproblem (RP_k) can be expressed in a closed form.

(1) Problem (QP):

$$x_j^k = (a_j - \lambda^k b_j)/d_j, \quad j \in J^k,$$
$$\lambda^k = \frac{\sum_{j \in J^k}(b_j a_j / d_j) - b^k}{\sum_{j \in J^k}(b_j^2 / d_j)}.$$

(2) Problem (MCP):

$$x_j^k = \gamma_j + \sqrt{\lambda^k b_j \gamma_j / c_j}, \quad j \in J^k,$$

$$\lambda^k = \left(\frac{\sum_{j \in J^k} \sqrt{b_j \gamma_j c_j}}{b^k} \right)^2.$$

(3) Problem $(SAMP)$:

$$x_j^k = \sqrt{d_j / (\lambda^k b_j)}, \quad j \in J^k,$$

$$\lambda^k = \left(\frac{\sum_{j \in J^k} \sqrt{b_j d_j}}{b^k} \right)^2.$$

6.2 0-1 Linearization Method

In this section, we consider the convex case of (NKP), i.e., f_j is a concave function and g_j is a convex function on $[l_j, u_j]$ for all $j = 1, \ldots, n$ (Assumption 6.1 (ii)). Without loss of generality, we assume $l_j = 0$ and $f_j(0) = g_j(0) = 0$ for $j = 1, \ldots, n$. It turns out that problem (NKP) can be converted into a 0-1 linear integer programming problem by piecewise linear approximation on the integer grid of X. The converted equivalent problem can be then dealt with by techniques developed for 0-1 knapsack problem.

6.2.1 0-1 linearization

As shown in Figures 6.1 and 6.2, the concave function f_j and the convex function g_j can be approximated on $0 \le x_j \le u_j$ by their piecewise linear underestimation and piecewise linear overestimation, respectively.

Let $x_j = \sum_{i=1}^{u_j} x_{ij}$, $p_{ij} = f_j(i) - f_j(i-1)$, $a_{ij} = g_j(i) - g_j(i-1)$, $i = 1, \ldots, u_j$, $j = 1, \ldots, n$. By the monotonicity, it holds $p_{ij} \ge 0$ and $a_{ij} \ge 0$. Consider the following 0-1 linear knapsack problem:

$$(LKP) \qquad \max\ \psi(x) = \sum_{j=1}^{n} \sum_{i=1}^{u_j} p_{ij} x_{ij} \tag{6.2.1}$$

$$\text{s.t. } \phi(x) = \sum_{j=1}^{n} \sum_{i=1}^{u_j} a_{ij} x_{ij} \le b,$$

$$x_{ij} \in \{0, 1\}, i = 1, \ldots, u_j, \ j = 1, \ldots, n.$$

We have the following equivalence result.

THEOREM 6.2 *Under Assumption 6.1 (ii), (NKP) and (LKP) are equivalent under the transformation $x_j = \sum_{i=1}^{u_j} x_{ij}$.*

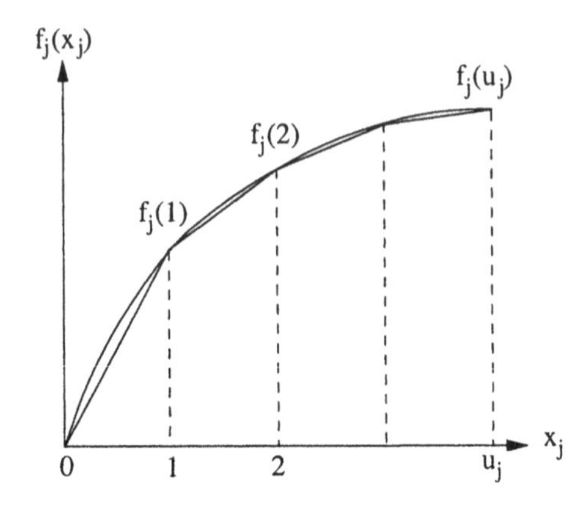

Figure 6.1. Linear approximation of $f_j(x_j)$.

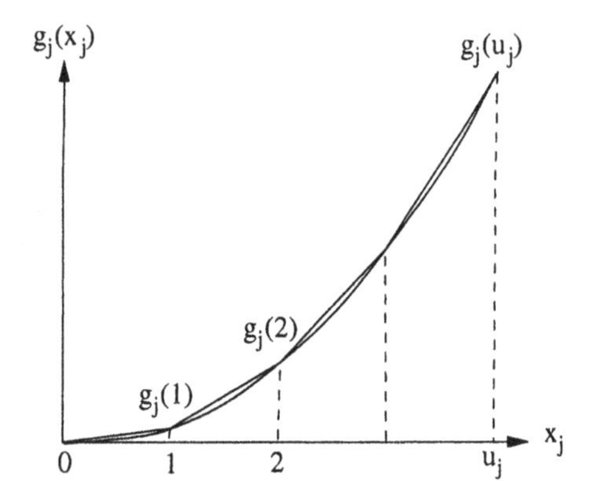

Figure 6.2. Linear Approximation of $g_j(x_j)$.

Proof. Notice that under transformation $x_j = \sum_{i=1}^{u_j} x_{ij}$, the functions ψ and ϕ take the same values as $f(x)$ and $g(x)$, respectively, on the integer points of X if for each j, there is no 1 after 0's in the 0-1 sequence $\{x_{ij}\}$. Thus, (LKP) is a relaxation of (NKP). By the monotonicity and concavity of f_j, we have $p_{1j} \geq p_{2j} \geq \cdots \geq p_{u_j j} \geq 0$ for $j = 1, \ldots, n$. Similarly, by the convexity of g_j, we have $0 \leq a_{1j} \leq a_{2j} \leq \cdots \leq a_{u_j j}$ for $j = 1, \ldots, n$. Thus, for the optimal solution x_{ij}^*, there must be no 1 after 0's in the sequence $\{x_{1j}^*, \ldots, x_{u_j j}^*\}$ for $j = 1, \ldots, n$. Therefore, (LKP) and (NKP) are equivalent. □

The property that there is no 1's after 0 in the optimal 0-1 sequence $\{x_{ij}^*\}$ for each j was exploited in [154][155] to derive a reduction process which can be used to compute tight lower and upper bounds on the integer variable x_i in (NKP).

EXAMPLE 6.1

$$\begin{aligned} \max\ & f(x) = 4x_1 - x_1^2 + 2x_2 + 7x_3 - x_3^2 \\ \text{s.t.}\ & g(x) = 2x_1^2 + x_2^2 + x_2 + 3x_3^2 \le 17, \\ & x \in X = \{x \in \mathbb{Z}^3 \mid 0 \le x_j \le 2,\ j = 1, 2, 3\}. \end{aligned}$$

We have in this example $p_{11} = 3$, $p_{21} = 1$, $p_{12} = 2$, $p_{22} = 2$, $p_{13} = 6$, $p_{23} = 4$, $a_{11} = 2$, $a_{21} = 6$, $a_{12} = 2$, $a_{22} = 4$, $a_{13} = 3$, $a_{23} = 9$. Therefore, the problem can be transformed into the following 0-1 linear knapsack problem:

$$\begin{aligned} \max\ & 3x_{11} + x_{21} + 2x_{12} + 2x_{22} + 6x_{13} + 4x_{23} \\ \text{s.t.}\ & 2x_{11} + 6x_{21} + 2x_{12} + 4x_{22} + 3x_{13} + 9x_{23} \le 17, \\ & x_{ij} \in \{0, 1\},\ i = 1, 2,\ j = 1, 2, 3. \end{aligned} \tag{6.2.2}$$

6.2.2 Algorithms for 0-1 linear knapsack problem

For the sake of simplicity, we rewrite (LKP) by

$$\begin{aligned} (LKP) \qquad \max\ & \sum_{j=1}^{N} p_j w_j \\ \text{s.t.}\ & \sum_{j=1}^{N} a_j w_j \le b, \\ & w_j \in \{0, 1\},\ j = 1, \ldots, N. \end{aligned}$$

There are two basic methods for solving the 0-1 knapsack problems (LKP): branch-and-bound method and dynamic programming method.

6.2.2.1 Branch-and-bound method

Assume that the variables have been ordered such that

$$p_1/a_1 \ge p_2/a_2 \ge \cdots \ge p_N/a_N. \tag{6.2.3}$$

Let s be the maximum index k such that

$$\sum_{j=1}^{k} a_j \le b. \tag{6.2.4}$$

The following theorem is due to Dantzig [49].

THEOREM 6.3 *The optimal solution to the continuous relaxation of* (LKP) *is*

$$w_j = 1,\ j = 1, \ldots, s,$$
$$w_j = 0,\ j = s+2, \ldots, N,$$
$$w_{s+1} = (b - \sum_{j=1}^{s} a_j)/a_{s+1}.$$

If p_j, $j = 1, \ldots, N$, are positive integers, then an upper bound of the optimal value of (LKP) is given by

$$UB = \sum_{j=1}^{s} p_j + \lfloor (b - \sum_{j=1}^{s} a_j) p_{s+1}/a_{s+1} \rfloor, \tag{6.2.5}$$

where $\lfloor x \rfloor$ denotes the largest integer less than or equal to x. Several improvements of the upper bound in (6.2.5) can be found in [152][153][158]. The following branch-and-bound method uses the depth-first search and finds an upper bound by using Theorem 6.3.

ALGORITHM 6.1 (BRANCH-AND-BOUND METHOD FOR (LKP))

Step 1 (Initialization). Set $p_{N+1} = 0$, $a_{N+1} = \infty$, $f_{opt} = f = 0$, $w_{opt} = w = (0, \ldots, 0)^T$, $W = b$, $i = 1$.

Step 2 (Test heuristic). If $a_i \le W$, find the largest s such that $\sum_{j=i}^{s} a_j \le W$, set $z = \sum_{j=i}^{s} p_j + (W - \sum_{j=i}^{s} a_j) p_{s+1}/a_{s+1}$. If $a_i > W$, set $s = i - 1$ and $z = W p_s / a_s$. If $f_{opt} \ge \lfloor z \rfloor + f$, go to Step 5.

Step 3 (New feasible solution). If $a_i \le W$ and $i \le N$, set $W := W - a_i$, $f := f + p_i$, $w_i = 1$, $i := i + 1$, repeat Step 3; otherwise, if $i \le N$, set $w_i = 0$, $i := i + 1$. If $i < N$, go to Step 2; if $i = N$, repeat Step 3; if $i > N$, go to Step 4.

Step 4 (Updating incumbent). If $f_{opt} < f$, set $f_{opt} = f$, $w_{opt} = w$. Set $i = N$, if $w_N = 1$, set $W := W + a_N$, $f := f - p_N$, $w_N = 0$.

Step 5 (Backtracking). Find the largest $k < i$ such that $w_k = 1$. If there is no such a k, stop and the current w_{opt} is the optimal solution. Otherwise, set $W := W + a_k$, $f := f - p_k$, $w_k = 0$, $i = k + 1$ and go to Step 2.

EXAMPLE 6.2 Consider the reformulation of the linear 0-1 knapsack problem (6.2.2) in Example 6.1:

$$\begin{aligned} \max\ & 6w_1 + 3w_2 + 2w_3 + 2w_4 + 4w_5 + w_6 \\ \text{s.t.}\ & 3w_1 + 2w_2 + 2w_3 + 4w_4 + 9w_5 + 6w_6 \le 17, \\ & w_j \in \{0, 1\},\ j = 1, \ldots, 6, \end{aligned}$$

where $(w_1, w_2, w_3, w_4, w_5, w_6)$ is corresponding to $(x_{13}, x_{11}, x_{12}, x_{22}, x_{23}, x_{21})$. Note that $\{p_j/a_j\}$ is in a decreasing order with

$$(p_j) = (6, 3, 2, 2, 4, 1),$$
$$(a_j) = (3, 2, 2, 4, 9, 6).$$

The process of Algorithm 6.1 is described as follows.

Step 1. Set $p_7 = 0$, $a_7 = \infty$, $f_{opt} = f = 0$, $w_{opt} = w = (0, 0, 0, 0, 0, 0)^T$, $W = 17$, $i = 1$.

Step 2. $s = 4$, $z = 13 + (17 - 11) \times 4/9 = 15.6667$. $f_{opt} < 15 + 0$.

Step 3. $W = 17 - 3 = 14$, $f = 0 + 6 = 6$, $w_1 = 1$, $i = 2$.

Step 3. $W = 14 - 2 = 12$, $f = 6 + 3 = 9$, $w_2 = 1$, $i = 3$.

Step 3. $W = 12 - 2 = 10$, $f = 9 + 2 = 11$, $w_3 = 1$, $i = 4$.

Step 3. $W = 10 - 4 = 6$, $f = 11 + 2 = 13$, $w_4 = 1$, $i = 5$.

Step 3. $a_5 > 6$, $i = 5 < N$, set $w_5 = 0$, $i = 6 = N$.

Step 3. $W = 6 - 6 = 0$, $f = 13 + 1 = 14$, $i = 6 = N$, set $w_6 = 1$, $i = 7 > N$.

Step 4. Set $f_{opt} = f = 14$, $w_{opt} = (1, 1, 1, 1, 0, 1)^T$; $i = 6$, $W = 0 + 6 = 6$, $f = 14 - 1 = 13$, $w_6 = 0$.

Step 5. $k = 4$, $W = 6 + 4 = 10$, $f = 13 - 2 = 11$, $w_4 = 0$, $i = 5$.

Step 2. $s = 5$, $z = 4 + (10 - 9) \times 1/6 = 4.1667$. $f_{opt} < 4 + 11$.

Step 3. $W = 10 - 9 = 1$, $f = 11 + 4 = 15$, $i = 5 < N$, set $w_5 = 1$, $i = 6 = N$.

Step 3. $a_6 > 1$, $i = 6 = N$, set $w_6 = 0$, $i = 7 > N$.

Step 4. $f_{opt} = f = 15$, $w_{opt} = (1, 1, 1, 0, 1, 0)^T$; $i = 6$.

Step 5. $k = 5$, $W = 1 + 9 = 10$, $f = 15 - 4 = 11$, $w_5 = 0$, $i = 6$.

Step 2. $s = 6$, $z = 1 + 0 = 1$. $f_{opt} > 1 + 11$.

Step 5. $k = 3$, $W = 10 + 2 = 12$, $f = 11 - 2 = 9$, $w_3 = 0$, $i = 4$.

Step 2. $s = 4$, $z = 2 + (12 - 4) \times 4/9 = 5.5556$. $f_{opt} > 5 + 9$.

Step 5. $k = 2$, $W = 12 + 2 = 14$, $f = 9 - 3 = 6$, $w_2 = 0$, $i = 3$.

Step 2. $s = 4$, $z = 4 + (14 - 6) \times 4/9 = 7.5556$. $f_{opt} > 7 + 6$.

Step 5. $k = 1$, $W = 14 + 3 = 17$, $f = 6 - 6 = 0$, set $w_1 = 0$, $i = 2$.

Step 2. $s = 5$, $z = 11 + (17 - 17) \times 1/6 = 11$. $f_{opt} > 11 + 0$.

Step 5. There exists no k such that $w_k = 1$. Stop and $w_{opt} = (1, 1, 1, 0, 1, 0)^T$ is the optimal solution to the problem. By Theorem 6.2, the optimal solution to Example 6.1 is $x^* = (1, 1, 2)^T$ with $f(x^*) = 15$.

6.2.2.2 Dynamic programming method

Dynamic programming approach is applicable to (LKP) if certain integrality conditions of the coefficients hold. We first assume that the coefficients a_j $(j = 1, \ldots, N)$ are positive integers. If the 0-1 problem (LKP) is transformed from (NKP), then one sufficient condition for this condition to hold is that g_j $(j = 1, \ldots, n)$ are integer valued.

For each $m = 1, \ldots, N$ and $z = 1, \ldots, b$, define

$$P_m(z) = \max\{\sum_{j=1}^{m} p_j w_j \mid \sum_{j=1}^{m} a_j w_j \le z,\ (w_1, \ldots, w_m) \in \{0,1\}^m\}.$$

The recursive equation at the m-th stage is

$$P_m(z) = \begin{cases} P_{m-1}(z), & 0 \le z < a_m \\ \max\{P_{m-1}(z), P_{m-1}(z - a_m) + p_m\}, & a_m \le z \le b \end{cases}$$

with the initial condition:

$$P_1(z) = \begin{cases} 0, & 0 \le z < a_1 \\ p_1, & a_1 \le z \le b. \end{cases}$$

Under the condition that a_j $(j = 1, \ldots, N)$ are positive integers, a dynamic programming algorithm constructs a table of dimension $N \times (b+1)$ and calculates the entries $P_m(z)$ $(m = 1, \ldots, N,\ z = 0, \ldots, b)$ in a bottom-up fashion. An optimal solution can be found by backtracking through the table once the optimal value $P_N(b)$ is obtained. The complexity of this dynamic programming algorithm is $O(Nb)$.

EXAMPLE 6.3 Let's consider again the reformulation of the linear 0-1 knapsack problem (6.2.2) in Example 6.1:

$$\begin{aligned} \max\ & 3w_1 + w_2 + 2w_3 + 2w_4 + 6w_5 + 4w_6 \\ \text{s.t.}\ & 2w_1 + 6w_2 + 2w_3 + 4w_4 + 3w_5 + 9w_6 \le 17, \\ & w_j \in \{0,1\},\ j = 1, \ldots, 6, \end{aligned}$$

where $(w_1, w_2, w_3, w_4, w_5, w_6)$ is corresponding to $(x_{11}, x_{21}, x_{12}, x_{22}, x_{13}, x_{23})$. Table 6.1 illustrates the dynamic programming solution process of calculating $P_m(z)$. The optimal value is $P_6(17) = 15$. The optimal solution $w^* = (1,0,1,0,1,1)^T$ can be found using backtracking.

6.3 Convergent Lagrangian and Domain Cut Algorithm

The solution methods discussed so far in the previous sections have been confined themselves in singly-constrained convex knapsack problems. We discuss in this section a novel convergent Lagrangian and domain cut method which is applicable to all types of multiply-constrained nonlinear knapsack problems.

As we have seen in Chapter 3, for general convex integer programming problems, the bound produced by the Lagrangian relaxation and dual search is never worse than the bound generated by the continuous relaxation. However, the optimal solutions to the Lagrangian relaxation problem corresponding to the

Table 6.1. Values of $P_m(z)$ in dynamic programming for Example 6.3.

	$m = 1$	2	3	4	5	6
$z = 0$	0	0	0	0	0	0
1	0	0	0	0	0	0
2	3	3	3	3	3	3
3	3	3	3	3	6	6
4	3	3	5	5	6	6
5	3	3	5	5	9	9
6	3	3	5	5	9	9
7	3	3	5	5	11	11
8	3	4	5	7	11	11
9	3	4	5	7	11	11
10	3	4	6	7	11	11
11	3	4	6	7	13	13
12	3	4	6	7	13	13
13	3	4	6	7	13	13
14	3	4	6	8	13	13
15	3	4	6	8	13	13
16	3	4	6	8	13	15
17	3	4	6	8	14	15

optimal multiplier do not necessarily solve the primal problem – a duality gap may exist even for linear or convex integer programming problems. The existence of the duality gap has been a major obstacle in the use of the Lagrangian dual method as an exact method for solving integer programming problems(see [17][56][57][75][192]).

In this section we will develop a convergent Lagrangian and domain cut method for problem (NKP). The algorithm will be then extended to deal with multi-dimensional nonlinear knapsack problems.

Let $\alpha, \beta \in \mathbb{Z}^n$. Denote by $\langle \alpha, \beta \rangle$ the set of integer points in $[\alpha, \beta]$,

$$\langle \alpha, \beta \rangle = \prod_{i=1}^{n} \langle \alpha_i, \beta_i \rangle = \langle \alpha_1, \beta_1 \rangle \times \langle \alpha_2, \beta_2 \rangle \cdots \times \langle \alpha_n, \beta_n \rangle.$$

The set $\langle \alpha, \beta \rangle$ is called an integer box or subbox. For convenience, we define $\langle \alpha, \beta \rangle = \emptyset$ if $\alpha \not\leq \beta$.

6.3.1 Derivation of the algorithm

To motivate the method, we consider an illustrative example as follows.

EXAMPLE 6.4

$$\begin{aligned} \max\ & f(x) = \frac{1}{2}x_1^2 + 5x_1 + 6x_2 \\ \text{s.t. } & g(x) = 6x_1 + x_2^2 \le 23, \\ & x \in X = \{x \in \mathbb{Z}^2 \mid 1 \le x_i \le 5,\ i = 1,2\}. \end{aligned}$$

The optimal solution of this example is $x^* = (3,2)^T$ with $f(x^*) = 31.5$.

The domain X and the perturbation function $z = w(y)$ of this example are illustrated in Figures 6.3 and 6.4, respectively. It is easy to check that the optimal Lagrangian multiplier is $\lambda^0 = 1.3333$ with dual value 34.8333. The duality gap is 3.3333. The Lagrangian problem

$$\max_{x \in X} \ [f(x) - 1.3333(g(x) - 23)]$$

has a feasible solution $x^0 = (1,2)^T$ with $f(x^0) = 17.5$ and an infeasible solution $y^0 = (5,2)^T$. In Figure 6.4, points A, B, C correspond to x^0, y^0

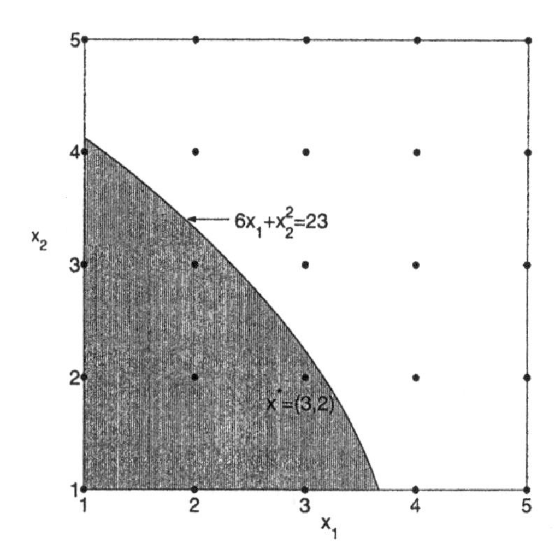

Figure 6.3. Domain X and the feasible region of Example 6.4.

and x^* in $(g(x), f(x))$ plane, respectively. We observe that if points A and B are removed from the plot of the perturbation function, then the duality gap of the revised problem will be smaller than the original duality gap and thus the "hidden" point C can be hopefully exposed on the concave envelope of the revised perturbation function after repeating such a process. The monotonicity of f and g motivates us to cut integer points satisfying $x \le x^0$ and integer points satisfying $x \ge y^0$ from box X. It is easy to see that cutting such integer points from X does not remove any better feasible point than x^0. Denote by X^1 the

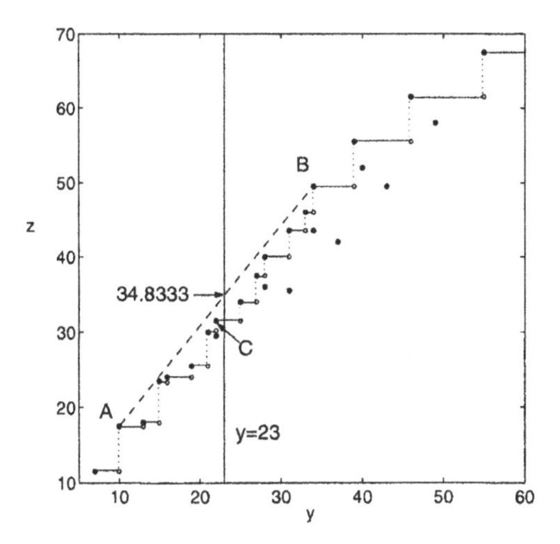

Figure 6.4. Perturbation function $z = w(y)$ with domain X of Example 6.4.

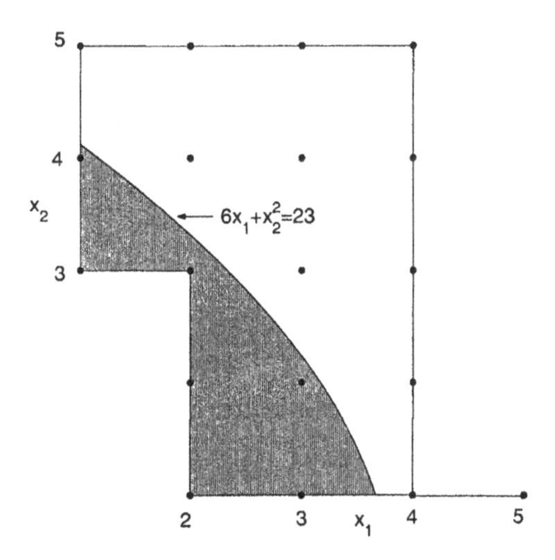

Figure 6.5. Domain X^1 in Example 6.4.

revised domain of integer points after such a cut. Figures 6.5 and 6.6 show the integer points in X^1 and the perturbation function corresponding to the revised problem by replacing X by X^1. The optimal Lagrangian multiplier for this revised problem is $\lambda^* = 1.2692$ with $d(\lambda^1) = 33.6538$. The Lagrangian problem

$$\max_{x \in X^1} [f(x) - 1.2692(g(x) - 23)]$$

has a feasible solution $x^1 = (1, 3)^T$ with $f(x^1) = 23.5$ and an infeasible solution $y^1 = (4, 2)^T$. We observe that X^1 can be partitioned into three integer

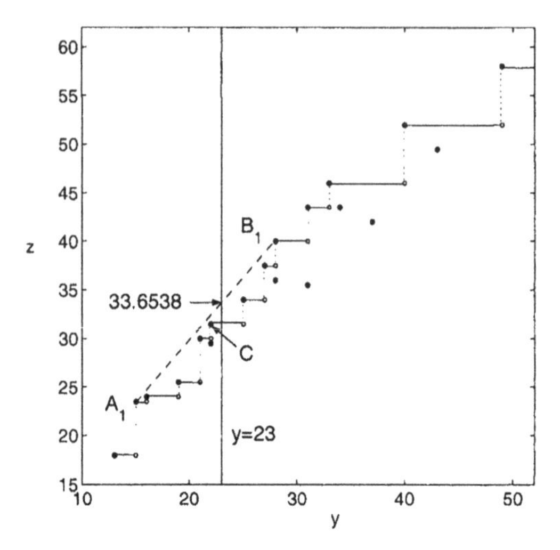

Figure 6.6. Perturbation function with domain X^1 of Example 6.4.

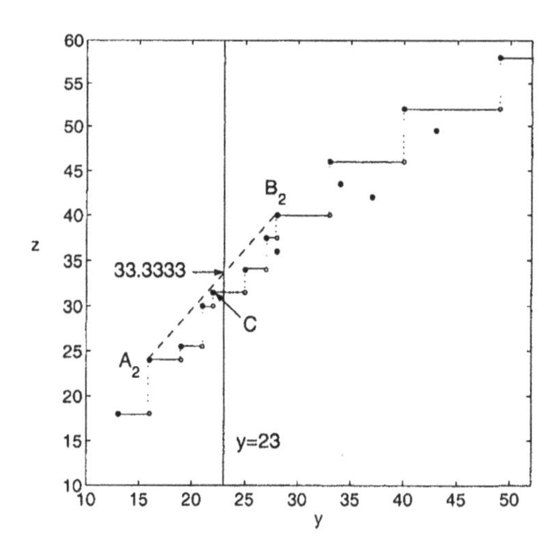

Figure 6.7. Perturbation function with domain X_1^1 of Example 6.4.

subboxes X_1^1 and X_2^1 and X_3^1,

$$\begin{aligned} X^1 &= X_1^1 \cup X_2^1 \cup X_3^1 \\ &= \langle (2,1)^T, (4,5)^T \rangle \cup \langle (1,3)^T, (1,5)^T \rangle \cup \langle (5,1)^T, (5,1)^T \rangle. \end{aligned}$$

Since the single point $(5,1)^T$ is infeasible, we can remove X_3^1 from X^1. Performing dual search on X_1^1 and X_2^1 separately, we may get a better upper bound than $d(\lambda^1)$. Figures 6.7 and 6.8 show the perturbation functions on X_1^1 and X_2^1, respectively. The dual values on X_1^1 and X_2^1 are 33.3333 and 30.1666, respec-

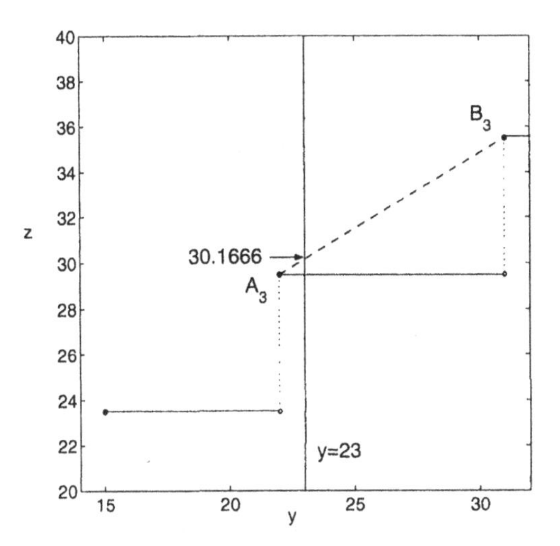

Figure 6.8. Perturbation function with domain X_2^1 of Example 6.4.

tively. Notice that both dual values on X_1^1 and X_2^1 are less than the dual value on X^1. On the other hand, the dual values on X_1^1 and X_2^1 are upper bounds of the optimal values on X_1^1 and X_2^1, respectively. Thus, the larger one of the dual values on X_1^1 and X_2^1, 33.3333, provides a better upper bound on X^1, which is smaller than the dual value on X^1, 33.6538. The feasible and infeasible solutions of X_1^1 and X_2^1 obtained in the dual search are $x^2 = (2,2)^T$, $y^2 = (4,2)^T$ and $x^3 = (1,4)^T$, $y^3 = (1,5)^T$, respectively. The incumbent is updated by $x^3 = (1,4)^T$ with $f(x^3) = 29.5$. Since the latest incumbent is generated from X_1^1, we choose X_1^1 to partition and obtain three integer subboxes:

$$X_1^2 = \langle (3,1)^T, (3,5)^T \rangle,\ X_2^2 = \langle (2,3)^T, (2,5)^T \rangle,\ X_3^2 = \langle (4,1)^T, (4,1)^T \rangle.$$

The single point in X_3^2 is infeasible. Thus, X_3^2 is discarded from further consideration. The feasible and infeasible solutions of X_1^2 and X_2^2 obtained in the dual search are $x^4 = (3,2)^T$, $y^4 = (3,3)^T$ and $x^5 = (2,3)^T$, $y^5 = (2,4)^T$, respectively. The feasible solution in X_1^2, $x^4 = (3,2)^T$, is the new incumbent with $f(x^4) = 31.5$. Domain X_2^1 is fathomed because its upper bound 30.1666 $< f(x^4)$. Further applying the cutting process to X_1^2 and X_2^2, respectively, yields empty sets. We therefore claim that $x^4 = (3,2)^T$ is the optimal solution to this example.

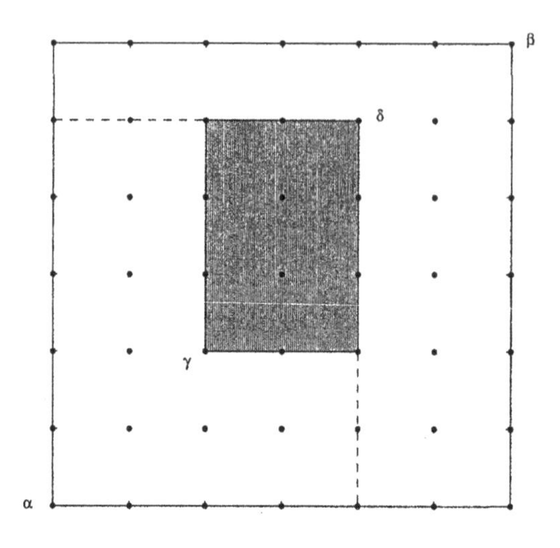

Figure 6.9. Partition of $A \setminus B$.

6.3.2 Domain cut

A key issue in implementing the above idea of Lagrangian dual and domain cut is to partition the non-rectangular domain, such as X^1 in Example 6.4, into a union of integer subboxes so that the Lagrangian relaxation on the revised domain can be decomposed.

LEMMA 6.1 *Let $A = \langle \alpha, \beta \rangle$ and $B = \langle \gamma, \delta \rangle$, where α, β, γ, $\delta \in \mathbb{Z}^n$ and $\alpha \le \gamma \le \delta \le \beta$. Then $A \setminus B$ can be partitioned into at most $2n$ integer boxes.*

$$\begin{aligned} A \setminus B \;=\; & \left\{ \cup_{j=1}^n \left(\prod_{i=1}^{j-1} \langle \alpha_i, \delta_i \rangle \times \langle \delta_j + 1, \beta_j \rangle \times \prod_{i=j+1}^{n} \langle \alpha_i, \beta_i \rangle \right) \right\} \\ & \cup \left\{ \cup_{j=1}^n \left(\prod_{i=1}^{j-1} \langle \gamma_i, \delta_i \rangle \times \langle \alpha_j, \gamma_j - 1 \rangle \times \prod_{i=j+1}^{n} \langle \alpha_i, \delta_i \rangle \right) \right\}. \end{aligned} \tag{6.3.1}$$

Proof. As illustrated in Figure 6.9, $A \setminus B$ can be expressed as

$$A \setminus B = \langle \alpha, \beta \rangle \setminus \langle \gamma, \delta \rangle = (\langle \alpha, \beta \rangle \setminus \langle \alpha, \delta \rangle) \cup (\langle \alpha, \delta \rangle \setminus \langle \gamma, \delta \rangle). \tag{6.3.2}$$

Let $C = \langle \alpha, \delta \rangle$. Then, by (6.3.2), we have

$$A \setminus B = (A \setminus C) \cup (C \setminus B). \tag{6.3.3}$$

For $j = 0, 1, \ldots, n-1$, define

$$A_j = \prod_{i=j+1}^{n} \langle \alpha_i, \beta_i \rangle,$$

$$C_j = \prod_{i=j+1}^{n} \langle \alpha_i, \delta_i \rangle.$$

Then

$$\begin{aligned} A_{j-1} \setminus C_{j-1} &= \prod_{i=j}^{n} \langle \alpha_i, \beta_i \rangle \setminus \prod_{i=j}^{n} \langle \alpha_i, \delta_i \rangle \\ &= \Bigg\{ (\langle \alpha_j, \delta_j \rangle \times \prod_{i=j+1}^{n} \langle \alpha_i, \beta_i \rangle) \\ &\qquad \cup (\langle \delta_j + 1, \beta_j \rangle \times \prod_{i=j+1}^{n} \langle \alpha_i, \beta_i \rangle) \Bigg\} \setminus \prod_{i=j}^{n} \langle \alpha_i, \delta_i \rangle \\ &= \Bigg\{ (\langle \alpha_j, \delta_j \rangle \times \prod_{i=j+1}^{n} \langle \alpha_i, \beta_i \rangle) \setminus \prod_{i=j}^{n} \langle \alpha_i, \delta_i \rangle \Bigg\} \\ &\qquad \cup (\langle \delta_j + 1, \beta_j \rangle \times \prod_{i=j+1}^{n} \langle \alpha_i, \beta_i \rangle) \end{aligned}$$

$$\begin{aligned} &= \Bigg\{ \langle \alpha_j, \delta_j \rangle \times (\prod_{i=j+1}^{n} \langle \alpha_i, \beta_i \rangle \setminus \prod_{i=j+1}^{n} \langle \alpha_i, \delta_i \rangle) \Bigg\} \\ &\qquad \cup (\langle \delta_j + 1, \beta_j \rangle \times \prod_{i=j+1}^{n} \langle \alpha_i, \beta_i \rangle) \\ &= \{ \langle \alpha_j, \delta_j \rangle \times (A_j \setminus C_j) \} \cup (\langle \delta_j + 1, \beta_j \rangle \times \prod_{i=j+1}^{n} \langle \alpha_i, \beta_i \rangle). \end{aligned}$$

Using the above partition formulation recursively for $j = 1, \ldots, n-1$, and noting that $A = A_0$, $C = C_0$, $A_{n-1} \setminus C_{n-1} = \langle \alpha_n, \beta_n \rangle \setminus \langle \alpha_n, \delta_n \rangle = \langle \delta_n + 1, \beta_n \rangle$, we get

$$A \setminus C = \cup_{j=1}^{n} \left(\prod_{i=1}^{j-1} \langle \alpha_i, \delta_i \rangle \times \langle \delta_j + 1, \beta_j \rangle \times \prod_{i=j+1}^{n} \langle \alpha_i, \beta_i \rangle \right). \quad (6.3.4)$$

Similarly, we have

$$C \setminus B = \cup_{j=1}^{n} \left(\prod_{i=1}^{j-1} \langle \gamma_i, \delta_i \rangle \times \langle \alpha_j, \gamma_j - 1 \rangle \times \prod_{i=j+1}^{n} \langle \alpha_i, \delta_i \rangle \right). \quad (6.3.5)$$

Combining (6.3.3) with (6.3.4) and (6.3.5), we obtain (6.3.1). □

COROLLARY 6.1 *Let $A = \langle \alpha, \beta \rangle$, $B = \langle \alpha, \gamma \rangle$ and $C = \langle \gamma, \beta \rangle$, where $\alpha \leq \gamma \leq \beta$. Then both $A \setminus B$ and $A \setminus C$ can be partitioned into at most n new integer subboxes:*

$$A \setminus B = \cup_{i=1}^{n} \left(\prod_{k=1}^{i-1} \langle \alpha_k, \gamma_k \rangle \times \langle \gamma_i + 1, \beta_i \rangle \times \prod_{k=i+1}^{n} \langle \alpha_k, \beta_k \rangle \right), \quad (6.3.6)$$

$$A \setminus C = \cup_{i=1}^{n} \left(\prod_{k=1}^{i-1} \langle \gamma_k, \beta_k \rangle \times \langle \alpha_i, \gamma_i - 1 \rangle \times \prod_{k=i+1}^{n} \langle \alpha_k, \beta_k \rangle \right). \quad (6.3.7)$$

The above corollary shows that the revised domain resulted from cutting two subboxes from an integer box can be partitioned into at most $2n - 1$ integer subboxes.

As an example, let us consider the domain cutting process in Example 6.4. Using (6.3.6), we have

$$\begin{aligned} X \setminus \langle l, x^0 \rangle &= \{\langle 1,5 \rangle \times \langle 1,5 \rangle\} \setminus \{\langle 1,1 \rangle \times \langle 1,2 \rangle\} \\ &= \{\langle 2,5 \rangle \times \langle 1,5 \rangle\} \cup \{\langle 1,1 \rangle \times \langle 3,5 \rangle\}. \end{aligned}$$

Further removing $\langle y^0, u \rangle$ by using (6.3.7), we get

$$\begin{aligned} X^1 &= (X \setminus \langle l, x^0 \rangle) \setminus \langle y^0, u \rangle \\ &= (\{\langle 2,5 \rangle \times \langle 1,5 \rangle\} \setminus \{\langle 5,5 \rangle \times \langle 2,5 \rangle\}) \cup \{\langle 1,1 \rangle \times \langle 3,5 \rangle\} \\ &= \{\langle 2,4 \rangle \times \langle 1,5 \rangle\} \cup \{\langle 5,5 \rangle \times \langle 1,1 \rangle\} \cup \{\langle 1,1 \rangle \times \langle 3,5 \rangle\} \\ &= \langle (2,1)^T, (4,5)^T \rangle \cup \langle (5,1)^T, (5,1)^T \rangle \cup \langle (1,3)^T, (1,5)^T \rangle. \end{aligned}$$

We will refer the process of cutting nonpromising integer boxes and partitioning a revised domain into integer subboxes as *domain cut*. The domain cut is based on the monotone properties of f and g_i. Specifically, we have the following property for (NKP):

LEMMA 6.2 *Let $x, y \in \langle \alpha, \beta \rangle$. Suppose that x is feasible to (NKP) and y is infeasible to (NKP). Then $\langle \alpha, x \rangle$ and $\langle y, \beta \rangle$ can be cut from the $\langle \alpha, \beta \rangle$,* *without missing any optimal solution of (NKP) after recording the feasible solution x.*

Notice that the above property holds as well for cases with multiple constraints.

Based on Theorem 3.3, when the problem domain can be expressed as a union of sub-domains, the dual search should be performed separately on all individual sub-domains, since it will provide a better dual value than performing the dual search globally on the entire domain.

6.3.3 The main algorithm

Based on the above discussion, a convergent Lagrangian dual and domain cut algorithm can be developed by combining the Lagrangian relaxation with the domain cut. Let $X^0 = X$. Initially, a dual search procedure is applied to (NKP) to produce an optimal dual value $d(\lambda^*)$ together with a feasible optimal solution x^0 and an infeasible optimal solution y^0 to (L_{λ^*}). Suppose that at the k-th iteration, an integer subbox is selected from X^k according to some rule, where X^k is the set of all integer boxes that have not been fathomed. The domain cut process as stated in Lemma 6.2 is performed on that integer subbox to generate at most $2n-1$ new integer subboxes. A Lagrangian dual search is then applied to each newly generated integer subbox to produce the dual value together with a feasible solution and an infeasible solution. The current best feasible solution is recorded as the incumbent solution and all integer subboxes whose upper bound is less than or equal to the function value of the incumbent are removed. The process repeats until there is no integer subbox in X^k and the incumbent solution is the optimal solution to (NKP) when the algorithm terminates.

We now describe the algorithm in details.

ALGORITHM 6.2 (CONVERGENT LAGRANGIAN AND DOMAIN CUT ALGORITHM)

Step 0 (Initialization). If $x = l$ is infeasible, then the problem has no feasible solution, or if u is feasible, then u is the optimal solution, stop. Apply the dual search procedure to (NKP) and obtain the dual value f_{dual} as an upper bound, a feasible solution x^0 and an infeasible solution y^0. Set $x_{opt} = x^0$, $f_{opt} = f(x_{opt})$, $X^0 = X$, $k = 0$.

Step 1 (Sub-Domain Selection). Select an integer subbox $\langle \alpha, \beta \rangle$ from X^k according to one of the following rules:

(a) $\langle \alpha, \beta \rangle$ is the integer subbox with the highest dual value among all subboxes in the revised domain;

(b) $\langle \alpha, \beta \rangle$ is the integer subbox with the maximum function value of a feasible solution among all subboxes in the revised domain;

(c) $\langle\alpha,\beta\rangle$ is selected according to a natural order in the formulas given in (6.3.6) and (6.3.7).

Step 2 (Domain Cut). Let x^k, $y^k \in \langle\alpha,\beta\rangle$ be the feasible solution and infeasible solution, respectively.

(i) Cut $\langle y^k,\beta\rangle$ from $\langle\alpha,\beta\rangle$, and partition the relative complement set $Y^{k+1} = \langle\alpha,\beta\rangle \setminus \langle y^k,\beta\rangle$ into integer subboxes by (6.3.7). Remove $\langle\alpha,\beta\rangle$ from X^k. Apply Step 3 for each new integer subbox.

(ii) If x^k is included in $\langle\gamma,\delta\rangle$, one of the remaining subboxes of Y^{k+1}, set $Z^{k+1} = \langle\gamma,\delta\rangle \setminus \langle\alpha,x^k\rangle$ and partition it into integer subboxes by (6.3.6). Apply Step 3 for each new integer subbox. Remove $\langle\gamma,\delta\rangle$ from Y^{k+1}.

(iii) Update x_{opt} and f_{opt} if one feasible solution found in the dual search is better than x_{opt}. Set X^{k+1} to be the set of integer subboxes by adding all integer subboxes remaining in Y^{k+1} and Z^{k+1} to X^k. Go to Step 4.

Step 3 (Dual Search and Fathoming).

(i) Remove the integer subbox $\langle\tilde{\alpha},\tilde{\beta}\rangle$ with $\tilde{\alpha}$ infeasible or $\tilde{\beta}$ feasible to problem (NKP). Update x_{opt} and f_{opt} if $\tilde{\beta}$ is feasible and $f(\tilde{\beta}) > f_{opt}$.

(ii) Apply the dual search procedure to the integer subbox to obtain its dual value, a feasible solution and an infeasible solution to problem (NKP).

(iii) Remove any integer subbox if its dual value is less than or equal to f_{opt}.

Step 4 (Termination). If X^{k+1} is empty, stop and x_{opt} is an optimal solution to (NKP). Otherwise, set $k := k+1$, go to Step 1.

REMARK 6.1 The three sub-domain selection rules in Step 1 will be compared in our computational experiments.

REMARK 6.2 Theorem 3.15 and Corollary 3.2 guarantee that at least one feasible solution and one infeasible solution can be found by a finite convergent dual search. In implementation, a feasible solution and an infeasible solution can be obtained by identifying the optimal solutions to the one-dimensional problem (3.1.8) with minimum and maximum values of g_j, respectively. There is no need to find out the entire solution set of the Lagrangian relaxation problem at its optimal multiplier.

REMARK 6.3 After performing Step 3 (i), only those newly generated integer boxes that cross the boundary of the feasible region will be left.

REMARK 6.4 Algorithm 6.2 can be interpreted as an extension of the traditional branch-and-bound method in a wide sense. It uses both monotonicity and Lagrangian bound to prune nodes. The process of domain cut is essentially

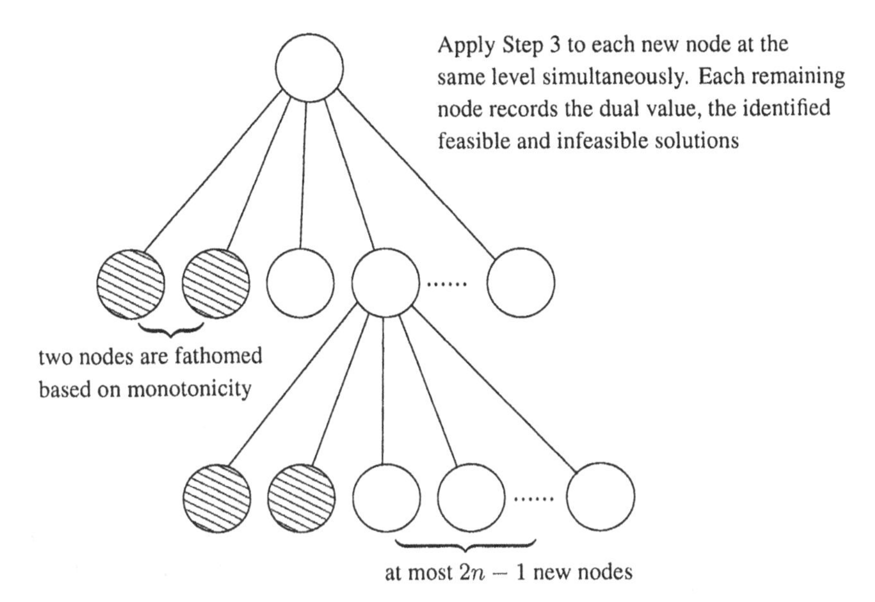

Figure 6.10. Structural diagram of the convergent Lagrangian and domain cut method under a branch-and-bound framework.

a branch process. At each level of the search tree, a parent node is branched into at most $2n + 1$ new nodes, among which two nodes are fathomed immediately based on the monotonicity. The dual procedure is applied to the remaining $2n - 1$ nodes simultaneously after removing in Step 3 (i) integer boxes that only contain feasible integer points or only contain infeasible points. Three items of information: the dual value and the identified feasible and infeasible solutions are recorded for each node. The nodes with the dual value equal to or less than the objective value of the incumbent are fathomed in the process. According to one of the selection rules in Step 1, a node from the current active node-list will be selected for further branch. A structural diagram in Figure 6.10 illustrates the convergent Lagrangian and domain cut method under a branch-and-bound framework.

REMARK 6.5 The concept behind Algorithm 6.2 also has a similarity to the traditional cutting plane method for linear integer program. Both of them aim at eliminating duality gap by reshaping the feasible region. While the revised domain in the traditional cutting plane method becomes more irregular when adding more cutting planes, Algorithm 6.2 keeps the revised domain as a union of boxes, thus maintaining the decomposability of the revised domain.

THEOREM 6.4 (i) *Let d^k denote the maximum upper bound of all the integer subboxes in X^k. Then $\{d^k\}$ is nonincreasing.*

(ii) *Algorithm 6.2 finds an optimal solution of (NKP) after finite steps of iterations.*

Proof. (i) Let $d(a,b)$ denote the Lagrangian bound on $\langle a,b\rangle$. For any integer subbox $\langle\gamma,\delta\rangle$ of X^{k+1}, there exists an integer subbox $\langle\alpha,\beta\rangle$ of X^k such that $\langle\gamma,\delta\rangle \subseteq \langle\alpha,\beta\rangle$. Thus, we have

$$d^{k+1} = \max_{\langle\gamma,\delta\rangle\in X^{k+1}} \min_{\lambda\geq 0} \max_{x\in\langle\gamma,\delta\rangle} L(x,\lambda) \leq \max_{\langle\alpha,\beta\rangle\in X^{k}} \min_{\lambda\geq 0} \max_{x\in\langle\alpha,\beta\rangle} L(x,\lambda) = d^k.$$

Hence $\{d^k\}$ is nonincreasing.

(ii) By Lemma 6.2 and the weak duality, the domain cut process in Step 2 and the fathoming process in Step 3 do not remove any solution better than the incumbent x_{opt}. Also, by the monotone property of f and g, all points in $\langle\alpha,\beta\rangle$ are infeasible when α is infeasible, thus cutting $\langle\alpha,\beta\rangle$ from X^k in Step 3 (i) does not remove any feasible point in X^k. Therefore, at each iteration, either x_{opt} is already the optimal solution or there is an optimal solution in X^k.

The finite termination of the algorithm is obvious by noting the finiteness of X and the fact that at least two integer boxes are cut from X at each iteration. □

6.3.4 Multi-dimensional nonlinear knapsack problems

We consider the following multi-dimensional nonlinear knapsack problem:

$$(MNKP) \qquad \max\ f(x) = \sum_{j=1}^{n} f_j(x_j)$$
$$\text{s.t. } g_i(x) = \sum_{j=1}^{n} g_{ij}(x_j) \leq b_i,\ i = 1,\ldots,m,$$
$$x \in X = \{x \in \mathbb{Z}^n \mid l_j \leq x_j \leq u_j,\ j = 1,\ldots,n\},$$

where all f_j's and all g_{ij}'s are nondecreasing functions on $[l_j, u_j]$ for $j = 1,\ldots,n$, $i = 1,\ldots,m$, with $l_j < u_j$ and l_j and u_j being integer numbers for $j = 1,\ldots,n$.

In this section we discuss how to extend Algorithm 6.2 to deal with problems $(MNKP)$ by using a surrogate technique. Let (SP) denote the subproblem of $(MNKP)$ by replacing X with an integer subbox $\langle\alpha,\beta\rangle \subseteq X$. In order to adopt the algorithmic framework developed in the previous sections, we relax the feasible region of (SP) by using a surrogate constraint technique ([51][176]) discussed in Chapter 4. For $\mu \in \mathbb{R}^m_+$ with $\mu \neq 0$, let $g^\mu(x) = \sum_{i=1}^m \mu_i g_i(x)$ and $b^\mu = \sum_{i=1}^m \mu_i b_i$. The surrogate constraint formulation of (SP) can be expressed as follows:

$$s(\mu) = \max \sum_{j=1}^{n} f_j(x_j) \tag{6.3.8}$$
$$\text{s.t. } g^{\mu}(x) = \sum_{j=1}^{n} g_j^{\mu}(x_j) \le b^{\mu},$$
$$x \in \langle \alpha, \beta \rangle,$$

where $g_j^{\mu}(x_j) = \sum_{i=1}^{m} \mu_i g_{ij}(x_j)$. Notice that both the separability and the monotonicity in $(MNKP)$ are still retained in the surrogate constraint formulation (6.3.8). It is easy to see that the Lagrangian relaxation of the surrogate constraint formulation (6.3.8) still provides an upper bound on the optimal value of (SP). The optimal surrogate multiplier vector for (6.3.8) is the vector μ^* that minimizes $s(\mu)$ over all $\mu \ge 0$:

$$(SD) \quad s(\mu^*) = \min_{\mu \ge 0} s(\mu).$$

Since μ^* is usually very expensive to obtain, we will use the optimal Lagrangian multiplier vector, which is much cheaper to calculate, as the surrogate multiplier vector. Consider the Lagrangian dual of (SP):

$$\min_{\mu \ge 0} v(\mu), \tag{6.3.9}$$

where

$$v(\mu) := \max_{x \in \langle \alpha, \beta \rangle} \left[f(x) - \sum_{i=1}^{m} \mu_i (g_i(x) - b_i) \right]. \tag{6.3.10}$$

As we discussed in Section 3.2, the optimal solution to (6.3.9) can be computed efficiently by the subgradient method or the outer Lagrangian linearization method.

Algorithm 6.2 can be extended to solve problem $(MNKP)$ with some modifications in the domain cut process and in the computation of the dual value on the integer subboxes of X^k. Now, the computation of the dual value on each integer subbox $\langle \alpha, \beta \rangle$ includes two steps. First, a surrogate multiplier vector $\tilde{\mu}$ for (6.3.8) is computed by solving problem (6.3.9)–(6.3.10) via certain dual search method; If $\tilde{\mu}$ is an exact solution to (6.3.9), set $\lambda^* = 1$. If $\tilde{\mu}$ is an approximate solution to (6.3.9), then a finite convergent dual search procedure (e.g., Procedure 3.3) for singly constrained problems is applied to the dual problem of the surrogate problem (6.3.8) with $\mu = \tilde{\mu}$:

$$(SCD) \quad \min_{\lambda \ge 0} d_{sc}(\lambda)$$

where

$$d_{sc}(\lambda) = \max_{x \in X} \left[f(x) - \lambda \sum_{i=1}^{m} \tilde{\mu}_i (g_i(x) - b_i) \right]. \qquad (6.3.11)$$

Let λ^* be the optimal solution to (SCD). In either case of dual search, we can obtain two optimal solutions to (6.3.11) with $\lambda = \lambda^*$: x^k with $g^{\tilde{\mu}}(x^k) \leq b^{\tilde{\mu}}$ and y^k with $g^{\tilde{\mu}}(y^k) > b^{\tilde{\mu}}$.

It is clear y^k is also infeasible for $(MNKP)$. Since x^k is not necessarily feasible for $(MNKP)$, a modification is needed for Step 2(ii) of Algorithm 6.2 to give a correct domain cut.

Step 2 (ii)$'$ Let $x^k \in \langle \gamma, \delta \rangle$, one of the subboxes in Y^{k+1}. If x^k is feasible for $(MNKP)$, then cut $\langle \gamma, x^k \rangle$ from $\langle \gamma, \delta \rangle$. Set $Z^{k+1} = \langle \gamma, \delta \rangle \setminus \langle \gamma, x^k \rangle$ and partition it into integer subboxes by (6.3.6). Otherwise, cut $\langle x^k, \delta \rangle$ from $\langle \gamma, \delta \rangle$. Set $Z^{k+1} = \langle \gamma, \delta \rangle \setminus \langle x^k, \delta \rangle$ and partition it into integer subboxes by (6.3.7). Remove $\langle \gamma, \delta \rangle$ from Y^{k+1}.

The finite convergence of the extended algorithm and the optimality of x_{opt} when the algorithm stops can be proved similarly as in Algorithm 6.2. Now we illustrate the extended algorithm by an illustrative example with two constraints.

EXAMPLE 6.5

$$\begin{aligned} \max\ & f(x) = x_1^2 + 1.5x_2 \\ \text{s.t.}\ & g_1(x) = 6x_1 + x_2^2 \leq 23, \\ & g_2(x) = 4x_1 + x_2 \leq 12.5, \\ & x \in X = \{x \in \mathbb{Z}^2 \mid 1 \leq x_i \leq 4,\ i = 1, 2\}. \end{aligned}$$

The optimal solution is $x^* = (2,3)^T$ with $f(x^*) = 8.5$. Rule (a) in Step 1 of Algorithm 6.2 is used for this example to select the integer subbox in Step 1.

We first use the subgradient method to generate the surrogate multiplier.

Initial Iteration

Step 0. Let $X^0 = \{X\}$. Solving the Lagrangian dual problem of the example by using the subgradient method, we obtain an approximate solution $\mu = (0.07054, 1.1433)^T$. The surrogate problem is

$$\begin{aligned} \max\ & f(x) = x_1^2 + 1.5x_2 \\ \text{s.t.}\ & g_\mu(x) = 5x_1 + 0.07054x_2^2 + 1.1433x_2 \leq 15.9242, \\ & x \in X. \end{aligned} \qquad (6.3.12)$$

Figure 6.11 depicts the domain and the feasible regions of both the primal problem and the surrogate problem (6.3.12). Applying the dual search procedure to (6.3.12), we obtain a dual value 12.3571, a feasible solution $x^0 = (1,3)^T$

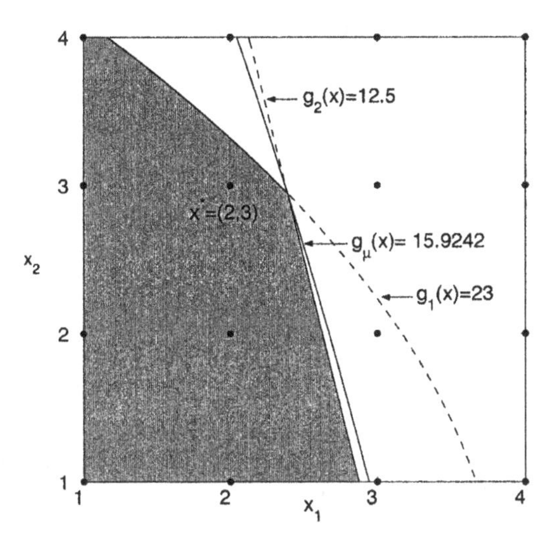

Figure 6.11. Domain X and the feasible region of Example 6.5.

and an infeasible solution $y^0 = (4,3)^T$ to (6.3.12). Since x^0 is also feasible to the original problem, set $x_{opt} = (1,3)^T$, $f_{opt} = 5.5$.

Iteration 1

Step 1. Select X to partition.

Step 2. Cutting $\langle y^0, u\rangle$ from X results in

$$\begin{aligned} Y^1 &= \langle (1,1)^T, (4,4)^T\rangle \setminus \langle (4,3)^T, (4,4)^T\rangle \\ &= \langle (1,1)^T, (3,4)^T\rangle \cup \langle (4,1)^T, (4,2)^T\rangle = \{\tilde{X}_1, \tilde{X}_2\}. \end{aligned}$$

Since $(4,1)^T$ is infeasible, $\tilde{X}_2$ is removed. The Lagrangian bound on $\tilde{X}_1$ is $10.9643 > f_{opt}$. Since $x^0 \in \tilde{X}_1$ and x^0 is feasible to the original problem, $\langle (1,1)^T, (1,3)^T\rangle$ is cut from $\tilde{X}_1$. We have

$$\begin{aligned} Z^1 &= \langle (1,1)^T, (3,4)^T\rangle \setminus \langle (1,1)^T, (1,3)^T\rangle \\ &= \langle (2,1)^T, (3,4)^T\rangle \cup \langle (1,4)^T, (1,4)^T\rangle = \{\tilde{X}_3, \tilde{X}_4\}. \end{aligned}$$

Computing a Lagrangian bound on $\tilde{X}_3$, we obtain the dual value 8.9286, a feasible solution $(2,3)^T$ and an infeasible solution $(3,3)^T$ to the corresponding surrogate problem. Since $(2,3)^T$ is also feasible to the original problem and $f((2,3)^T) = 8.5 > 5.5 = f_{opt}$, set $x_{opt} = (2,3)^T$ and $f_{opt} = 8.5$. For $\tilde{X}_4$, since $(1,4)^T$ is feasible to the original problem and $f((1,4)^T) = 7 < f_{opt}$, $\tilde{X}_4$ is removed. Set $X^1 = \{\tilde{X}_3\}$.

Iteration 2

Step 1. Select $\tilde{X}_3$.

Step 2. Cut $\langle (3,3)^T, (3,4)^T \rangle$ from $\tilde{X}_3$. We have

$$\begin{aligned} Y^2 &= \langle (2,1)^T, (3,4)^T \rangle \setminus \langle (3,3)^T, (3,4)^T \rangle \\ &= \langle (2,1)^T, (2,4)^T \rangle \cup \langle (3,1)^T, (3,2)^T \rangle = \{\tilde{X}_5, \tilde{X}_6\}. \end{aligned}$$

The Lagrangian bound on $\tilde{X}_5$ is $8.9286 > f_{opt}$ with a feasible solution $(2,3)^T$ and an infeasible solution $(2,4)^T$. Since $(3,1)^T$ is infeasible to the original problem, $\tilde{X}_6$ is removed. Since $(2,3)^T \in \tilde{X}_5$ and $(2,3)^T$ is feasible, $\langle (2,1)^T, (2,3)^T \rangle$ is cut from $\tilde{X}_5$. We have

$$Z^2 = \langle (2,1)^T, (2,4)^T \rangle \setminus \langle (2,1)^T, (2,3)^T \rangle = \langle (2,4)^T, (2,4)^T \rangle = \{\tilde{X}_7\}.$$

Since $(2,4)^T$ is infeasible, $\tilde{X}_7$ is removed. Now, the remaining domain, X^2, is empty.

Step 4. The algorithm stops at an optimal solution $x^1 = (2,3)^T$.

Next, we re-solve the example using the outer Lagrangian linearization method as the dual search procedure for (6.3.9). The algorithm process is described as follows.

Initial Iteration

Step 0. Set $X^0 = \{X\}$. Solving the Lagrangian dual problem of the example by the outer Lagrangian linearization method, we obtain an exact dual solution $\mu = (0.07143, 1.14286)^T$ with a dual value 11.3571. The surrogate problem is

$$\begin{aligned} &\max\ f(x) = x_1^2 + 1.5x_2 \qquad (6.3.13)\\ &\text{s.t. } g^{\mu}(x) = 5.00002x_1 + 0.07143x_2^2 + 1.14286x_2 \le 15.92864,\\ &\quad x \in X. \end{aligned}$$

Solving the Lagrangian relaxation of (6.3.13) with $\lambda^* = 1$, we obtain a feasible solution $x^0 = (1,1)^T$ and an infeasible solution $y^0 = (3,4)^T$ to (6.3.13). Since x^0 is also feasible to the original problem, set $x_{opt} = (1,1)^T$ and $f_{opt} = 2.5$.

Iteration 1

Step 1. Select X to partition.

Step 2. Cutting $\langle y^0, u \rangle$ from X results in

$$\begin{aligned} Y^1 &= \langle (1,1)^T, (4,4)^T \rangle \setminus \langle (3,4)^T, (4,4)^T \rangle \\ &= \langle (1,1)^T, (2,4)^T \rangle \cup \langle (3,1)^T, (3,3)^T \rangle \\ &= \{\tilde{X}_1, \tilde{X}_2\} \end{aligned}$$

Since $(3,1)^T$ is infeasible, $\tilde{X}_2$ is removed. The Lagrangian bound on $\tilde{X}_1$ is $11.56476 > f_{opt}$. Since $x^0 \in \tilde{X}_1$ and x^0 is feasible to the original problem, $\langle (1,1)^T, (1,1)^T \rangle$ is cut from $\tilde{X}_1$. We have

$$\begin{aligned} Z^1 &= \langle (1,1)^T, (2,4)^T \rangle \setminus \langle (1,1)^T, (1,1)^T \rangle \\ &= \langle (1,2)^T, (1,4)^T \rangle \cup \langle (2,1)^T, (2,4)^T \rangle \\ &= \{\tilde{X}_3, \tilde{X}_4\} \end{aligned}$$

For $\tilde{X}_3$, since $(1,4)^T$ is feasible to the original problem and $f((1,4)^T) = 7 < f_{opt}$, $\tilde{X}_3$ is removed. Applying the outer Lagrangian linearization method to the dual problem on $\tilde{X}_4$, we obtain a Lagrangian dual bound 12.75, a feasible solution $(2,3)^T$ and an infeasible solution $(2,4)^T$ to the corresponding surrogate problem. Since $(2,3)^T$ is also feasible to the original problem and $f((2,3)^T) = 8.5 > 2.5 = f_{opt}$, set $x_{opt} = (2,3)^T$ and $f_{opt} = 8.5$. Set $X^1 = \tilde{X}_4$.

Iteration 2

Step 1. Select $\tilde{X}_4$.

Step 2. Cut $\langle (2,1)^T, (2,3)^T \rangle$ from $\tilde{X}_4$. We have

$$Y^2 = \langle (2,1)^T, (2,4)^T \rangle \setminus \langle (2,1)^T, (2,3)^T \rangle = \langle (2,4)^T, (2,4)^T \rangle = \{\tilde{X}_5\}.$$

Since $(2,4)^T$ is infeasible, $\tilde{X}_5$ is removed. $X^2 = \emptyset$.

Step 3. The algorithm stops at an optimal solution $x^1 = (2,3)^T$.

6.4 Concave Nonlinear Knapsack Problems

We consider in this section an efficient solution algorithm for a special class of nonlinear knapsack problems: The *concave* knapsack problem with a linear constraint. The problem is in the following form:

$$\begin{aligned} (CCKP) \qquad & \max\ f(x) = \sum_{j=1}^{n} f_j(x_j) \\ & \text{s.t. } g(x) = \sum_{j=1}^{n} b_j x_j \le b, \\ & x \in X = \{x \in \mathbb{Z}^n \mid l_j \le x_j \le u_j,\ j = 1, \ldots, n\}, \end{aligned}$$

where f_j, $j = 1, \ldots, n$, are increasing convex functions on $\mathbb{R}$, $b_j > 0$, $j = 1, \ldots, n$, and l_j and u_j are integer lower and upper bounds of x_j with $u_j > l_j \ge 0$, $j = 1, \ldots, n$.

The convergent Lagrangian and domain cut method developed in Section 6.3 is applicable to $(CCKP)$. However, the special structure of $(CCKP)$ allows a development of a more efficient solution scheme which combines the domain cut idea with a linear approximation method.

6.4.1 Linear approximation

A natural way to overcome the nonconcavity of the objective function in $(CCKP)$ is to overestimate each f_j by a linear function. Let $\langle \alpha, \beta \rangle \subseteq X$ be a nonempty integer box. Consider the following subproblem of $(CCKP)$:

$$(SP) \qquad \max\ f(x) = \sum_{j=1}^{n} f_j(x_j)$$
$$\text{s.t. } g(x) = \sum_{j=1}^{n} b_j x_j \leq b,$$
$$x \in \langle \alpha, \beta \rangle.$$

Denote by $v(\cdot)$ the optimal value of problem $(\cdot)$. The linear overestimating function of $f(x) = \sum_{j=1}^{n} f_j(x_j)$ over box $[\alpha, \beta]$ can be expressed as:

$$L(x) = \sum_{j=1}^{n} L_j(x_j),$$

where $L_j(x_j) = f_j(\alpha_j) + a_j(x_j - \alpha_j)$ with

$$a_j = \begin{cases} \frac{f_j(\beta_j) - f_j(\alpha_j)}{\beta_j - \alpha_j}, & \alpha_j < \beta_j, \\ 0, & \alpha_j = \beta_j. \end{cases}$$

By the convexity of f_j $(j = 1, \ldots, n)$, we have $L(x) \geq f(x)$ for all $x \in \langle \alpha, \beta \rangle$ and $L(x) = f(x)$ for all the extreme points of $\langle \alpha, \beta \rangle$. The linear approximation of (SP) is:

$$(LSP) \qquad \max\ L(x) = a_0 + \sum_{j=1}^{n} a_j x_j$$
$$\text{s.t. } g(x) = \sum_{j=1}^{n} b_j x_j \leq b,$$
$$x \in \langle \alpha, \beta \rangle,$$

where a_j is the coefficient of x_j in $L(x)$ and $a_0 = \sum_{j=1}^{n} [f_j(\alpha_j) - a_j \alpha_j]$ is the constant term of $L(x)$. Since f_j, $j = 1, \ldots, n$, are increasing functions, we have $a_j \geq 0$ for $j = 1, \ldots, n$. Without loss of generality, we assume that

$$\frac{a_1}{b_1} \geq \frac{a_2}{b_2} \geq \cdots \geq \frac{a_n}{b_n}.$$

Let

$$\xi_j = (b - \sum_{i=1}^{j-1} b_i \beta_i - \sum_{i=j+1}^{n} b_i \alpha_i)/b_j, \ j = 1, \ldots, n. \tag{6.4.1}$$

Let k be the largest index j satisfying $\xi_j > \alpha_j$. By Theorem 6.3, the optimal solution of the continuous relaxation of (LSP) is

$$x^R = (\beta_1, \ldots, \beta_{k-1}, \xi_k, \alpha_{k+1}, \ldots, \alpha_n)^T. \tag{6.4.2}$$

A feasible solution can be derived from x^R by rounding down ξ_k:

$$x^F = (\beta_1, \ldots, \beta_{k-1}, \tau_k, \alpha_{k+1}, \ldots, \alpha_n)^T, \tag{6.4.3}$$

where $\tau_k = \lfloor \xi_k \rfloor$ is the largest integer less than or equal to ξ_k. From (6.4.1) and (6.4.3), we infer that if $\xi_k = \tau_k$, then $x^F = x^R$ is an optimal solution to (LSP). Suppose that $\xi_k < \beta_k$. Let

$$x^I = (\beta_1, \beta_2, \ldots, \beta_{k-1}, \tau_k + 1, \alpha_{k+1}, \ldots, \alpha_n)^T. \tag{6.4.4}$$

It follows that $x^I \in \langle \alpha, \beta \rangle$ and x^I is infeasible. Let $(\overline{LSP})$ denote the continuous relaxation problem of (LSP). Then, from the above discussion, we have

$$L(x^R) = v(\overline{LSP}) \geq v(LSP) \geq v(SP) \geq f(x^F).$$

Therefore, by solving $(\overline{LSP})$, we can get an upper bound $L(x^R)$ and a lower bound $f(x^F)$ of the subproblem (SP).

It is interesting to compare $L(x^R)$ with the upper bound provided by Lagrangian dual problem of (SP). The Lagrangian dual problem of (SP) is

$$(SD) \qquad \min_{\lambda \geq 0} d(\lambda),$$

where $d(\lambda)$ is the dual function defined by

$$d(\lambda) = \max_{x \in \langle \alpha, \beta \rangle} l(x, \lambda) = f(x) - \lambda(\sum_{j=1}^{n} b_j x_j - b). \tag{6.4.5}$$

The following theorem shows that $L(x^R)$ coincides with the optimal Lagrangian dual value of problem (SP).

THEOREM 6.5 $v(SD) = v(\overline{LSP}) = L(x^R)$.

Proof. Since $f(x)$ is convex, the Lagrangian function $l(x, \lambda)$ in (6.4.5) is a convex function of x for any $\lambda \geq 0$. Thus, it always achieves its maximum over $[\alpha, \beta]$ at one of the extreme points of $\langle \alpha, \beta \rangle$. On the other hand, $f(x)$ takes the same values as $L(x)$ over all the extreme points of box $[\alpha, \beta]$. Therefore, we have

$$\begin{aligned} v(SD) &= \min_{\lambda \geq 0} d(\lambda) \\ &= \min_{\lambda \geq 0} \max_{x \in \langle \alpha, \beta \rangle} \{f(x) - \lambda(\sum_{j=1}^{n} b_j x_j - b)\} \\ &= \min_{\lambda \geq 0} \max_{x \in [\alpha, \beta]} \{L(x) - \lambda(\sum_{j=1}^{n} b_j x_j - b)\} \\ &= \max\{L(x) \mid \sum_{j=1}^{n} b_j x_j \leq b,\ x \in [\alpha, \beta]\} \\ &= v(\overline{LSP}) = L(x^R). \end{aligned}$$

The fourth equation above is due to the duality theorem of linear programming. □

Theorem 6.5 shows that the upper bound obtained by solving $(\overline{LSP})$ is the same as the Lagrangian bound to (SP). We observe that computing the solution of $(\overline{LSP})$ is much easier than that of (SD).

6.4.2 Domain cut and linear approximation method

Let $A = \langle \alpha, \beta \rangle$, $B = \langle \alpha, x^F \rangle$ and $C = \langle x^I, \beta \rangle$, where x^F and x^I are defined in (6.4.3) and (6.4.4), respectively. By the monotonicity of $f(x)$ and $g(x)$, cutting integer box B does not remove any feasible solution better than x^F from A. Moreover, cutting integer box C does not remove any feasible solution from A. Let $\Omega = (A \setminus B) \setminus C$. The following result shows that Ω can be partitioned into a union of at most $n-1$ integer boxes. A lower bound and an upper bound on Ω can be then calculated by using the linear approximation approach.

PROPOSITION 6.1 *The set $\Omega = (A \setminus B^F) \setminus B^I$ can be partitioned into at most $n - 1$ integer boxes:*

$$\begin{aligned} \Omega = & \left\{ \cup_{j=1}^{k-1} \left(\prod_{i=1}^{j-1} \langle \beta_i, \beta_i \rangle \times \langle \alpha_j, \beta_j - 1 \rangle \times \prod_{i=j+1}^{k-1} \langle \alpha_i, \beta_i \rangle \times \langle \tau_k + 1, \beta_k \rangle \right.\right. \\ & \left.\left. \times \prod_{i=k+1}^{n} \langle \alpha_i, \beta_i \rangle \right) \right\} \cup \left\{ \cup_{j=k+1}^{n} \left(\prod_{i=1}^{k-1} \langle \alpha_i, \beta_i \rangle \times \langle \alpha_k, \tau_k \rangle \right.\right. \\ & \left.\left. \times \prod_{i=k+1}^{j-1} \langle \alpha_i, \alpha_i \rangle \times \langle \alpha_j + 1, \beta_j \rangle \times \prod_{i=j+1}^{n} \langle \alpha_i, \beta_i \rangle \right) \right\}. \end{aligned} \quad (6.4.6)$$

Proof. The partition formula (6.4.6) can be obtained by applying Lemma 6.1. □

As we have seen from Corollary 6.1, partitioning the set $(A \setminus B) \setminus C$ in general situations generates at most $2n-1$ new integer subboxes. The property that x^F and x^I are neighboring integer points on the boundary of $\langle \alpha, \beta \rangle$ leads to a partition of Ω with at most $n-1$ new integer subboxes.

We now describe the algorithm.

ALGORITHM 6.3

Step 0 (Initialization). Let $l = (l_1, \ldots, l_n)^T$ and $u = (u_1, \ldots, u_n)^T$. If l is infeasible, then problem (P) has no feasible solution, stop. Otherwise, set $x_{opt} = l$, $f_{opt} = f(x_{opt})$, $X^1 = \langle l, u \rangle$, $Y^1 = X^1$, $Z^1 = \emptyset$. Set $k = 1$.

Step 1 (Linear approximation). For each $\langle \alpha, \beta \rangle \in Y^k$, do the following:

(i) If $g(\alpha) > b$, then remove $\langle \alpha, \beta \rangle$ from Y^k, repeat Step 1.

(ii) Compute the linear approximation function $L(x)$ and rank the sequence $\{a_j/b_j\}_{j=1}^n$ in a decreasing order. Calculate the continuous optimal solution x^R by (6.4.2). If ξ_k is an integer, then $x^F = x^R$ is an optimal solution to the corresponding subproblem (LSP), set $f_{opt} = f(x^F)$ and $x_{opt} = x^F$ if $f(x^F) > f_{opt}$, remove $\langle \alpha, \beta \rangle$ from Y^k. Otherwise, go to (iii).

(iii) Calculate x^F and x^I by (6.4.3) and (6.4.4), respectively. Set $\tau_k = \lfloor \xi_k \rfloor$. Determine τ_j for $j = k+1, \ldots, n$ by

$$\tau_j = \min\{\beta_j, \lfloor (b - \sum_{i=1}^{k-1} b_i \beta_i - \sum_{i=k}^{j-1} b_i \tau_i - \sum_{i=j+1}^{n} b_i \alpha_i)/b_j \rfloor\}.$$

Let

$$\bar{x}^F = (\beta_1, \beta_2, \ldots, \beta_{k-1}, \tau_k, \tau_{k+1}, \ldots, \tau_n)^T.$$

Set $f_{opt} = f(\bar{x}^F)$ and $x_{opt} = \bar{x}^F$ if $f(\bar{x}^F) > f_{opt}$, repeat Step 1.

Step 2 (Fathoming). Let $r(\alpha, \beta)$ denote the upper bound $L(x^R)$, the optimal value of $(\overline{LSP})$ on $\langle \alpha, \beta \rangle$. Let $T^k = Y^k \cup Z^k$. For each $\langle \alpha, \beta \rangle \in T^k$, remove $\langle \alpha, \beta \rangle$ from T^k if $r(\alpha, \beta) \le f_{opt}$.

Step 3 (Partition). If $T^k = \emptyset$, stop and x_{opt} is an optimal solution to (P). Otherwise, find the integer box $\langle \alpha^k, \beta^k \rangle$ with maximum value of $r(\alpha, \beta)$:

$$f_k = r(\alpha^k, \beta^k) = \max_{\langle \alpha, \beta \rangle \in T^k} r(\alpha, \beta).$$

Set $Z^{k+1} = T^k \setminus \{\langle \alpha^k, \beta^k \rangle\}$ and

$$Y^{k+1} = (\langle \alpha^k, \beta^k \rangle \setminus \langle \alpha^k, x^F \rangle) \setminus \langle x^I, \beta^k \rangle,$$

where x^F and x^I were calculated in Step 1 (iii) for the integer box $\langle \alpha^k, \beta^k \rangle$. Partition Y^{k+1} into a union of integer boxes by using the formula (6.4.6). Set $X^{k+1} = Y^{k+1} \cup Z^{k+1}$, $k := k + 1$, go to Step 1.

A few remarks about the algorithm are as follows.

REMARK 6.6 In the algorithm, $X^k = Y^k \cup Z^k$ represents all the active integer boxes, where Y^k is the set of newly generated integer boxes on each of which a lower bound and an upper bound will be calculated in Step 1, and Z^k is the set of old integer boxes inherited from X^{k-1}. After executing Step 1, each integer box in X^k is associated with an upper bound $L(x^R)$, a feasible solution x^F and an infeasible solution x^I. The incumbent x_{opt} and the corresponding best function value f_{opt} are obtained by comparing the last incumbent with the maximum of lower bounds achieved by feasible solutions identified from the integer boxes in Y^k.

REMARK 6.7 Calculating $\bar{x}^F$ in Step 2 (iii) is to improve the feasible solution x^F by filling the slack of constraint at x^F. Since x^F is feasible, it follows that $\bar{x}^F$ is also feasible and $f(\bar{x}^F) \geq f(x^F)$.

THEOREM 6.6 *The algorithm generates a strictly decreasing sequence of upper bounds $\{f_k\}$ and terminates at an optimal solution of $(CCKP)$ within a finite number of iterations.*

Proof. For each integer box $\langle \alpha, \beta \rangle$ of $X^{k+1} = Y^{k+1} \cup Z^{k+1}$, it is either identical to an integer box in X^k or a subset of an integer box in X^k. Thus, the linear overestimation of $f(x)$ on $\langle \alpha, \beta \rangle$ majorizes that on the corresponding integer box of X^k. Moreover, from Step 3, the continuous optimal solution x^R corresponding to the maximum upper bound f_k is excluded in X^{k+1}. Therefore, $f_{k+1} \leq f_k$ for $k \geq 1$. The finite termination of the algorithm is obvious from the finiteness of X and the fact that at least the feasible solution x^F and infeasible solution x^I corresponding to the maximum upper bound f_k are cut from X^k and excluded from X^{k+1}. Since the fathoming rule in Step 2 and the domain cutting process in Step 3 do not remove from X^k any feasible solution better than x_{opt}, the feasible solution x_{opt} must be an optimal solution to $(CCKP)$ when the algorithm stops at Step 3 with no integer boxes left in T^k. □

To illustrate the algorithm, let us consider a small-size numerical example:

EXAMPLE 6.6

$$\begin{aligned}\max\ f(x) &= 5x_1^2 + 15x_1 + 4x_2^2 + 6x_2 + 2x_3^2 + 4x_3 + x_4^2 + 9x_4 \\ &\quad + 2x_5^2 + 18x_5 \\ \text{s.t. } g(x) &= 7x_1 + x_2 + 5x_3 + 4x_4 + 2x_5 \leq 47.5, \\ & x \in X = \{x \in \mathbb{Z}^5 \mid 0 \leq x_j \leq 5,\ j = 1,2,3,4,5\}.\end{aligned}$$

The optimal solution of this problem is $x^* = (4,5,0,1,5)^T$ with $f(x^*) = 420$. The algorithm terminates at the 4-th iteration with the optimal solution x^* achieved. The iterative process is described as follows.

Initial Iteration:

Step 0. Set $l = (0,0,0,0,0)^T$, $u = (5,5,5,5,5)^T$, $X^1 = \{\langle l,u\rangle\}$, $Y^1 = X^1$, $Z^1 = \emptyset$, $x_{opt} = (0,0,0,0,0)^T$, $f_{opt} = 0$, $k = 1$.

Iteration 1:

Step 1. For box $\langle l,u\rangle$, we have

$$\begin{aligned}&x^R = (4.64,5,0,0,5)^T,\ L(x^R) = 445.71,\ x^F = (4,5,0,0,5)^T, \\ &x^I = (5,5,0,0,5)^T, \bar{x}^F = (4,5,0,1,5)^T, \\ &x_{opt} = \bar{x}^F = (4,5,0,1,5)^T,\ f_{opt} = 420.\end{aligned}$$

Step 2. $T^1 = \{\langle l,u\rangle\}$.

Step 3. Integer box $\langle l,u\rangle$ is chosen to partition. $Z^2 = \emptyset$. Using (6.4.6),

$$Y^2 = (\langle l,u\rangle \setminus \langle (0,0,0,0,0)^T, (4,5,0,0,5)^T\rangle) \setminus \langle (5,5,0,0,5)^T, (5,5,5,5,5)^T\rangle$$

is partitioned into 4 integer subboxes:

$$\begin{aligned}&Y_1^2 = \langle (0,0,1,0,0)^T, (4,5,5,5,5)^T\rangle,\ Y_2^2 = \langle (0,0,0,1,0)^T, (4,5,0,5,5)^T\rangle, \\ &Y_3^2 = \langle (5,0,0,0,0)^T, (5,4,5,5,5)^T\rangle,\ Y_4^2 = \langle (5,5,0,0,0)^T, (5,5,5,5,4)^T\rangle.\end{aligned}$$

Thus, $X^2 = Y^2 \cup Z^2 = \{Y_1^2, Y_2^2, Y_3^2, Y_4^2\}$. Set $k = 2$ and go to Step 1.

Iteration 2:

Step 1. (1) For box Y_1^2, we have $x^R = (3.93,5,1,0,5)^T$ and $L(x^R) = 413.52 < f_{opt}$. Remove Y_1^2 from Y^2.

(2) For box Y_2^2, we have $x^R = (4,5,0,1.13,5)^T$, $L(x^R) = 421.87 > f_{opt}$, $x^F = (4,5,0,1,5)^T$, $x^I = (4,5,0,2,5)^T$, $\bar{x}^F = x^F$.

(3) For box Y_3^2, we have $x^R = (5,4,0,0,4.25)^T$ and $L(x^R) = 407 < f_{opt}$. Remove Y_3^2 from Y^2.

(4) For box Y_4^2, we have $x^R = (5,5,0,0,3.75)^T$, $L(x^R) = 427.5 > f_{opt}$, $x^F = (5,5,0,0,3)^T$, $x^I = (5,5,0,0,4)^T$, $\bar{x}^F = x^F$.

Step 2. $T^2 = \{Y_2^2, Y_4^2\}$.

Step 3. Integer box Y_4^2 is chosen to partition. $Z^3 = \{Y_2^2\}$. Using (6.4.6),

$$Y^3 = (Y_4^2 \setminus \langle (5,5,0,0,0)^T, (5,5,0,0,3)^T\rangle) \setminus \langle (5,5,0,0,4)^T, (5,5,5,5,4)^T\rangle$$

is partitioned into 2 integer subboxes:

$$Y_1^3 = \langle (5,5,1,0,0)^T, (5,5,5,5,4)^T \rangle,\ Y_2^3 = \langle (5,5,0,1,0)^T, (5,5,0,5,4)^T \rangle.$$

Thus, $X^3 = Y^3 \cup Z^3 = \{Y_2^2, Y_1^3, Y_2^3\}$. Set $k = 3$, go to Step 1 .

Iteration 3:

Step 1. (1) For Y_1^3, we have $x^R = (5,5,1,0,1.25)^T$ and $L(x^R) = 368.5 < f_{opt}$. Remove Y_1^3 from Y^3.

(2) For Y_2^3, we have $x^R = (5,5,0,1,1.75)^T$ and $L(x^R) = 385.5 < f_{opt}$. Remove Y_2^3 from Y^3.

Step 2. $T^3 = \{Y_2^2\}$.

Step 3. Integer box Y_2^2 is chosen to partition. $Z^4 = \emptyset$. Using (6.4.6),

$$Y^4 = (Y_2^2 \setminus \langle (0,0,0,1,0)^T, (4,5,0,1,5)^T \rangle) \setminus \langle (4,5,0,2,5)^T, (4,5,0,5,5)^T \rangle$$

is partitioned into 3 integer subboxes:

$$Y_1^4 = \langle (0,0,0,2,0)^T, (3,5,0,5,5)^T \rangle,\ Y_2^4 = \langle (4,5,0,2,0)^T, (4,5,0,5,4)^T \rangle,$$
$$Y_3^4 = \langle (4,0,0,2,0)^T, (4,4,0,5,5)^T \rangle.$$

Thus, $X^4 = Y^4 \cup Z^4 = \{Y_1^4, Y_2^4, Y_3^4\}$. Set $k = 4$ and go to Step 1.

Iteration 4:

Step 1. (1) For Y_1^4, we have $x^R = (3,5,0,2.87,5)^T$ and $L(x^R) = 396 < f_{opt}$. Remove Y_1^4 from Y^4.

(2) For Y_2^4, we have $x^R = (4,5,0,2,3.25)^T$ and $L(x^R) = 376.5 < f_{opt}$. Remove Y_2^4 from Y^4.

(3) For Y_3^4, we have $x^R = (4,4,0,2,3.75)^T$ and $L(x^R) = 355 < f_{opt}$. Remove Y_3^4 from Y^4.

Step 2. $T^4 = \emptyset$.

Step 3. Stop, $x_{opt} = (4,5,0,1,5)^T$ is the optimal solution.

6.5 Reliability Optimization in Series-Parallel Reliability Networks

We now consider a special class of multi-dimensional nonlinear knapsack problems arising from series-parallel reliability systems. Consider a series system shown in Figure 6.12. The system is functioning if and only if all its n independent components are functioning. In order to improve the overall reliability of the system, one can use more reliable components. However, the expense and more often the technological limits may prohibit an adoption of this strategy. An alternative method is to add redundant components as shown in Figure 6.13.

We now consider the constrained redundancy optimization problem in series systems (see [217][219]). The goal of the problem is to determine an optimal

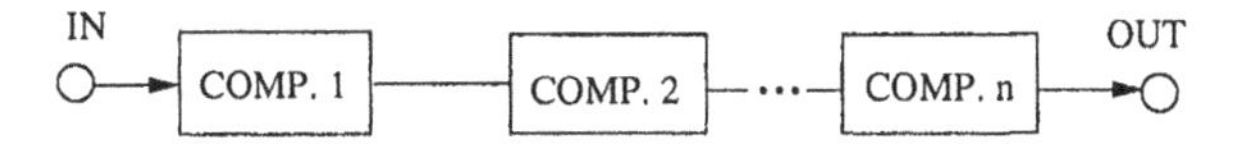

Figure 6.12. A series system with n components.

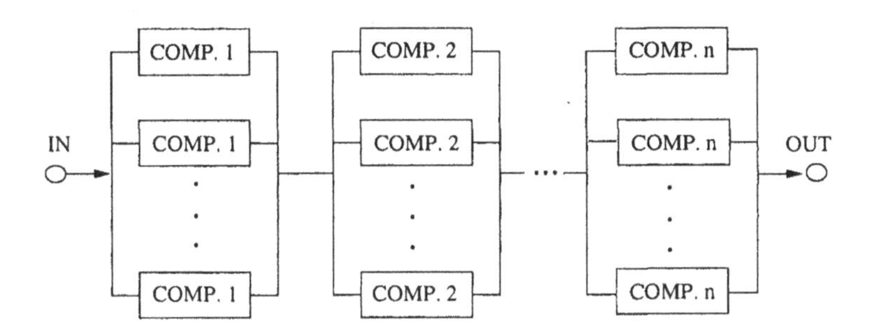

Figure 6.13. A series system with redundancies.

redundancy assignment so as to maximize the overall system reliability under certain limited resource constraints. This kind of problem is often encountered in the design of various engineering systems. The components with redundancy in a series setting can be independent subsystems or basic elements in an overall system. The components in Figure 6.12, for example, can represent electronic parts in a section of circuits, coolers and filters in a lubrication system, valves in a pipeline (see, e.g.,[25][209]) or subsystems of a complicated communication network. Typically, the adding of redundant components is constrained to cost, volume and weight limitations.

The mathematical model of the constrained redundancy optimization problem can be formulated as follows:

$$(CROP) \qquad \max\ R(x) = \prod_{j=1}^{n} R_j(x_j)$$

$$\text{s.t.}\ \ C_i(x) = \sum_{j=1}^{n} c_{ij}(x_j) \le b_i,\ i = 1, \ldots, m,$$

$$x \in X \subseteq \mathbb{Z}^n_+,$$

where $R_j(x_j) = 1-(1-r_j)^{x_j}$ is the reliability of the jth subsystem when having x_j identical components, $r_j \in (0, 1)$ is the component reliability in the jth subsystem in series, $x = (x_1, x_2, \ldots, x_n)$ represents a redundancy assignment, $R(x)$ is the overall systems reliability when adopting redundancy assignment x, $c_{ij}(x_j)$ is an increasing function of x_j that represents the ith resource consumed in the jth subsystem, b_i is the ith total available resource, and X is a subset of $\mathbb{Z}^n_+$, the positive integer vector set in $\mathbb{R}^n_+$. Denote by S the feasible region of the problem, $(CROP)$. Without loss of generality, we assume that $r_1 \le r_2 \le \cdots \le r_n$. Notice that problem $(LCROP)$ discussed in Section 6.1.1.2 is a special case of $(CROP)$ with a single linear constraint.

A closely related problem is the cost minimization in reliability systems (see [52][217][219]). The problem is to minimize the cost of a series-parallel system under a minimum overall reliability requirement. The problem can be modelled as:

$$(COST) \quad \min\ c(x) = \sum_{j=1}^{n} c_j(x_j)$$
$$\text{s.t. } R(x) = \prod_{j=1}^{n} R_j(x_j) \ge R_0,$$
$$x \in X = \{x \in \mathbb{Z}^n \mid l_j \le x_j \le u_j,\ j = 1, \ldots, n\},$$

where $c_j(x_j)$ is an increasing convex function on $[l_j, u_j]$, $R_j(x_j)$ is defined the same as in $(CROP)$, $R_0 \in (0, 1)$ is the given minimum reliability level. The above problem can be rewritten as a maximization problem by letting $y_j = u_j - x_j$:

$$\max\ f(y) = \sum_{j=1}^{n} [-c_j(u_j - y_j)]$$
$$\text{s.t. } g(y) = \sum_{j=1}^{n} [-\ln(R_j(u_j - y_j))] \le -\ln(R_0).$$
$$y \in Y = \{y \in \mathbb{Z}^n \mid 0 \le y_j \le u_j - l_j,\ j = 1, \ldots, n\}.$$

Let $f_j(y_j) = -c_j(u_j - y_j)$ and $g_j(y_j) = -\ln(R_j(u_j - y_j))$. Then $f_j(y_j)$ is an increasing function of y_j and $g_j(y_j)$ is a convex and increasing function of y_j on $[0, u_j - l_j]$ for $j = 1, \ldots, n$. We notice that problems $(CROP)$ and $(COST)$ are convex knapsack problems. When $c_j(x_j)$ is linear, problem $(COST)$ reduces to the convex knapsack problem $(LCOST)$ in Section 6.1.1.2.

6.5.1 Maximal decreasing property

The following is always true in constrained redundancy optimization. Increasing the number of parallel paths in one subsystem, while keeping all other subsystems unchanged, will increase both the overall systems reliability and the resources consumed. This point can be enhanced by observing that all $R(x)$ and $C_i(x)$, $i = 1, \ldots, m$, are strictly increasing functions of x. Some researchers, for example, Misra and Sharma [164], have noticed that an optimal redundancy assignment of $(CROP)$ always locates close to the boundary of the feasible region due to the monotonicity of $R(x)$ and $C_i(x)$, $i = 1, 2, \ldots, m$.

A feasible redundancy assignment x is said to be *noninferior* if there exists no other feasible $y \in S$ such that $y_i \geq x_i$, $(i = 1, \ldots, n)$, with at least one strict inequality. It is easy to see that x is noninferior iff there does not exist a j such that $x + e_j \in S$, where e_j denotes the jth unit vector in $\mathbb{R}^n$.

PROPOSITION 6.2 *Any optimal redundancy assignment of* $(CROP)$ *must be noninferior.*

Proof. The proof can be easily achieved by contradiction, as in [131]. □

The search of the optimal redundancy assignment of $(CROP)$ can now be confined in the set of noninferior redundancy assignments. This represents a significant reduction in the search space. We will proceed to achieve further reduction by claiming that only those noninferior solutions with certain properties need to be considered.

Let $x = (x_1, \ldots, x_{i-1}, x_i, \ldots, x_j, x_{j+1}, \ldots, x_n) \in S$. Notice that we have already ranked the subsystem reliability in an increasing order, $r_1 \leq r_2 \leq \ldots \leq r_n$. Consider a transformation of x by adding redundancy in subsystems with smaller reliability and reducing redundancy in subsystems with larger reliability. Redundancy assignment $u_{(i,j)}(x) = (x_1, \ldots, x_{i-1}, x_i + 1, \ldots, x_j - 1, x_{j+1}, \ldots, x_n)$ is said to be a *unit decreasing transformation* of x (on i, j) if $i < j$ and $x_j \geq x_i + 1$. The following proposition shows that a unit decreasing transformation can be used to improve the overall systems reliability.

PROPOSITION 6.3 $R(u_{(i,j)}(x)) > R(x)$ *if* $i < j$, $r_i < r_j$, *and* $x_j \geq x_i + 1$.

To prove Proposition 6.3, we need the following lemma.

LEMMA 6.3 (i) *If* $0 < a < b < 1$ *and* $0 < p < q$, *then*

$$(1 - b^q)(1 - a^p) > (1 - b^p)(1 - a^q). \tag{6.5.1}$$

(ii) *If* $0 < a < b < 1$ *and* $1 < p + 1 \leq q$, *then*

$$(1 - b^{p+1})(1 - a^{q-1}) > (1 - b^p)(1 - a^q). \tag{6.5.2}$$

Proof. (i) Inequality (6.5.1) is equivalent to

$$\frac{1 - b^q}{1 - b^p} > \frac{1 - a^q}{1 - a^p}.$$

Let $\phi(t) = (1 - t^q)/(1 - t^p)$. It suffices to prove $\phi(t)$ to be a strictly increasing function in $(0, 1)$ or equivalently, $\phi'(t) > 0$ for $t \in (0, 1)$ when $q > p$. Note that

$$\phi'(t) = \frac{t^{p-1}[p - qt^{q-p} + (q - p)t^q]}{(1 - t^p)^2}. \tag{6.5.3}$$

Let $\phi_1(t) = p - qt^{q-p} + (q-p)t^q$. Since $\phi_1'(t) = q(q-p)t^{q-1}(1 - t^{-p}) < 0$ for $t \in (0,1)$, $\phi_1(t)$ is a strictly decreasing function on $(0,1]$. Thus, $\phi_1(t) > \phi_1(1) = 0$ for $t \in (0,1)$. From (6.5.3), we obtain $\phi'(t) > 0$ for $t \in (0,1)$.

(ii) Inequality (6.5.2) is equivalent to

$$\frac{1 - b^{p+1}}{1 - b^p} > \frac{1 - a^q}{1 - a^{q-1}}. \tag{6.5.4}$$

Set $p = q - 1$ in (6.5.1), we obtain

$$\frac{1 - b^q}{1 - b^{q-1}} > \frac{1 - a^q}{1 - a^{q-1}}. \tag{6.5.5}$$

We now prove that

$$\frac{1 - b^{p+1}}{1 - b^p} \geq \frac{1 - b^q}{1 - b^{q-1}}. \tag{6.5.6}$$

Let $\psi(t) = (1 - b^t)/(1 - b^{t-1})$. We have

$$\psi'(t) = \frac{b^{t-1}(1-b)\ln(b)}{(1 - b^{t-1})^2} < 0, \quad t > 1.$$

It then follows that $\psi(t)$ is a strictly decreasing function on $(1, \infty)$. Since $p + 1 \leq q$, we imply that (6.5.6) holds. Upon combining (6.5.5) with (6.5.6), we obtain (6.5.4). □

Proof of Proposition 6.3.

By the definition of $u_{(i,j)}(x)$, $R(x)$ and $R(u_{(i,j)}(x))$ are different only in their ith and jth factors. Thus $R(u_{(i,j)}(x)) > R(x)$ is equivalent to

$$[1 - (1 - r_i)^{x_i+1}][1 - (1 - r_j)^{x_j-1}] > [1 - (1 - r_i)^{x_i}][1 - (1 - r_j)^{x_j}]. \tag{6.5.7}$$

Since $r_i < r_j$ and $x_i + 1 \leq x_j$, we imply (6.5.7) by applying Lemma 6.3(ii) with $a = 1 - r_j$, $b = 1 - r_i$, $p = x_i$, $q = x_j$. □

The significance of Proposition 6.3 is that for a feasible redundancy assignment x with $i < j$, $r_i < r_j$, and $x_i < x_j$ ($x_i + 1 \leq x_j$ by the integrality of x_i and x_j), if the unit decreasing transformation is feasible, i.e., $u_{(i,j)}(x) \in S$, then the overall systems reliability can get higher via replacing x by $u_{(i,j)}(x)$.

A feasible redundancy assignment x is said to be *maximal decreasing* if there does not exist a feasible unit decreasing transformation of x. We therefore obtain, from Propositions 6.2 and 6.3, the following theorem.

THEOREM 6.7 *An optimal redundancy assignment of $(CROP)$ must be both noninferior and maximal decreasing.*

In the linearly constrained cases, we have the following interesting result that coincides with intuition.

COROLLARY 6.2 *Assume that $C_i(x) = \sum_{j=1}^{n} c_{ij}x_j$, $r_1 < r_2 < \cdots < r_n$, and $0 \le c_{i1} \le c_{i2} \le \cdots \le c_{in}$ for $i = 1, \ldots, m$. Then an optimal redundancy assignment $x^* = (x_1^*, x_2^*, \ldots, x_n^*)$ of problem $(CROP)$ must be noninferior and satisfy*

$$x_1^* \ge x_2^* \ge \cdots \ge x_n^*. \tag{6.5.8}$$

Proof. For an optimal redundancy assignment x^*, if $x_i^* < x_j^*$ for some i, j with $i < j$ and $r_i < r_j$, then by Proposition 6.3, $u_{(i,j)}(x^*)$ has a higher overall system reliability than that of x^*. Moreover, $u_{(i,j)}(x^*)$ is feasible by the ordering of c_{ij}'s. This contradicts the optimality of x^*. □

In the situations with a single linear resource constraint, if the resource consumption of an additional parallel component is the same for all subsystems, the decreasing property derived from Corollary 6.2 states that in order to achieve the maximum overall systems reliability, more redundant components should be placed into a subsystem with lower reliability.

To verify that a redundancy assignment x is *not* maximal decreasing, one can show that there exist i and j with $i < j$, $r_i < r_j$, and $x_i < x_j$ such that $u_{(i,j)}(x) \in S$, i.e.,

$$C_k(u_{(i,j)}(x)) \le b_k, \quad \forall k \in \{1, \ldots, m\}.$$

Let $s_k(x) = b_k - C_k(x)$ be the slack at the k-th constraint. Then, the above inequality is equivalent to:

$$s_k(x) \ge [c_{ki}(x_i + 1) - c_{ki}(x_i)] + [c_{kj}(x_j - 1) - c_{kj}(x_j)] \tag{6.5.9}$$

for $k \in \{1, \ldots, m\}$. In the linearly constrained cases, (6.5.9) is equivalent to

$$s_k(x) \ge c_{ki} - c_{kj}, \quad \forall k \in \{1, \ldots, m\}. \tag{6.5.10}$$

As shown in the following illustrative example, the necessary condition being both noninferior and maximal decreasing significantly facilitates the elimination of non-optimal redundancy assignments from among the set of noninferior redundancy assignments.

Consider a series-parallel system with 4 subsystems, $r = (0.65, 0.70, 0.75, 0.80)$, and two linear constraints, $C_1(x) = 6x_1 + 4x_2 + 3x_3 + 2x_4 \le b_1 = 30$, and $C_2(x) = 9x_1 + 4x_2 + 4x_3 + 3x_4 \le b_2 = 40$. Table 6.2 lists all 22 noninferior

Table 6.2. List of noninferior points for a 4-subsystem problem.

No.	x	Maximal decreasing	$R(x)$	$C_1(x)$	$C_2(x)$
1	(3, 1, 1, 1)	*	0.4020	27	38
2	(2, 3, 1, 1)	*	0.5123	29	37
3	(1, 4, 2, 1)	*	0.4836	30	36
4	(1, 3, 3, 1)	*	0.4981	29	36
5	(1, 1, 6, 1)	*	0.3639	30	40
6	(1, 4, 1, 2)		0.4642	29	35
7	(2, 2, 2, 2)	*	0.7187	30	40
8	(2, 1, 3, 2)		0.5805	29	40
9	(1, 2, 4, 2)	*	0.5656	30	39
10	(1, 1, 5, 2)		0.4364	29	39
11	(2, 2, 1, 3)		0.5941	29	39
12	(2, 1, 2, 3)		0.5713	28	39
13	(1, 3, 2, 3)	*	0.5882	30	38
14	(1, 2, 3, 3)		0.5776	29	38
15	(1, 1, 4, 3)		0.4496	28	38
16	(2, 1, 1, 4)		0.4600	27	38
17	(1, 3, 1, 4)		0.4736	29	37
18	(1, 2, 2, 5)	*	0.5544	30	40
19	(1, 1, 3, 5)		0.4477	29	40
20	(1, 2, 1, 6)		0.4436	29	39
21	(1, 1, 2, 6)		0.4265	28	39
22	(1, 1, 1, 7)		0.3412	27	38

solutions of this example, among which only 9 solutions (marked by *) are maximal decreasing.

The necessary optimality condition stated in Theorem 6.7 can be used as a fathoming criteria for $(CROP)$. Similar condition can be derived for problem $(COST)$ using Proposition 6.3.

6.5.1.1 Fathoming condition

The maximal decreasing property can be used to derive an additional fathoming condition in an enumeration method for solving $(CROP)$ and $(COST)$. We will only consider the convex cases of $(CROP)$ and $(COST)$.

Using this fathoming condition may significantly speed up the convergence of the algorithm by further eliminating certain nodes that do not generate an optimal solution of $(CROP)$. Let $N = (z, \alpha, \beta)$ denote a node in an enumeration algorithm, where z is the optimal objective function value of the continuous relaxation at the parent node, and vectors α, $\beta \in \mathbb{R}^n$ are the lower bound and upper bound on the decision variables, respectively.

PROPOSITION 6.4 (i) *Let $N = (z, \alpha, \beta)$ be a node in a branch-and-bound method for solving $(CROP)$. If there exist i and j ($i < j$), $r_i < r_j$ such that $\beta_i < \alpha_j$ and*

$$[c_{ki}(x_i + 1) - c_{ki}(x_i)] + [c_{kj}(x_j - 1) - c_{kj}(x_j)] \leq 0 \qquad (6.5.11)$$

for $k \in \{1, \ldots, m\}$ and $x \in X_N = \{x \in X \mid \alpha \leq x \leq \beta\}$, then the node N can be fathomed from further consideration.

(ii) *Let $N = (z, \alpha, \beta)$ be a node in a branch-and-bound method for solving $(COST)$. If there exist i and j ($i < j$), $r_i < r_j$ such that $\beta_i < \alpha_j$ and $c_i(x_i + 1) - c_i(x_i) \leq c_j(x_j - 1) - c_j(x_j)$, then the node N can be fathomed from further consideration.*

Proof. (i) For any possible optimal integer solution x^* of a subproblem corresponding to the node N or nodes branched out from N, the condition of $x_i^* \leq \beta_i < \alpha_j \leq x_j^*$ must hold. On the other hand, since $s_k(x^*) = b_k - C_k(x^*) \geq 0$, $\forall\, k \in \{1, 2, \ldots, m\}$, due to the feasibility of x^*, (6.5.11) implies that (6.5.9) holds. Thus, x^* is not a maximal decreasing redundancy assignment and hence, by Theorem 6.7, is not an optimal solution of $(CROP)$.

Part (ii) can be proved similarly. □

We note that in the linearly constrained cases, (6.5.11) is equivalent to

$$c_{ki} \leq c_{kj},\ i < j, \quad \forall k \in \{1, \ldots, m\}. \qquad (6.5.12)$$

The monotone condition (6.5.12) can be interpreted as follows: A component with lower reliability consumes less resources than the one with higher reliability.

Theorem 6.7 can be used to improve incumbent solutions, too. Whenever an integer optimal solution x^* of a continuous relaxation subproblem is found, we check the maximal decreasing property for x^*. If x^* is not maximal decreasing, we can immediately identify a feasible solution to $(CROP)$ with a higher reliability or a feasible solution to $(COST)$ with a lower cost by making use of the unit decreasing transformations. If the resulting feasible solution is not noninferior, we add certain redundant components to certain subsystems until the feasible solution is both maximal decreasing and noninferior.

6.6 Implementation and Computational Results

We present in this section the implementation issues and computational results for the following algorithms:

- Algorithm 6.2 and its extension for multiply constrained problems;
- Algorithm 6.3 for concave knapsack problems.

Comparison results with other methods will be also reported. The algorithms were coded by Fortran 90 and run on a Sun Workstation (Blade 2000).

6.6.1 Test problems

The first set of test problems for Algorithm 6.2 and its extension for multiply constrained problems includes the following six classes of test problems. Except for Problem 6.6, all constraint functions are linear.

PROBLEM 6.1 Convex quadratic knapsack problems (QP_1).

$$\begin{aligned} \max\ & f(x) = \sum_{j=1}^{n}(c_j x_j - d_j x_j^2) \\ \text{s.t. }\ & g(x) = Ax \le b, \\ & x \in X = \{x \in \mathbb{Z}^n \mid l_j \le x_j \le u_j,\ j = 1,\ldots,n\}, \end{aligned}$$

where $c_j \ge 0, d_j > 0, u_j \le c_j/(2d_j)$ for $j = 1,\ldots,n$, and $A = (a_{ij})_{m\times n}$ with $a_{ij} \ge 0$ for $i = 1,\ldots,m$, $j = 1,\ldots,n$. The function $f_j(x_j) = c_j x_j - d_j x_j^2$ is concave on $[l_j, u_j]$ for $j = 1,\ldots,n$. The condition of $u_j \le c_j/(2d_j)$, $j = 1,\ldots,n$, is imposed to guarantee that f is nondecreasing with respect to all x_j, $j = 1,\ldots,n$.

PROBLEM 6.2 Concave quadratic knapsack problems (QP_2).

$$\begin{aligned} \max\ & f(x) = \sum_{j=1}^{n}(c_j x_j + d_j x_j^2) \\ \text{s.t. }\ & g(x) = Ax \le b, \\ & x \in X = \{x \in \mathbb{Z}^n \mid l_j \le x_j \le u_j,\ j = 1,\ldots,n\}, \end{aligned}$$

where $c_j \ge 0$, $d_j > 0$, $0 \le l_j < u_j$ for $j = 1,\ldots,n$, and $A = (a_{ij})_{m\times n}$ with $a_{ij} \ge 0$ for $i = 1,\ldots,m$, $j = 1,\ldots,n$. The function $f_j(x_j) = c_j x_j + d_j x_j^2$ is nondecreasing and convex on $[l_j, u_j]$ for $j = 1,\ldots,n$.

PROBLEM 6.3 Polynomial knapsack problems $(POLY)$.

$$\begin{aligned} \max\ & f(x) = \sum_{j=1}^{n}[c_j x_j + d_j (x_j - e_j)^3] \\ \text{s.t. }\ & g(x) = Ax \le b, \\ & x \in X = \{x \in \mathbb{Z}^n \mid l_j \le x_j \le u_j,\ j = 1,\ldots,n\}, \end{aligned}$$

where $c_j \ge 0$, $d_j > 0$, $e_j \in (l_j, u_j)$ for $j = 1,\ldots,n$, and $A = (a_{ij})_{m\times n}$ with $a_{ij} \ge 0$ for $i = 1,\ldots,m$, $j = 1,\ldots,n$. We notice that function $f_j(x_j) =$

$c_jx_j + d_j(x_j - e_j)^3$ is nondecreasing but not necessarily convex or concave on $[l_j, u_j]$ for $j = 1, \ldots, n$.

PROBLEM 6.4 Optimal sample allocation in stratified sampling ($SAMP$).

$$\max\ f(x) = -\sum_{j=1}^{n} d_j/x_j$$
$$\text{s.t. } g(x) = Ax \leq b,$$
$$x \in X = \{x \in \mathbb{Z}^n \mid l_j \leq x_j \leq u_j,\ j = 1, \ldots, n\},$$

where $d_j > 0$, $A = (a_{ij})_{m\times n}$ with $a_{ij} \geq 0$ for $i = 1, \ldots, m$, $j = 1, \ldots, n$. The function $f_j(x_j) = -d_j/x_j$ is a concave and nondecreasing function on $[l_j, u_j]$ for $j = 1, \ldots, n$.

PROBLEM 6.5 Linearly constrained redundancy problems in reliability systems ($LCROP$) (see Section 6.5).

PROBLEM 6.6 Linear cost minimization problem in reliability systems ($LCOST$) (see Section 6.5).

The data in the above testing problems are randomly generated from uniform distributions. In all the test problems, $l_j = 1$ and $u_j = 5$ for $j = 1, \ldots, n$. The parameters are set as follows:

- (QP_1): $c_j \in [100, 300]$, $d_j \in (0, 10]$, $a_{ij} \in [1, 50]$ for $i = 1, \ldots, m$, $j = 1, \ldots, n$, $b = 0.7A \times u$.
- (QP_2): $c_j \in [1, 50]$, $d_j \in [1, 10]$, for $j = 1, \ldots, n$, $a_{ij} \in [1, 50]$ for $i = 1, \ldots, m$, $j = 1, \ldots, n$; and $b = 0.7A \times u$.
- $(POLY)$: $c_j \in [1, 50]$, $d_j \in [1, 10]$, $e_j \in [1, 5]$ for $j = 1, \ldots, n$; $a_{ij} \in [1, 50]$ for $i = 1, \ldots, m$, $j = 1, \ldots, n$; and $b = 0.7A \times u$.
- $(SAMP)$: $d_j \in [1, 20]$ for $j = 1, \ldots, n$, $a_{ij} \in [1, 50]$ for $i = 1, \ldots, m$, $j = 1, \ldots, n$; and $b = 0.7A \times u$.
- $(LCROP)$: $r_j \in [0.8, 0.98]$ for $j = 1, \ldots, n$, $a_{ij} \in [1, 50]$ for $i = 1, \ldots, m$, $j = 1, \ldots, n$; and $b = 0.7A \times u$.
- $(LCOST)$: $c_j \in [1, 50]$ for $j = 1, \ldots, n$, $a_{ij} \in [1, 50]$ for $i = 1, \ldots, m$, $j = 1, \ldots, n$; and $b = 0.7A \times u$.

The second set of test problems is the concave knapsack problem.

PROBLEM 6.7 Concave knapsack problem $(CCKP)$.

$$\max\ f(x) = \sum_{j=1}^{n}(c_j x_j^3 + d_j x_j^2 + e_j x_j)$$

$$\text{s.t. } g(x) = \sum_{j=1}^{n} b_j x_j \le b,$$

$$x \in X = \{x \in \mathbb{Z}^n \mid l_j \le x_j \le u_j,\ j = 1, \ldots, n\},$$

where c_j, d_j, e_j and b_j are positive real numbers. For each n, 20 test problems are randomly generated by uniform distribution with $c_j \in [0, 1]$, $d_j \in [1, 10]$, $e_j \in [1, 20]$, and $b_j \in [1, 40]$. In all the test problems, $l_j = 1$, $u_j = 5$ and $b = \sum_{j=1}^{n} b_j l_j + 0.5(\sum_{j=1}^{n} b_j(u_j - l_j))$.

6.6.2 Heuristics for feasible solutions

Due to the monotonicity of f_j and g_j in (NKP), the performance of Algorithm 6.2 and its extension for multidimensional nonlinear knapsack problems can be improved significantly by using certain heuristics. The greedy method can be used to generate a good initial feasible point and to improve feasible solutions obtained in the dual search.

For general problems, Procedure 6.2 in Subsection 6.1.1.3 can be applied. Procedure 6.3 is suitable for convex (NKP). For concave knapsack problems, all f_j's are convex and all g_j's are concave. Since for any fixed j, the ratio in (6.1.24) is nondecreasing on k, we can replace (6.1.24) by

$$x + k_{j_0} e_{j_0} = \arg \max_{j=1,\ldots,n} \frac{f_j(x_j + k_j) - f_j(x_j)}{g_j(x_j + k_j) - g_j(x_j)}, \tag{6.6.1}$$

where $k_j = \max\{k \in \mathbb{Z}^+ \mid g(x + ke_j) \le b\}$.

PROCEDURE 6.5 (HEURISTIC FOR CONCAVE KNAPSACK PROBLEMS)

Step 1. If there exists $j_0 \in \{1, \ldots, n\}$ such that (6.6.1) holds, then set $x := x + k_{j_0} e_{j_0}$.

Step 2. Repeat Step 1 until there is no $j \in \{1, \ldots, n\}$ satisfying $g(x + e_j) \le b$.

Similar to singly constrained cases, we can use a simple heuristic in Step 0 of the extended algorithm for $(MNKP)$ to generate a good initial feasible point and to improve feasible solutions generated in the dual search process.

PROCEDURE 6.6 (A GENERAL HEURISTIC FOR $(MNKP)$)

Given a feasible solution $x = (x_1, x_2, \ldots, x_n)^T$ to $(MNKP)$.

Step 1. For $j = 1, 2, \ldots, n$, if there is j such that

$$b_i - g_i(x) \geq g_{ij}(x_j + 1) - g_{ij}(x_j), \forall i = 1, \ldots, m, \tag{6.6.2}$$

then set $x := x + e_j$, where e_j is the j-th unit vector of $\mathbb{R}^n$.

Step 2. Repeat Step 1 until (6.6.2) does not hold for any j.

Based on the results in Section 6.5, heuristics of finding better feasible solutions to reliability problem can be developed.

PROCEDURE 6.7 (A SPECIAL HEURISTIC FOR $(LCROP)$)
Given a feasible solution $x = (x_1, x_2, \ldots, x_n)^T$ to problem $(LCROP)$.

Step 1. For each $j \in \{1, \ldots, n\}$, set $x := x + e_j$ if $g(x + e_j) \leq b$.

Step 2. If there exists a pair (i, j) with $i < j$ such that $x_i < x_j$ and

$$b_k - g_k(x) \geq a_{ki} - a_{kj}, \ \forall k = 1, \ldots, m, \tag{6.6.3}$$

then set $x := x + e_i - e_j$.

Step 3. Repeat Steps 1-2 until there is no such pair (i, j) satisfying (6.6.3).

PROCEDURE 6.8 (A SPECIAL HEURISTIC FOR $(LCOST)$)
Given a feasible solution $x = (x_1, x_2, \ldots, x_n)^T$ to problem $(LCOST)$.

Step 1. For each $j \in \{1, \ldots, n\}$, set $x := x - e_j$ if $R(x - e_j) \geq R_0$.

Step 2. If there exists a pair (i, j), with $i < j$ such that

$$x_i < x_j \text{ and } c_i \leq c_j, \tag{6.6.4}$$

then set $x := x + e_i - e_j$.

Step 3. Repeat Steps 1-2 until there is no such pair (i, j) satisfying (6.6.4).

6.6.3 Numerical results of Algorithm 6.2 for singly constrained cases

In order to compare the computational effects of the different sub-domain selection rules in Step 1 of Algorithm 6.2, we first tested the algorithm for 4 types of test problems with $n = 200$ and $m = 1$ when using different selection rules. The results are reported in Table 6.3. It is obvious from Table 6.3 that the algorithm with Rule (a) outperformed the ones with Rules (b) or (c) for all the test problems. All the following numerical results were obtained by using Rule (a).

Tables 6.4–6.9 summarize the numerical results of Algorithm 6.2 for Problems 6.1–6.6 with a single constraint. We see that the Lagrangian and domain cut method can solve different kinds of large-scale singly constrained nonlinear knapsack problems efficiently. The results in Tables 6.4–6.9 also indicate that the algorithm is most efficient in solving problem (QP_2) in terms of the total integer boxes generated by the algorithm and the average CPU time. This could be due to, in part, the fact that the Lagrangian relaxation always achieves its optimal solution at one of the extreme points of the integer box. Additional fathoming rules based on the results in Section 6.5 are used in the algorithm for problems $(LCROP)$ and $(LCOST)$. Comparing the results in Tables 6.4–6.9, we can see that the efficiency of the convergent Lagrangian and domain cut method does not depend significantly on the convexity of the problems.

Table 6.3. Comparison of node selection rules ($n = 200$, $m = 1$).

Problem	Rule (a) Average CPU Seconds	Rule (b) Average CPU Seconds	Rule (c) Average CPU Seconds
QP_1	2.5	4.4	4.7
QP_2	0.9	1.4	1.4
$PLOY$	2.7	5.2	3.9
$SAMP$	3.8	5.0	8.2
$LCROP$	5.1	5.7	6.5
$LCOST$	8.8	12.6	15.9

Table 6.4. Numerical results for (QP_1) with single constraint.

n	Average Number of Integer Boxes	Average CPU Time (Seconds)
400	23044	13.9
600	85951	81.1
1000	139275	233.9
1500	411982	1165.2

Table 6.5. Numerical results for (QP_2) with single constraint.

n	Average Number of Integer Boxes	Average CPU Time (Seconds)
500	15444	12.4
1000	80127	132.8
2000	207011	785.7
2500	276592	1431.9

Table 6.6. Numerical results for $(POLY)$ with single constraint.

n	Average Number of Integer Boxes	Average CPU Time (Seconds)
500	30017	30.7
1000	97212	214.5
1500	171976	614.9
2000	269877	1382.1

Table 6.7. Numerical results for $(SAMP)$ with single constraint.

n	Average Number of Integer Boxes	Average CPU Time (Seconds)
400	38817	34.5
600	67096	93.7
1000	187280	447.6
1500	344526	1329.5

6.6.4 Numerical results of Algorithm 6.2 for multiply constrained cases

Two versions of the extended Algorithm 6.2 for multiply constrained problems are programmed using the *outer Lagrangian linearization method* and the *subgradient method* respectively, as dual search procedures for solving (6.3.9). The stepsize in the subgradient method is taken as $t_k = 1/(2k)$. The maximum number of iterations to terminate the subgradient method is set to be 500. The numerical results for Problems 6.1–6.5 with multiple constraints are summarized in Tables 6.10–6.14, where OLL and SG stand for the algorithms using the

Table 6.8. Numerical results for $(LCROP)$ with single constraint.

n	Average Number of Integer Boxes	Average CPU Time (Seconds)
400	19169	45.9
600	23917	90.1
1000	67020	481.7
1500	112432	1295.4

Table 6.9. Numerical results for $(LCOST)$ with single constraint.

n	Average Number of Integer Boxes	Average CPU Time (Seconds)
400	26381	61.2
600	47118	173.9
1000	107423	727.2
1200	163399	1399.2

outer Lagrangian linearization method and the subgradient method as the dual search procedures, respectively, NS denotes the situation where the algorithm did not find the solutions for 20 test problems in 24 CPU hours, and

$$Ratio = \frac{\text{Average CPU Seconds Used by OLL}}{\text{Average CPU Seconds Used by SG}}.$$

From Tables 6.10–6.14, we see that Algorithm 6.2 can find the exact solutions of large-scale multi-dimensional nonlinear knapsack problems within reasonable computation time. Comparing the results in Tables 6.10–6.14, we observe that the algorithm using the outer Lagrangian linearization method is 3-5 times faster than that using the subgradient method.

6.6.5 Numerical results of Algorithm 6.3

Table 6.15 summarizes the numerical results for Algorithm 6.3, where min, max and avg stand for minimum, maximum and average, respectively. From Table 6.15, we can see that the linear approximation and partition method can find exact optimal solutions of concave knapsack problems with up to 1200 integer variables in reasonable computation time.

Table 6.10. Numerical results for (QP_1) with multiple constraints.

$n \times m$	Average CPU Times (Seconds)		Average Number of Integer Subboxes		*Ratio*
	OLL	SG	OLL	SG	
20×5	3.2	12.3	1304	1122	0.26
50×5	156.3	425.48	26530	16023	0.37
100×5	1832.1	NS	147057	–	–
30×10	43.3	948.8	8812	49195	0.05
30×20	176.6	NS	23501	–	–
30×30	350.7	NS	43502	–	–

Table 6.11. Numerical results for (QP_2) with multiple constraints.

$n \times m$	Average CPU Times (Seconds)		Average Number of Integer Subboxes		*Ratio*
	OLL	SG	OLL	SG	
20×5	5.7	19.4	1072	2255	0.29
40×5	273.2	644.0	26477	42286	0.42
60×5	2184.6	NS	37937	–	–
30×3	10.4	27.1	2324	2438	0.39
30×10	195.8	NS	13483	–	–
30×30	1930.4	NS	37915	–	–

Table 6.12. Numerical results for $(POLY)$ with multiple constraints.

$n \times m$	Average CPU Times (Seconds)		Average Number of Integer Subboxes		*Ratio*
	OLL	SG	OLL	SG	
20×5	2.8	9.2	901	688	0.31
40×5	43.8	167.2	7042	6845	0.26
60×5	417.0	NS	40342	–	–
30×10	34.6	613.5	5064	30925	0.06
30×20	82.8	NS	9251	–	–
30×30	228.6	NS	25320	–	–

6.6.6 Comparison results

We have compared the performance of the Algorithm 6.2 and its extension with the following methods:

Table 6.13. Numerical results for $(SAMP)$ with multiple constraints.

$n \times m$	Average CPU Time (Seconds)		Average Number of Integer Subboxes		*Ratio*
	OLL	SG	OLL	SG	
30×5	73.5	206.88	8681	12414	0.36
40×5	289.7	1213.83	24228	57230	0.24
50×5	1273.3	NS	83995	–	–
30×3	24.2	70.40	4833	4103	0.34
30×10	251.1	NS	15815.5	–	–
30× 30	392.1	NS	15427.1	–	–

Table 6.14. Numerical results for $(LCROP)$ with multiple constraints.

n	m	Average Number of Integer Subboxes	Average CPU Time (Seconds)
50	5	2034	11.6
100	5	25895	284.6
150	5	67773	1255.9
90	10	24770	394.3
90	30	33617	869.8
90	50	61766	2455.7

Table 6.15. Numerical results for $(CCKP)$.

n	Number of Iterations			Number of Integer Subboxes			CPU Seconds		
	Min	Max	Avg	Min	Max	Avg	Min	Max	Avg
200	15	128	50	1664	13386	5481	1.31	10.51	4.25
400	16	261	106	4516	48558	22782	9.61	104.09	47.11
600	18	262	94	7046	72811	32373	28.61	280.22	127.23
800	17	517	193	8675	201691	76347	57.0	1214.76	464.57
1000	10	691	231	5407	360897	119775	53.52	3398.96	1116.32
1200	27	679	237	20245	523844	162477	255.21	7157.14	2183.64

- **0-1 Linearization**: 0-1 linearization method of Hochbaum (see Section 6.2)
- **B & B**: Pegging method of Brettauer and Shetty (see Section 6.1)

- **Hybrid Method**: Hybrid method of Marstern and Morin (see Section 7.2).

Note that the 0-1 linearization and the pegging method (branch-and-bound) of Brettauer and Shetty can be only applied to singly constrained convex separable integer programming problems.

The first set of test problems is for singly constrained *convex* knapsack problems: (QP_1), $(SAMP)$, $(LCROP)$ and $(LCOST)$. Comparison results with n ranged from 50 to 150 are reported in Table 6.16, where Domain Cut represents a version of Algorithm 6.2 for convex (NKP). NS denotes the situation where the algorithm did not find the exact solution in 24 hours for the 20 problems.

Table 6.16. Comparison results for convex knapsack problems.

Problem	n	Domain Cut	0-1 Linearization	B&B	Hybrid Method
		Average CPU Seconds	Average CPU Seconds	Average CPU Seconds	Average CPU Seconds
	50	0.05	< 0.01	0.32	10.3
QP_1	100	0.3	< 0.01	16.5	243.7
	150	1.3	0.01	485.1	NS
	40	0.07	< 0.01	1071.8	4.1
$SAMP$	100	0.6	0.01	2367.1	183.0
	150	1.7	0.02	NS	NS
	50	0.2	< 0.01	1541.8	31.1
$LCROP$	100	0.8	0.01	NS	180.6
	150	2.4	0.02	NS	NS
	50	0.09	< 0.01	623.5	8.8
$LCOST$	100	1.4	0.01	NS	212.5
	150	3.6	0.03	NS	NS

The average CPU time in Table 6.16 indicates that the domain cut algorithm is much more efficient than other methods except for the 0-1 linearization method. One theoretical reason behind the out-performance of the convergent Lagrangian and domain cut method to the continuous relaxation-based branch-and-bound method could be that for convex integer programming problems, the Lagrangian bound is never worse than the continuous bound, as stated in Theorem 2.4. For 0-1 linearization method, the greedy method for the transformed 0-1 linear knapsack problem (0-1KP) generates high-quality feasible solutions and thus making the branch-and-bound method for (0-1KP) very efficient.

The second set of test problems is for singly constrained nonconvex knapsack problems: (QP_2) and $(POLY)$ with n ranging from 100 to 200. For this set of problems, only the convergent Lagrangian and domain cut method and the hybrid method are applicable. Table 6.17 summarizes the comparison results.

Table 6.17. Comparison results for nonconvex knapsack problems.

Problem	n	Domain Cut	Hybrid Method
		Average CPU Seconds	Average CPU Seconds
	100	0.16	26.6
QP_2	150	0.50	131.0
	200	0.89	397.0
	100	0.25	64.3
$POLY$	150	1.2	378.4
	200	2.7	NS

It is clear from Table 6.17 that the domain cut algorithm uses much less CPU time than the hybrid method in finding the exact solution of singly constrained nonconvex knapsack problems.

The third set of test problems is for multidimensional knapsack problems. Again, only Algorithm 6.2 and the hybrid method are applicable to this set of problems. The comparison results are reported in Table 6.18. From Table 6.17, it is clear that Algorithm 6.2 outperforms significantly over the hybrid method. Part of the reason is that the dynamic programming is inefficient for multiply constrained problems in generating efficient solutions due to the "curse of dimensionality."

6.7 Notes

Problems (NKP) and $(MNKP)$ are natural extensions of the classical 0-1 knapsack problems and bounded knapsack problems which have been extensively studied in the literature (see the book of Martello and Toth [153] and the recent book of Kellerer, Pferschy and Pisinger [117]).

Resource allocation problems, which can be viewed as a special class of nonlinear knapsack problems with a *packing* constraint $\sum_{j=1}^{n} x_j = b$, have also been well studied. The algorithms for various resource allocation problems were summarized in Ibaraki and Katoh's book [106].

Algorithms for the continuous version of convex (NKP) and branch-and-bound methods based on the continuous relaxation of the convex case of (NKP) were studied in [34][36][122][159]. The 0-1 linearization methods for convex

Table 6.18. Comparison results for multidimensional knapsack problems.

Problem	n	m	Domain Cut	Hybrid Method
			Average CPU Seconds	Average CPU Seconds
QP_1	10	3	0.29	25.5
	15	3	1.8	368.7
	20	3	4.8	NS
QP_2	10	3	0.08	9.7
	15	3	0.43	51.6
	20	3	2.5	376.9
$POLY$	10	3	0.40	6.2
	15	3	0.92	76.6
	20	3	1.6	374.0
$SAMP$	10	3	0.42	15.9
	15	3	1.9	212.0
	20	3	6.4	NS
$LCROP$	10	3	0.16	7.4
	15	3	0.80	46.6
	20	3	1.2	173.7

(NKP) were presented in [101][154][155]. Using a surrogate technique, the 0-1 linearization method was extended in [51] to deal with quadratic multidimensional knapsack problems.

The Lagrangian dual and domain cut method in Section 6.3 was presented in [141] (see also [139]). The optimality conditions for reliability problems $(CROP)$ and $(COST)$ were derived in [205]. The algorithm in Section 6.4 for concave knapsack problems was proposed in [208].

Chapter 7

SEPARABLE INTEGER PROGRAMMING

In this chapter, we consider the following general class of separable integer programming problems:

$$(P) \qquad \min\ f(x) = \sum_{j=1}^{n} f_j(x_j)$$
$$\text{s.t. } g_i(x) = \sum_{j=1}^{n} g_{ij}(x_j) \le b_i,\ i = 1,\ldots,m,$$
$$x \in X = X_1 \times X_2 \times \cdots \times X_n,$$

where f_j and g_{ij}'s are defined on $\mathbb{R}$, and all X_j's are finite integer sets in $\mathbb{R}$. Let $g(x) = (g_1(x), g_2(x), \ldots, g_m(x))^T$ and $b = (b_1, b_2, \ldots, b_m)^T$. Problem (P) covers very general situations of nonlinear integer programming problems as no additional property such as convexity, concavity, monotonicity or differentiability is assumed in (P).

In Section 7.1 we discuss the conventional dynamic programming method for solving (P). In Section 7.2, a hybrid method that combines solution strategies of branch-and-bound, domination and surrogate with dynamic programming is discussed to partially overcome the difficulty caused by the "curse of dimensionality." In Section 7.3, a novel convergent Lagrangian and objective level cut method is discussed for (P), which is an exact solution scheme and is efficient in implementation by retaining the decomposability of (P).

7.1 Dynamic Programming Method

Dynamic programming has been widely used in discrete optimization. The separability of both the objective function f and constraint functions g_i's makes dynamic programming method an ideal technique to solve (P). A key assumption for an efficient implementation of a dynamic programming method for (P) is the integrality of g_i's.

ASSUMPTION 7.1 *Function g_{ij} is integer-valued, for all $j = 1, \ldots, n$ and $i = 1, \ldots m$.*

To apply dynamic programming, we first introduce a stage variable k, $0 \leq k \leq n$, and a state vector at stage k, $s_k \in \mathbb{R}^m$, satisfying the following recursive equation:

$$s_{k+1} = s_k + g^k(x_k), \; k = 1, \ldots, n-1, \tag{7.1.1}$$

with an initial condition $s_1 = 0$, where

$$g^k(x_k) = (g_{1k}(x_k), \ldots, g_{mk}(x_k))^T.$$

Since the constraints are integer-valued, we only need to consider integer points in the state space. Furthermore, the feasible region of the state vector at stage k with $2 \leq k \leq n+1$ can be confined as follows:

$$\underline{s}_k \leq s_k \leq \bar{s}_k,$$

where

$$\underline{s}_k = \begin{bmatrix} \sum_{t=1}^{k-1} \min_{x_t \in X_t} g_{1t}(x_t) \\ \vdots \\ \sum_{t=1}^{k-1} \min_{x_t \in X_t} g_{mt}(x_t) \end{bmatrix} \tag{7.1.2}$$

and

$$\bar{s}_k = \begin{bmatrix} \min\{\sum_{t=1}^{k-1} \max_{x_t \in X_t} g_{1t}(x_t), b_1 - \sum_{t=k}^{n} \min_{x_t \in X_t} g_{1t}(x_t)\} \\ \vdots \\ \min\{\sum_{t=1}^{k-1} \max_{x_t \in X_t} g_{mt}(x_t), b_m - \sum_{t=k}^{n} \min_{x_t \in X_t} g_{mt}(x_t)\} \end{bmatrix}. \tag{7.1.3}$$

Dynamic programming can be applied to solve problem (P) either by a backward recursion or by a forward recursion.

7.1.1 Backward dynamic programming

For a given state s at stage k, $1 \leq k \leq n$, define the cost-to-go function as follows,

$$\hat{t}_k(s) = \min \sum_{j=k}^{n} f_j(x_j),$$

$$\text{s.t. } s + \sum_{j=k}^{n} g^j(x_j) \leq b,$$

$$x_j \in X_j, \; j = k, \ldots, n.$$

It is obvious that

$$v(P) = \hat{t}_1(0).$$

Based on Bellman's principle of optimality, the cost-to-go function satisfies the following backward recursive relation for $k = n-1, n-2, \ldots, 1$,

$$\hat{t}_k(s) = \min_{x_k \in X_k} \{f_k(x_k) + \hat{t}_{k+1}(s + g^k(x_k))\}$$

with boundary condition

$$\hat{t}_n(s) = \min_{x_n \in X_n} \{f_n(x_n) \mid s + g^n(x_n) \le b\}.$$

Define

$$x_n^*(s) = \arg \min_{x_n \in X_n} \{f_n(x_n) \mid s + g^n(x_n) \le b\},$$

$$x_k^*(s) = arg \min_{x_k \in X_k} \{f_k(x_k) + \hat{t}_{k+1}(s + g^k(x_k))\}, \ k = n-1, \ldots, 1.$$

The backward dynamic programming starts at $k = n-1$ and moves backwards, $k = n-2, \ldots, 1$. It calculates the cost-to-go recursively for every s at stage k between $\underline{s}_k$ and $\bar{s}_k$ and finally stops at $s_1 = 0$. The tracing process is then carried out in a forward way to identify the optimal solution of (P). Starting from $x_1^*(0)$, the optimal state at stage 2 is obtained as $s_2^* = g^1(x_1^*(0))$. The algorithm then identifies the optimal solution at stage 2, $x_2^*(s_2^*)$, which yields the optimal state at stage 3, $s_3^* = s_2^* + g^2(x_2^*(s_2^*))$. The process terminates when it reaches s_n^* and finds out $x_n^*(s_n^*)$.

7.1.2 Forward dynamic programming

For a given state s at stage k, $2 \le k \le n+1$, define the cost-to-accumulate function as follows,

$$\begin{aligned} \tilde{t}_k(s) = \min & \sum_{j=1}^{k-1} f_j(x_j), \\ \text{s.t.} & \sum_{j=1}^{k-1} g^j(x_j) \le s, \\ & x_j \in X_j, \ j = 1, \ldots, k-1. \end{aligned}$$

It is obvious that

$$v(P) = \min\{\tilde{t}_{n+1}(s) \mid s \le b\}.$$

Based on Bellman's principle of optimality, the cost-to-accumulate function satisfies the following forward recursive relation for $k = 3, \ldots n+1$,

$$\tilde{t}_k(s) = \min_{x_{k-1} \in X_{k-1}} \{f_{k-1}(x_{k-1}) + \tilde{t}_{k-1}(s - g^{k-1}(x_{k-1}))\},$$

with boundary condition

$$\tilde{t}_2(s) = \min_{x_1 \in X_1} \{f_1(x_1) \mid g^1(x_1) \le s\}.$$

Define

$$x_1^*(s) = \arg \min_{x_1 \in X_1} \{f_1(x_1) \mid g^1(x_1) \le s\},$$

$$x_{k-1}^*(s) = \arg \min_{x_{k-1} \in X_{k-1}} \{f_{k-1}(x_{k-1}) + \tilde{t}_{k-1}(s - g^{k-1}(x_{k-1}))\},$$

$$k = 2, \dots, n+1.$$

The forward dynamic programming starts at $k = 2$ and moves forward, $k = 3, \dots, n+1$. It calculates the cost-to-accumulate recursively for every s at stage k between $\underline{s}_k$ and $\bar{s}_k$ and finally stops at stage $n+1$. Let

$$s_{n+1}^* = \arg\min\{\tilde{t}_{n+1}(s) \mid s \le b\}.$$

The tracing process is then carried out in a backward way to identify the optimal solution of (P). Starting from $x_n^*(s_{n+1}^*)$, the optimal state at stage n is obtained as $s_n^* = s_{n+1}^* - g^n(x_n^*(s_{n+1}^*))$. The algorithm then identifies the optimal solution at stage n, $x_{n-1}^*(s_n^*)$, which yields the optimal state at stage $n-1$, $s_{n-1}^* = s_n^* - g^{n-1}(x_{n-1}^*(s_n^*))$. The process terminates when it reaches s_2^* and finds out $x_1^*(s_2^*)$.

EXAMPLE 7.1

$$\begin{aligned} \min\ & f(x) = 3x_1^2 - 4x_2^3 + 5x_3 \\ \text{s.t.}\ & g_1(x) = x_1^3 - x_2 - x_3^2 \le 0, \\ & g_2(x) = -x_1 + x_2^3 + x_3^2 \le 1, \\ & x_i \in \{-1, 0, 1\},\ i = 1, 2, 3. \end{aligned}$$

The optimal solution is $x^* = (1, 1, -1)^T$ with $f(x^*) = -6$.

Using the formulas in (7.1.2) and (7.1.3), the feasible regions of the state vector can be found as follows for $k = 2$, 3, and 4,

$$\begin{bmatrix} -1 \\ -1 \end{bmatrix} \le s_2 \le \begin{bmatrix} \min\{1,2\} \\ \min\{1,2\} \end{bmatrix},$$

$$\begin{bmatrix} -2 \\ -2 \end{bmatrix} \le s_3 \le \begin{bmatrix} \min\{2,1\} \\ \min\{2,1\} \end{bmatrix},$$

$$\begin{bmatrix} -3 \\ -2 \end{bmatrix} \le s_4 \le \begin{bmatrix} \min\{2,0\} \\ \min\{3,1\} \end{bmatrix}.$$

Table 7.1 gives the solution processes using backward dynamic programming.

Table 7.1. Solution process for Example 7.1 using backward dynamic programming.

s_1	$x_1^*(s_1)/\hat{t}_1(s_1)$	s_2	$x_2^*(s_2)/\hat{t}_2(s_2)$	s_3	$x_3^*(s_3)/\hat{t}_3(s_3)$
$(0,0)^T$	1/−6	$(-1,-1)^T$	1/−9	$(-2,-2)^T$	−1/−5
		$(-1,0)^T$	0/−5	$(-2,-1)^T$	−1/−5
		$(-1,1)^T$	−1/−1	$(-2,0)^T$	−1/−5
		$(0,-1)^T$	1/−9	$(-2,1)^T$	0/0
		$(0,0)^T$	0/−5	$(-1,-2)^T$	−1/−5
		$(0,1)^T$	−1/−1	$(-1,-1)^T$	−1/−5
		$(1,-1)^T$	1/−9	$(-1,0)^T$	−1/−5
		$(1,0)^T$	0/−5	$(-1,1)^T$	0/0
		$(1,1)^T$	infeasible/∞	$(0,-2)^T$	−1/−5
				$(0,-1)^T$	−1/−5
				$(0,0)^T$	−1/−5
				$(0,1)^T$	0/0
				$(1,-2)^T$	−1/−5
				$(1,-1)^T$	−1/−5
				$(1,0)^T$	−1/−5
				$(1,1)^T$	infeasible/∞

The solution process using backward dynamic programming starts from stage 3. For each possible s_3, the optimal decision $x_3^*(s_3)$ is found and the corresponding optimal cost-to-go $\hat{t}_3(s_3)$ is recorded. For example, at $s_3 = (-1,1)^T$, both $x_3 = 1$ and $x_3 = -1$ are infeasible. The optimal decision $x_3^*((-1,1)^T)$ is found to be 0 and the corresponding $\hat{t}_3((-1,1)^T)$ is 0. If there does not exist a feasible solution at s_3, $x_3^*(s_3)$ is set as ∞. Then, we move back to stage 2. At each possible s_2, we compare $f_2(x_2) + \hat{t}_3(s_2 + g^2(x_2))$ for $x_2 = -1$, 0 and 1 and find out $x_2^*(s_2)$ and the corresponding optimal cost-to-go $\hat{t}_2(s_2)$. For example, at $s_2 = (-1,1)^T$, comparison of $-4(-1)^3+\hat{t}_3((0,0)^T) = -1$, $-4(0)^3+\hat{t}_3((-1,1)^T) = 0$, and $-4(1)^3+\hat{t}_3((-2,2)^T) = \infty$ yields $x_2^*((-1,1)^T) = -1$ and $\hat{t}_2((-1,1)^T) = -1$. Finally, we move back to stage 1. Checking $f_1(x_1) + \hat{t}_2((0,0)^T + g^1(x_1))$ for $x_1 = -1$, 0 and 1 gives $x_1^*(s_1 = (0,0)^T) = 1$ and $\hat{t}_1(s_1 = (0,0)^T) = -6$. Tracing back, we find the optimal solution for the example problem: $x_1 = x_2 = 1$ and $x_3 = -1$.

Next we examine how the forward dynamic programming is used to solve Example 7.1. Table 7.2 summarizes the solution process.

The solution process using forward dynamic programming starts from stage 2 and ends at stage 4. Minimizing $\tilde{t}_4$ with respect to $s_4 \le (0,1)^T$ finds out the optimal value of the example problem $\tilde{t}_4((-1,1)^T) = -6$. Tracing back identifies optimal solution: $x_3^* = -1$, $x_2^* = x_1^* = 1$.

Table 7.2. Solution process for Example 7.1 using forward dynamic programming.

s_2	$x_1^*(s_2)/\tilde{t}_2(s_2)$	s_3	$x_2^*(s_3)/\tilde{t}_3(s_3)$	s_4	$x_3^*(s_4)/\tilde{t}_4(s_4)$
$(-1,-1)^T$	infeasible/∞	$(-2,-2)^T$	infeasible /∞	$(0,1)^T$	$-1/-5$
$(-1,0)^T$	infeasible /∞	$(-2,-1)^T$	infeasible/∞	$(-1,1)^T$	$-1/-6$
$(-1,1)^T$	$-1/3$	$(-2,0)^T$	infeasible/∞	$(-2,1)^T$	infeasible/∞
$(0,-1)^T$	infeasible/∞	$(-2,1)^T$	infeasible/∞	$(-3,1)^T$	infeasible/∞
$(0,0)^T$	0/0	$(-1,-2)^T$	infeasible/∞	$(0,0)^T$	$-1/-2$
$(0,1)^T$	0/0	$(-1,-1)^T$	infeasible/∞	$(-1,0)^T$	infeasible/∞
$(1,-1)^T$	1/3	$(-1,0)^T$	infeasible/∞	$(-2,0)^T$	infeasible/∞
$(1,0)^T$	0/0	$(-1,1)^T$	$1/-4$	$(-3,0)^T$	infeasible/∞
$(1,1)^T$	0/0	$(0,-2)^T$	infeasible/$-\infty$	$(0,-1)^T$	infeasible/∞
		$(0,-1)^T$	infeasible/∞	$(-1,-1)^T$	infeasible/∞
		$(0,0)^T$	$1/-1$	$(-2,-1)^T$	infeasible/∞
		$(0,1)^T$	$1/-4$	$(-3,-1)^T$	infeasible/∞
		$(1,-2)^T$	infeasible/∞	$(0,-2)^T$	infeasible/∞
		$(1,-1)^T$	0/3	$(-1,-2)^T$	infeasible/∞
		$(1,0)^T$	0/0	$(-2,-2)^T$	infeasible/∞
		$(1,1)^T$	0/0	$(-3,-2)^T$	infeasible/∞

Determining the feasible region could become a tedious task in applying dynamic programming. This difficulty can be alleviated to certain degree when the following assumption is satisfied.

ASSUMPTION 7.2 *For all $j = 1,\ldots,n$ and $i = 1, \ldots m$, function g_{ij} is integer-valued and is nonnegative for all $x_j \in X_j$.*

When Assumption 7.2 is satisfied, the range of s_k at stage k, for $k = 2, 3, \ldots, n, n+1$, can be simply determined by $[(0,\ldots,0)^T, (b_1,\ldots,b_m)^T]$.

If the nonnegativity assumption does not hold for some g_{ij}, then we can subtract $\min_{x_j \in X_j} g_{ij}(x_j)$ from both g_{ij} and b_i at the same time. Repeating this equivalent transformation for all g_{ij}'s that do not possess the nonnegativity property such that Assumption 7.2 holds for the transformed problem. The range of $(s_k)_i$ at stage k for $k = 2, 3, \ldots, n, n+1$ can be then given by $[0, b_i - \sum_{j \in I_i} \min_{x_j \in X_j} g_{ij}]$, where $I_i = \{j = 1,\ldots,n \mid \min_{x_j \in X_j} g_{ij} < 0\}$. The price to perform such a transformation is an enlargement of the feasible region of the state space which affects an efficient implementation of dynamic programming.

It is evident that the number of the possible states increases exponentially with respect to the number of constraints. Thus, although dynamic programming is conceptually an ideal solution scheme for separable integer programming, the "curse of dimensionality" prevents its direct application to multiply constrained cases of (P) when m is large. Dynamic programming, however, remains as an efficient solution scheme for separable integer programming problems when m is small, especially for singly constrained cases.

7.1.3 Singly constrained case

Consider the singly constrained case of (P):

$$
\begin{aligned}
(P_1) \qquad & \min\ f(x) = \sum_{j=1}^{n} f_j(x_j) \\
& \text{s.t. } g(x) = \sum_{j=1}^{n} g_j(x_j) \le b, \\
& \quad x \in X = X_1 \times X_2 \times \cdots \times X_n,
\end{aligned}
$$

where $X_j = \{x_j \in \mathbb{Z} \mid l_j \le x_j \le u_j\}$ with l_j and u_j being integers. We assume $g_j(x_j) \ge 0$ on X_j for all $j = 1, \ldots, n$.

For adopting backward dynamic programming, the cost-to-go function is defined first as follows,

$$
\begin{aligned}
\hat{t}_k(s) &= \min \sum_{j=k}^{n} f_j(x_j), \\
\text{s.t. } & s + \sum_{j=k}^{n} g_j(x_j) \le b, \\
& x_j \in X_j,\ j = k, \ldots, n,
\end{aligned}
$$

for $k = 1, \ldots, n-1$, $s = 0, \ldots, b$. The *backward* recursive equation is

$$
\begin{aligned}
& \hat{t}_k(s) = \min\{f_k(x_k) + \hat{t}_{k+1}(s + g_k(x_k))\} \\
& \text{s.t. } s + g_k(x_k) \le b, \\
& \quad x_k = l_k, \ldots, u_k,
\end{aligned}
$$

for $k = n-1, \ldots, 1$, $s = 0, \ldots, b$, with boundary conditions

$$
\begin{aligned}
\hat{t}_k(s) &= +\infty, \ \text{for } s < 0,\ k = 1, \ldots, n, \\
\hat{t}_n(s) &= \min\{f_n(x_n) \mid s + g_n(x_n) \le b,\ x_n = l_n, l_n + 1, \ldots, u_n\}, \\
& \quad s = 0, \ldots, b.
\end{aligned}
$$

For adopting forward dynamic programming, we define the following cost-to-accumulate function,

$$\begin{aligned}\tilde{t}_k(s) = \min & \sum_{j=1}^{k-1} f_j(x_j),\\ \text{s.t.} & \sum_{j=1}^{k-1} g_j(x_j) \le s,\\ & x_j \in X_j,\ j = 1,\ldots,k-1.\end{aligned}$$

The *forward* recursive equation is

$$\begin{aligned}\tilde{t}_k(s) &= \min f_k(x_k) + \tilde{t}_{k-1}(s - g_k(x_k))\\ \text{s.t.}\ & g_k(x_k) \le s,\\ & x_k = l_k, l_k+1,\ldots,u_k,\end{aligned}$$

for $k = 3,\ldots,n$, $s = 0,\ldots,b$, with boundary conditions

$$\begin{aligned}\tilde{t}_j(s) &= +\infty,\ \text{for } s < 0,\ j = 1,\ldots,n,\\ \tilde{t}_2(s) &= \min\{f_1(x_1) \mid g_1(x_1) \le s,\ x_1 = l_1, l_1+1,\ldots,u_1\},\\ & \quad s = 0,\ldots,b.\end{aligned}$$

The dynamic programming table has a size of $n \times (b+1)$.

EXAMPLE 7.2

$$\begin{aligned}\min\ & f(x) = -2\sqrt{x_1} - 2x_2 - x_3^2 - (1/2)x_4^3\\ \text{s.t.}\ & g(x) = 3x_1 - x_1^2 + x_2 + x_3^2 + x_4 \le 5,\\ & x \in [0,2]^4 \cap \mathbb{Z}^4.\end{aligned}$$

The optimal solution is $x^* = (0,2,1,2)^T$ with $f(x^*) = -9$.

Table 7.3 shows the process of the forward dynamic programming for this example, where $w_k(s) = s - g_k(x_k^*(s))$.

Thus $\tilde{t}_4(5)$ is the optimal value and the optimal solution can be obtained by backtracking out through the table:

$$\begin{aligned}& s_5^* = 5, x_4^* = 2 \Longrightarrow s_5^* - g_4(x_4^*) = 3\\ & s_4^* = 3, x_3^* = 1 \Longrightarrow s_4^* - g_3(x_3^*) = 2\\ & s_3^* = 2, x_2^* = 2 \Longrightarrow s_3^* - g_2(x_2^*) = 0\\ & s_2^* = 0, x_1^* = 0.\end{aligned}$$

Therefore the optimal solution is $x^* = (0,2,1,2)^T$.

Table 7.3. Dynamic programming table for Example 7.2.

s	$\tilde{t}_2(s)/x_1^*(s)$	$\tilde{t}_3(s)/x_2^*(s)/w_2(s)$	$\tilde{t}_4(s)/x_3^*(s)/w_3(s)$	$\tilde{t}_5(s)/x_4^*(s)/w_4(s)$
0	0/0	0/0/0	0/0/0	0/0/0
1	0/0	−2/1/0	−2/0/1	−2/0/1
2	−2.8284/2	−4/2/0	−4/0/2	−4/2/0
3	−2.8284/2	−4.8284/1/2	−5/1/2	−6/2/1
4	−2.8284/2	−6.8284/2/2	−6.8284/0/4	−8/2/2
5	−2.8284/2	−6.8284/1/2	−7.8284/1/4	−9/2/3

7.2 Hybrid Method

In this section, we introduce a hybrid method for (P) which combines the dynamic programming with dominance rules and branch-and-bound method. The purpose of the hybrid method is to partially overcome the curse of dimensionality and the basic idea of the method is to recursively generate the efficient feasible solutions of the problem and to remove in the solution process the inefficient feasible solutions by dominance rules. Branch-and-bound strategy is employed to remove nonpromising incomplete solutions during the recursion. We assume in this section that X_j has the form: $X_j = \{0, 1, 2, \ldots, K_j\}$, $j = 1, \ldots, n$. For convenience, we consider problem (P) in the following maximization form:

$$(P_2) \qquad \max\ f(x) = \sum_{j=1}^{n} f_j(x_j)$$

$$\text{s.t. } g_i(x) = \sum_{j=1}^{n} g_{ij}(x_j) \le b_i, \ i = 1, \ldots, m,$$

$$x \in X = X_1 \times X_2 \times \cdots \times X_n.$$

We need the following assumption about (P_2):

ASSUMPTION 7.3 *For all $j = 1, \ldots, n$, the function $f_j(x_j)$ is nonnegative on X_j. For all $i = 1, \ldots, m$, b_i is nonnegative and $g_{ij}(x_j)$ is nonnegative on X_j for all $j = 1, \ldots, n$.*

7.2.1 Dynamic programming procedure

Consider the following k-stage subproblem of (P_2):

$$(SP_k) \qquad s_k(b) = \max f^k(x_1, \ldots, x_k) := \sum_{j=1}^{k} f_j(x_j)$$

$$\text{s.t. } g_i^k(x_1, \ldots, x_k) := \sum_{j=1}^{k} g_{ij}(x_j) \leq b_i, \ i = 1, \ldots, m,$$

$$x_j \in X_j, \ j = 1, \ldots, k.$$

Obviously, $f^* = s_n(b)$. Let S_k be a subset of the set of all partial feasible solutions of (SP_k):

$$S_k \subseteq \{(x_1, \ldots, x_k) \in X_1 \times X_2 \times \cdots X_k \mid g_i^k(x_1, \ldots, x_k) \leq b_i, \ i = 1, \ldots, m\}.$$

DEFINITION 7.1 A partial solution $x^1 \in S_k$ is said to be *dominated* by $x^2 \in S_k$ if $g_i^k(x^2) \leq g_i^k(x^1)$, $i = 1, \ldots, m$, and $f^k(x^2) \geq f^k(x^1)$ with at least one strict inequality. A partial solution $x \in S_k$ is said to be *efficient* with respect to S_k if it is not dominated by any other partial feasible solutions in S_k.

Let $S_1^0 = X_1$ and

$$S_1^f = \{x_1 \in X_1 \mid g_{i1}(x_1) \leq b_i, \ i = 1, \ldots, m\},$$
$$S_1^e = \{x_1 \in S_1^f \mid x_1 \text{ is efficient with respect to } S_1^f\}.$$

It holds $S_1^e \subseteq S_1^f \subseteq S_1^0$. For $k = 2, \ldots, n$, define the following recursively:

$$S_k^0 = \{(x_1, \ldots, x_{k-1}, x_k) \mid (x_1, \ldots, x_{k-1}) \in S_{k-1}^e, \ x_k \in X_k\}, \qquad (7.2.1)$$
$$S_k^f = \{(x_1, \ldots, x_k) \in S_k^0 \mid g_i^k(x_1, \ldots, x_k) \leq b_i, \ i = 1, \ldots, m\}, \qquad (7.2.2)$$
$$S_k^e = \{(x_1, \ldots, x_k) \in S_k^f \mid (x_1, \ldots, x_k) \text{ is efficient with respect to } S_k^f\}. \qquad (7.2.3)$$

It is clear that set S_k^e includes all efficient solutions of (SP_k) and set S_n^e is the set of all efficient solutions of (P_2). We can compute set S_n^e using (7.2.1)–(7.2.3) recursively.

PROCEDURE 7.1 (DP PROCEDURE FOR GENERATING S_n^e)

Step 0. Set $k = 1$, $S_1^0 = X_1$.

Step 1. Compute S_k^f by eliminating all infeasible solutions in S_k^0.

Step 2. Compute S_k^e by eliminating all dominated solutions in S_k^f.

Step 3. If $k = n$, stop. Otherwise, set $S_{k+1}^0 = S_k^e \times X_{k+1}$, $k := k+1$, go to Step 1.

Let x^* be an optimal solution to (P_2). It can be proved that $x^* \in S_n^e$ and for $0 \le y \le b$,

$$s_n(y) = \max\{\sum_{j=1}^n f_j(x_j) \mid x \in S_n^e, \ \sum_{j=1}^n g_{ij}(x_j) \le y_i, \ i = 1, \ldots, m\}.$$

7.2.2 Incorporation of elimination procedure

Now, we consider to further incorporate an elimination procedure into the above dynamic programming framework. For any $x = (x_1, \ldots, x_k) \in S_k^e$, let

$$\beta = \sum_{j=1}^k g^j(x_j),$$

where $g^j = (g_{1j}(x_j), \ldots, g_{mj}(x_j))^T$. It is clear that β represents the resource consumed by a partial solution $x = (x_1, \ldots, x_k)^T$. Define the following *residual* subproblem:

$$(RSP_k) \qquad \tilde{s}_{k+1}(b - \beta) = \max \sum_{j=k+1}^n f_j(x_j)$$

$$\text{s.t.} \ \sum_{j=k+1}^n g_{ij}(x_j) \le b_i - \beta_i, \ i = 1, \ldots, m,$$

$$x_j \in X_j, \ j = k+1, \ldots, n.$$

Thus, $\tilde{s}_{k+1}(b - \beta)$ represents the maximum return of the remaining $n - k + 1$ stages after resource β has been consumed in the first k stages. Let $UB_{k+1}(b - \beta)$ be an upper bound function of $\tilde{s}_{k+1}(b - \beta)$:

$$\tilde{s}_{k+1}(b - \beta) \le UB_{k+1}(b - \beta), \quad 0 \le \beta \le b.$$

UB_{k+1} can be computed by certain relaxation of (RSP_k), Lagrangian relaxation of (RSP_k), for example.

Let LB be a lower bound of $s_n(b)$ or (P_2), which could be determined by $f(x^*)$ with x^* being the incumbent of (P_2). A partial solution $(x_1, \ldots, x_k)$ can be eliminated from further consideration if

$$UB(x_1, \ldots, x_k) := \sum_{j=1}^k f_j(x_j) + UB_{k+1}(b - \beta) \le LB \qquad (7.2.4)$$

because no completion of x can be better than the incumbent. Let S_k^s denote the set of all efficient partial solutions at stage k after eliminating x that satisfies (7.2.4),

$$S_k^s = \{(x_1, \ldots, x_k) \in S_k^e \mid UB(x_1, \ldots, x_k) > LB\}. \tag{7.2.5}$$

Then, only the solutions in S_k^s are needed to generate potential solution at stage $k+1$. In order to incorporate this bound elimination process into the dynamic programming framework, we can redefine S_k^0 as $S_{k-1}^s \times X_k$ for $k = 2, \ldots, n$ and calculate S_k^f and S_k^e accordingly. Now, the hybrid method can be described as follows.

ALGORITHM 7.1 (HYBRID METHOD FOR (P_2))

Step 0. Choose an accuracy $\epsilon \geq 0$ and an integer $N \geq 1$ that controls the maximum number of solutions in S_k^s when an upper bound is computed and updated. Set $k = 1$, $S_1^0 = X_1$ and $UB = UB_1(b)$. Compute an initial feasible solution x^0 by certain heuristic method and set $LB = f(x^0)$.

Step 1. Compute S_k^f by eliminating all infeasible solutions in S_k^0.

Step 2. If $k = n$, stop and either S_n^f contains an optimal solution or the incumbent is optimal. Otherwise, compute S_k^e by eliminating all dominated solutions in S_k^f.

Step 3. If $|S_k^e| \leq N$, set $S_k^s = S_k^e$ and go to Step 8, where $|S_k^e|$ denotes the cardinality of set S_k^e.

Step 4. Compute S_k^s by (7.2.5).

Step 5. Calculate

$$UB' = \max\{\sum_{j=1}^{k} f_j(x_j) + UB_{k+1}(b - \beta) \mid x = (x_1, \ldots, x_k) \in S_k^s\},$$

where $\beta = \sum_{j=1}^{k} g^j(x_j)$. Set $UB = \min(UB', UB)$.

Step 6. Update the lower bound and incumbent if a better feasible solution is obtained during the computation of $UB_{k+1}(b - \beta)$ or by some heuristic method. Update S_k^s if necessary.

Step 7. If $(UB - LB)/UB \leq \epsilon$, then stop and the incumbent is an approximate optimal solution to (P_2).

Step 8. Set $S_{k+1}^0 = S_k^s \times X_{k+1}$ and $k := k + 1$, go to Step 1.

REMARK 7.1 Using the monotonicity of f_j and the nonnegativity of g_{ij} in problem (P_2), heuristic methods can be derived to obtain the feasible solution and lower bound LB in the algorithm (see Section 6.6). Since $|S_k^e| > 1$ holds in most situations, setting $N = 1$ leads to computing upper bound and lower bound at every stage. When ϵ is set to be 0, the algorithm finds an exact solution to (P_2). The finite convergence of Algorithm 7.1 is evident by observing that S_k^0, S_k^f and S_k^s are finite sets and at most n stages are executed by the algorithm.

7.2.3 Relaxation of (RSP_k)

Due to the separable structure of the residual subproblem (RSP_k), the Lagrangian relaxation method discussed in Chapter 3 can be used to get an upper bound of $\tilde{s}_{k+1}(b-\beta)$.

Let

$$UB_{k+1}(b-\beta) = \min_{\lambda \geq 0} d_k(\lambda), \tag{7.2.6}$$

where

$$\begin{aligned} d_k(\lambda) = \quad & \max \sum_{j=k+1}^{n} f_j(x_j) - \sum_{i=1}^{m} \lambda_i [\sum_{j=k+1}^{n} g_{ij}(x_j) - (b_i - \beta_i)] \\ & \text{s.t. } x_j \in X_j, \; j = k+1, \ldots, n. \end{aligned} \tag{7.2.7}$$

As discussed in Section 3.1, the Lagrangian relaxation problem (7.2.7) with a separable structure can be solved efficiently. Thus the efficient dual search procedures in Section 3.2 can be adopted to search for an optimal solution to (7.2.6).

An alternative way of computing an upper bound of $\tilde{s}_{k+1}(b-\beta)$ is by linear programming. For each $x_j \in X_j = \{0, 1, \ldots, K_j\}$, introduce a 0-1 variable y_{jl}, $l = 0, 1, \ldots, K_j$,

$$y_{jl} = \begin{cases} 1, & \text{if } x_j = l, \\ 0, & \text{otherwise.} \end{cases}$$

Then x_j takes exactly one value from $0, 1, \ldots, K_j$, if an additional constraint $\sum_{l=1}^{K_j} y_{jl} = 1$ is imposed. Let

$$\begin{aligned} & f_{jl} = f_j(l), \; l = 0, 1, \ldots, K_j, \; j = 1, \ldots, n, \\ & g_{ijl} = g_{ij}(l), \; l = 0, 1 \ldots, K_j, \; j = 1, \ldots, n, \; i = 1, \ldots, m. \end{aligned}$$

Then problem (RSP_k) can be written as

$$\tilde{s}_{k+1}(b-\beta) = \max \sum_{j=k+1}^{n} \sum_{l=0}^{K_j} f_{jl} y_{jl}$$

$$\text{s.t.} \quad \sum_{j=k+1}^{n} \sum_{l=0}^{K_j} g_{ijl} y_{jl} \leq b_i - \beta_i, \ i = 1, \ldots, m,$$

$$\sum_{l=0}^{K_j} y_{jl} = 1, \ j = k+1, \ldots, n,$$

$$y_{jl} \in \{0, 1\}, \ l = 0, \ldots, K_j, \ j = k+1, \ldots, n.$$

Relaxing $y_{jl} \in \{0,1\}$ by $0 \leq y_{jl} \leq 1$ in the above 0-1 integer linear programming leads to a linear programming:

$$UB_{k+1}(b-\beta) = \max \sum_{j=k+1}^{n} \sum_{l=0}^{K_j} f_{jl} y_{jl} \tag{7.2.8}$$

$$\text{s.t.} \quad \sum_{j=k+1}^{n} \sum_{l=0}^{K_j} g_{ijl} y_{jl} \leq b_i - \beta_i, \ i = 1, \ldots, m, \tag{7.2.9}$$

$$\sum_{l=0}^{K_j} y_{jl} \leq 1, \ j = k+1, \ldots, n, \tag{7.2.10}$$

$$0 \leq y_{jl}, \ l = 0, \ldots, K_j, \ j = k+1, \ldots, n. \tag{7.2.11}$$

Notice in (7.2.8)–(7.2.11) that we have replaced $\sum_{l=1}^{K_j} y_{jl} = 1$ with $\sum_{l=1}^{K_j} y_{jl} \leq 1$. This is because the optimal solution to the problem (7.2.8)–(7.2.11) is always binding at the constraint (7.2.10). Moreover, due to the presence of constraint (7.2.10), $y_{jl} \leq 1$ can be omitted in (7.2.11). The upper bound $UB_{k+1}(b-\beta)$ can be computed by either solving (7.2.8)–(7.2.11) directly or solving its dual problem.

To illustrate the hybrid method, let's consider the following example:

EXAMPLE 7.3

$$\begin{aligned}
\max \ & f(x) = x_1 + 4x_1^2 + 2x_2 + x_2^2 + 9x_3 + x_3^2 \\
\text{s.t.} \ & g_1(x) = 7x_1 + 7x_2 + 2x_3 \leq 38, \\
& g_2(x) = 7x_1 + 6x_2 + 3x_3 \leq 38, \\
& x \in X = \{x \in \mathbb{Z}^3 \mid 0 \leq x_i \leq 4, \ i = 1, 2, 3\}.
\end{aligned}$$

The optimal solution of this example is $x^* = (4, 0, 3)^T$ with $f(x^*) = 104$. In our implementation, Procedure 3.2 is used to obtain the upper bound $UB_{k+1}(b-$

β) via solving the Lagrangian dual (7.2.6). The best feasible solution found during Procedure 3.2 is used in Step 6 of the algorithm to update the incumbent and S_k^s. The solution process of Algorithm 7.1 for this example is described as follows.

Initialization

Step 0. Choose $\epsilon = 0$, $N = 1$. Set $k = 1$. $S_1^0 = \{0,1,2,3,4\}$. $UB = UB_1(b) = 115\frac{1}{7}$. A feasible solution $x_{opt} = (4,1,1)^T$ is found. Set $LB = f(x_{opt})$=81.

Iteration 1

Step 1. $S_1^f = \{0,1,2,3,4\}$.

Step 2. Table 7.4 shows the values of g^1 for all the partial solutions in S_1^f. We see that no partial solution is dominated. Thus, $S_1^e = \{0,1,2,3,4\}$.

Table 7.4. Domination and upper bounds at Iteration 1 of Algorithm 7.1 for Example 7.3.

(x_1)	$g^1(x_1)$	$f^1(x_1)$	$UB_2(b - g^1(x_1))$	$UB(x_1)$
(0)	$(0,0)^T$	0	76	76
(1)	$(7,7)^T$	5	71	76
(2)	$(14,14)^T$	18	64	82
(3)	$(21,21)^T$	39	57	96
(4)	$(28,28)^T$	68	$43\frac{1}{3}$	$111\frac{1}{3}$

Step 4. The upper bounds, UB, of all partial solutions in S_1^e are given in Table 7.4. Using the fathoming rule in (7.2.4), 0 and 1 are removed from S_1^e. We have $S_1^s = \{2,3,4\}$.

Step 5. $UB' = 111\frac{1}{3}$, $UB := \min(UB', UB) = \min(115\frac{1}{7}, 111\frac{1}{3}) = 111\frac{1}{3}$.

Step 6. A new feasible solution $x = (4,0,3)^T$ is found. Set $LB = f((4,0,3)^T) = 104$. Since $UB(2)$ and $UB(3)$ are less than 104, 2 and 3 are removed from S_1^s. Set $S_1^s = \{4\}$.

Step 8. Set $S_2^0 = S_1^s \times X_2 = \{(4,j)^T \mid j = 0,1,2,3,4\}$ and $k = 2$.

Iteration 2

Step 1. Since the partial solutions $(4,2)^T$, $(4,3)^T$, and $(4,4)^T$ are infeasible, we have

$$S_2^f = \{(4,0)^T, (4,1)^T\}.$$

Step 2. The values of g^2, f^2 and the upper bound of the two partial solutions in S_2^f are given in Table 7.5. No domination occurs.

$$S_2^e = \{(4,0)^T, (4,1)^T\}.$$

Step 4. Since $UB((4,1)^T) < 104$, the partial solution $(4,1)^T$ is removed

Table 7.5. Domination and upper bounds at Iteration 2 of Algorithm 7.1 for Example 7.3.

(x_1, x_2)	$g^2(x_1, x_2)$	$f^2(x_1, x_2)$	$UB_3(b - g^2(x_1, x_2))$	$UB(x_1, x_2)$
$(4, 0)$	$(28, 28)^T$	68	$43\frac{1}{3}$	$111\frac{1}{3}$
$(4, 1)$	$(35, 34)^T$	71	$17\frac{12}{3}$	$88\frac{1}{3}$

from S_2^e. $S_2^s = \{(4, 0)^T\}$.

Step 5. $UB = UB' = 111\frac{1}{3}$.

Step 6. $LB = 104$.

Step 8. Set $S_3^0 = \{(4, 0, j) \mid j = 0, 1, 2, 3, 4\}$ and $k = 3$.

Iteration 3

Step 1. Eliminating infeasible solutions from S_3^0, we obtain

$$S_3^f = \{(4, 0, 0)^T, (4, 0, 1)^T, (4, 0, 2)^T, (4, 0, 3)^T\}.$$

Step 2. Calculating the objective values of the feasible solutions in S_3^f, we get $f((4, 0, 0)^T) = 68$, $f((4, 0, 1)^T) = 78$, $f((4, 0, 2)^T) = 90$, $f((4, 0, 3)^T) = 104$. Thus, $x = (4, 0, 3)^T$ is the optimal solution to the example.

7.3 Convergent Lagrangian and Objective Level Cut Method

As already witnessed from our earlier discussion, the "curse of dimensionality" prevents dynamic programming as well as its improved versions, such as the hybrid method, from their successful execution in multiply constrained cases of (P) when m is large. When there is no convexity assumption, branch-and-bound-type methods may fail to solve (P) due to lack of an ability in identifying a global optimal solution to nonconvex continuous relaxation subproblems. Although the conventional Lagrangian method makes an efficient use of the separable structure of (P) in its solution process, it is often unable to find an exact solution of (P) due to the existence of a duality gap.

Stimulated by the relationship between the duality gap and the geometry of the perturbation function, we discuss in this section a convergent Lagrangian and objective level cut algorithm for (P). In this section, we need the following assumption for problem (P):

ASSUMPTION 7.4 *For each $j = 1, \ldots, n$, f_j is integer-valued.*

7.3.1 Motivation

To motivate the solution algorithm, let us consider the following example:

EXAMPLE 7.4

$$\begin{aligned} &\min\ f(x) = -2x_1^2 - x_2 + 3x_3^2 \\ &\text{s.t. } 5x_1 + 3x_2^2 - \sqrt{3}x_3 \le 7, \\ &\qquad x \in X = \{x \in \mathbb{Z}^2 \mid 0 \le x_i \le 2,\ i = 1,2,3\}. \end{aligned}$$

The optimal solution of this example is $x^* = (1,0,0)^T$ with $f^* = f(x^*) = -2$. The perturbation function of this problem is illustrated in Figure 7.1. From

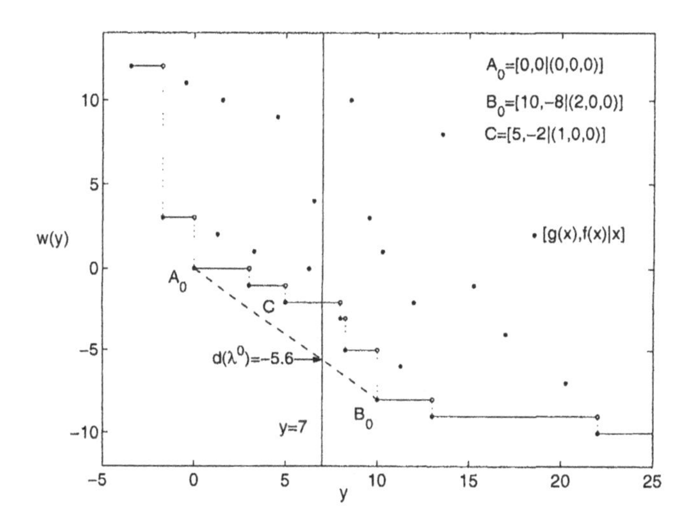

Figure 7.1. The perturbation function of Example 7.4.

Figure 7.1 we can see that point C that corresponds to the optimal solution x^* "hides" above the convex envelope of the perturbation function and therefore there is no optimal generating multiplier for x^*. In other words, it is impossible for x^* to be found by the conventional Lagrangian dual method. The optimal solution to (D) is $\lambda^0 = 0.8$ with $d(\lambda^0) = -5.6$. Thus, the duality gap is $f(x^*) - d(\lambda^0) = -2 + 5.6 = 3.6$. A key observation of the perturbation function is that point C can be exposed to the convex envelope or the convex hull of the perturbation function by adding an objective cut. As a matter of fact, since A_0 corresponds to a feasible solution $x^0 = (0,0,0)^T$, the function value $f(x^0) = 0$ is an upper bound of f^*. Moreover, by the weak duality, the dual value $d(\lambda^0) = -5.6$ is a lower bound of f^*. The current duality bound is $0-(-5.6) = 5.6$. Therefore, adding an objective cut of $-5.6 \le f(x) \le 0$ to the original problem does not exclude the optimal solution while the perturbation function will be reshaped. Since the objective function is integer-valued, we can set a stronger objective cut of $-5 \le f(x) \le -1$ after storing the current best feasible solution x^0 as the incumbent. The modified problem then has the following form:

$$\min\ f(x) = -2x_1^2 - x_2 + 3x_3^2 \tag{7.3.1}$$
$$\text{s.t. } 5x_1 + 3x_2^2 - \sqrt{3}x_3 \le 7,$$
$$x \in X_1 = X \cap \{x \mid -5 \le f(x) \le -1\}.$$

The perturbation function of problem (7.3.1) is shown in Figure 7.2. The optimal

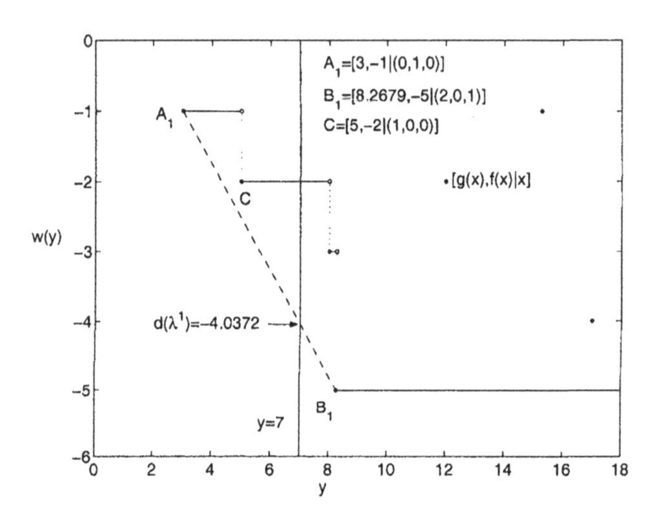

Figure 7.2. The perturbation function of problem (7.3.1).

dual multiplier to (7.3.1) is $\lambda^1 = 0.7593$ with dual value $d(\lambda^1) = -4.0372$. Since $x^1 = (0,1,0)^T$ corresponding to A_1 is feasible, the duality gap bound is now reduced to $f(x^1) - (-4.0372) = -1 + 4.037 = 3.0372$. Again we can add an objective cut $-4 \le f(x) \le f(x^1) - 1 = -2$ to (7.3.1) and obtain the following problem:

$$\min\ f(x) = -2x_1^2 - x_2 + 3x_3^2 \tag{7.3.2}$$
$$\text{s.t. } 5x_1 + 3x_2^2 - \sqrt{3}x_3 \le 7,$$
$$x \in X_2 = X \cap \{x \mid -4 \le f(x) \le -2\}.$$

The perturbation function of problem (7.3.2) is shown in Figure 7.3. The optimal dual multiplier is $\lambda^2 = 0.3333$ with dual value $d(\lambda^2) = -2.6667$. Now point C corresponding to x^* is exposed to the convex hull of the perturbation function and the duality bound is reduced to $f(x^*) - (-2.6667) = -2 + 2.6667 = 0.6667 < 1$. Since the objective function is integer-valued, we claim that $x^* = (1,0,0)^T$ is the optimal solution to the original problem.

This example clearly illustrates a procedure of eliminating the duality bound and thus the duality gap by using objective cuts. The convergent Lagrangian and objective level cut method exposes an optimal solution of (P) to the convex

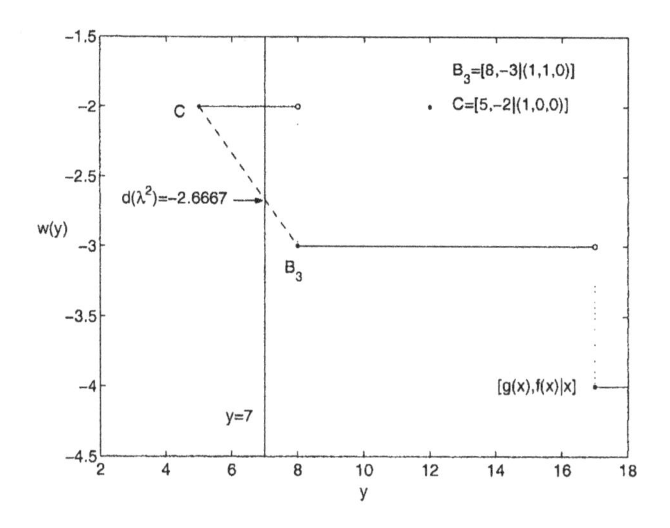

Figure 7.3. The perturbation function of problem (7.3.2).

hull of the revised perturbation function by successively using objective cuts. The algorithm starts with a lower bound derived from the dual value by the conventional dual search and an upper bound by a feasible solution generated in the dual search (if any). The lower level cut and upper level cut are imposed to (P) such that the duality bound (duality gap) is forced to shrink. The objective cut is updated successively with the distance between the upper cut and the lower cut monotonically decreasing. The algorithm terminates in finite iterations, either reaching an optimal solution to (P) or reporting an infeasibility of (P).

One crucial issue to an efficient implementation of this solution idea is how to solve the relaxation problems of the revised problems such as the Lagrangian relaxations in (7.3.1) and (7.3.2). Since the objective function is integer-valued, dynamic programming can be used to search for optimal solutions to this kind of problems quite efficiently.

7.3.2 Algorithm description

Consider the following modified version of (P) by imposing a lower cut l and an upper cut u:

$$(P(l,u)) \qquad \begin{aligned} &\min\ f(x) \\ &\text{s.t. } g_i(x) \le b_i,\ i = 1, \ldots, m, \\ &\qquad x \in X(l,u) = \{x \in X \mid l \le f(x) \le u\}. \end{aligned} \tag{7.3.3}$$

It is obvious that $(P(l, u))$ is equivalent to (P) if $l \leq f^* \leq u$. The Lagrangian relaxation of $(P(l, u))$ is:

$$(L_\lambda(l, u)) \qquad d(\lambda, l, u) = \min_{x \in X(l,u)} L(x, \lambda), \tag{7.3.4}$$

where $\lambda \in \mathbb{R}^m_+$ and $L(x, \lambda) = f(x) + \sum_{i=1}^m \lambda_i(g_i(x) - b_i)$. The Lagrangian dual problem of $(P(l, u))$ is then given as

$$(D(l, u)) \quad \max_{\lambda \in \mathbb{R}^m_+} d(\lambda, l, u). \tag{7.3.5}$$

Notice that $L(x, \lambda) = \sum_{j=1}^n \theta_j(x_j, \lambda) - \alpha(\lambda)$, where $\theta_j(x_j, \lambda) = f_j(x_j) + \sum_{i=1}^m \lambda_i g_{ij}(x_j)$ and $\alpha(\lambda) = \sum_{i=1}^m \lambda_i b_i$. Problem $(L_\lambda(l, u))$ can be explicitly written as:

$$d(\lambda, l, u) = \min \sum_{j=1}^n \theta_j(x_j, \lambda) - \alpha(\lambda) \tag{7.3.6}$$

$$\text{s.t. } l \leq \sum_{j=1}^n f_j(x_j) \leq u,$$

$$x \in X.$$

It is clear that $(L_\lambda(l, u))$ is a separable integer programming problem with a lower bound and upper bound constraint for $f(x)$. By the assumptions in (P), each $f_j(x_j)$ is integer-valued for all $x_j \in X_j$. Therefore, $(L_\lambda(l, u))$ can be efficiently solved by dynamic programming. Let

$$s_k = \sum_{j=1}^{k-1} f_j(x_j), \quad k = 2, \ldots, n+1, \tag{7.3.7}$$

with an initial condition $s_1 = 0$. Then $(L_\lambda(l, u))$ can be solved by the following dynamic programming formulation:

$$(DP) \qquad \min\ s_{n+1} + \sum_{j=1}^n \sum_{i=1}^m \lambda_i g_{ij}(x_j) \tag{7.3.8}$$

$$\text{s.t. } s_{j+1} = s_j + f_j(x_j),\ j = 1, 2, \ldots, n,$$

$$s_1 = 0,$$

$$l \leq s_{n+1} \leq u,$$

$$x_j \in X_j,\ j = 1, 2, \ldots, n.$$

The state in the above dynamic programming formulation takes finite values at each stage. All the solutions to $(L_\lambda(l, u))$ can be generated using the conventional dynamic programming technique.

Since $d(\lambda)$ or $d(\lambda, l, u)$ is a nonsmooth concave function of λ, the subgradient method or the outer Lagrangian linearization method can be used to solve (D), the dual problem of (P), or $(D(l, u))$ (see, e.g., [56][176][192]). In practice, the subgradient method terminates at an approximate solution when certain stopping criteria are met.

We now describe the algorithm as follows.

ALGORITHM 7.2 (CONVERGENT LAGRANGIAN AND OBJECTIVE LEVEL CUT ALGORITHM)

Step 0 (Initialization).

(i) Solve the dual problem (D) by using the subgradient method or by the outer Lagrangian linearization method. Let λ^0 be the best dual vector found. Set $d^0 = d(\lambda^0)$.

(ii) Let x^* denote the current best feasible solution (if there is one) and set $v^0 = f(x^*)$. The initial feasible solution can either be found during the dual search or by certain heuristic method. If $v^0 - d^0 < 1$, stop and x^* is an optimal solution to (P); Otherwise, set $l_0 = \lceil d^0 \rceil$, $u_0 = v^0 - 1$ and $k = 0$, where $\lceil x \rceil$ is the minimum integer number larger than or equal to x.

(iii) When no feasible solution is found, set v^0 to be equal to an upper bound of $f(x)$ over X. If $v^0 - d^0 < 0$, stop and there is no feasible solution to (P); Otherwise, set $l_0 = \lceil d^0 \rceil$, $u_0 = v^0$ and $k = 0$.

Step 1 (Finding feasible solution). If $l_k = u_k$, go to Step 3. Otherwise, solve the following problem using dynamic programming,

$$(P_f) \qquad \min\ g_{\lambda^k}(x) = \sum_{j=1}^{n}\sum_{i=1}^{m} \lambda_i^k g_{ij}(x_j)$$
$$\text{s.t. } l_k \le f(x) \le u_k$$
$$x \in X.$$

Let C^k be the set of optimal solutions to the above problem.

(i) If there is a feasible solution in C^k, then set the incumbent $x^* = \arg\min\{f(x) \mid x \in C^k \cap S\}$ and $v^k = f(x^*)$, where S is the feasible region of (P). If $v^k - l_k < 1$, stop and the current incumbent x^* is an optimal solution to (P). Otherwise, set $u_{k+1} = v^k - 1$, $l_{k+1} = l_k$, $\lambda^{k+1} = \lambda^k$ and $k := k + 1$, and go to Step 1.

(ii) If for any $x \in C^k$, $g_{\lambda^k}(x) > \sum_{i=1}^m \lambda_i^k b_i$ holds, stop. The current incumbent x^* is an optimal solution to (P) or there is no feasible solution to (P) if no incumbent has been found.

Step 2 (Dual search with objective cut). Solve $(D(l_k, u_k))$ by the subgradient method or the outer Lagrangian linearization method, while the Lagrangian relaxation problem $(L_\lambda(l_k, u_k))$ is solved by using dynamic programming. The subgradient method terminates when the algorithm is not able to increase the dual value after a given number of iterations. Let λ^k be the dual vector that generates the highest dual value in the dual search process. Set $d^k = d(\lambda^k, l_k, u_k)$.

(i) If there is a feasible solution x^* found during the dual search process, replace the incumbent by x^*, set $v^k = f(x^*)$, $u_{k+1} = v^k - 1$, $l_{k+1} = \max\{l_k, \lceil d^k \rceil\}$, $k := k + 1$, and go to Step 1.

(ii) If no feasible solution is found and $d^k > l_k$, set $l_{k+1} = \lceil d^k \rceil$, $u_{k+1} = u_k$, $k := k + 1$, and go to Step 1.

Step 3 (Finding feasible solution when $\lambda = 0$). Solve the following dynamic programming problem

$$(DP_0) \qquad \begin{array}{ll} \min & s_{n+1} \\ \text{s.t.} & s_{j+1} = s_j + f_j(x_j),\ j = 1, 2, \ldots, n \\ & s_1 = 0, \\ & l_k \le s_{n+1} \le u_k, \\ & x_j \in X_j,\ j = 1, 2, \ldots, n. \end{array} \tag{7.3.9}$$

(i) If there is a feasible optimal solution x^* to (DP_0), stop. The incumbent x^* is the optimal solution to (P).

(ii) Set $u_{k+1} = u_k$, $l_{k+1} = v(DP_0) + 1$. If $v^k - l_{k+1} < 1$, stop. The incumbent x^* is an optimal solution to (P) or there is no feasible solution to (P) if no incumbent has been found. Otherwise, set $k := k + 1$ and go to Step 1.

Step 1 in the above algorithm is adopted to speed up the convergence of the algorithm. When the objective level cut is updated, solving (P_f) could sometimes identify a feasible solution of (P) with an objective level less than u_k. As we learned from Section 3.5, there exist multiply constrained cases where more than one points $(g(x), f(x))$ with $g(x) \not\le b$ surrounding the axis $y = b$ and span a horizontal plane (corresponding to $\lambda = 0$) with the same f value (being the lowest objective value over the defined domain). In such a situation, the dual search method will fail to raise the dual value higher than the lowest objective value. Step 3 of the above algorithm deals with this kind of situations.

Next, we discuss the properties of the algorithm and its finite convergence. We need the following lemma.

LEMMA 7.1 (i) *Let $\lambda^*(l,u)$ denote the optimal solution to $(D(l,u))$. The optimal dual value $d(\lambda^*(l,u),l,u)$ is a nondecreasing function of l.*

(ii) *If $l \leq f^* \leq u$, then $d(\lambda^*) \leq d(\lambda^*(l,u),l,u) \leq f^*$. Moreover, let $\sigma = \max\{f(x) \mid f(x) < f^*, x \in X \setminus S\}$. If $\sigma < l \leq f^*$, then $\lambda^*(l,u) = 0$ and $d(\lambda^*(l,u),l,u) = f^*$.*

(iii) *For $l < f^*$, we have $d(\lambda^*(l,u),l,u) \geq l$.*

Proof. (i) If $l_1 \leq l_2$, then $d(\lambda,l_1,u) \leq d(\lambda,l_2,u)$ for all $\lambda \in \mathbb{R}^m_+$. Thus,

$$d(\lambda^*(l_1,u),l_1,u) = \max_{\lambda \in \mathbb{R}^m_+} d(\lambda,l_1,u) \leq \max_{\lambda \in \mathbb{R}^m_+} d(\lambda,l_2,u) = d(\lambda^*(l_2,u),l_2,u).$$

(ii) Since $X(l,u) \subseteq X$, we have

$$d(\lambda) = \min_{x \in X} L(x,\lambda) \leq \min_{x \in X(l,u)} L(x,\lambda) = d(\lambda,l,u), \quad \forall \lambda \in \mathbb{R}^m_+.$$

Thus, $d(\lambda^*) \leq d(\lambda^*(l,u),l,u)$. If $l \leq f^* \leq u$, then $S^* \subseteq X(l,u)$, where S^* is the set of optimal solutions to (P). For any $\lambda \in \mathbb{R}^m_+$, we have

$$\begin{aligned} d(\lambda,l,u) &= \min_{x \in X(l,u)} L(x,\lambda) \\ &\leq \min_{x \in S^*} L(x,\lambda) \\ &\leq \min_{x \in S^*} f(x) \\ &= f^*. \end{aligned}$$

Therefore $d(\lambda^*(l,u),l,u) \leq f^*$. Suppose that $\sigma < l \leq f^* \leq u$, then there is no infeasible point x in $X(l,u)$ with $f(x) < f^*$. Thus

$$d(0,l,u) = \min_{x \in X(l,u)} f(x) = \min_{x \in S^*} f(x) = f^* \geq \min_{x \in S^*} L(x,\lambda) \geq d(\lambda,l,u)$$

for all $\lambda \in \mathbb{R}^m_+$. Thus, $\lambda = 0$ solves $(D(l,u))$ and $d(0,l,u) = f^*$.

(iii) Consider the perturbation function of $(P(l,u))$. The set of corner points of it is a subset of Φ_c satisfying $l \leq f_k \leq u$. Thus, applying (3.3.8) and Theorem 3.10, we infer that there exist an index set $I(l,u) \subset \{1,2,\ldots,K\}$ and $\mu_k^*(l,u) > 0$, $k \in I(l,u)$, such that

$$d(\lambda^*(l,u),l,u) = \sum_{k \in I(l,u)} \mu_k^*(l,u) f_k, \tag{7.3.10}$$

$$\sum_{k \in I(l,u)} \mu_k^*(l,u) c_k \leq b,$$

$$\sum_{k \in I(l,u)} \mu_k^*(l,u) = 1,$$

$$l \leq f_k \leq u, \; k \in I(l,u).$$

Since for each $k \in I(l,u)$, $f_k \geq l$, the above conditions imply that $d(\lambda^*(l,u),l,u) \geq l$. □

LEMMA 7.2 *If* $d(\lambda^*(l,u),l,u) < v(D(l,u)) = f^*$, *then*

$$\min\{f(x) \mid x \in T(\lambda^*(l,u),l,u) \setminus S\} \leq d(\lambda^*(l,u),l,u),$$

where $T(\lambda^*(l,u),l,u)$ *is the solution set to problem* $(L_\lambda(l,u))$ *with* $\lambda = \lambda^*(l,u)$.

Proof. From (7.3.10), we have

$$\sum_{k \in I(l,u)} \mu_k^*(l,u)(f_k - d(\lambda^*(l,u),l,u)) = 0.$$

If there is a k such that f_k is not equal to $d(\lambda^*(l,u),l,u)$, then there must be a k_1 such that f_{k_1} is strictly greater than $d(\lambda^*(l,u),l,u)$ and there must be a k_2 such that f_{k_2} is strictly smaller than $d(\lambda^*(l,u),l,u)$. From the weak duality, the solution corresponding to f_{k_2} must be infeasible in (P). If all f_k's are equal to $d(\lambda^*(l,u),l,u)$, then all solutions in $T(\lambda^*(l,u),l,u)$ must be infeasible from the assumption of $d(\lambda^*(l,u),l,u) < f^*$. □

Lemma 7.2 implies that at least one infeasible solution will be removed when placing a cut higher than $d(\lambda^*(l,u),l,u)$.

THEOREM 7.1 *Algorithm 7.2 either finds an optimal solution of* (P) *or reports an infeasibility of* (P) *in at most* $u_0 - l_0 + 1$ *iterations.*

Proof. First, from the algorithm and Lemma 7.1, it always holds $l_k \leq f^*$. It is clear that (P) is infeasible if the algorithm stops at Step 0 (iii), Step 1 (ii) or Step 3 (ii) when the incumbent is empty. The optimality of the incumbent x^* is obvious when the algorithm stops at Step 0 (ii) or Step 1 (i). If the algorithm stops at Step 1 (ii), then there is no feasible solution x satisfying $l_k \leq f(x) \leq u_k$. Thus, from the algorithm, if the incumbent is x^*, then $f(x^*) = u_k + 1$ and x^* is an optimal solution to (P). If the algorithm stops at Step 3 (i), then $\lambda = 0$ is the dual optimal solution to $(P(l_k,u_k))$ and $f(x^*) \geq l_k$. By Theorem 3.17, x^* must be an optimal solution to (P). If the algorithm stops at Step 3 (ii), then there is no feasible solution x satisfying $l_k \leq f(x) \leq u_k$ and the stopping condition $v^k - l_{k+1} < 1$ implies that there is no better feasible solution than the incumbent x^*.

Suppose that the algorithm does not stop at iteration k, then by the algorithm, either $u_{k+1} \leq u_k - 1$ or $l_{k+1} \geq l_k + 1$. Notice that for any k, $l_k \leq f^* \leq u_k + 1$ holds. Therefore, in at most $u_0 - l_0$ iterations, $u_k = l_k$ will be satisfied. If the algorithm does not stop before $u_0 - l_0 + 1$ iterations, then the algorithm will stop in $(u_0 - l_0 + 1)$-th iteration either at Step 3 (i) or at Step 3 (ii), reaching an optimal solution or reporting an infeasibility of (P). □

7.3.3 Implementation of dynamic programming

We now discuss several implementation issues of dynamic programming. Three techniques will be developed to facilitate an efficient use of dynamic programming: partition of objective cut, reduction of state space and feasibility check of (DP_0).

The magnitude of the initial duality bound $u_0 - l_0$ at Step 0 of Algorithm 7.2 has a great effect on the efficiency of dynamic programming when solving $(P(l_k, u_k))$. As a matter of fact, if the initial duality bound is very large then the dynamic programming can be very time-consuming and inefficient due to a large range of the state space. In order to reduce the range without losing any optimal solution, a partition scheme of the objective cut is proposed to divide the range $[l_0, u_0]$ at Step 0 into q smaller non-overlapping blocks such that

$$[l_0, u_0] = \cup_{s=1}^{q} [l_0^s, u_0^s],$$

where $l_0^1 = l_0$, $u_0^q = u_0$ and $l_0^{s+1} = u_0^s + 1$. The original problem can be then divided into q subproblems with $s = 1, 2, \ldots, q$:

$$\begin{aligned} (P^s) \qquad & \min\ f(x) \\ & \text{s.t. } g_i(x) \le b_i,\ i = 1, \ldots, m, \\ & \quad l_0^s \le f(x) \le u_0^s,\ x \in X. \end{aligned} \tag{7.3.11}$$

These q problems will be solved successively from $s = 1$ to $s = q$. If an optimal solution x^* is found in Problem (P^s) for $1 \le s \le q$, then x^* is also an optimal solution to (P) and there is no need to solve (P^{s+1}), (P^{s+2}), ... If all problems (P^s) are infeasible, then we claim that the original problem is infeasible.

Next, we discuss the strategy for reducing state space. Let $\bar{s}_j$, $\underline{s}_j$ denote the upper bound and lower bound of the range of state variable s_j, respectively. Let

$$\overline{f}_j = \max_{l_j \le x_j \le u_j} f_j(x_j),$$

$$\underline{f}_j = \min_{l_j \le x_j \le u_j} f_j(x_j).$$

With the initial condition $\overline{s}_1^F = \underline{s}_1^F = 0$, the range s_j^F of the state variable s_j at stage j can be determined by a forward recursive formulation,

$$\overline{s}_{j+1}^F = \overline{s}_j^F + \overline{f}_j, \quad \text{for } j = 1, \ldots, n,$$

$$\underline{s}_{j+1}^F = \underline{s}_j^F + \underline{f}_j, \quad \text{for } j = 1, \ldots, n.$$

With the initial condition $\overline{s}_{n+1}^B = u_k$, $\underline{s}_{n+1}^B = l_k$, the range s_j^B of the state variable s_j at stage j can be determined by a backward recursive formulation,

$$\overline{s}_j^B = \overline{s}_{j+1}^B - \underline{f}_j, \quad \text{for } j = n, \ldots, 1,$$
$$\underline{s}_j^B = \underline{s}_{j+1}^B - \overline{f}_j, \quad \text{for } j = n, \ldots, 1.$$

Therefore, the exact expression of the state range can be given as follows:

$$[\underline{s}_j, \overline{s}_j] = \begin{cases} [0,0], & \text{for } j = 1, \\ [\underline{s}_j^B, \overline{s}_j^B] \cap [\underline{s}_j^F, \overline{s}_j^F], & \text{for } j = 2, \ldots, n, \\ [l^k, u^k], & \text{for } j = n+1. \end{cases} \tag{7.3.12}$$

If any $[\underline{s}_j, \overline{s}_j]$ is empty, then $P(l_k, u_k)$ has no feasible solution. In general, the state space of dynamic programming can be significantly reduced by (7.3.12).

Now we discuss the implementation of solving (DP_0) at Step 3 of Algorithm 7.2, a situation when λ is set to be zero in the dual search. Since there may exist a large number of optimal solutions to (DP_0), an efficient ordering of the optimal solutions by certain rules is crucial to the feasibility check process. For given $\mu \geq 0$ and $\mu \neq 0$, consider the following surrogate constraint:

$$g^\mu(x) = \sum_{i=1}^m \mu_i g_i(x) \leq \sum_{i=1}^m \mu_i b_i = b^\mu.$$

Let $S_\mu = \{x \in X \mid g^\mu(x) \leq b^\mu\}$. It is clear that $S \subseteq S_\mu$. Suppose that the set of optimal solutions to (DP_0) is T_0. Rank the points in T_0 from the smallest to the largest in terms of the value of $g^\mu(x)$:

$$T_0 = \{x^1, x^2, \ldots, x^N\}.$$

Let t be such that $g^\mu(x^t) \leq b^\mu$ and $g^\mu(x^{t+1}) > b^\mu$. The point x^t is called a "turning point." When solving (DP_0) by dynamic programming, we generate and calculate $g^\mu(x^k)$ for $k = 1, 2, \ldots$, till a feasible solution to (P) is found or a turning point is met. In the latter case there is no feasible solution in T_0. In the worst case, checking feasibility of T_0 requires generating $t+1$ optimal solutions in T_0.

Finally, we point out that although the objective function is assumed to be integer-valued in the algorithm, a rational objective function can be also handled by multiplying a suitable number.

To illustrate Algorithm 7.2, we consider the following small-size example.

Table 7.6. Iteration process of Example 7.5.

Iteration	λ^k	d^k	x^*	$f(x^*)$	l_k	u_k
0	$(0.853, 0, 0.915)^T$	-548.526			-548	113
1	$(0.853, 0, 0.915)^T$	-548.526	$(-1, -4, 5, 4, 5)^T$	-367	-548	-368
2	$(0.853, 0, 0.915)^T$	-548.526	$(-2, -4, 5, 4, 5)^T$	-373	-548	-374
3	$(0.853, 0, 0.915)^T$	-548.526	$(-3, -4, 5, 4, 5)^T$	-385	-548	-386
4	$(0.853, 0, 0.915)^T$	-548.526	$(-1, -5, 5, 5, 5)^T$	-400	-548	-401
5	$(0.853, 0, 0.915)^T$	-548.526	$(-2, -5, 5, 5, 5)^T$	-406	-548	-407
6	$(0.853, 0, 0.915)^T$	-548.526	$(-3, -5, 5, 5, 5)^T$	-418	-548	-419
7	$(0.246, 0, 0.385)^T$	-540.492	$(-3, -5, 5, 5, 5)^T$	-418	-540	-419
8	$(0, 0, 0)^T$	-540.000	$(-3, -5, 5, 5, 5)^T$	-418	-539	-419
9	$(0.140, 0, 0.151)^T$	-530.359	$(-3, -5, 5, 5, 5)^T$	-418	-530	-419
10	$(0, 0, 0)^T$	-530.000	$(-3, -5, 5, 5, 5)^T$	-418	-529	-419
11	$(0.047, 0, 0.047)^T$	-528.899	$(-3, -5, 5, 5, 5)^T$	-418	-528	-419
12	$(0, 0, 0)^T$	-528.000	$(-3, -5, 5, 5, 5)^T$	-418	-527	-419
13	$(0, 0, 0)^T$	-527.000	$(-4, -5, 5, 2, 5)^T$	-526	-527	-527
14	$(0, 0, 0)^T$	-527.000	$(-4, -5, 5, 2, 5)^T$	-526		

EXAMPLE 7.5

$$
\begin{aligned}
\min\ & -3x_1 - 3x_1^2 + 8x_2 - 7x_2^2 - 5x_3 - 3x_3^2 + 2x_4 + 4x_4^2 - 4x_5 - 7x_5^2 \\
\text{s.t. } & 7x_1 + 7x_1^2 + 4x_2 + 4x_2^2 - 8x_3 - 7x_3^2 - 7x_4 + 2x_4^2 - 5x_5 + 2x_5^2 \le -6, \\
& 8x_1 - 5x_1^2 + 4x_2 - 7x_2^2 - 4x_3 + 8x_3^2 + 7x_4 - 6x_4^2 - 2x_5 - 7x_5^2 \le -2, \\
& -x_1 - 3x_1^2 - 2x_2 + x_2^2 - 2x_3 + 8x_3^2 - 5x_4 - 3x_4^2 + 5x_5 - 7x_5^2 \le 9, \\
& \quad x \in X = \{x \in \mathbb{Z}^5 \mid -5 \le x_i \le 5,\ i = 1, 2, 3, 4, 5\}.
\end{aligned}
$$

It can be verified that the optimal solution of Example 7.5 is $x^* = (-4, -5, 5,2, 5)^T$ with $f(x^*) = -526$.

The initial dual value is $d^0 = -548.526$ and an upper bound of $f(x)$ is $v^0 = 113$. Therefore, the initial interval of objective cut is $[-548, 113]$. A partition scheme is used to divide the initial interval of objective cut into smaller ones with an interval length of 200. The algorithm finds the optimal solution x^* at iteration 13. The dual search at iteration 14 finds a zero optimal dual solution and there is no feasible solution in the set of optimal solutions to the corresponding Lagrangian relaxation problem. The algorithm thus terminates and reports x^* as an optimal solution. Table 7.6 summaries the iteration process of the algorithm.

7.3.4 Computational experiment

We report in this section the computational results in testing Algorithm 7.2. The efficiency of Algorithm 7.2 is tested by five classes of randomly generated separable integer programming problems.

PROBLEM 7.1 3rd polynomial integer programming problem (PIP):

$$f_j(x_j) = \sum_{k=1}^{3} c_{jk}x_j^k,\ j = 1,\ldots,n,$$
$$g_{ij}(x_j) = \sum_{k=1}^{3} a_{ijk}x_j^k,\ i = 1,\ldots,m,\ j = 1,\ldots,n.$$

Coefficients c_{ik} are integer numbers with $c_{i1} \in [-20, 20]$, $c_{i2} \in [-10, 10]$ and $c_{i3} \in [-5, 5]$. Coefficients a_{ijk} are of real values with $a_{ij1} \in [-20, 20]$, $a_{ij2} \in [-10, 10]$ and $a_{ij3} \in [-5, 5]$.

PROBLEM 7.2 Convex quadratic integer programming problem with convex quadratic constraints (QIP_1):

$$f_j(x_j) = c_{j1}x_j^2 + c_{j2}x_j,\ j = 1,\ldots,n,$$
$$g_{ij}(x_j) = a_{ij1}x_j^2 + a_{ij2}x_j,\ i = 1,\ldots,m,\ j = 1,\ldots,n.$$

Coefficients c_{j1} and c_{j2}, $j = 1,\ldots,n$, are integer numbers taken from [1,10] and $[-100, 20]$, respectively. Coefficients a_{ij1} and a_{ij2}, $i = 1,\ldots,m$, $j = 1,\ldots,n$, are of real values taken from $[1, 10]$ and $[100, 220]$, respectively.

PROBLEM 7.3 Convex quadratic integer programming problem with linear constraints $(QIPL_1)$:

$$f_j(x_j) = c_{j1}x_j^2 + c_{j2}x_j,\ j = 1,\ldots,n,$$
$$g_{ij}(x_j) = a_{ij}x_j,\ i = 1,\ldots,m,\ j = 1,\ldots,n.$$

Coefficients c_{j1} and c_{j2}, $j = 1,\ldots,n$, are integer numbers taken from [1,10] and $[-100, 20]$, respectively. Coefficient a_{ij}, $i = 1,\ldots,m$, $j = 1,\ldots,n$, are of real values taken from $[20, 60]$.

PROBLEM 7.4 Concave quadratic integer programming problem with convex quadratic constraints (QIP_2):

$$f_j(x_j) = c_{j1}x_j^2 + c_{j2}x_j,\ j = 1,\ldots,n,$$
$$g_{ij}(x_j) = a_{ij1}x_j^2 + a_{ij2}x_j,\ i = 1,\ldots,m,\ j = 1,\ldots,n.$$

Coefficients c_{j1} and c_{j2}, $j = 1, \ldots, n$, are integer numbers taken from $[-10, -1]$ and $[-20, 60]$, respectively. Coefficients a_{ij1} and a_{ij2}, $i = 1, \ldots, m$, $j = 1, \ldots, n$, are of real values taken from $[1, 10]$ and $[100, 220]$, respectively.

PROBLEM 7.5 Concave quadratic integer programming problem with linear constraints $(QIPL_2)$:

$$f_j(x_j) = c_{j1}x_j^2 + c_{j2}x_j, \ j = 1, \ldots, n,$$
$$g_{ij}(x_j) = a_{ij}x_j, \ i = 1, \ldots, m, \ j = 1, \ldots, n.$$

Coefficients c_{j1} and c_{j2}, $j = 1, \ldots, n$, are integer numbers taken from $[-10, -1]$ and $[-20, 60]$, respectively. Coefficient a_{ij}, $i = 1, \ldots, m$, $j = 1, \ldots, n$, are of real values taken from $[20, 80]$.

All the coefficients in the above problems are taken uniformly and independently. The finite integer set X_j's are of the following form:

$$X_j = \{x_j \in \mathbb{Z} \mid 1 \leq x_j \leq 5\}, \quad j = 1, \ldots, n.$$

The right-hand side b in the above problems is generated according to the following rule. Let $0 < r < 1$. Set

$$b_i = \underline{g}_i + r(\bar{g}_i - \underline{g}_i), \quad i = 1, \ldots, m, \tag{7.3.1}$$

where $\bar{g}_i = \max_{x \in X} g_i(x)$ and $\underline{g}_i = \min_{x \in X} g_i(x)$. The ratio r is used to control the size of the feasible regions of the test problems and the degree of difficulty of the problems. As we will see in the numerical results, the smaller the value of r, the more difficult the problem. A similar rule of determining the right-hand side was used in generating test problems in [34][36].

Algorithm 7.2 has been coded by Fortran 90 and runs on a Sun Workstation (Blade 2000). The computational results for the five classes of test problems are reported in Tables 7.7–7.11. All the results are obtained by running the algorithm for 20 randomly generated problems. The following notations are used in the tables:

- n=number of variables;
- m=number of constraints;
- r=ratio defining the right-hand side b in (7.3.1);
- Duality Bound=initial duality bound $u_0 - l_0$, where l_0 and u_0 are defined in Algorithm 7.2.

Table 7.7. Numerical results for (PIP) $(r = 0.62)$.

n	m	Average Duality Bound	Average Number of Iterations	Average CPU Seconds
50	5	29.6	4	0.5
50	10	442.0	12	5.8
50	15	20.4	5	1.9
50	20	842.2	17	24.2
50	30	1874.2	15	223.4

Table 7.8. Numerical results for (QIP_1) $(r = 0.62)$.

n	m	Average Duality Bound	Average Number of Iterations	Average CPU Seconds
50	5	685.3	5	67.6
50	10	773.5	5	5.3
50	20	795.7	13	334.7
50	25	1007.7	7	126.7
50	30	986.1	8	194.3

Table 7.9. Numerical results for $(QIPL_1)$ $(r = 0.65)$.

n	m	Average Duality Bound	Average Number of Iterations	Average CPU Seconds
50	10	5.7	3	113.1
50	15	84.4	6	51.1
50	20	29.4	2	2.5
50	30	18.9	3	381.3

7.4 Notes

The principle of optimality and the first dynamic programming algorithm were presented in [18]. Dynamic programming methods for integer programming were discussed in many books (see e.g. [50][106] [117][153][168]). The hybrid method of dynamic programming and branch-and-bound was proposed in [151]. The strategy of combining dynamic programming and branch-and-

Table 7.10. Numerical results for (QIP_2) $(r = 0.70)$.

n	m	Average Duality Bound	Average Number of Iterations	Average CPU Seconds
50	5	533.5	3	47.4
50	8	765.6	4	5.5
50	10	791.4	6	8.8
50	20	1624.6	10	266.3

Table 7.11. Numerical results for $(QIPL_2)$ $(r = 0.70)$.

n	m	Average Duality Bound	Average Number of Iterations	Average CPU Seconds
50	5	70.2	2	0.1
50	10	57.9	6	86.1
50	15	122.0	7	26.8
50	20	132.6	5	1027.2

bound was also used in [123]. The objective level cut method in Section 7.3 was developed in [142] (see also [139]).

Chapter 8

NONLINEAR INTEGER PROGRAMMING WITH A QUADRATIC OBJECTIVE FUNCTION

In this chapter, we consider the following nonlinear integer programming problem with a quadratic objective function:

$$(QIP) \qquad \min\ q(x) = \sum_{j=1}^{n}(\frac{1}{2}c_j x_j^2 + d_j x_j)$$

$$\text{s.t. } g_i(x) = \sum_{j=1}^{n} g_{ij}(x_j) \le b_i,\ i = 1,\ldots,m,$$

$$x \in X = \{x \in \mathbb{Z}^n \mid l_j \le x_j \le u_j,\ j = 1,\ldots,n\},$$

where g_{ij}'s are continuous functions and l_j and u_j are integer lower and upper bounds of x_j for $j = 1,\ldots,n$. Problem (QIP) is a special class of the separable integer programming problems discussed in Chapter 7. The special geometry of the quadratic objective function can be exploited to derive more efficient algorithms for (QIP).

Two cases of quadratic objective functions are considered first: (a) $q(x)$ is a convex function, i.e., $c_j > 0$ for $j = 1,\ldots,n$, and (b) $q(x)$ is a concave function, i.e., $c_j < 0$ for $j = 1,\ldots,n$. Problems with an indefinite quadratic objective function will be considered later in this chapter as an extension.

8.1 Quadratic Contour Cut

In this section, we establish a domain cut and partition scheme by exploiting the geometry of the quadratic contour of the objective function $q(x)$. The domain cut and partition technique will be used later on to develop an exact solution method for solving (QIP).

8.1.1 Ellipse of quadratic contour

Let $q(x)$ be the quadratic function defined in (QIP). Let $\tau = -\sum_{j=1}^{n} d_j^2/(2c_j)$. Consider the ellipse contour of $q(x)$:

$$\sum_{j=1}^{n}[(1/2)c_j x_j^2 + d_j x_j] = v, \tag{8.1.1}$$

where $v \geq \tau$ when $c_j > 0$ $(j = 1, \ldots, n)$ and $v \leq \tau$ when $c_j < 0$ $(j = 1, \ldots, n)$. The center of ellipse (8.1.1) is

$$o = (-d_1/c_1, \ldots, -d_n/c_n)^T. \tag{8.1.2}$$

The length of the j-th axis of ellipse (8.1.1) is

$$2r_j = 2\sqrt{|2(v - \tau)/c_j|}. \tag{8.1.3}$$

Let $E(v)$ denote the ellipsoid formed by the contour (8.1.1). Then

$$E(v) = \begin{cases} \{x \in \mathbb{R}^n \mid q(x) \leq v\}, & \text{if } q(x) \text{ is convex,} \\ \{x \in \mathbb{R}^n \mid q(x) \geq v\}, & \text{if } q(x) \text{ is concave.} \end{cases} \tag{8.1.4}$$

The minimum rectangle that encloses the ellipsoid $E(v)$ is $[a, b]$ with

$$\begin{aligned} a &= (o_1 - r_1, \ldots, o_n - r_n)^T, \\ b &= (o_1 + r_1, \ldots, o_n + r_n)^T, \end{aligned}$$

where o is defined in (8.1.2) and r_j is defined in (8.1.3). Let $\lfloor t \rfloor$ denote the maximum integer less than or equal to t and $\lceil t \rceil$ the minimum integer greater than or equal to t. Then the minimum integer box containing all the integer points in the ellipsoid $E(v)$ can be expressed as $M(v) = \langle \alpha, \beta \rangle$, where

$$\alpha = (\lceil o_1 - r_1 \rceil, \ldots, \lceil o_n - r_n \rceil)^T, \tag{8.1.5}$$

$$\beta = (\lfloor o_1 + r_1 \rfloor, \ldots, \lfloor o_n + r_n \rfloor)^T. \tag{8.1.6}$$

Let $\tilde{x}$ be an integer point inside the ellipsoid $E(v)$. Let $N(\tilde{x})$ denote the integer subbox inside $E(v)$ with $\tilde{x}$ being one of its corner points. By the symmetry of $E(v)$, we have $N(\tilde{x}) = \langle \gamma, \delta \rangle$, where

$$\gamma = (\lceil o_1 - |\tilde{x}_1 - o_1| \rceil, \ldots, \lceil o_n - |\tilde{x}_n - o_n| \rceil)^T, \tag{8.1.7}$$

$$\delta = (\lfloor o_1 + |\tilde{x}_1 - o_1| \rfloor, \ldots, \lfloor o_n + |\tilde{x}_n - o_n| \rfloor)^T. \tag{8.1.8}$$

Notice that if $q(\tilde{x}) = v$, then $\langle \gamma, \delta \rangle$ is the maximum integer box inside $E(v)$ that passes through $\tilde{x}$.

8.1.2 Contour cuts of quadratic function

Consider the singly constrained case of (QIP):

$$(P_s) \qquad \min\ q(x) = \sum_{j=1}^{n}((1/2)c_j x_j^2 + d_j x_j)$$
$$\text{s.t. } g(x) = \sum_{j=1}^{n} g_j(x_j) \le b,$$
$$x \in X. \tag{8.1.9}$$

A subproblem (SP) of (P_s) is formed by replacing X by a subset $\widetilde{X} = \langle \tilde{l}, \tilde{u}\rangle \subseteq X$. Assume that $\widetilde{X} \cap S \neq \emptyset$ and $\widetilde{X} \setminus S \neq \emptyset$, where S is the feasible region of (P_s). Let q_s denote the optimal value of (SP). Let $\lambda^* > 0$ be the dual optimal solution to (SP). Suppose that the duality gap of (SP) is nonzero, i.e., $d(\lambda^*) < q_s$. By Theorems 3.15 and 3.16, Procedure 3.3 for dual search can find two optimal solutions, $\tilde{x} \in S$ and $\tilde{y} \in \widetilde{X} \setminus S$, to the Lagrangian relaxation problem (L_{λ^*}). The following always holds:

$$q(\tilde{y}) < d(\lambda^*) < q_s \le q(\tilde{x}). \tag{8.1.10}$$

In the following we will show that cutting certain integer boxes from $\widetilde{X}$ will not remove any optimal solution of (SP) after recording $\tilde{x}$. We consider the contour cut for the two cases where $q(x)$ is convex or is concave.

Case (a): $q(x)$ is convex, i.e., $c_j > 0$, $j = 1, \ldots, n$. Let $v_1 = q(\tilde{x})$ and $v_2 = d(\lambda^*)$. By (8.1.10) and the convexity of q, either $\tilde{x}$ is the optimal solution of (SP) or the optimal solution still lies in the set

$$\Omega = (\widetilde{X} \cap E(v_1)) \setminus E(v_2), \tag{8.1.11}$$

where $E(v_1)$ and $E(v_2)$ are defined by (8.1.4). In other words, removing sets $\widetilde{X} \setminus E(v_1)$ and $E(v_2)$ from $\widetilde{X}$ will not miss any optimal solution to (SP) after we record $\tilde{x}$. Since both $E(v_1)$ and $E(v_2)$ are ellipsoids, it is difficult to calculate Ω in (8.1.11). We instead outer-approximate Ω using integer boxes. More specifically, we consider a union of boxes of which Ω is a subset. Note that set Ω is a finite set containing only integer points. The following is true,

$$\widetilde{X} \cap M(v_1) \supset \widetilde{X} \cap E(v_1), \tag{8.1.12}$$

where $M(v_1)$ is the minimum integer box enclosing all the integer points in $E(v_1)$. Let $B(v_1) = \widetilde{X} \cap M(v_1)$. Then $B(v_1) = \langle \bar{\alpha}, \bar{\beta}\rangle$, where

$$\bar{\alpha} = (\max(\tilde{l}_1, \alpha_1), \ldots, \max(\tilde{l}_n, \alpha_n))^T, \tag{8.1.13}$$
$$\bar{\beta} = (\min(\tilde{u}_1, \beta_1), \ldots, \min(\tilde{u}_n, \beta_n))^T, \tag{8.1.14}$$

with α and β defined in (8.1.5) and (8.1.6), respectively.

By (8.1.10), the infeasible point $\tilde{y}$ is contained in the ellipsoid $E(v_2)$. So, the integer box $N(\tilde{y}) = \langle \gamma, \delta \rangle$ is also contained in $E(v_2)$, where γ and δ can be found using (8.1.7)–(8.1.8). This, combined with (8.1.12), implies that

$$B(v_1) \setminus N(\tilde{y}) \supset \Omega. \tag{8.1.15}$$

We further would like to cut $\tilde{x}$ from $\widetilde{X}$ if $\tilde{x} \in B(v_1)$ after recording $\tilde{x}$. Let $T(\tilde{x}) = \langle \tilde{\alpha}, \tilde{\beta} \rangle$ be the integer box with i) $\tilde{x}$ being one of its corner points and ii) all edges starting from $\tilde{x}$ being leaving the ellipsoid $E(v_1)$ and being towards the boundaries of $B(v_1)$. Specifically, $T(\tilde{x})$ can be determined by

$$\tilde{\alpha}_j = \begin{cases} \min(\tilde{x}_j, \bar{\alpha}_j), & \tilde{x}_j \le o_j \\ \min(\tilde{x}_j, \bar{\beta}_j), & \tilde{x}_j > o_j \end{cases} \tag{8.1.16}$$

$$\tilde{\beta}_j = \begin{cases} \max(\tilde{x}_j, \bar{\alpha}_j), & \tilde{x}_j \le o_j \\ \max(\tilde{x}_j, \bar{\beta}_j), & \tilde{x}_j > o_j \end{cases} \tag{8.1.17}$$

where o is defined in (8.1.2) and $\bar{\alpha}$ and $\bar{\beta}$ are defined in (8.1.13) and (8.1.14), respectively. Since $\tilde{x}$ is on the boundary of $E(v_1)$, we can cut $T(\tilde{x})$ from $B(v_1)$. We have

$$\tilde{\Omega} = [B(v_1) \setminus N(\tilde{y})] \setminus T(\tilde{x}) \supset \Omega \setminus \{\tilde{x}\}. \tag{8.1.18}$$

Figure 8.1 illustrates the contour cut process for case (a).

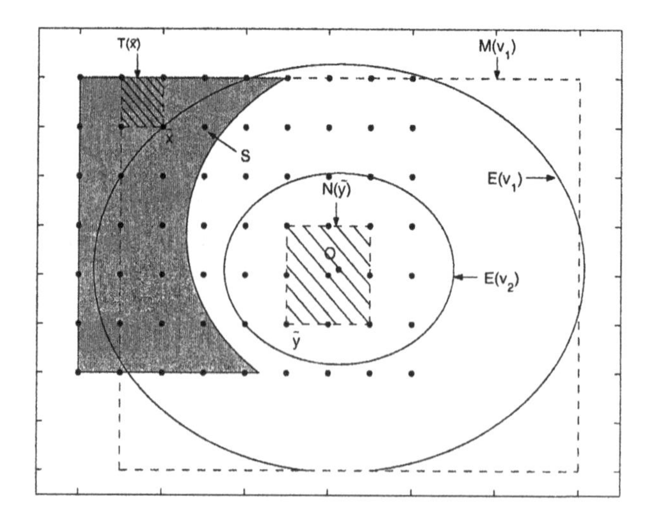

Figure 8.1. Contour cuts for case (a).

Case (b): $q(x)$ is concave, i.e., $c_j < 0$, $j = 1, \ldots, n$. Let $v_1 = d(\lambda^*)$ and $v_2 = q(\tilde{x})$. Then, by (8.1.10) and the concavity of q, the optimal solution of (SP) must lie in the set Ω defined in (8.1.11). Similar to case (a), we have

$$B(v_1) \setminus N(\tilde{x}) \supset \Omega. \tag{8.1.19}$$

Since $q(\tilde{y}) < d(\lambda^*) = v_1$, $\tilde{y}$ is outside the ellipsoid $E(v_1)$. If $\tilde{y}$ is contained in $B(v_1)$, then we can cut $T(\tilde{y})$ from $B(v_1)$, where $T(\tilde{y}) = \langle \tilde{\alpha}, \tilde{\beta} \rangle$, $\tilde{\alpha}$ and $\tilde{\beta}$ are defined in (8.1.16)–(8.1.17) with $\tilde{x}$ replaced by $\tilde{y}$. Therefore, we have

$$\tilde{\Omega} = [B(v_1) \setminus N(\tilde{x})] \setminus T(\tilde{y}) \supset \Omega. \tag{8.1.20}$$

Figure 8.2 illustrates the contour cut process for case (b).

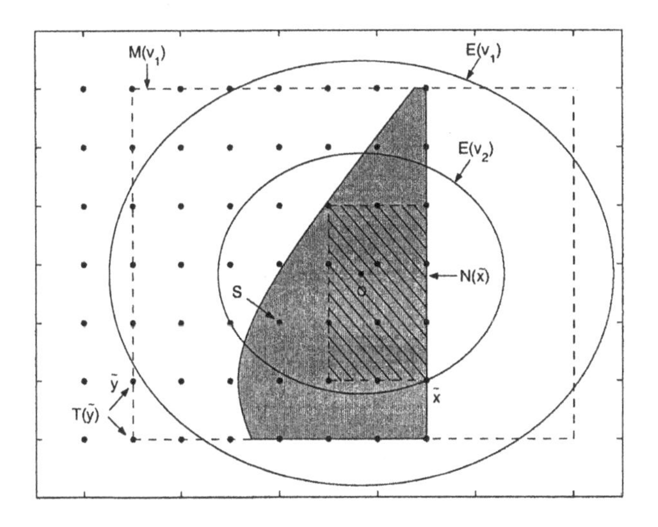

Figure 8.2. Contour cuts for case (b).

One clear conclusion is that after recording the feasible solution $\tilde{x}$, we can reduce the domain of (SP) from $\widetilde{X}$ to $\tilde{\Omega}$ without missing any optimal solution to (SP). This domain reduction process will improve the quality of the dual search, as seen in the following sections.

8.2 Convergent Lagrangian and Objective Contour Cut Method

In this section, we develop a convergent Lagrangian and contour cut method for the singly constrained problem (P_s). The method will be extended in Section 8.3 to handle multiple constraints. We first demonstrate the method by an example and then describe the method formally.

To motivate the method, let us consider a two-dimensional example with a concave quadratic objective function.

EXAMPLE 8.1

$$\begin{aligned} \min\ & q(x) = -1.5x_1^2 + 2x_1 - 2x_2^2 + 8x_2 \\ \text{s.t.}\ & g(x) = 3x_1^2 - 2x_1 + 2x_2^2 - 6x_2 \le 35, \\ & x \in X = \{x \in \mathbb{Z}^2 \mid -1 \le x_1 \le 5,\ 0 \le x_2 \le 6\}. \end{aligned}$$

The optimal solution of this problem is $x^* = (-1, 5)^T$ with $q(x^*) = -13.5$. The perturbation function of the example is illustrated in Figure 8.3. It can be observed from Figure 8.3 that the point C that corresponds to the optimal solution x^* is "hidden" above the convex envelope of the perturbation function and thus the traditional Lagrangian dual method will fail to find the optimal solution x^*.

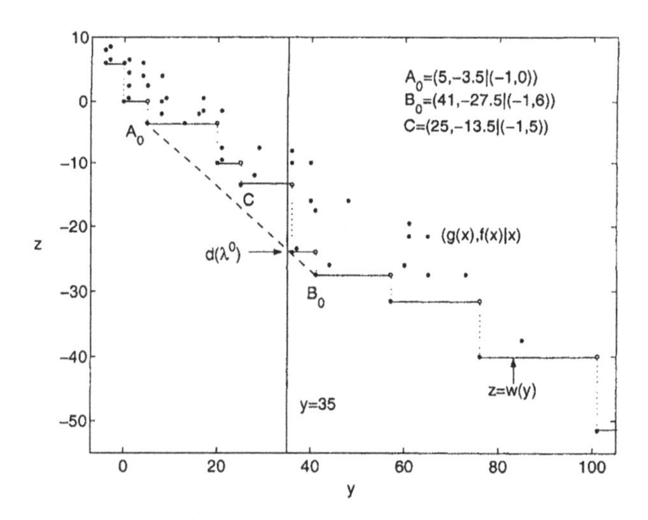

Figure 8.3. Perturbation function of Example 8.1.

Solving the dual problem of the example, we obtain the optimal multiplier $\lambda^0 = 0.6667$ with $d(\lambda^0) = -23.5$. The optimal solutions to (L_{λ^0}) are $x^0 = (-1, 0)^T$ and $y^0 = (-1, 6)^T$. The current duality bound is $q(x^0) - d(\lambda^0) = -3.5 + 23.5 = 20$.

Now, let $v_1^0 = d(\lambda^0) = -23.5$, $v_2^0 = q(x^0) = -3.5$. Applying the contour cut scheme in Section 8.1 to the example by using (8.1.20), we obtain a revised domain

$$X^1 = [B(v_1^0) \setminus N(x^0)] \setminus T(y^0),$$

where

$$\begin{aligned} B(v_1^0) &= X \cap M(v_1^0) = \langle(-1,0)^T,(5,6)^T\rangle \cap \langle(-3,-2)^T,(5,6)^T\rangle \\ &= \langle(-1,0)^T,(5,6)^T\rangle, \\ N(x^0) &= \langle(-1,0)^T,(2,4)^T\rangle,\ T(y^0) = \{(-1,6)^T\}. \end{aligned}$$

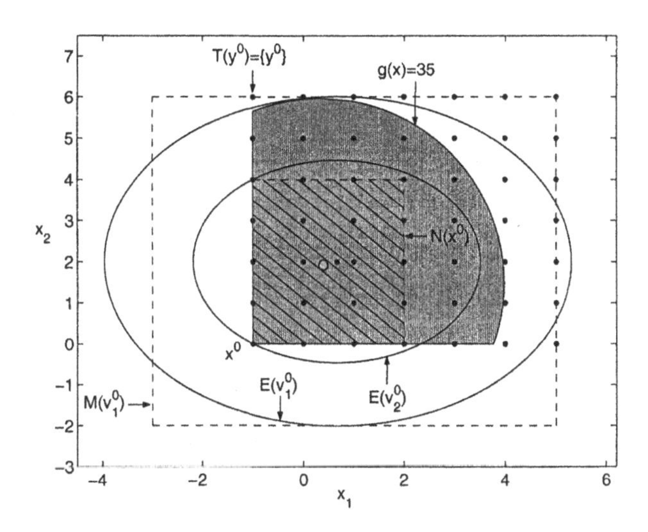

Figure 8.4. Domain X and the objective contour cuts.

The ellipsoids $E(v_1^0)$, $E(v_2^0)$, the integer boxes $M(v_1^0)$, $N(x^0)$ and $T(y^0)$ are illustrated in Figure 8.4. It can be seen from Figs. 8.3 and 8.4 that cutting sets $N(x^0)$ and $T(y^0)$ from the domain X will remove the corner points A_0 and B_0 in the plot of the perturbation function and thus raising the dual value. The revised domain X^1 and the corresponding perturbation function are shown in Figure 8.5 and Figure 8.6, respectively. The optimal dual value of the revised problem is $d(\lambda^1) = -23.125$ and the feasible and infeasible solutions of (L_{λ^1}) are: $x^1 = (0,5)^T, y^1 = (0,6)^T$. The dual bound is reduced to $q(x^1) - d(\lambda^1) = -10 + 23.125 = 13.125$. Let $v_1^1 = d(\lambda^1) = -23.125$ and $v_2^1 = q(x^1) = -10$. The ellipsoids $E(v_1^1)$, $E(v_2^1)$, the integer boxes $M(v_1^1)$, $N(x^1)$ and $T(y^1)$ are illustrated in Figure 8.5.

The above discussion reveals that the contour cut scheme described in Section 8.1 will reduce the duality bound and thus the duality gap and will eventually expose the "hidden" optimal point to the convex envelope of the perturbation function. In fact, as we can foresee from Figure 8.6, one more contour cut will make the point C lie on the convex envelope of the revised perturbation function, thus enabling the dual search to find the optimal solution x^*.

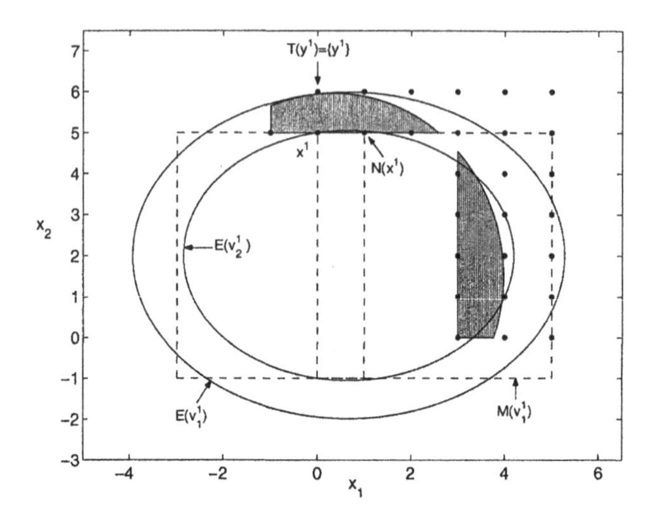

Figure 8.5. The revised domain X^1.

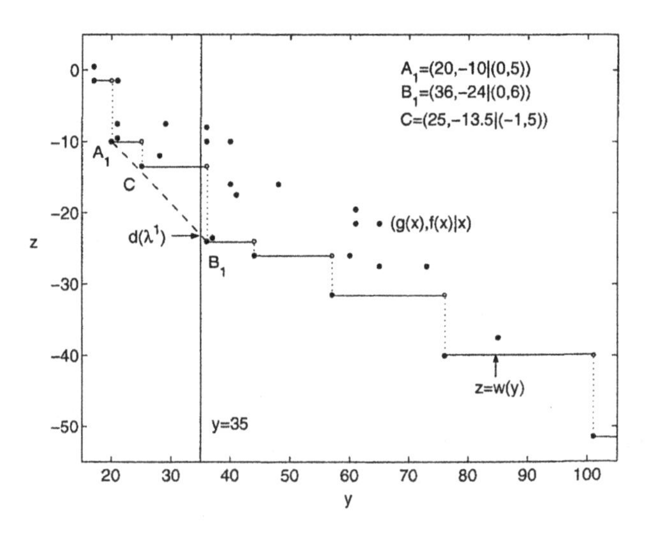

Figure 8.6. Perturbation function of the revised problem on X^1.

Based on the above discussion, a convergent Lagrangian and contour cut algorithm can be developed by combining the Lagrangian relaxation with the domain cut and partition scheme. Let $X^0 = \{X\}$. Initially, a dual search procedure is applied to (P_s) to produce an optimal dual value $d(\lambda^0)$ together with a feasible optimal solution x^0 and an infeasible optimal solution y^0 to (L_{λ^0}). The optimal dual value $d(\lambda^0)$ gives a lower bound of the problem and x^0 is set to be the incumbent. At the k-th iteration, the integer subbox with the minimum dual value is selected from X^k, the set of all integer subboxes which have not been fathomed. The domain cut and partition scheme is then applied to that integer subbox. For each newly generated integer subbox, Procedure 3.3 is applied to determine its dual value together with a feasible solution and an infeasible solution. The current best feasible solution is recorded as the incumbent solution and all integer subboxes whose dual value is greater than or equal to the objective function value of the incumbent are removed. The process repeats until there is no integer subbox in X^k and the incumbent solution is the optimal solution to (P_s) when the algorithm terminates.

We now formally present the algorithm.

ALGORITHM 8.1 (CONVERGENT LAGRANGIAN AND CONTOUR CUT METHOD FOR (P_s))

Step 0 (Initialization). Apply Procedure 3.3 to (P_s) and obtain the dual value $d(\lambda^0)$, a feasible solution x^0 and an infeasible solution y^0. Set $LB = d(\lambda^0)$ as the lower bound, $x_{opt} = x^0$, $f_{opt} = q(x_{opt})$, $X^0 = X$, $k = 0$.

Step 1. Select the integer subbox $\langle \alpha^k, \beta^k \rangle$ from X^k that yields the minimum lower bound LB. Let x^k, $y^k \in \langle \alpha^k, \beta^k \rangle$ be the feasible and infeasible solutions generated by Procedure 3.3, respectively.

Step 2 (Contour cut and partition).

Case (a): q is a convex function. Set $v_1 = q(x^k)$, $v_2 = LB$; Calculate integer boxes $B(v_1)$, $N(y^k)$ and $T(x^k)$. Use (6.3.1) to partition the set

$$Y^{k+1} = [B(v_1) \setminus N(y^k)] \setminus T(x^k). \tag{8.2.21}$$

Case (b): q is a concave function. Set $v_1 = LB$, $v_2 = q(x^k)$; Calculate integer boxes $B(v_1)$, $N(x^k)$ and $T(y^k)$; Use (6.3.1) to partition the set

$$Y^{k+1} = [B(v_1) \setminus N(x^k)] \setminus T(y^k). \tag{8.2.22}$$

Step 3 (Dual Search).

(i) Apply Procedure 3.3 to each integer subbox $\langle \alpha, \beta \rangle \in Y^{k+1}$ with X replaced by $\langle \alpha, \beta \rangle$. Let

$$\tilde{x}^0 \in \arg \min_{x \in \langle \alpha, \beta \rangle} g(x), \ \tilde{y}^0 \in \arg \min_{x \in \langle \alpha, \beta \rangle} q(x).$$

One of the following three cases happens: (a) If $g(\tilde{x}^0) > b$, then remove $\langle \alpha, \beta \rangle$ from Y^{k+1}; (b) If $g(\tilde{y}^0) \le b$, then, set $x_{opt} = \tilde{y}^0$ and $f_{opt} = q(\tilde{y}^0)$ if $q(\tilde{y}^0) < f_{opt}$, and remove $\langle \alpha, \beta \rangle$ from Y^{k+1}; (c) If $g(\tilde{x}^0) \le b$ and $g(\tilde{y}^0) > b$, then Procedure 3.3 generates a dual value on the integer box, a feasible solution and an infeasible solution. If the dual value is greater than or equal to f_{opt}, then remove $\langle \alpha, \beta \rangle$ from Y^{k+1}. Compute the objective function value of the feasible solution and update x_{opt} and f_{opt} if necessary.

(ii) Set $X^{k+1} = Y^{k+1} \cup (X^k \setminus \{\langle \alpha^k, \beta^k \rangle\})$. If the dual value of any integer box in X^{k+1} is greater than or equal to f_{opt}, remove this integer box from X^{k+1}.

Step 4 (Termination). If X^{k+1} is empty, stop and x_{opt} is an optimal solution to (P_s). Otherwise, set $k := k + 1$, go to Step 1.

THEOREM 8.1 *Algorithm 8.1 stops within a finite number of iterations with either an optimal solution to (P_s) being found or an infeasibility of (P_s) being reported.*

Proof. The finite convergence is obvious by noting that X is a finite integer set and at each iteration, x^k and y^k are cut from X^k in Step 2 and are not included in X^{k+1}. From the discussion in Section 8.1, no feasible solution better than x^k will be cut from X^k in Step 2. Also, by weak duality, no feasible solution better than x^k will be cut from X^k in Step 3. Thus, at each iteration, either x_{opt} is already the optimal solution or there is an optimal solution in X^k. Therefore, x_{opt} must be an optimal solution to the original problem when the algorithm stops at Step 4. □

8.3 Extension to Problems with Multiple Constraints

The algorithm developed in Section 8.2 can be extended to deal with multiply constrained cases of (QIP). Consider a subproblem (SP) of (QIP) with X replaced by an integer subbox $\widetilde{X} \subseteq X$. The Lagrangian dual of (SP) is:

$$\max_{\lambda \in \mathbb{R}^m_+} d(\lambda), \tag{8.3.23}$$

where

$$d(\lambda) := \min_{x \in \widetilde{X}} \left[q(x) + \sum_{i=1}^{m} \lambda_i (g_i(x) - b_i) \right]. \tag{8.3.24}$$

From the weak duality, $d(\lambda) \leq q(x)$ for any feasible solution $x \in \widetilde{X}$. Therefore, $d(\lambda)$ provides a lower bound of the optimal value of (SP). Let λ^* be an optimal solution to (8.3.23). Then, $LB = d(\lambda^*)$ is the best lower bound generated by the Lagrangian relaxation (8.3.24).

Since $d(\lambda)$ is a concave piecewise linear function, the subgradient method is an efficient method to compute an approximate solution to (8.3.23). Alternatively, we can use the outer Lagrangian linearization method to compute an exact solution to (8.3.23) when an initial feasible solution to (QIP) is available.

Consider the following surrogate constraint problem:

$$\min\ q(x) = \sum_{j=1}^{n}(\frac{1}{2}c_j x_j^2 + d_j x_j) \tag{8.3.25}$$

$$\text{s.t. } g_{\lambda^*}(x) = \sum_{i=1}^{m} \lambda_i^* g_i(x) \le \sum_{i=1}^{m} \lambda_i^* b_i,$$

$$x \in \tilde{X}.$$

Let $b_{\lambda^*} = \sum_{i=1}^{m} \lambda_i^* b_i$. Denote by $\underline{g}_{\lambda^*}$ and $\overline{g}_{\lambda^*}$ the minimum value and maximum value of $g_{\lambda^*}(x)$ over $\widetilde{X}$, respectively. Without loss of generality, we can assume that

$$\underline{g}_{\lambda^*} \le b_{\lambda^*} < \overline{g}_{\lambda^*}. \tag{8.3.26}$$

Suppose that λ^* is an exact solution to (8.3.23). It is easy to see that (8.3.25) and (SP) have the same dual value and the optimal solution to the dual problem of problem (8.3.25) is 1. Moreover, by Theorems 3.15 and 3.16, there exist a feasible solution $\tilde{x}$ and an infeasible solution $\tilde{y}$ to problem (8.3.25) that solve the Lagrangian relaxation (8.3.24) with $\lambda = \lambda^*$.

If $\tilde{\lambda}$ is an approximate solution to (8.3.23), then we can apply Procedure 3.3 to search for an exact dual solution μ^* to problem (8.3.25) with λ^* replaced by $\tilde{\lambda}$. Set $\lambda^* = \mu^* \tilde{\lambda}$. Again, by Theorems 3.15 and 3.16, there exist a feasible solution $\tilde{x}$ and an infeasible $\tilde{y}$ to problem (8.3.25) that solve the Lagrangian relaxation (8.3.24) with $\lambda = \lambda^*$.

Now we are ready to extend Algorithm 8.1 to multiply constrained case of (QIP). Notice that

$$q(\tilde{y}) < d(\lambda^*) \le q(\tilde{x}). \tag{8.3.27}$$

Moreover, $\tilde{y}$ is infeasible to (QIP) while $\tilde{x}$ is not necessarily feasible to (QIP). Therefore, the contour cutting process in Step 2 of Algorithm 8.1 has to be modified for situations where $\tilde{x}$ is infeasible to (QIP). More specifically, we need the following modifications in Algorithm 8.1.

Step 2'. Case (a): q is a convex function. If $\tilde{x}$ is feasible to (P), set $v_1 = q(\tilde{x})$ and compute Y^{k+1} by

$$Y^{k+1} = [B(v_1) \setminus T(\tilde{x})] \setminus N(\tilde{y}).$$

Otherwise, if $\tilde{x}$ is infeasible to (P), then compute Y^{k+1} by

$$Y^{k+1} = [\langle \tilde{l}, \tilde{u}\rangle \setminus \{\tilde{x}\}] \setminus N(\tilde{y}).$$

Case (b): q is a concave function. Set $v_1 = LB$. If $\tilde{x}$ is feasible to (P), then compute Y^{k+1} by

$$Y^{k+1} = [B(v_1) \setminus N(\tilde{x})] \setminus T(\tilde{y}).$$

Otherwise, if $\tilde{x}$ is infeasible to (QIP), compute Y^{k+1} by

$$Y^{k+1} = [B(v_1) \setminus \{\tilde{x}\}] \setminus T(\tilde{y}).$$

We also need to replace the dual search procedure used in Step 0 and Step 3 (i) of Algorithm 8.1 with an exact dual search method or an approximate method for (8.3.23). When the dual problems (8.3.23) in Step 0 and Step 3 (i) are solved approximately, Procedure 3.3 is applied to the surrogate problem (8.3.25) to search for the lower bound together with a feasible solution and infeasible solution to (8.3.25). Finally, two special cases have to be considered in the algorithm when (8.3.26) does not hold. If $\underline{g}_{\lambda^*} > b_{\lambda^*}$, then there is no feasible solution in $\widetilde{X}$ and $\widetilde{X}$ can be removed from further consideration. If $\overline{g}_{\lambda^*} \leq b_{\lambda^*}$, then solving (8.3.25) using the dual search will yield a zero dual solution and an optimal solution $\tilde{x}$ which is feasible to (8.3.25). If $\tilde{x}$ is also feasible to (QIP), discard $\tilde{X}$ from further consideration after updating x_{opt} and f_{opt} if $q(\tilde{x}) < f_{opt}$. Otherwise, remove $\tilde{x}$ from $\tilde{X}$.

The finite convergence of the extended algorithm for multiply constrained problems and the optimality of x_{opt} when the algorithm stops can be proved similarly as in Theorem 8.1.

An important observation from Step 2′ is that in multiply constrained situations, we are not always able to find a feasible solution to the primal problem during the dual search procedure, which constitutes a major difference between multiply constrained problems and singly constrained problems. The unavailability of feasible solutions to the primal problem affects the efficiency of the contour cut algorithm for multiply constrained problems, as witnessed from our computational experiences. Specifically, a guaranteed two-direction cutting process (cutting the outside of a bigger ellipse and the inside of a smaller ellipse) in singly constrained situations often becomes a one-direction cutting process in multiply constrained situations when a feasible solution is not available. Nevertheless, in some situations, certain heuristics can be used to search for a feasible solution which does not necessarily solve problem (8.3.24). This may improve the efficiency of the contour cutting process. For example, if the constraint functions are nondecreasing, as is the case in nonlinear knapsack problems, then the lower bound point $\tilde{l}$ of $\widetilde{X}$ is always feasible to (SP) in nontrivial cases.

We now illustrate the extended algorithm for multiply constrained problems by a two-dimensional example with a concave quadratic objective function, a convex constraint and a nonconvex constraint.

EXAMPLE 8.2

$$\begin{aligned} \min\ & q(x) = -1.5x_1^2 + 2x_1 - 2x_2^2 + 8x_2 \\ \text{s.t.}\ & g_1(x) = 3x_1^2 - 2x_1 + 2x_2^2 - 6x_2 \le 66, \\ & g_2(x) = -x_1^2 - x_1 + x_2^2 - 2x_2 \le -3.5, \\ & x \in X = \{x \in \mathbb{Z}^2 \mid -1 \le x_1 \le 5,\ 0 \le x_2 \le 6\}. \end{aligned}$$

The optimal solution is $x^* = (5,0)^T$ with $q(x^*) = -27.5$.

For this example, we use the subgradient method to solve the dual problem (8.3.23). The iterative process is described as follows.

Iteration 0

Step 0. Solving (8.3.23) with $\widetilde{X} = X$, we get $\lambda^* = (0.5145, 0.2284)^T$. Applying Procedure 3.3 to the surrogate constraint problem (8.3.25), we obtain the dual value $LB = -34.0771$ and two optimal solutions $x^0 = (-1,6)^T$ and $y^0 = (5,6)^T$. An initial feasible solution $(5,0)^T$ is also obtained during the dual search. Set $x_{opt} = (5,0)^T$ and $f_{opt} = q(x_{opt}) = -27.5$. Notice that both x^0 and y^0 are infeasible to the example. Set $X^0 = X$ and $k = 0$.

Iteration 1

Step 1. Select X to generate new integer boxes.

Step 2. Set $v_1 = LB = -34.0771$. We have

$$B(v_1) = M(v_1) \cap X = \langle(-4,-2)^T, (6,6)^T\rangle \cap X = X$$

and

$$Z^1 = B(v_1) \setminus \{x^0\} = \langle(0,0)^T, (5,6)^T\rangle \cup \langle(-1,0)^T, (-1,5)^T\rangle = Z_1^1 \cup Z_2^1.$$

Since the dual value on Z_2^1 is $-13.5 > -27.5 = f_{opt}$, we can remove Z_2^1 from Z^1. We have $T(y^0) = \langle(5,6)^T, (5,6)^T\rangle$. Thus,

$$Y^1 = Z^1 \setminus T(y^0) = \langle(0,0)^T, (4,6)^T\rangle \cup \langle(5,0)^T, (5,5)^T\rangle = Y_1^1 \cup Y_2^1.$$

For Y_1^1, the dual value is -33.1476 with two solutions $(0,6)^T$ and $(4,6)^T$; For Y_2^1, the dual value is -32.2875 with two solutions $(5,0)^T$ and $(5,5)^T$.

Step 3. Set $X^1 = Y^1$, $k = 1$.

Iteration 2

Step 1. Select Y_1^1 from X^1 to generate new integer boxes. Set $x^1 = (0,6)^T$ and $y^1 = (4,6)^T$. Notice that x^1 is infeasible to the example.

Step 2. Set $v_1 = -33.1476$. Calculate $B(v_1) = M(v_1) \cap Y_1^1 = \langle(-4,-2)^T, (5,6)^T\rangle \cap Y_1^1 = Y_1^1$. We have

$$Z^2 = B(v_1) \setminus \{x^1\} = \langle(1,0)^T, (4,6)^T\rangle \cup \langle(0,0)^T, (0,5)^T\rangle = Z_1^2 \cup Z_2^2.$$

Since the dual value on Z_2^2 is $-1.5966 > -27.5 = f_{opt}$, we can remove Z_2^2 from Z^2. We have $T(y^1) = \langle (4,6)^T, (4,6)^T \rangle$. Thus

$$Y^2 = Z^2 \setminus T(y^1) = \langle (1,0)^T, (3,6)^T \rangle \cup \langle (4,0)^T, (4,5)^T \rangle = Y_1^2 \cup Y_2^2.$$

The dual value on Y_1^2 is -20.7748 and the dual value on Y_2^2 is -26.0. Since both of them are greater than f_{opt}, we can remove Y_1^2 and Y_2^2 from Y^2.

Step 3. Set $X^2 = \{Y_2^1\}$, $k = 2$.

Iteration 3.

Step 1. Select Y_2^1 to generate the new integer subboxes. Set $x^2 = (5,0)^T$ and $y^2 = (5,5)^T$. Note that x^2 is feasible to the example.

Step 2. Set $v_1 = -32.2875$. Calculate $B(v_1) = M(v_1) \cap Y_2^1 = \langle (-4,-2)^T, (5,6)^T \rangle \cap Y_2^1 = Y_2^1$ and $N(x^2) = \langle (5,0)^T, (5,4)^T \rangle$. We have

$$Z^2 = B(v_1) \setminus N(x^2) = \{(5,5)^T\}.$$

Thus

$$Y^2 = Z^2 \setminus \{y^2\} = \emptyset.$$

Step 3. $X^3 = \emptyset$.

Step 4. Stop and $x_{opt} = (5,0)^T$ is an optimal solution to the example.

8.4 Extension to Problems with Indefinite q

The contour cut method developed in the previous sections can be extended to handle problems with an indefinite quadratic objective function. We describe the main idea of this extension in this section. Let's first consider the singly constrained problem (P_s) where some of the coefficients c_j's are positive and all others are negative.

We can always express $q(x)$ as the sum of a convex quadratic function and a concave quadratic function: $q(x) = q_1(x) + q_2(x)$ with $q_1(x) = \sum_{j=1}^n (\frac{1}{2} c_j^1 x_j^2 + d_j x_j)$ and $q_2(x) = -\sum_{j=1}^n \frac{1}{2} c_j^2 x_j^2$, where all c_j^1 and c_j^2, $j = 1, 2, \ldots, n$, are positive. Note that the expression of $q(x)$ is not unique. The subproblem (SP) of problem (P_s) can be expressed as follows:

$$\begin{aligned} \min\ & q(x) = q_1(x) + q_2(x) \\ \text{s.t. } & g(x) = \sum_{j=1}^n g_j(x_j) \le b, \\ & x \in \widetilde{X} = \{x \in \mathbb{Z}^n \mid \tilde{l}_j \le x_j \le \tilde{u}_j,\ j = 1, \ldots, n\}, \end{aligned}$$

where $\widetilde{X} \subseteq X$. Consider the following two problems associated with (SP):

$$(SP^1) \qquad \min\ q_1(x) = \sum_{j=1}^{n} (\frac{1}{2} c_j^1 x_j^2 + d_j x_j)$$
$$\text{s.t. } g(x) = \sum_{j=1}^{n} g_j(x_j) \le b,$$
$$x \in \widetilde{X} = \{x \in \mathbb{Z}^n \mid \tilde{l}_j \le x_j \le \tilde{u}_j,\ j = 1, \ldots, n\},$$

and

$$(SP^2) \qquad \min\ q_2(x) = -\sum_{j=1}^{n} \frac{1}{2} c_j^2 x_j^2$$
$$\text{s.t. } g(x) = \sum_{j=1}^{n} g_j(x_j) \le b,$$
$$x \in \widetilde{X} = \{x \in \mathbb{Z}^n \mid \tilde{l}_j \le x_j \le \tilde{u}_j,\ j = 1, \ldots, n\}.$$

Obviously, (SP^1) and (SP^2) are nonlinear integer programming problems with a convex quadratic objective function and a concave quadratic objective function, respectively. Let $f_i^* = \min_{x \in S \cap \widetilde{X}} q_i(x)$, $i = 1, 2$, where S is the feasible region of (P_s). Further define the following Lagrangian relaxation for (SP^1) and (SP^2), respectively, for $\lambda \ge 0$,

$$(L_\lambda^i) \qquad d_i(\lambda) = \min_{x \in \widetilde{X}} [q_i(x) + \lambda(g(x) - b)], \quad i = 1, 2.$$

Let λ_i^* be the optimal solutions to the dual problems $\max_{\lambda \ge 0} d_i(\lambda)$ for $i = 1, 2$, respectively. Let $\tilde{x} \in S \cap \widetilde{X}$. By the weak duality, we have,

$$d_1(\lambda_1^*) + d_2(\lambda_2^*) \le f_1^* + f_2^* \le f^* \le q_1(\tilde{x}) + q_2(\tilde{x}). \tag{8.4.28}$$

Let

$$C_1 = \{x \in \widetilde{X} \mid q_i(x) < d_i(\lambda_i^*),\ i = 1, 2\},$$
$$C_2(\tilde{x}) = \{x \in \widetilde{X} \mid q_i(x) \ge q_i(\tilde{x}),\ i = 1, 2\}.$$

It is easy to see from (8.4.28) and the weak duality that sets C_1 and $C_2(\tilde{x})$ can be cut off from $\widetilde{X}$ without removing the optimal solution after recording $\tilde{x}$. Let $\tilde{x}_i$ and $\tilde{y}_i$ be the feasible and infeasible optimal solutions to $(L_{\lambda_i^*}^i)$ $(i = 1, 2)$, respectively. Notice that $q_i(\tilde{y}_i) \le d_i(\lambda_i^*)$, $i = 1, 2$. Let $v_i = q_1(\tilde{x}_i)$, $i = 1, 2$ and $w = d_2(\lambda_2^*)$. Similar to Section 8.1, we define sets $B_i(\cdot)$ and $N_i(\cdot)$ for functions q_i, $i = 1, 2$, respectively. Then, we have

$$Q_1 = N_1(\tilde{y}_1) \cap [\widetilde{X} \setminus B_2(w)] \subseteq C_1 \cap \widetilde{X},$$
$$Q_2(\tilde{x}_i) = [\widetilde{X} \setminus B_1(v_i)] \cap N_2(\tilde{x}_i) \subseteq C_2(\tilde{x}_i) \cap \widetilde{X}, \quad i = 1, 2.$$

Thus, cutting both Q_1 and $Q_2(\tilde{x}_i)$ $(i = 1, 2)$ from $\widetilde{X}$ will not remove any optimal solution to the primal problem after recording the current best feasible solution as the incumbent. Note Q_1 and/or $Q_2(\tilde{x}_i)$ could be empty in certain circumstances. In the cutting process, points $\tilde{x}_i$, $i = 1, 2$, will be removed from $\widetilde{X}$ after updating the incumbent.

Replacing Step 2 of Algorithm 8.1 with the above contour cutting process, we can then deal with (P_s) with an indefinite quadratic objective function. Similar to Section 8.3, we can further extend the algorithm to solve the multiply constrained case of (QIP) with an indefinite objective function.

Now, let's demonstrate the above solution idea by an illustrative example.

EXAMPLE 8.3

$$\begin{aligned} \min\ & q(x) = -1.75x_1^2 - 1.75x_1 + x_2^2 - 12x_2 \\ \text{s.t.}\ & g(x) = 4(x_1 - 1)^2 + 9(x_2 - 2.5)^2 \le 10, \\ & x \in X = \{x \in \mathbb{Z}^2 \mid 0 \le x_i \le 4,\ i = 1, 2\}. \end{aligned}$$

The optimal solution of the example is $x^* = (2, 3)^T$ with $q(x^*) = -37.5$.

Decompose the above example into the following two associated problems, of which the first has a convex quadratic objective function and the second has a concave quadratic objective function,

$$\begin{aligned} \min\ & q_1(x) = 0.25x_1^2 - 1.75x_1 + 3x_2^2 - 12x_2 \\ \text{s.t.}\ & g(x) = 4(x_1 - 1)^2 + 9(x_2 - 2.5)^2 \le 10, \\ & x \in X = \{x \in \mathbb{Z}^2 \mid 0 \le x_i \le 4,\ i = 1, 2\} \end{aligned} \tag{8.4.29}$$

and

$$\begin{aligned} \min\ & q_2(x) = -2x_1^2 - 2x_2^2 \\ \text{s.t.}\ & g(x) = 4(x_1 - 1)^2 + 9(x_2 - 2.5)^2 \le 10, \\ & x \in X = \{x \in \mathbb{Z}^2 \mid 0 \le x_i \le 4,\ i = 1, 2\}. \end{aligned} \tag{8.4.30}$$

Iteration 0

Step 0. Solving the dual problem of (8.4.29) yields a dual value, $d_1 = -14.6563$, and two solutions, $\tilde{x}_1 = (2, 2)^T$ and $\tilde{y}_1 = (3, 2)^T$. Solving the dual problem of (8.4.30) yields a dual value, $d_2 = -29.1250$, and two solutions, $\tilde{x}_2 = (2, 3)^T$ and $\tilde{y}_2 = (3, 3)^T$. Thus, the lower bound is $LB = d_1 + d_2 =$

$-14.6563 - 29.1250 = -43.7813$ and the incumbent is $x_{opt} = (2,3)^T$ with $f_{opt} = q((2,3)^T) = -37.5$. Set $X^0 = \{\langle(0,0)^T, (4,4)^T\rangle\}$ and $k = 0$.

Iteration 1

Step 1. Select the unique integer subbox in X^0.

Steps 2-3. Since $N_1(\tilde{y}_1) = \langle(3,2)^T, (4,2)^T\rangle$ and $B_2(d_2) = \langle(0,0)^T, (3,3)^T\rangle$, we have $Q_1 = N_1(\tilde{y}_1) \cap [X \setminus B_2(d_2)] = \{(4,2)^T\}$. Thus,

$$\begin{aligned} Z^1 &= X \setminus Q_1 = \langle(0,3)^T, (4,4)^T\rangle \cup \langle(0,0)^T, (3,2)^T\rangle \cup \langle(4,0)^T, (4,1)^T\rangle \\ &= Z_1^1 \cup Z_2^1 \cup Z_3^1. \end{aligned}$$

For Z_1^1, we have $d_1 + d_2 = -11.6563 - 29.1250 = -40.7813 < -37.5 = f_{opt}$. For Z_2^1, we have $d_1 + d_2 = -14.6563 - 19.1250 = -33.7813 > -37.5 = f_{opt}$. So Z_2^1 is removed from Z^1. Since there is no feasible solution in Z_3^1, Z_3^1 is also removed from Z^1. Set $Y^1 = \{Z_1^1\}$. Figure 8.7 illustrates the set $Z^1 = X \setminus Q_1$.

Let $v_1 = q_1(\tilde{x}_1) = -14.5$. Since $N_2(\tilde{x}_1) = \langle(0,0)^T, (2,2)^T\rangle$ and $B_1(v_1) = \langle(2,2)^T, (4,2)^T\rangle$, we have

$$\begin{aligned} Q_2(\tilde{x}_1) &= [X \setminus B_1(v_1)] \cap N_2(\tilde{x}_1) = N_2(\tilde{x}_1) \setminus \{(2,2)^T\} \\ &= \langle(0,0)^T, (2,2)^T\rangle \setminus \{(2,2)^T\}. \end{aligned}$$

Notice $Q_2(\tilde{x}_2)$ is an empty set. Since $Q_2(\tilde{x}_1) \cap Y^1 = \emptyset$, a revised domain X^1 is generated from cutting $\tilde{x}_2$ from Y^1 (see Figure 8.8). Decompose X^1 as

$$X^1 = \langle(3,3)^T, (4,4)^T\rangle \cup \langle(0,4)^T, (2,4)^T\rangle \cup \langle(0,3)^T, (1,3)^T\rangle = X_1^1 \cup X_2^1 \cup X_3^1.$$

Since there is no feasible solution in X_1^1 and X_2^1, they can be removed from X^1. For X_3^1, we have $d_1 + d_2 = -10.5 - 20.0 = -30.5 > f_{opt}$, so X_3^1 is also removed. Therefore, $X^2 = \emptyset$.

Step 4. Stop and $x_{opt} = (2,3)^T$ is an optimal solution.

In computational implementation with an indefinite q, we can also solve the dual problem of (SP) directly to obtain a dual value $d(\lambda^*)$ and use

$$\max\{d(\lambda^*), d_1(\lambda_1^*) + d_2(\lambda_2^*)\}$$

as the lower bound to identify unpromising subboxes to be fathomed. Let $\tilde{x}$ and $\tilde{y}$ be the feasible and infeasible optimal solutions to the Lagrangian relaxation problem of (SP) with λ set as λ^*, respectively. Instead of cutting $Q_2(\tilde{x}_i)$, $i = 1, 2$, we cut $Q_2(\tilde{x})$ in the algorithm.

8.5 Computational Results

In this section, we present the computational results of the algorithms in Section 8.2 and its extensions in Sections 8.3 and 8.4. The algorithms were programmed in FORTRAN 90 and run on a SUN Workstation (Blade 2000). Comparison results with other methods in the literature will be also presented.

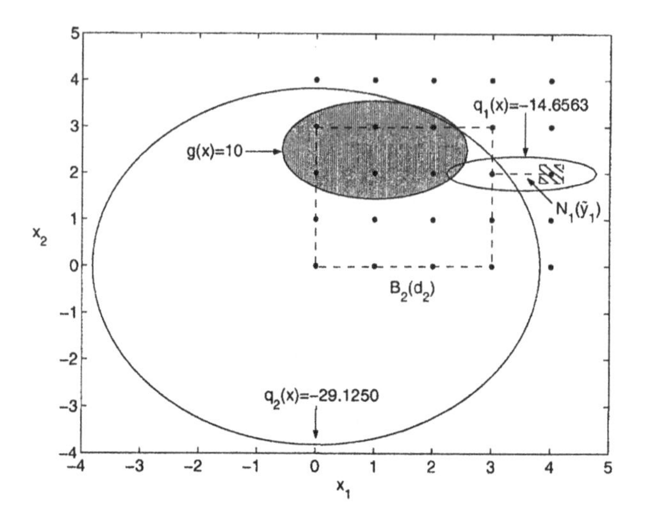

Figure 8.7. Set $Z^1 = X \setminus Q_1$.

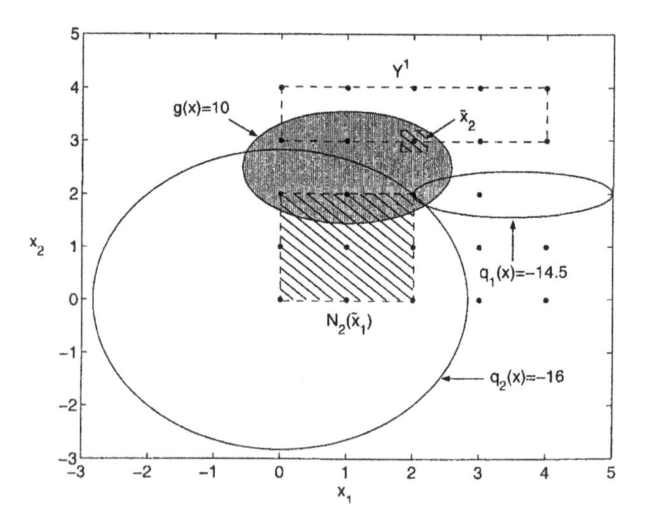

Figure 8.8. Sets $X^1 = Y^1 \setminus \{\tilde{x}_2\}$ and $Q_2(\tilde{x}_1)$.

8.5.1 Test problems

Two sets of test problems are considered in our computational experiments. The first set of test problems consists of 12 problems with different types of objective functions and constraint functions. The second set of test problems is a class of convex quadratic integer programming problems arising in portfolio optimization. All the coefficients in the test problems are randomly generated from uniform distributions.

In the first set of test problems, three types of objective functions in the form of $q(x) = \sum_{j=1}^{n}(\frac{1}{2}c_j x_j^2 + d_j x_j)$ are generated using the following data:

- $q(x)$ is convex quadratic with $c_j \in [2, 20]$ and $d_j \in [-100, -50]$;
- $q(x)$ is concave quadratic with $c_j \in [-20, -2]$ and $d_j \in [-10, 40]$;
- $q(x)$ is indefinite quadratic with $c_j \in [-10, 10]$ and $d_j \in [-40, 10]$.

The constraint functions in the test problems are in the following form:

$$g_i(x) = \sum_{j=1}^{n}(\alpha_{ij}x_j + \beta_{ij}x_j^2 + \gamma_{ij}x_j^3), \quad i = 1, \ldots, m.$$

Table 8.1 describes the ranges of coefficients in g_i's for singly constrained test problems and multiply constrained test problems, where Type 1 denotes the linear constraints, Type 2 the convex quadratic constraints, Type 3 the concave quadratic constraints and Type 4 the 3rd polynomial constraints.

Table 8.1. Coefficients in the test problems for Algorithm 8.1 and its extensions.

Type	Single Constraint			Multiple Constraints		
	α_{1j}	β_{1j}	γ_{1j}	α_{ij}	β_{ij}	γ_{ij}
1	$[-10, 40]$	0	0	$[1, 40]$	0	0
2	$[-10, 30]$	$[1, 20]$	0	$[10, 50]$	$[1, 10]$	0
3	$[100, 200]$	$[-20, -1]$	0	$[-10\beta_{ij}, -10\beta_{ij} + 5]$	$[-15, -5]$	0
4	$[-10, 20]$	$[5, 25]$	$[-2, 8]$	$[10, 50]$	$[1, 10]$	$[1, 5]$

In the first set of test problems, we take $l_j = 1$ and $u_j = 5$, $j = 1, \ldots, n$, and the right-hand side b is taken as $b = g_{min} + r \times (g_{max} - g_{min})$, where g_{min} and g_{max} are the minimum and maximum values of $g(x)$ over X, respectively, and $r \in (0, 1)$.

The second set of problems arises from portfolio optimization. It has been shown in [193][194] that the Markowitz's mean-variance portfolio selection model can be reformulated as a simplified model which is a separable convex quadratic programming problem with linear constraints. The discrete version of the simplified portfolio selection problem [210] can be expressed as

$$(SMV) \qquad \min\ q(x) = \sum_{j=1}^{n} (\frac{1}{2} c_j x_j^2 + d_j x_j)$$
$$\text{s.t. } Ax \le b,$$
$$x \in X = \{x \in \mathbb{Z}^n \mid l_j \le x_j \le u_j,\ j = 1, \ldots, n\},$$

where $c_j > 0$ for all j and $A = (a_{ij})$ is an $m \times n$ matrix. Obviously, problem (SMV) is a special case of (QIP). In our testing, the data in (SMV) are taken as the same as in [210] where additional dependency relationships are considered. The ranges of coefficients in (SMV) are: $c_j \in [10, 50]$, $d_j \in [-3000, -1000]$, $a_{ij} \in [1, 5]$, $l_j \in [0, 40]$ and $u_j = l_j + 5$. The right-hand side b is taken as $b = A \times [l + r \times (u - l)]$, where $l = (l_1, \ldots, l_n)^T, u = (u_1, \ldots, u_n)^T$ and $r \in (0, 1)$.

8.5.2 Computational results

The computational results of the convergent Lagrangian and contour cut method for the first set of test problems are summarized in Tables 8.2–8.4. The following notations are used in the numerical results:

- n = number of variables;
- m = number of constraints;
- N_{iter} = average number of iterations of the algorithm for 20 test problems;
- N_{box} = average number of the total integer boxes examined during the algorithm for 20 test problems;
- T_{cpu} = average CPU seconds measured on a SUN Workstation (Blade 2000) for 20 test problems.

In our implementation of the algorithms for multiply constrained problems, the outer Lagrangian linearization method is used to solve the dual problem (8.3.23). The results in Tables 8.2–8.4 show that the convergent Lagrangian and contour cut methods are efficient and robust for solving large-scale quadratic integer problems with convex, concave and indefinite objective functions and different types of constraint functions. Comparing results in Tables 8.2–8.4, we can see that the algorithm is more efficient for problems with a concave objective function. We can also see that the efficiency of the algorithm is not sensitive to the convexity of the constraint functions. This is partially due to the fact that the domain cut and partition scheme does not depend on the property of the constraints.

Table 8.2. Numerical results for convex $q(x)$ ($r = 0.6$).

Type of Constraint	Single Constraint				Multiple Constraints				
	n	N_{iter}	N_{box}	T_{cpu}	n	m	N_{iter}	N_{box}	T_{cpu}
Linear	500	162	42928	52.9	30	10	598	10732	74.9
	1000	253	123197	332.0	40	10	1074	23765	211.0
	1500	538	404297	1716.6	50	10	5313	129684	1901.4
Convex Quadratic	500	155	46853	86.1	30	10	123	2426	8.0
	1000	242	138212	521.5	40	10	204	4952	34.3
	1500	354	286512	1746.4	50	10	432	13568	80.9
Concave Quadratic	500	186	43213	76.9	30	10	161	1875	4.8
	1000	508	189918	669.0	50	10	340	6521	27.6
	1500	555	338348	2075.4	70	10	674	16870	107.4
3rd Polynomial	500	111	31924	77.2	30	10	45	927	2.4
	1000	155	80996	427.9	50	10	140	4314	16.8
	1500	199	156222	1301.4	70	10	344	14296	78.9

Table 8.3. Numerical results for concave $q(x)$ ($r = 0.6$).

Type of Constraint	Single Constraint				Multiple Constraints				
	n	N_{iter}	N_{box}	T_{cpu}	n	m	N_{iter}	N_{box}	T_{cpu}
Linear	500	32	8748	13.4	50	10	32	765	2.3
	1000	53	27622	97.1	100	10	112	4580	24.8
	2000	58	60408	464.1	150	10	268	17602	173.8
Convex Quadratic	500	43	9388	20.0	100	10	33	1407	5.9
	1000	57	23858	114.1	150	10	77	4973	31.2
	2000	149	120776	1294.3	200	10	110	9739	79.5
Concave Quadratic	500	18	4334	9.8	100	10	26	1268	4.5
	1000	70	34094	163.0	150	10	65	4659	24.4
	2000	108	105606	1085.8	200	10	237	21910	169.1
3rd Polynomial	500	47	8943	27.6	100	10	53	2302	9.9
	1000	76	30419	196.6	150	10	113	6701	45.7
	2000	104	75337	1080.5	200	10	215	17852	188.2

The computational results for portfolio selection problems (SMV) are presented in Table 8.5, where N_{iter}, N_{box} and T_{cpu} are obtained by running the code for 10 test problems. We see from Table 8.5 that the problem becomes more difficult as the ratio of right-hand side, r, decreases.

Table 8.4. Numerical results for indefinite $q(x)$ ($r = 0.6$).

Type of Constraint	Single Constraint				Multiple Constraints				
	n	N_{iter}	N_{box}	T_{cpu}	n	m	N_{iter}	N_{box}	T_{cpu}
	200	183	23743	4.2	30	10	121	4081	29.7
Linear	600	447	184493	166.1	40	10	179	8188	73.9
	1000	744	482480	997.1	50	10	601	38933	414.3
	200	288	36056	7.6	30	10	65	2403	15.9
Convex Quadratic	600	896	349573	174.2	50	10	103	6107	58.7
	1000	1145	739706	572.7	70	10	238	20093	249.8
	200	191	22386	5.0	30	10	42	1469	9.3
Concave Quadratic	600	776	272840	157.1	50	10	74	4184	36.9
	1000	2386	1376007	1806.0	70	10	126	9958	114.1
	200	121	17283	5.9	30	10	39	1566	12.9
3rd Polynomial	600	836	341505	246.9	50	10	72	4713	55.1
	1000	1059	756976	857.0	70	10	75	7570	101.7

Table 8.5. Numerical results for problem (SMV).

r	n	m	N_{iter}	N_{box}	T_{cpu}
	30	5	243	3274	9.6
0.5	50	5	2191	46787	345.5
	80	5	4265	128091	860.0
	30	5	326	4653	11.5
0.6	50	5	1437	32564	143.0
	80	5	7888	232647	1292.2
	30	5	95	1523	3.6
0.7	50	5	361	7889	32.7
	80	5	596	22140	106.4

8.5.3 Comparison with other methods

To compare the convergent Lagrangian and contour cut method with other existing methods, we implemented two exact methods which are applicable to (QIP):

- Branch-and-bound method of Brettauer and Shetty (see Section 6.1) which is applicable to singly constrained convex (QIP).

- Hybrid method of Marstern and Morin (see Section 7.2) which is applicable to general (QIP).

We have implemented the above two methods by FORTRAN 90 and tested for two sets of test problems for comparison. The first set of test problems is a convex instance of (QIP) with a single linear constraint. Both the branch-and-bound method and hybrid method are applicable to this set of test problems. The ranges of the parameters of $q(x)$ are: $c_j \in [1, 10]$ and $d_j \in [-100, -300]$. The linear constraint is $g(x) = \sum_{j=1}^{n} \alpha_j x_j$ with $\alpha_j \in [1, 50]$. The ratio of the right-hand side b is taken as $r = 0.7$, and $l_j = 1$, $u_j = 5$, $j = 1, \ldots, n$. Table 8.6 summarizes the average CPU time of the convergent Lagrangian and contour cut (CLCC) method, the branch-and-bound method and the hybrid method for 20 randomly generated test problems in the first set.

Table 8.6. Comparison results for convex problems.

n	CLCC Method	Branch-and-Bound Method	Hybrid Method
	T_{cpu}	T_{cpu}	T_{cpu}
50	0.10	0.32	8.0
100	0.88	16.5	152.1
150	2.0	485.1	833.6

The second set of test problems for comparison is a concave instance of (QIP) with a single linear constraint. Note that only the hybrid method is applicable to this kind of nonconvex problems. The ranges of the parameters of $q(x)$ are: $c_j \in [-10, -1]$ and $d_j \in [-50, -1]$. The ranges of the coefficients in the linear constraint are: $\alpha_j \in [1, 50]$. The ratio of the right-hand side b is taken as $r = 0.7$, and $l_j = 1$, $u_j = 5$, $j = 1, \ldots, n$. The comparison results for test problems with different n are reported in Table 8.7, where the average CPU time is obtained by running the algorithms for 20 randomly generated test problems.

The average CPU time in Tables 8.6 and 8.7 indicates that the convergent Lagrangian and contour cut method is much more efficient than the branch-and-bound method and the hybrid method for both convex and nonconvex problems. Part of the theoretical reason for the out-performance of contour cut methods over the continuous relaxation-based branch-and-bound method is that the Lagrangian bound of a convex integer programming problem is better than or equal to the continuous bound. Moreover, cutting certain integer boxes from the domain at each iteration in the domain cut and partition scheme of the convergent Lagrangian and contour cut method speeds up the convergence of the algorithm significantly. We also notice that it is difficult for dynamic pro-

Table 8.7. Comparison results for nonconvex problems.

n	CLCC Method	Hybrid Method
	T_{cpu}	T_{cpu}
100	0.4	26.6
150	2.0	131.0
200	1.6	397.0

gramming in the hybrid method to exploit the special structure of the problems in generating efficient feasible solutions and it is thus not efficient to find an exact solution of the original problem.

8.6 Note

The convergent Lagrangian and contour cut method for problem (QIP) was proposed in [138]. Surveys for general quadratic programming problems can be found in [62][216].

Integer programming models with a convex quadratic objective function have various applications, including capital budgeting [126][155], capacity planning [34], optimization problems from graph theory [15][125]. An important class of applications of problem (QIP) arises in portfolio selection models with discrete features (see [14][21][108][140]). It was shown in [193][194] that the Markowitz's mean-variance model [150] can be simplified to a separable problem formulation of (QIP) by using market indices together with some additional variables and constraints. A method for reformulating general nonlinear programs to separable forms was discussed in [165].

Concave quadratic cost functions are often encountered in real-world integer programming models involving economies of scale (see [62][183]). It corresponds to the economic phenomenon of "decreasing marginal cost." The continuous version of problem (QIP) with $q(x)$ being concave and $g_i(x)$ linear or convex quadratic was extensively studied, for example, in [29][47][183] [112][195][221] and was served as the standard test problems in concave minimization. These methods exploit the special structures of quadratic functions and the extreme point property of concave programming that the minimum of a concave function over a polyhedron is always achieved at one of its extreme points. Branch-and-bound methods based on continuous relaxation and convex underestimating were proposed in [19][20][32][34][37] for solving concave integer problems over a polyhedron. Solution methods for general quadratic integer programming problems were also studied in [215].

Chapter 9

NONSEPARABLE INTEGER PROGRAMMING

Consider the following general nonlinear integer programming problem:

$$
\begin{aligned}
(P) \qquad & \min\ f(x) \\
& \text{s.t. } g_i(x) \le b_i,\ i = 1, \ldots, m, \\
& \quad h_k(x) = c_k,\ k = 1, \ldots, l, \\
& \quad x \in X = \{x \in \mathbb{Z}^n \mid l_j \le x_j \le u_j,\ j = 1, \ldots, n\}.
\end{aligned}
$$

In this chapter, we will focus on situations of (P) where at least one function in (P) is nonseparable. Evidently, we expect nonseparable problem (P) to be much more difficult to solve than separable nonlinear integer programming problems.

In Section 9.1, we will investigate a general continuous relaxation-based branch-and-bound method for solving the convex case of (P). A Lagrangian decomposition method for linearly constrained convex case of (P) will be discussed in Section 9.2, along with its integration with a domain cut scheme in implementation. In Section 9.3, we study the monotone case of (P). We first describe a discrete polyblock method. We then investigate the relationship between convexity and monotonicity. Finally, we demonstrate how to combine convexification with the discrete polyblock method to improve the algorithm efficiency.

9.1 Branch-and-Bound Method based on Continuous Relaxation

In this section, we focus on the general branch-and-bound methodology for solving convex nonlinear integer programming based on continuous relaxation.

Consider the continuous relaxation problem of (P):

$$
\begin{array}{lll}
(\overline{P}) & \min & f(x) \\
& \text{s.t.} & g_i(x) \le b_i,\ i = 1,\dots,m, \\
& & h_k(x) = c_k,\ k = 1,\dots,l, \\
& & x \in \overline{X} = \{x \in \mathbb{R}^n \mid l_j \le x_j \le u_j,\ j = 1,\dots,n\}.
\end{array}
$$

A subproblem of $(\overline{P})$ is obtained by replacing l_j with α_j and u_j with β_j, where $l_j \le \alpha_j < \beta_j \le u_j$ for $j = 1,\dots,n$.

To guarantee that all the subproblems can be solved to the global optimality correctly by using nonlinear programming methods, we require that functions f and g_i, $i = 1,\dots,m$, be convex, functions h_k, $k = 1,\dots,l$, be linear, and certain constraint qualifications be satisfied for all subproblems of $(\overline{P})$.

The branch-and-bound method using lower bound generated by continuous relaxation can be outlined as follows. The algorithm starts by finding an optimal solution x^* of the continuous relaxation problem of (P). Let x^0 denote the optimal solution to $(\overline{P})$ with $\alpha = l$ and $\beta = u$. If x^0 is integral, then it is also optimal to (P). Otherwise, let x_i^0 be a fractional variable of x^0. Two new subproblems are generated by adding variable constraints $x_j \le \lfloor x_j^0 \rfloor$ and $x_j \ge \lfloor x_j^0 \rfloor + 1$, respectively, where $\lfloor x_j^0 \rfloor$ denotes the maximum integer less than or equal to x_j^0. At the k-th iteration, one of the generated subproblems is chosen to be solved next. If its optimal solution is integral and its objective value is better than that of the incumbent, then it becomes the new incumbent. The subproblem is fathomed or pruned from further consideration if one of the following three conditions holds: (a) the corresponding continuous relaxation subproblem generates an optimal integer solution, (b) the optimal value of the continuous relaxation is larger than or equal to the upper bound associated with the current incumbent, or (c) the continuous relaxation problem is infeasible. Otherwise the subproblem is divided again, and the process is repeated until no subproblem remains to be solved. The above process can fit into the framework of Algorithm 2.1, while in the current methodological framework only one subproblem is selected in Step 1, the lower bounds are generated by the continuous relaxation and a fractional variable of the continuous solution is branched to form the two subproblems. Suppose that the subproblem at the k-th node is selected to solve. Then this subproblem is called "parent subproblem." Let x^k denote the optimal solution of this subproblem and x_j^k is a fractional variable. The new subproblems generated by adding $x_j \le \lfloor x_j^k \rfloor$ and $x_j \ge \lfloor x_j^k \rfloor + 1$ are called "left son subproblem" and "right son subproblem," respectively.

The overall performance of the above branch-and-bound algorithm for (P) is significantly affected by the following three factors:

- The efficiency of the nonlinear programming solver for solving subproblems of $(\overline{P})$;

- The rules to select a variable to branch upon;
- The rules to select a node for generating new subproblems.

There are many different choices for selecting nonlinear programming solvers. These solvers are developed based on different solution methods including penalty method, generalized gradient method, sequential quadratic programming method and trust region method. It was shown in [87] that solvers based on generalized gradient method are significantly superior to the others as evidenced in the numerical experience in [87].

9.1.1 Branching variables

Suppose that a node is selected and the optimal solution to the corresponding subproblem is $x = (x_1, \ldots, x_n)^T$. Let I denote the index set of fractional variables of x. There are three commonly used branching rules.

1. Most fractional integer variable. This rule selects the variable x_j which has the most fractional part,

$$j = \arg \max_{i \in I} \{\min(x_i - \lfloor x_i \rfloor, \lceil x_i \rceil - x_i)\}.$$

It is the intention of selecting such a j to produce the largest difference between the objective function values of the new subproblems so that an earlier fathoming may take place and hence more nodes can be pruned.

2. *Lowest-index-first.* In many situations, some decision variables x_i's play more important roles in the model than others. Therefore it is reasonable to branch variables in terms of their importance. The rule of lowest-index-first orders the index set I in decreasing priorities and selects the first variable in I to branch.

3. Pseudo-costs. The idea underlying the pseudo-cost branching rule is to determine a priority of the variables in terms of the change in the optimal objective value of the continuous subproblem per unit change of x_j. This is accomplished by ordering the differential of the optimal value of the subproblems before and after adding a new constraint. Let x_j^k be the variable that is selected to branch at the k-th node. Let f_k be the optimal objective value of the continuous subproblem at the k-th node. Denote by f_L and f_R the optimal objective values of the two son subproblems after adding the constraint $x_j \leq \lfloor x_j^k \rfloor$ and $x_j \geq \lfloor x_j^k \rfloor + 1$, respectively. Define the pseudo-costs of the left and right son subproblems as follows:

$$c_j^L = \frac{f_L - f_k}{x_j^k - \lfloor x_j^k \rfloor},$$

$$c_j^R = \frac{f_R - f_k}{\lceil x_j^k \rceil - x_j^k}.$$

Since c_j^L and c_j^R are only available after solving the two son subproblems, it is reasonable to compute them only once and use them in all remaining nodes. The pseudo-costs are computed in the course of the tree search. We can use the most fractional integer variable to be the branching variable before all c_j^L and c_j^R are computed. Let x^s be the optimal solution to the subproblem at the s-th node. Let the pseudo-cost be defined as follows for $j = 1, \ldots, n$,

$$v_j = \min\{c_j^L(x_j^s - \lfloor x_j^s \rfloor), c_j^R(\lceil x_j^s \rceil - x_j^s)\}. \tag{9.1.1}$$

Suppose that $v_{j_0} = \max_{j=1,\ldots,n} v_j$. Then x_{j_0} is selected to be the branching variable to generate two subproblems at the s-th node.

9.1.2 Branching nodes

There are three commonly used rules for selecting a branching node. Suppose that the list of active nodes is $\{i_1, i_2, \ldots, i_N\}$.

Node with lowest bound. Suppose that the lower bounds of the active nodes are $\{f_{i_1}, f_{i_2}, \ldots, f_{i_N}\}$. The next node to branch is selected to be the node with minimum f_{i_k}.

Newest node. The node list is ordered in a way of last-in first-out. The newest node is selected to branch next. Since there are two son subproblems, the node corresponding to the left son subproblem is given a preference over the right son subproblem.

Estimation. At node k, the pseudo-costs v_j ($j = 1, \ldots, n$) defined in (9.1.1) are added to the lower bound f_k to form an estimation of the best objective function value for the descendants of node k.

$$E_k = f_k + \sum_{j=1}^{n} v_j.$$

The quantity E_k is computed for all unfathomed nodes. The node with the lowest E_k is chosen to branch next.

It is also useful to combine all or some of the above strategies in a branch-and-bound method. A typical combination, for example, is to use the rule of newest node until a node is pruned and then to backtrack to the node with the lowest bound.

9.2 Lagrangian Decomposition Method

We consider in this section the following convex knapsack problem:

$$(CVKP) \qquad \begin{aligned} &\min\ f(x) \\ &\text{s.t. } Ax \le b, \\ &\qquad x \in X = \{x \in \mathbb{Z}^n \mid l_j \le x_j \le u_j,\ j = 1, \ldots, n\}, \end{aligned}$$

where f is a nonincreasing convex function on $conv(X)$ and $A = (a_{ij})_{m \times n}$ with all $a_{ij} \geq 0$. Problem $(CVKP)$ is a special case of (P_l) studied in Section 3.6.

Lagrangian decomposition method discussed in Section 3.6 provides an alternative way to compute the lower bounds in a branch-and-bound method for solving $(CVKP)$. Since the Lagrangian decomposition produces a tighter lower bound than the continuous relaxation, it is more reasonable to solve the dual problem of $(CVKP)$ at each node to give a lower bound, instead of solving the continuous relaxation.

By the Lagrangian decomposition scheme discussed in Section 3.6, the Lagrangian bound of $(CVKP)$ is given by solving the following dual problem:

$$(D_{CVKP}) \qquad \max_{\mu \in \mathbb{R}^n} \ell(\mu) = \ell_1(\mu) + \ell_2(\mu),$$

where

$$\ell_1(\mu) = \min\{f(y) - \mu^T y \mid Ay \leq b,\ y \in conv(X)\}, \tag{9.2.1}$$
$$\ell_2(\mu) = \min\{\mu^T x \mid Ax \leq b,\ x \in X\}. \tag{9.2.2}$$

A subgradient procedure can be developed to search for the optimal solution to the dual problem (D_{CVKP}).

PROCEDURE 9.1 (SUBGRADIENT METHOD FOR (D_{CVKP}))

Step 0. *Choose the tolerance parameters* $\sigma_1 > 0$ *and* $\sigma_2 > 0$. *Set* $i = 0$, $\mu^0 = 0$, $L^0 = -\infty$, $U^0 = +\infty$.

Step 1. *Solve* (9.2.1) *and* (9.2.2) *to obtain their optimal solutions* y^i *and* x^i, *respectively. Set* $L^{i+1} := \max(L^i, \ell(\mu^i))$ *and* $U^{i+1} := \min(U^i, f(x^i))$.

Step 2. *If* $\|x^i - y^i\| \leq \sigma_1$ *or* $U^{i+1} - L^{i+1} \leq \sigma_2$, *then stop.*

Step 3. *Set* $\mu^{i+1} = \mu^i + t^i(x^i - y^i)$, *where* $t^i > 0$ *is the stepsize.*

Step 4. *Set* $i := i + 1$, *go to Step 1.*

Procedure 9.1 converges to the optimal solution of problem (D_{CVKP}) if σ_1 and σ_2 are set to zero and the stepsize t^i satisfies certain rules (see Section 3.2.1). Note that $L^1 = \ell(0)$ corresponds to the continuous bound of $(CVKP)$. In practice, the procedure can be terminated after a given number of iterations or a satisfactory improvement of the dual value, $\ell(\mu) - \ell(0)$, is achieved. Notice that x^i is feasible to $(CVKP)$. An infeasible solution can be easily found by the monotonicity of constraint $Ax \leq b$, for example, by increasing the coordinate of a feasible point successively. Therefore, Procedure 9.1 produces a lower bound L^i and an upper bound U^i together with a feasible solution x^i and an infeasible solution z^i when it is terminated at the i-th iteration.

We now discuss a convergent Lagrangian decomposition algorithm for problem $(CVKP)$ by integrating the Lagrangian decomposition method with the domain cut scheme in Section 6.3. Initially, a lower bound is computed on the initial integer box $\langle l, u \rangle$ by solving the dual problem (D_{CVKP}). The feasible solution and the infeasible solution generated by the dual search procedure are used to partition the domain into a union of subboxes using Lemma 6.1. For each new subbox, we apply Procedure 9.1 to compute a lower bound of the objective function on the subbox, together with a feasible solution and an infeasible solution. A feasible solution better than the incumbent is used to update the upper bound and to replace the incumbent. As the same as in Algorithm 6.2, certain integer subboxes are fathomed and the remaining subboxes are added to the node list. The integer subbox with the minimum bound is chosen to partition further and the above process repeats until there is no integer subbox left in the node list.

ALGORITHM 9.1

Step 0. (Initialization) If $x = l$ is infeasible, then problem $(CVKP)$ has no feasible solution, stop. If $x = u$ is feasible, then $x = u$ is the optimal solution of $(CVKP)$, stop. Otherwise, apply Procedure 9.1 to (D_{CVKP}) and obtain a lower bound LB^0, a feasible solution x^0 and an infeasible solution z^0. Set $x_{opt} = x^0$, $f_{opt} = f(x_{opt})$, $X^0 = \langle l, u \rangle$. Set $k = 0$.

Step 1. Choose the integer subbox $\langle \alpha^k, \beta^k \rangle$ from X^k with the minimum lower bound. Let x^k and z^k be the feasible and infeasible solutions on $\langle \alpha^k, \beta^k \rangle$ found by Procedure 9.1, respectively. Set $X^k := X^k \setminus \langle \alpha^k, \beta^k \rangle$.

Step 2. Partition $\langle \alpha^k, \beta^k \rangle \setminus (\langle \alpha^k, x^k \rangle \cup \langle z^k, \beta^k \rangle)$ into a union of integer boxes by using the formula (6.3.1). Let Z^k be the set of the newly generated integer subboxes.

Step 3. For each integer subbox $\langle \alpha, \beta \rangle$ in Z^k, apply Procedure 9.1 to find a lower bound $LB_{\langle \alpha, \beta \rangle}$ and a feasible solution $x_{\langle \alpha, \beta \rangle}$. Starting from $x_{\langle \alpha, \beta \rangle}$, increase the value of $(x_{\langle \alpha, \beta \rangle})_i$ coordinately for $i = 1, \ldots, n$ to search for an infeasible $z_{\langle \alpha, \beta \rangle}$.

Step 4. (Fathoming and Updating). For each integer subbox $\langle \alpha, \beta \rangle \in X^k \cup Z^k$, check the following:

(i) If β is feasible, then remove $\langle \alpha, \beta \rangle$; Update x_{opt} and f_{opt} if $f(\beta) < f_{opt}$.

(ii) If α is infeasible, then remove $\langle \alpha, \beta \rangle$.

(iii) If $LB_{\langle \alpha, \beta \rangle} \geq f_{opt}$, then remove $\langle \alpha, \beta \rangle$ based on the weak duality.

Step 5. Let X^{k+1} be the set of integer subboxes of $X^k \cup Z^k$ after carrying out Step 4. If $X^{k+1} = \emptyset$, then stop, x_{opt} and f_{opt} are the optimal solution and the optimal function value of $(CVKP)$, respectively. Otherwise, set $k := k + 1$, go to Step 1.

THEOREM 9.1 *Algorithm 9.1 terminates at an optimal solution of* $(CVKP)$ *within a finite number of iterations.*

Proof. Similar to the proof of Theorem 6.4. □

To illustrate Algorithm 9.1, let us consider the following quadratic knapsack example.

EXAMPLE 9.1

$$\begin{aligned} \min\ f(x) &= -2x_1^2 - x_1x_2 - x_1x_3 - x_1x_4 - 2x_2^2 - x_2x_3 - x_2x_4 \\ &\quad - 5/2x_3^2 - x_3x_4 - 3x_4^2 \\ \text{s.t.}\ g(x) &= x_1 + 2x_2 + x_3 + 3x_4 \leq 25.2, \\ x &\in X = \{x \in \mathbb{Z}^4 \mid 1 \leq x_j \leq 5,\ j = 1, 2, 3, 4\}. \end{aligned}$$

The optimal solution of this problem is $x^* = (5, 3, 5, 3)^T$ with $f(x^*) = -363.5$.

The iterative solution process can be described as follows.

Initial iteration: $X^0 = \langle \alpha, \beta \rangle$, where $\alpha = (1, 1, 1, 1)^T$, $\beta = (5, 5, 5, 5)^T$. Applying Procedure 9.1, we obtain a feasible solution $x^0 = (5, 5, 5, 1)^T$ with $f(x^0) = -339.5$, a lower bound $LB^0 = -364.631$ and an infeasible solution $z^0 = (5, 5, 5, 2)^T$. Set $x_{opt} = x^0$ and $f_{opt} = -339.5$. $k = 0$.

Iteration 1: Cutting $\langle \alpha, x^0 \rangle$ and $\langle z^0, \beta \rangle$ from X^0 generates: $Z^0 = \{Z_1^0, Z_2^0, Z_3^0\}$, where

$$\begin{aligned} Z_1^0 &= \langle (1, 1, 1, 2)^T, (4, 5, 5, 5)^T \rangle, \\ Z_2^0 &= \langle (5, 1, 1, 2)^T, (5, 4, 5, 5)^T \rangle, \\ Z_3^0 &= \langle (5, 5, 1, 2)^T, (5, 5, 4, 5)^T \rangle. \end{aligned}$$

The feasible solution x, the infeasible solution z, the upper bound $f(x)$ and the lower bound LB for the three integer subboxes in Z^0 are:

$$\begin{aligned} Z_1^0 &: x = (4, 5, 5, 2)^T,\ z = (4, 5, 5, 3)^T,\ f(x) = -355.5,\ LB = -363.476; \\ Z_2^0 &: x = (5, 4, 5, 2)^T,\ z = (5, 4, 5, 3)^T,\ f(x) = -355.5,\ LB = -364.562; \\ Z_3^0 &: x = (5, 5, 4, 2)^T,\ z = (5, 5, 4, 3)^T,\ f(x) = -355.0,\ LB = -356.197. \end{aligned}$$

Since $-355.5 < f_{opt}$, update $x_{opt} = (5, 4, 5, 2)^T$ and $f_{opt} = -355.5$. $X^1 = Z^0$. Box Z_2^0 is chosen to partition since it has the minimum lower bound. $k = 1$.

Iteration 2: Cutting $\langle (5, 1, 1, 2)^T, (5, 4, 5, 2)^T \rangle$ and $\langle (5, 4, 5, 3)^T, (5, 4, 5, 5)^T \rangle$ from Z_2^0 generates two subboxes in Step 2: $Z^1 = \{Z_1^1, Z_2^1\}$, where

$$\begin{aligned} Z_1^1 &= \langle (5, 1, 1, 3)^T, (5, 3, 5, 5)^T \rangle, \\ Z_2^1 &= \langle (5, 4, 1, 3)^T, (5, 4, 4, 5)^T \rangle. \end{aligned}$$

The feasible solution x, the infeasible solution z, the upper bound $f(x)$ and the lower bound LB for the two integer subboxes in Z^1 are:

$$Z_1^1 : x = (5,3,5,3)^T, \; z = (5,3,5,4)^T, \; f(x) = -363.5, \; LB = -364.456;$$
$$Z_2^1 : x = (5,4,3,3)^T, \; z = (5,4,4,3)^T, \; f(x) = -355.5, \; LB = -357.8.$$

Since $-363.5 < f_{opt} = -355.5$, update x_{opt} to $(5,3,5,3)^T$ and f_{opt} to -363.5. By Step 4, Z_2^1 is removed from Z^1 and Z_1^0 and Z_3^0 are removed from X^1. Set $X^2 = \{Z_1^1\}$. $k = 2$.

Iteration 3: Cutting $\langle(5,1,1,3)^T,(5,3,5,3)^T\rangle$ and $\langle(5,3,5,4)^T,(5,3,5,5)^T\rangle$ from $Z_1^1 \in X^2$ generates two subboxes: $Z^2 = \{Z_1^2, Z_2^2\}$, where

$$Z_1^2 = \langle(5,1,1,4)^T,(5,2,5,5)^T\rangle,$$
$$Z_2^2 = \langle(5,3,1,4)^T,(5,3,4,5)^T\rangle.$$

The feasible solution x, the infeasible solution z, the upper bound $f(x)$ and the lower bounds LB for the two integer subboxes in Z^2 are:

$$Z_1^2 : x = (5,2,4,4)^T, \; z = (5,2,5,4)^T, \; f(x) = -357, \; LB = -358.821;$$
$$Z_2^2 : x = (5,3,2,4)^T, \; z = (5,3,3,4)^T, \; f(x) = -343, \; LB = -346.3.$$

Both Z_1^2 and Z_2^2 are removed by Step 4. Thus, $X^3 = \emptyset$. The algorithm terminates with the optimal solution $x_{opt} = (5,3,5,3)^T$ and the optimal function value $f_{opt} = -363.5$.

9.3 Monotone Integer Programming

Consider the following monotone integer programming:

$$(MIP) \qquad \begin{aligned} & \max \; f(x) \\ & \text{s.t. } g_i(x) \le b_i, \; i = 1,\ldots,m, \\ & \quad x \in X = \{x \in \mathbb{Z}^n \mid l_j \le x_j \le u_j, \; j = 1,\ldots,n\}, \end{aligned}$$

where f and all g_i's are increasing functions of x_j on $[l_j, u_j]$ for $j = 1,\ldots,n$, $i = 1,\ldots,m$, l_j and u_j are integer numbers with $l_j < u_j$ for $j = 1,\ldots,n$. Functions f and g_i's are not necessarily convex or separable. Problem (MIP) is often referred as a *multi-dimensional nonseparable knapsack problem.*

The difficulty of designing a solution method for problem (MIP) lies in the nonconvexity and nonseparability of f and g_i's. Due to the nonconvexity and nonseparability, the classical branch-and-bound method and Lagrangian relaxation (decomposition) method are not applicable to problem (MIP).

In this section, we first discuss a discrete polyblock method for (MIP). The relationship between monotonicity and convexity will be then investigated. A branch-and-bound method that combines the polyblock method with the convexification method is finally developed.

9.3.1 Discrete polyblock method for (MIP)

Define

$$G(x) = \max_{i=1,\dots,m} \{g_i(x) - b_i\}. \tag{9.3.1}$$

The boundary of the constraints can then be expressed as $\Gamma = \{x \in X \mid G(x) = 0\}$. Let $S = \{x \in X \subset \mathbb{Z}^n \mid g_i(x) \le b_i,\ i = 1, \dots, m\}$. Let $\langle \alpha, \beta \rangle$ be an integer box in X with $\alpha \in S$ and $\beta \notin S$. Suppose also that $G(\alpha) < 0$. Let x_b be an intersection point of the line $x = \lambda\alpha + (1 - \lambda)\beta$, $0 \le \lambda \le 1$, and the boundary Γ. Since $G(\alpha) < 0$ and $G(\beta) > 0$, there must exist an x_b in X that satisfies $G(x_b) = 0$, i.e., $g_i(x_b) \le b_i$ for $i = 1, \dots, m$ and there exists at least one i such that $g_i(x_b) = b_i$.

Denote by $\lfloor x \rfloor$ the integer vector with its i-th component being the maximum integer less than or equal to x_i, $i = 1, \dots, n$, and denote by $\lceil x \rceil$ the integer vector with its i-th component being the minimum integer greater than or equal to x_i, $i = 1, \dots, n$. Let $x^F = \lfloor x_b \rfloor$ and $x^I = \lceil x_b \rceil$. Suppose that x_b is not integral (otherwise $x^F = x^I$). It is easy to see that x^F is a feasible point ($x^F \in S$) and x^I is infeasible ($x^I \notin S$). Consider the integer boxes $\langle \alpha, x^F \rangle$ and $\langle x^I, \beta \rangle$. By the monotonicity of f and g_i, there are no feasible points better than x^F in $\langle a, x^F \rangle$ and there are no feasible points in $\langle x^I, \beta \rangle$. Therefore, when searching for an optimal solution to (MIP), we can remove integer boxes $\langle \alpha, x^F \rangle$ and $\langle x^I, \beta \rangle$ from $\langle \alpha, \beta \rangle$ for further consideration after comparing x^F with the incumbent solution. Corollary 6.1 shows that the set of the integer points left in $\langle \alpha, \beta \rangle$ after removing $\langle \alpha, x^F \rangle$ and $\langle x^I, \beta \rangle$ can be partitioned into a union of at most $2n - 1$ smaller integer boxes.

Based on the above discussion, we can derive an exact method for searching for an optimal solution of (MIP). The algorithm consists of two main steps: finding a feasible point x^F and an infeasible point x^I and generating integer boxes using the formulas (6.3.6) and (6.3.7). The points x^F and x^I are obtained by first finding a boundary point on Γ and then rounding down and up, respectively, the boundary point. The best feasible solution obtained during the generation of integer boxes is kept as an *incumbent* solution. For the newly generated integer boxes at each iteration, only the ones across the boundary Γ are needed to be kept for further consideration. Moreover, by the monotonicity of the problem, an integer box with the function value of its upper bound point less than or equal to the function value of the incumbent can be discarded. The algorithm proceeds successively by refining the partition and removing integer boxes that do not contain an optimal solution, and finally terminates at an optimal solution in a finite number of iterations.

We now describe the algorithm in detail.

ALGORITHM 9.2

Step 0 (Initialization). Let $l = (l_1, \ldots, l_n)^T$, $u = (u_1, \ldots, u_n)^T$. If l is infeasible, then problem (MIP) has no feasible solution; If u is feasible, then u is the optimal solution to (MIP), stop; Otherwise, set $x_{opt} = l$, $f_{opt} = f(x_{opt})$, $X^1 = \{\langle l, u\rangle\}$, and set $k = 1$.

Step 1 (Box Selection and Finding Boundary Point). Select an integer box $\langle \alpha, \beta\rangle \in X^k$ by certain selection rule. Set $X^k := X^k \setminus \langle \alpha, \beta\rangle$. Finding the root λ^* of the following equation:

$$G(\lambda\alpha + (1-\lambda)\beta) = 0, \quad \lambda \in [0,1], \tag{9.3.2}$$

where G is defined in (9.3.1). Set $x_b = \lambda^*\alpha + (1-\lambda^*)\beta$. Set $x^F = \lfloor x_b \rfloor$. If $x^F = x_b$ then set $x^I = x_b + e_j$, where e_j is the j-th unit vector in $\mathbb{R}^n$ with $x_b + e_j \le \beta$. Otherwise, set $x^I = \lceil x_b \rceil$. If $f(x^F) > f_{opt}$, set $x_{opt} = x^F$ and $f_{opt} = f(x^F)$.

Step 2 (Partition and Remove).

(i) Apply the formula (6.3.7) to partition the set $\Omega_1 = \langle \alpha, \beta\rangle \setminus \langle x^I, \beta\rangle$ into a union of integer boxes. Let $x^F \in \langle \tilde{\alpha}, \tilde{\beta}\rangle \in \Omega_1$. Set $\Omega_1 := \Omega_1 \setminus \langle \tilde{\alpha}, \tilde{\beta}\rangle$.

(ii) Apply the formula (6.3.6) to partition set $\Omega_2 = \langle \tilde{\alpha}, \tilde{\beta}\rangle \setminus \langle \tilde{\alpha}, x^F\rangle$.

(iii) Set $Y^k = \Omega_1 \cup \Omega_2$.

(iv) Perform the following for each integer box $\langle \alpha, \beta\rangle$ generated in the above partition process:

(a) If β is feasible, remove $\langle \alpha, \beta\rangle$ from Y^k. Furthermore if $f(\beta) > f_{opt}$, set $x_{opt} = \beta$ and $f_{opt} = f(\beta)$;

(b) If α is infeasible, remove $\langle \alpha, \beta\rangle$ from Y^k;

(c) If $f(\beta) \le f_{opt}$, remove $\langle \alpha, \beta\rangle$ from Y^k;

(d) If α is feasible, β is infeasible and $f(\alpha) > f_{opt}$, set $x_{opt} = \alpha$ and $f_{opt} = f(\alpha)$.

Denote Z^k the set of integer boxes after the above removing process.

Step 3 (Updating Integer Boxes). Removing all integer boxes $\langle \alpha, \beta\rangle$ in X^k with $f(\beta) \le f_{opt}$. Set $X^{k+1} = X^k \cup Z^k$. If $X^{k+1} = \emptyset$, stop. Otherwise, set $k := k+1$ and go to Step 1.

REMARK 9.1 Two box-selection strategies can be used in Step 1. The first strategy is to select the integer box in X^k with the maximum objective function value of the upper bound point. The second strategy is to select the last integer box included in X^k. To find the boundary point x_b, bisection method or Newton's method can be used in searching the root of equation (9.3.2).

REMARK 9.2 Heuristics can be used in the algorithm to obtain a good initial feasible point or to improve the feasible solution obtained during the algorithm.

For example, the feasible point x^F in Step 1 may be improved by testing the feasibility of the trial point $x^F + e_j$ for $j = 1, \ldots, n$ and update $x^F := x^F + e_j$ when successful until an infeasible point is reached.

THEOREM 9.2 *Algorithm 9.2 stops at an optimal solution to (MIP) within a finite number of iterations.*

Proof. The finite convergence of the algorithm can be easily seen from the finiteness of X and the fact that at each iteration at least the integer points x^F and x^I are removed from X^k. Since the partition formulas (6.3.6) and (6.3.7) and the cutting process in Step 2 do not remove any integer point better than the incumbent x_{opt}, the algorithm terminates with an optimal solution to (MIP). □

To illustrate Algorithm 9.2, let's consider the following problem:

EXAMPLE 9.2

$$\begin{aligned} \min\ & f(x) = 3x_1x_2 - x_1 + 6x_2 \\ \text{s.t. } & g(x) = 5x_1x_2 - 4x_1 - 4.5x_2 \le 32, \\ & x \in X = \{x \in \mathbb{Z}^2 \mid 1 \le x_j \le 5, j = 1, 2\}. \end{aligned}$$

The optimal solution of this example is $x^* = (2, 5)^T$ with $f(x^*) = 58$. The feasible region of the example is shown in Figure 9.1. The iterations of the algorithm are described as follows.

Iteration 1

Step 0. $l = (1, 1)^T$, $u = (5, 5)^T$, $x_{opt} = (1, 1)^T$, $f_{opt} = 8$, $X^1 = \{\langle l, u\rangle\}$, $k = 1$.

Step 1. Select $\langle \alpha, \beta\rangle = \langle l, u\rangle$. Use bisection procedure to find $x_b = (3.5188, 3.5188)^T$. $x^F = (3, 3)^T$, $x^I = (4, 4)^T$. Since $f(x^F) = 42 > 8 = f_{opt}$, set $x_{opt} = (3, 3)^T$ and $f_{opt} = 42$.

Step 2. Partition $\Omega_1 = \langle \alpha, \beta\rangle \setminus \langle x^I, \beta\rangle$ into 2 integer boxes:

$$\Omega_1 = \{B_1, B_2\} = \{\langle (1, 1)^T, (3, 5)^T\rangle, \langle (4, 1)^T, (5, 3)^T\rangle\}.$$

Since $x^F \in B_1$, set $\Omega_1 = \{B_2\}$. $\Omega_2 = B_1 \setminus \langle (1, 1)^T, (3, 3)^T\rangle = \{\langle (1, 4)^T, (3, 5)^T\rangle\}$.

$$Y^1 = \Omega_1 \cup \Omega_2 = \{\langle (4, 1)^T, (5, 3)^T\rangle, \langle (1, 4)^T, (3, 5)^T\rangle\}.$$

We obtain $Z^1 = Y^1$.

Step 3. $X^2 = Z^1$.

The process of Iteration 1 is illustrated in Figure 9.2.

Iteration 2

Step 1. Select $\langle \alpha, \beta\rangle = \langle (1, 4)^T, (3, 5)^T\rangle$ from X^2 since $f((3, 5)^T) > f((5, 3)^T)$. Set $X^2 = \{\langle (4, 1)^T, (5, 3)^T\rangle\}$. The bisection procedure finds out

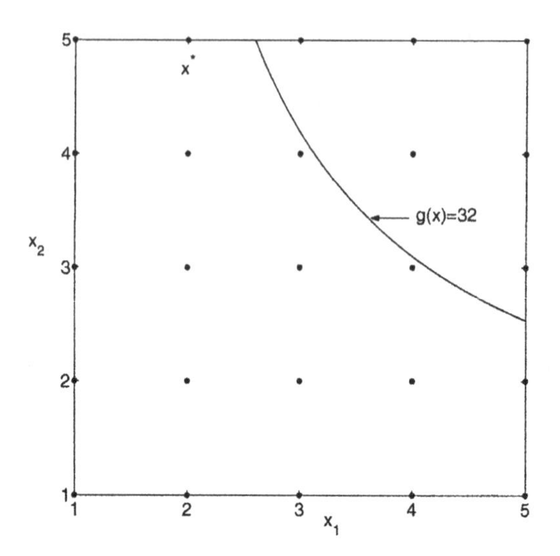

Figure 9.1. Feasible region of the Example 9.2.

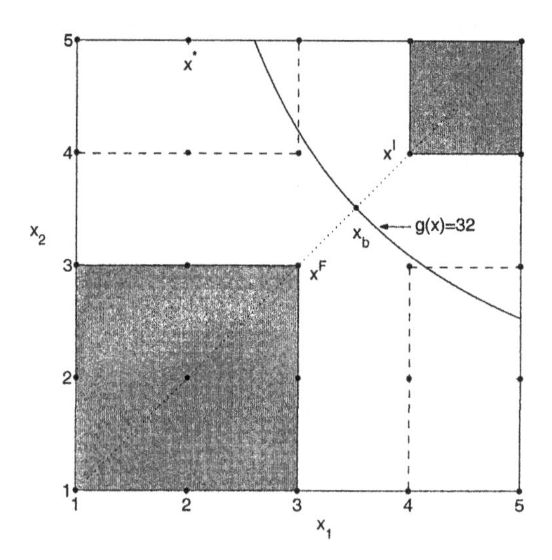

Figure 9.2. Illustration of Iteration 1 of Algorithm 9.2 for Example 9.2.

$x_b = (2.6655, 4.8327)^T$. $x^F = (2,4)^T$, $x^I = (3,5)^T$. Since $f(x^F) = 46 > 42 = f_{opt}$, set $x_{opt} = (2,4)^T$ and $f_{opt} = 46$.

Step 2. Partition $\Omega_1 = \langle \alpha, \beta \rangle \setminus \langle x^I, \beta \rangle$ into 2 integer boxes:

$$\Omega_1 = \{B_1, B_2\} = \langle (1,4)^T, (2,5)^T \rangle, \langle (3,4)^T, (3,4)^T \rangle \}.$$

Since $(2,5)^T$ is feasible and $f((2,5)^T) = 58 > 46 = f_{opt}$, set $x_{opt} = (2,5)^T$ and $f_{opt} = 58$. Remove B_1 from Ω_1. Since $f((3,4)^T) = 57 < f_{opt}$, remove B_2 from Ω_1. $\Omega_1 = \emptyset$. $Z^2 = Y^2 = \emptyset$.

Step 3. For $\langle (4,1)^T, (5,3)^T \rangle \in X^2$, since $f((5,3)^T) = 58 = f_{opt}$, remove it from X^2. Thus $X^3 = X^2 \cup Z^2 = \emptyset$. Stop and the incumbent $x_{opt} = (2,5)^T$ is an optimal solution to the problem with $f_{opt} = 58$.

9.3.2 Convexity and monotonicity

Due to the monotonicity of f and the g_i's, the optimal solution of the continuous relaxation of (MIP) always lies on the boundary of the feasible region. However, there may exist multiple local optimal solutions in the continuous relaxation of (MIP) since f is not necessarily concave and g_i's are not necessarily convex. Therefore, solution methods in nonlinear programming may fail to find the global solution to the continuous relaxation of (MIP). In order to apply the branch-and-bound strategy to (MIP), we need to develop global optimization methods for solving the continuous relaxation for subproblems of (MIP).

Convexity has been playing a key role in optimization theory and applications. An interesting question is: Is it possible to convert a nonconvex function into a convex function by certain transformation? In this section we discuss convexification schemes of a monotone function under a variable transformation.

ASSUMPTION 9.1 *Functions f and g_i $(i = 1,\ldots,m)$ in (MIP) are twice differentiable and strictly increasing on $\underline{X} = \{x \in \mathbb{R}^n \mid l_j \le x_j \le u_j,\ j = 1,\ldots,n\}$.*

A function t is called a strictly monotone function on its domain if it is either a strictly increasing function of all its variables on its domain or a strictly decreasing function of all its variables on its domain. Let $t : (y_1,\ldots,y_n) \mapsto (t_1(y_1),\ldots,t_n(y_n))$ be a separable one-to-one mapping. Let function h be defined on X. We introduce the following variable transformation for function h:

$$h_t(y) = h(t(y)). \tag{9.3.3}$$

The domain of h_t is:

$$Y^t = \prod_{j=1}^{n} Y_j = \prod_{j=1}^{n} t_j^{-1}([l_j, u_j]). \tag{9.3.4}$$

Define

$$\sigma = \min\{d^T \nabla^2 h(x) d \mid x \in \overline{X},\ \|d\|_2 = 1\}, \tag{9.3.5}$$

$$\eta = \min\{\frac{\partial h}{\partial x_j} \mid x \in \overline{X},\ j = 1, \ldots n\}. \tag{9.3.6}$$

We assume in the following that σ in (9.3.5) is strictly negative, since otherwise h is already convex. We have the following theorem.

THEOREM 9.3 *Let h be a twice continuously differentiable function on X with $\frac{\partial h}{\partial x_j} > 0$ for $j = 1, \ldots, n$. Assume that functions t_j ($j = 1, \ldots, n$) are strictly monotone functions and satisfy the following condition:*

$$\frac{t_j''(y_j)}{(t_j'(y_j))^2} \geq -\frac{\sigma}{\eta} > 0, \quad \forall\, y_j \in Y_j,\ j = 1, \ldots, n, \tag{9.3.7}$$

where σ and η are defined in (9.3.5) and (9.3.6), respectively. Then $h_t(y)$ is a convex function on Y^t.

Proof. Due to the twice continuous differentiability, it suffices to prove that the Hessian of $h_t(y)$ is a positive semidefinite matrix on Y^t. For any $y \in Y^t$, let $x = t(y)$. Then $x \in \overline{X}$. From (9.3.3), we have

$$\frac{\partial h_t}{\partial y_j} = t_j'(y_j) \frac{\partial h}{\partial x_j}, \quad j = 1, \ldots, n.$$

Furthermore,

$$\frac{\partial^2 h_t}{\partial y_i^2} = t_i''(y_i) \frac{\partial h}{\partial x_i} + (t_i'(y_i))^2 \frac{\partial^2 h}{\partial x_i^2}, \quad i = 1, \ldots, n, \tag{9.3.8}$$

$$\frac{\partial^2 h_t}{\partial y_i \partial y_j} = t_i'(y_i) t_j'(y_j) \frac{\partial^2 h}{\partial x_i \partial x_j}, \quad i \neq j,\ i, j = 1, \ldots, n. \tag{9.3.9}$$

Combining (9.3.8) with (9.3.9) gives the Hessian of h_t:

$$\nabla^2 h_t(y) = A(y)[\nabla^2 h(x) + B(x)]A(y), \tag{9.3.10}$$

where

$$A(y) = diag\left(t_1'(y_1), \ldots, t_n'(y_n)\right),$$
$$B(x) = diag\left(\frac{\partial h}{\partial x_1} \frac{t_1''(y_1)}{(t_1'(y_1))^2}, \ldots, \frac{\partial h}{\partial x_n} \frac{t_n''(y_n)}{(t_n'(y_n))^2}\right).$$

Let $C(x) = \nabla^2 h(x) + B(x)$. Since $t_j'(y_j) \neq 0$ for all $y_j \in t_j^{-1}([l_j, u_j])$, it is clear that $\nabla^2 h_t(y)$ is a positive semidefinite matrix for all $y \in Y^t$ if $C(x)$ is a positive semidefinite matrix for all $x \in \overline{X}$. Let $S^n = \{d \in \mathbb{R}^n \mid \|d\|_2 = 1\}$, the unit sphere in $\mathbb{R}^n$. By the definitions of σ and η, we have

$$d^T \nabla^2 h(x) d \geq \sigma, \quad \forall d \in S^n,$$
$$\frac{\partial h}{\partial x_j} \geq \eta > 0, \quad \forall x_j \in [l_j, u_j].$$

Now, for any $d \in S^n$, we have

$$\begin{aligned} d^T C(x) d &= d^T \nabla^2 h(x) d + d^T B(x) d \\ &\geq \sigma + \sum_{j=1}^{n} \frac{\partial h}{\partial x_j} \frac{t_j''(y_j)}{(t_j'(y_j))^2} d_j^2 \\ &\geq \sigma - \eta \times \frac{\sigma}{\eta} = 0. \end{aligned}$$

Therefore $\nabla^2 h_t(y)$ is a positive semidefinite matrix for all $y \in Y^t$. □

REMARK 9.3 Similar convexification results can be achieved for situations where h is a strictly decreasing function. Theorem 9.3 was generalized in [206] to convexify a class of nonsmooth functions.

The condition (9.3.7) in Theorem 9.3 is satisfied by many special convexification schemes (see [136][205]). In what follows, we give two typical convexification schemes.

COROLLARY 9.1 *Let $l_j > 0$ for $j = 1, \ldots, n$. Let h be a function satisfying the conditions in Theorem 9.3. Let*

$$t_j(y_j) = (1/p) \ln(1 - 1/y_j), \quad j = 1, \ldots, n. \tag{9.3.11}$$

Then there exists a $p_1 > 0$ such that $h_t(y)$ defined in (9.3.3) is a convex function on $Y^1 = \prod_{j=1}^{n} [1/(1 - \exp(pl_j)), 1/(1 - \exp(pu_j))]$ when $p \geq p_1$.

Proof. It suffices to show that condition (9.3.7) is satisfied. Notice that

$$t_j'(y_j) = \frac{1}{p} \frac{1}{y_j(y_j - 1)},$$
$$t_j''(y_j) = \frac{1}{p} \frac{(1 - 2y_j)}{y_j^2(y_j - 1)^2}.$$

Since $y_j < 0$ for $y_j \in [1/(1 - \exp(pl_j)), 1/(1 - \exp(pu_j))]$, we have

$$\frac{t_j''(y_j)}{(t_j'(y_j))^2} = p(1 - 2y_j) > p.$$

Obviously, condition (9.3.7) will be satisfied when $p \geq p_1 = \max\{0, -\sigma/\eta\}$.
□

COROLLARY 9.2 *Let h be a function satisfying the conditions in Theorem 9.3. Let*

$$t_j(y_j) = y_j^{-1/p}, \quad j = 1, \ldots, n. \tag{9.3.12}$$

Let $l_j > 0$ for $j = 1, \ldots, n$. Then there exists a $p_2 > 0$ such that $h_t(y)$ defined in (9.3.3) is a convex function on $Y^2 = \prod_{j=1}^n [u_j^{-p}, l_j^{-p}]$ when $p \geq p_2$.

Proof. To verify the condition (9.3.7), we calculate

$$t_j'(y_j) = -\frac{1}{p} y_j^{-1/p-1},$$
$$t_j''(y_j) = \frac{1}{p}(\frac{1}{p} + 1) y_j^{-1/p-2}.$$

For $y_j \in [u_j^{-p}, l_j^{-p}]$, we have

$$\frac{t_j''(y_j)}{(t_j'(y_j))^2} = (1+p) y_j^{1/p} \geq (1+p)/u_j.$$

Let $\bar{u} = \min_{1 \leq j \leq n} u_j$. Condition (9.3.7) will be satisfied when $p \geq p_2 = \max\{0, -(\bar{u}\sigma)/\eta - 1\}$. □

Note that function t_j in (9.3.11) is increasing, while function t_j in (9.3.12) is decreasing. To illustrate the convexification transformations in Corollaries 9.1 and 9.2, we consider a nonconvex function

$$h(x) = (1/3)(x-2)^3 + x, \quad x \in \overline{X} = [1, 3].$$

Figure 9.3 shows the plot of $h(x)$. Since $h'(x) = (x-2)^2 + 1 \geq 1$ and $h''(x) = 2(x-2) \geq -2$ for $x \in \overline{X}$, we can choose $p_1 = \max\{0, 2\} = 2$ in transformation (9.3.11) and $p_2 = \max\{0, 5\} = 5$ in transformation (9.3.12). Figures 9.4 and 9.5 show the convexified function $h_t(y)$.

The above results reveal that a real strictly monotone function can possess convexity in a transformed space. Because the variable transformation in (9.3.3) is a one-to-one monotone and continuous mapping, no minima or maxima of h on X will be lost in the new transformed set Y^t and no new minima or maxima will be created in Y^t.

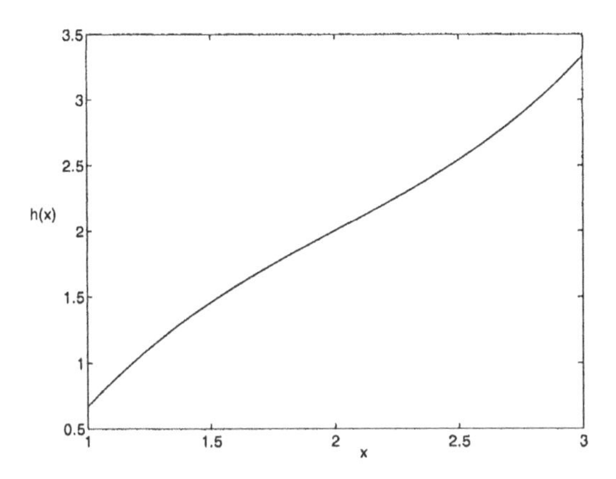

Figure 9.3. The nonconvex function $h(x)$.

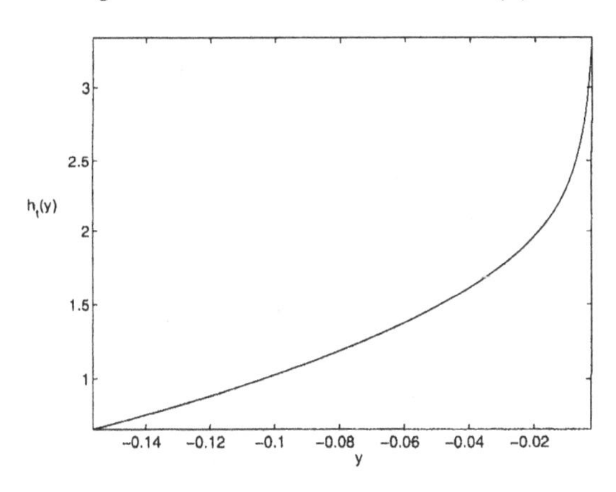

Figure 9.4. The convexified function $h_t(y)$ with t defined in (9.3.11) and $p = 2$.

9.3.3 Equivalent transformation using convexification

Consider the continuous relaxation of (MIP):

$$(\overline{MIP}) \qquad \max\ f(x)$$
$$\text{s.t. } g_i(x) \le b_i,\ i = 1, \ldots, m,$$
$$x \in \overline{X} = \{x \in \mathbb{R}^n \mid l_j \le x_j \le u_j,\ j = 1, \ldots, n\}.$$

For any one-to-one mapping t, problem $(\overline{MIP})$ is equivalent to the following transformed problem:

$$(\overline{MIP_t}) \qquad \max\ \phi(y) = f(t(y))$$
$$\text{s.t. } \psi_i(y) = g_i(t(y)) \le b_i,\ i = 1, \ldots, m,$$
$$y \in Y^t,$$

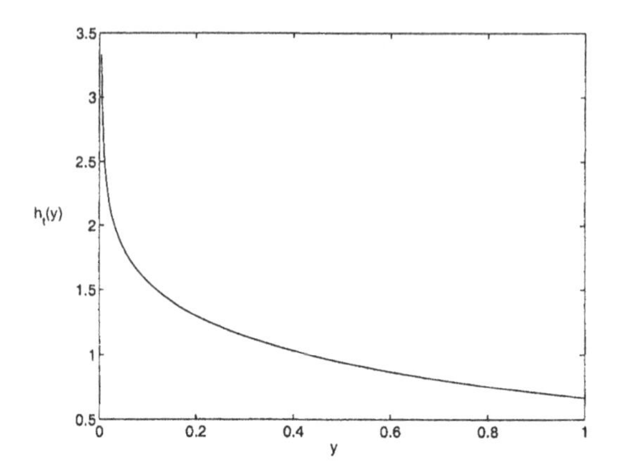

Figure 9.5. The convexified function $h_t(y)$ with t defined in (9.3.12) and $p = 5$.

where $Y^t = t^{-1}(\overline{X})$. Denote by S and S_t the feasible region of problems $(\overline{MIP})$ and $(\overline{MIP}_t)$, respectively, i.e.

$$S = \{x \in \overline{X} \mid g_i(x) \leq b_i,\ i = 1, \dots, m\}, \tag{9.3.13}$$

$$S_t = \{y \in Y^t \mid \psi_i(y) \leq b_i,\ i = 1, \dots, m\}. \tag{9.3.14}$$

If the mapping t in $(\overline{MIP}_t)$ satisfies the conditions in Theorem 9.3 for functions f and g_i's, then problem $(\overline{MIP}_t)$ is a convex maximization (or concave minimization) problem. Especially, when t_i takes the form of (9.3.11) or (9.3.12) and the parameter p is greater than certain threshold value, problem $(\overline{MIP}_t)$ is a convex maximization problem.

Concave minimization is a class of global optimization problems studied intensively in the literature. It is well-known that a convex function always achieves its maximum over a polyhedron at one of its vertices. Ranking the function values at all vertices of the polyhedron gives an optimal solution. For a convex maximization (or concave minimization) problem with a general convex feasible set, Hoffman [103] proposed an outer approximation algorithm. The convex objective function is successively maximized on a sequence of polyhedra that encloses the feasible region. At each iteration the current enclosing polyhedron is refined by adding a cutting plane tangential to the feasible region at a boundary point. The algorithm generates a nonincreasing sequence of upper bounds for the optimal value of $(\overline{MIP}_t)$ and terminates when the difference of the objective value of the current feasible solution and that of the optimal solution is within a given tolerance.

An outer approximation procedure for $(\overline{MIP}_t)$ can be described briefly as follows:

ALGORITHM 9.3 (POLYHEDRAL OUTER APPROXIMATION METHOD)

Step 1. Choose an initial polyhedron P_0 that contains S_t with vertex set V_0 and set $k = 0$.

Step 2. Compute v^k, the best vertex in the current enclosing polyhedron, and ϕ^k such that $\phi^k = \phi(v^k) = \max_{v \in V_k} \phi(v)$.

Step 3. Find a feasible point y^k on the boundary of S_t. Let i be such that $\psi_i(y^k) = b_i$. Form a new polyhedron P_{k+1} by adding a cutting plane inequality: $\xi_k^T(y - y^k) \leq 0$, where ξ_k is a subgradient of the binding constraint ψ_i at y^k.

Step 4. Calculate the vertex set V_{k+1} of P_{k+1}. Set $k := k + 1$, return to Step 2.

It can be proved that the above method converges to a global optimal solution to $(\overline{MIP_t})$. In implementation, the above procedure can be terminated when $\phi^k - \phi(y^k) \leq \epsilon$, where $\epsilon > 0$ is a given tolerance. There are many ways to generate the feasible point y^k in Step 3. A simple method is to find the (relative) boundary point of S_t on the line connecting v^k and a fixed (relative) interior point of S_t. Horst and Tuy [105] suggested projecting v^k onto the boundary of S_t and choosing z^k to be the projected point. Finding vertices of P_{k+1} is the major computational burden in the outer approximation method. After adding a cutting plane $\{y \mid \xi_k^T(y - y^k) = 0\}$, the new vertices can be generated by computing the intersection point of each edge of P_k with the new cutting plane.

Let us consider a small-size example to illustrate the convexification and outer approximation method.

EXAMPLE 9.3

$$\begin{aligned} \max\ & f(x) = 4.5(1 - 0.40^{x_1 - 1})(1 - 0.40^{x_2 - 1}) + 0.2\exp(x_1 + x_2 - 7) \\ \text{s.t.}\ & g_1(x) = 5x_1x_2 - 4x_1 - 4.5x_2 \leq 32, \\ & x \in \overline{X} = \{x \in \mathbb{R}^2 \mid 2 \leq x_1 \leq 6.2,\ 2 \leq x_2 \leq 6\}. \end{aligned}$$

It is clear that f and g_1 are strictly increasing functions on $\overline{X}$. The problem has three local optimal solutions: $x_{loc}^1 = (2.2692, 6)^T$ with $f(x_{loc}^1) = 3.7735$, $x_{loc}^2 = (3.4528, 3.5890)^T$ with $f(x_{loc}^2) = 3.857736$ and $x_{loc}^3 = (6.2, 2.1434)^T$ with $f(x_{loc}^3) = 3.6631$. Figure 9.6 shows the feasible region of the example. It is clear that the global optimal solution x_{loc}^2 is not on the convex hull of the nonconvex feasible region S. Take t to be the convexification transformation (9.3.11) with $p = 2$. The convexified feasible region is shown in Figure 9.7. Set $\epsilon = 10^{-4}$. The outer approximation procedure finds an approximate global optimal solution $y^* = (-0.21642, -0.19934)$ to the transformed problem $(\overline{MIP_t})$ after 17 iterations and generating 36 vertices. The point y^* corresponds to $x^* = (3.45290, 3.58899)^T$, an approximate optimal solution to Example 9.3 with $f(x^*) = 3.857736887$.

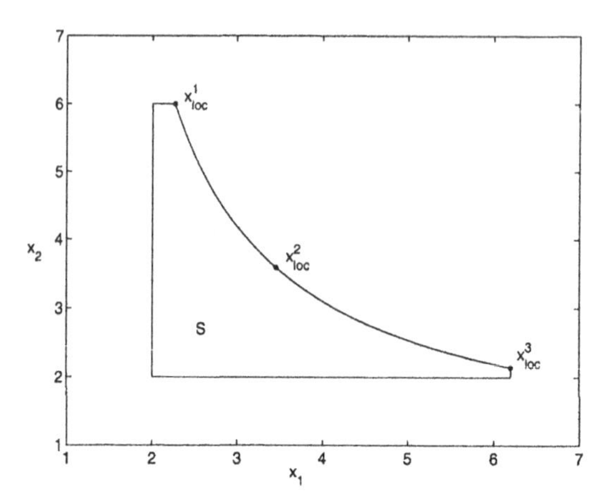

Figure 9.6. Feasible region of Example 9.3.

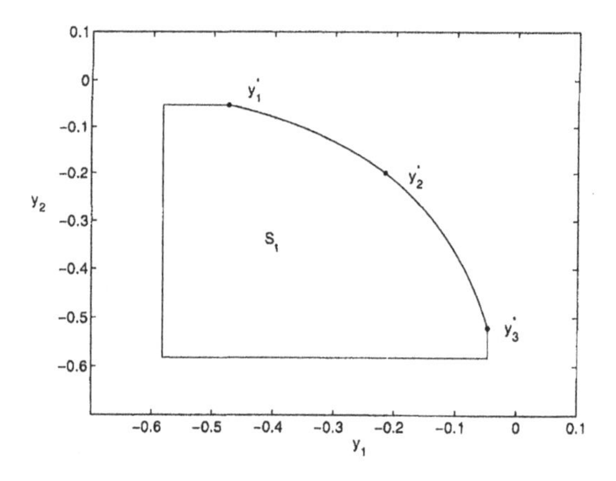

Figure 9.7. Convexified feasible region of Example 9.3.

9.3.4 Polyblock and convexification method for (MIP)

In Algorithm 9.2 of the discrete polyblock method, $f(\beta)$ is simply taken as the upper bound of $f(x)$ on $S \cap \langle \alpha, \beta \rangle$. Although this bound is easy to calculate, it could be a poor estimation of the optimal value of $f(x)$ on $S \cap \langle \alpha, \beta \rangle$. A much tighter upper bound can be obtained by using the convexification method discussed in Subsections 9.3.2 and 9.3.3.

Consider the continuous relaxation of the subproblem on integer $\langle \alpha, \beta \rangle$:

$$(\overline{MIP}(\alpha, \beta)) \qquad \begin{aligned} & \max\ f(x) \\ & \text{s.t. } g_i(x) \le b_i,\ i = 1, \ldots, m, \\ & \qquad x \in [\alpha, \beta]. \end{aligned}$$

Given a one-to-one mapping t that satisfies the conditions in Theorem 9.3, problem $(\overline{MIP}(\alpha, \beta))$ can be convexified into the following equivalent convex maximization problem:

$$(\overline{MIP}_t(\alpha, \beta)) \qquad \begin{array}{l} \max\ \phi(y) = f(t(y)) \\ \text{s.t. } \psi_i(y) = g_i(t(y)) \leq b_i,\ i = 1, \ldots, m, \\ \quad y \in t^{-1}([\alpha, \beta]). \end{array}$$

The outer approximation procedure (Algorithm 9.3) starts from $t^{-1}([\alpha, \beta])$ as the initial polyhedral approximation with the upper corner point $t^{-1}(\beta)$ as the initial solution. Adding cutting planes successively, the algorithm constructs a better and better polyhedral approximation to the feasible region of problem $(\overline{MIP}_t(\alpha, \beta))$, thus computing a better upper bound than $f(\beta)$. Since problem $(\overline{MIP}_t(\alpha, \beta))$ has to be solved many times in a branch-and-bound method, it is time-consuming to compute an approximate optimal solution to $(\overline{MIP}_t(\alpha, \beta))$ with high accuracy. Therefore, there is a trade-off between the tightness of the upper bound and the time to compute it. In practice, the procedure can be terminated either after given number of iterations or when a sufficient improvement of upper bound is achieved.

In the following paragraphs we describe some special properties of $(MIP(\alpha, \beta))$ that can be exploited to improve the efficiency of the branch-and-bound method.

Firstly, when solving the remaining relaxed subproblems, the current incumbent provides an extra criterion to stop the outer approximation method before the normal stopping rule is satisfied. In fact, suppose that the objective value of the incumbent is $\underline{\phi}$. If the condition $\phi(v^k) \leq \underline{\phi}$ holds at the k-th iteration of the outer approximation method, where v^k is the vertex with the maximum value of ϕ, then it is impossible for this subproblem to produce a feasible solution with function value greater than $\underline{\phi}$.

Secondly, the vertex information generated by the outer approximation method in solving a subproblem can be used to form a tight initial enclosing polyhedron for all its descendant subproblems. Suppose that the last polyhedron in solving a transformed relaxed subproblem $(\overline{MIP}_t(\alpha, \beta))$ is P and that its optimal solution is y^*. Let $x^* = t(y^*)$. Then x^* is an optimal solution to the relaxed subproblem $(\overline{MIP}(\alpha, \beta))$. Suppose that x_j^* is the branching variable. Then the initial enclosing polyhedron for the two transformed child subproblems of $(\overline{MIP}_t(\alpha, \beta))$ can be chosen as:

$$P^- = P \cap \{x \mid x_j \leq t_j^{-1}(\lfloor x_j^* \rfloor)\},$$

and

$$P^+ = P \cap \{x \mid x_j \geq t_j^{-1}(\lfloor x_j^* \rfloor + 1)\}.$$

The new vertices of P^- or P^+ can be easily obtained by computing the intersection points of the edges of P with the branching plane.

Thirdly, applying a convexification transformation and the outer approximation method is only necessary to subproblems for which the rectangular constraint set intersects with the boundary of the feasible region:

$$S = \{x \in X \mid g_i(x) \leq b_i,\ i = 1, \ldots, m\}.$$

In fact, if $\beta \in S$, then β is optimal to $(\overline{MIP}(\alpha, \beta))$. Moreover, if the left lower corner point $\alpha \notin S$, then we conclude that $(\overline{MIP}(\alpha, \beta))$ is infeasible.

Replacing the upper bound $f(\beta)$ in Algorithm 9.2 with an upper bound $UB_{\langle \alpha,\beta \rangle}$ obtained by the outer approximation method yields a combined polyblock and convexification algorithm that has a much better performance than the original Algorithm 9.2 for large-scale (MIP) as evidenced in the numerical results reported in the next section.

9.3.5 Computational results

In this section, we report computational results of Algorithm 9.2 discussed in Subsection 9.3.1 and its combination with the convexification method in Subsection 9.3.4. The algorithm was coded by Fortran 90 and run on a Sun Workstation (Blade 2000).

Four classes of nonseparable knapsack integer programming test problems will be considered. The objective functions of the test problems are described as follows.

- Polynomial function of the form

$$f(x) = \sum_{i=1}^{q} p_i \prod_{j \in N_i} x_j^{\alpha_{ij}},$$

  where q is a positive integer number, $p_i \in [0, 10]$, $N_i \subset \{1, \ldots, n\}$ with $1 \leq |N_i| \leq 3$, each element of N_i is randomly generated from $\{1, \ldots, n\}$, and α_{ij}'s are randomly generated from $\{1, 2, 3\}$. In our testing, q is taken to be n.

- Quadratic function $f(x) = x^T A x$, where $A = (a_{ij})_{n \times m}$ with a_{ij} randomly generated from $[0, 50]$, $i = 1, 2, \ldots, n$, $j = 1, 2, \ldots, n$.

- Minimax function

$$f(x) = \max_{j=1,\ldots,n} \min_{i=1,\ldots,n} a_{ij} x_i,$$

  where $a_{ij} (i = 1, 2, \ldots, n,\ j = 1, 2 \ldots, n)$ are randomly generated from $[0, 50]$.

- Reliability function of 12 variables in a 12-link complex network. (see Figure 9.8). Details of the expression of the reliability function can be found in [137].

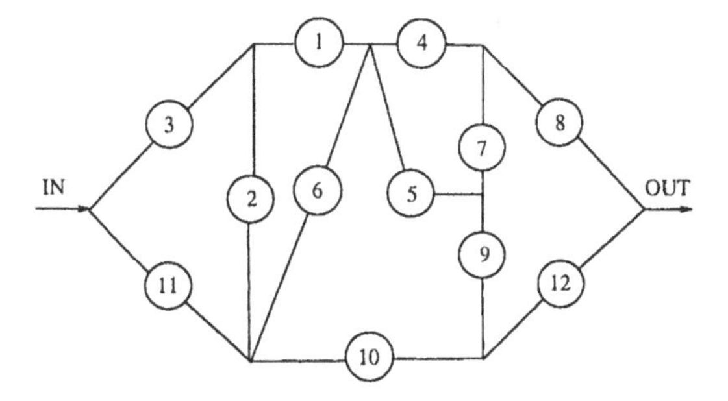

Figure 9.8. A 12-link complex reliability system.

Two types of constraint functions are considered for the test problems.

- Linear function $g_i(x) = \sum_{j=1}^{n} b_{ij}x_j$, where b_{ij} $(i = 1, \ldots, m,\ j = 1, \ldots, n)$ are randomly generated from $[100, 200]$.

- Polynomial function

$$g_l(x) = \sum_{i=1}^{q} b_{il} \prod_{j \in N_{il}} x_j^{\alpha_{ijl}}, \quad l = 1, \ldots, m,$$

  where $b_{il} \in [10, 50]$, $N_{il} \subset \{1, \ldots, n\}$ with $1 \le |N_{il}| \le 3$, each element of N_{il} is randomly generated from $\{1, \ldots, n\}$, and α_{ijl}'s are randomly generated from $\{1, 2, 3\}$. In our testing, q is set to be n.

For all the test problems, we set $l_j = 1, u_j = 5, j = 1, \ldots, n$. The right-hand side b_i $(i = 1, \ldots, m)$ affects the feasibility of the test problems and determines the degree of the difficulty of the test problems. In testing the algorithm, we have $b_i = g_i(l) + r(g_i(u) - g_i(l))$, $i = 1, \ldots, m$, where $r = 0.2$ for linear constraints and $r = 0.1$ for polynomial constraints.

In our implementation, the boundary point x_b in Step 1 is sought by the Bolzano's bisection method. The following two box selection rules are employed for choosing the next box for partition in Step 1 of the algorithm.

- **Selection Rule 1:** Select from X^k the box with the maximum objective function value of the upper bound to partition;

- **Selection Rule 2:** Select the last box included in X^k to partition, while the integer boxes in X^k are ordered based on the time they are generated.

Computational results of Algorithm 9.2 for two sets of test problems using the two different selection rules are summarized in Tables 9.1–9.4, where n is the number of variables, m the number of constraints and *box ratio* denotes the ratio of the total number of integer boxes generated by the algorithm to the total number of integer points in the domain X. The average CPU time (seconds) and the average box ratio are obtained by running the code 20 times.

Table 9.1. Numerical results for test problems with a polynomial objective function.

n	m	Constraint	Selection Rule 1		Selection Rule 2	
			Average CPU Time	Average Box Ratio (10^{-4})	Average CPU Time	Average Box Ratio (10^{-4})
10	5	Linear	0.34	2.09	0.28	2.41
10	5	Polynomial	3.85	5.25	1.7	2.20
14	5	Linear	3.7	0.03	6.6	0.06
14	5	Polynomial	1057.3	0.59	643.7	0.94

Table 9.2. Numerical results for test problems with a quadratic objective function.

n	m	Constraint	Selection Rule 1		Selection Rule 2	
			Average CPU Time	Average Box Ratio (10^{-4})	Average CPU Time	Average Box Ratio (10^{-4})
10	5	Linear	4.7	11.15	1.0	11.74
10	5	Polynomial	9.0	12.23	13.9	19.98
14	5	Linear	3462.7	1.59	121.6	1.64
14	5	Polynomial	2050.8	0.73	841.0	1.18

Table 9.3. Numerical results for test problems with a minimax objective function.

n	m	Constraint	Selection Rule 1		Selection Rule 2	
			Average CPU Time	Average Box Ratio (10^{-7})	Average CPU Time	Average Box Ratio (10^{-7})
15	5	Linear	1.9	2.78	1.9	2.23
15	5	Polynomial	18.2	5.00	31.6	8.49
20	5	Linear	154.3	0.019	60.8	0.017
20	5	Polynomial	316.2	0.008	600.8	0.035

Table 9.4. Numerical results for the reliability optimization problem of the 12-link complex network.

n	m	Constraint	Selection Rule 1		Selection Rule 2	
			Average CPU Time	Average Box Ratio (10^{-4})	Average CPU Time	Average Box Ratio (10^{-4})
12	5	Linear	6.9	1.44	4.6	1.34
12	5	Polynomial	55.1	1.99	51.4	2.25

To study the effect of the number of constraints to the efficiency of the algorithm, we have tested the algorithm for problems with a minimax objective function for different m. The comparison results are presented in Table 9.5. We conclude from Table 9.5 that for cases with linear constraints, the efficiency of the algorithm is not very sensitive to the number of constraints as evidenced by the fact that the number of integer boxes and the CPU time have the tendency to decrease as m increases. This is due to the decrease of the number of feasible points as m increases. The increase of CPU time for cases with nonlinear constraints as m increases accounts for the complexity of the feasible region and the significant increase of the computational time of evaluating nonlinear constraint functions in finding the root of $G(x)$ and checking the feasibility of the lower and upper bound points of integer boxes during the algorithm.

Table 9.5. Comparison results for problems with a minimax objective function ($n = 15$) for different m.

m	Constraint	Selection Rule 1		Selection Rule 2	
		Average CPU Time	Average Box Ratio (10^{-7})	Average CPU Time	Average Box Ratio (10^{-7})
1	Linear	5.6	6.85	2.9	5.34
5	Linear	3.0	4.27	2.6	4.06
15	Linear	2.0	2.65	1.94	2.24
1	Polynomial	59.2	18.4	17.9	17.1
5	Polynomial	18.2	5.0	31.6	8.49
15	Polynomial	275.2	22.79	368.7	35.5

Next, we discuss the implementation of Algorithm 9.2 when combined with a convexification method for upper bounding. Since the convexification and outer approximation method can improve the upper bound of f on each integer subbox, the number of integer subboxes examined in Algorithm 9.2 can be significantly decreased. However, additional computational time is needed to perform outer approximation on the transformed convex maximization subproblems. There is a trade-off between the quality of the upper bound and the computational time to obtain it. In our implementation, the outer approximation procedure is terminated whenever a sufficient improvement of the upper bound is achieved or after $2n$ cutting planes are generated.

Our numerical experiment shows that the discrete polyblock method outperforms its combination with the convexificaton method when the size of domain X is small, for example, $u_j - l_j \leq 10$. As the size of domain X increases, Algorithm 9.2 using the convexification and outer approximation bounding procedure becomes more efficient than Algorithm 9.2. To show this effect, we implement two versions of Algorithm 9.2 with and without using the convexification and outer approximation method, which are denoted by A_1 and A_2, respectively. Table 9.6 summarizes some comparison results for test problems with $l_i = 1$ and $u_i = 15$ $(i = 1, \ldots, n)$. The quadratic constraint functions are of the same form as the quadratic objective function.

Table 9.6. Comparison results of the discrete polyblock method.

Objective	Constraint	n	m	r	Average CPU Seconds	
					A_1	A_2
Polynomial	Quadratic	10	5	0.4	40.4	119.5
Polynomial	Quadratic	12	5	0.4	876.3	NS
Polynomial	Polynomial	10	5	0.1	186.4	584.4
Polynomial	Polynomial	12	5	0.1	881.5	NS
Quadratic	Quadratic	10	5	0.8	8.1	20.7
Quadratic	Quadratic	12	5	0.8	64.5	123.8
Quadratic	Polynomial	10	5	0.2	339.6	NS
Quadratic	Polynomial	12	5	0.2	573.7	NS
Minmax	Quadratic	10	5	0.2	118.2	381.5
Minmax	Quadratic	12	5	0.2	595.0	NS
Minmax	Polynomial	10	5	0.05	272.1	476.7
Minmax	Polynomial	12	5	0.05	92.0	235.2

9.4 Notes

A survey of the early works on general nonlinear integer programming can be found in [45]. Branch-and-bound methods for convex nonlinear integer programming were investigated in [87]. Lagrangian decomposition method for convex integer programming was proposed in [161].

Problem (MIP) is often encountered in optimization models of resource allocation problems ([106]), reliability optimization in complex systems ([217]) and optimal design ([173]). The continuous version of (MIP) is a global optimization problem and has been studied by various authors in the framework of monotone optimization. Rubinov and Tuy proposed a polyblock method for finding the continuous solution of (MIP) by using polyblock approximation to the continuous feasible region of (MIP) (see [186][218]). Convexification methods were introduced in [136][207] to convert the continuous version of (MIP) into a concave minimization problem which can then be solved by the outer approximation method.

Convexification methods for monotone optimization were presented in [136] [207]. Applications to reliability optimization in complex networks were discussed in [137]. The outer approximation method for concave minimization problems with general convex constraints was proposed in [103]. Techniques of computing new vertices resulted from an intersection of a polyhedron with a cutting plane were discussed in [41][104].

Chapter 10

UNCONSTRAINED POLYNOMIAL 0-1 OPTIMIZATION

Nonlinear programming in 0-1 variables plays an important role in many optimization models involving polynomial (multilinear) objective and constraint functions. The theory of nonlinear 0-1 programming or pseudo-Boolean optimization has been extensively studied during the last three decades. In this chapter, we study the theory and algorithms for unconstrained polynomial 0-1 programming.

This chapter is organized as follows. In Section 10.1, we introduce roof duality theory for unconstrained polynomial 0-1 programming. In Section 10.2, we discuss how to perform local search for an unconstrained polynomial 0-1 programming problem. In Section 10.3, we present a basic algorithm in searching for an optimal solution for an unconstrained polynomial 0-1 programming problem. In Section 10.4, we reveal the relationship between an unconstrained polynomial 0-1 programming problem and its continuous relaxation. We concentrate in Section 10.5, the last section in the chapter, on quadratic 0-1 programming problems.

10.1 Roof Duality

This section discusses the theory of roof duality which was first developed for unconstrained quadratic 0-1 optimization and was later extended to polynomial 0-1 programming. This section also examines the relation between the roof duality and other linearization approaches.

The unconstrained polynomial 0-1 optimization problem can be described as follows,

$$(0\text{-}1UPP) \qquad \max_{x\in\{0,1\}^n} f(x) = \sum_{i=1}^{n} c_i x_i + \sum_{k\in N} q_k \prod_{i\in S_k} x_i,$$

where N is an index set, $S_k \subseteq I = \{1, 2, \ldots, n\}$, $s_k = |S_k| \geq 2$.

10.1.1 Basic concepts

DEFINITION 10.1 *A linear function $p(x)$ is said to be an upper plane of $f(x)$ if $p(x) \geq f(x)$ for all $x \in \{0,1\}^n$. A local upper plane of the nonlinear term $f_k(x) = q_k \prod_{i \in S_k} x_i$ is a linear function with a form $p_k(x) = \lambda_k^0 + \sum_{i \in S_k} \lambda_k^i x_i$ that satisfies $p_k(x) \geq f_k(x)$ for all $x \in \{0,1\}^{s_k}$.*

It is easy to see that $p_k(x)$ is a local upper plane of $f_k(x)$ if and only if

$$\lambda_k^0 \geq 0, \tag{10.1.1}$$

$$\lambda_k^0 + \sum_{j \in S_k^i} \lambda_k^j \geq 0, \; i = 2, \ldots, 2^{s_k} - 1, \tag{10.1.2}$$

$$\lambda_k^0 + \sum_{j \in S_k} \lambda_k^j \geq q_k, \tag{10.1.3}$$

where S_k^i $(i = 2, \ldots, 2^{s_k} - 1)$ are all the possible nonempty proper subsets of S_k.

DEFINITION 10.2 *A paved upper plane of $f(x)$ is the sum of all local upper planes:*

$$\begin{aligned} p(x) &= \sum_{i=1}^{n} c_i x_i + \sum_{k \in N} p_k(x) \\ &= \sum_{i=1}^{n} c_i x_i + \sum_{k \in N} (\lambda_k^0 + \sum_{i \in S_k} \lambda_k^i x_i) \\ &= \sum_{k \in N} \lambda_k^0 + \sum_{i=1}^{n} (c_i + \sum_{k \in S^{-1}(i)} \lambda_k^i) x_i, \end{aligned} \tag{10.1.4}$$

where $S^{-1}(i) = \{k \in N \mid i \in S_k\}$ and λ_k^i satisfies (10.1.1)–(10.1.3) for $k \in N$.

Since $p(x) \geq f(x)$ for all $x \in \{0,1\}^n$, $\max_{x \in \{0,1\}^n} p(x)$ provides an upper bound for the maximum of $f(x)$ over $\{0,1\}^n$. Let $\mathcal{P}$ denote the set of all paved upper planes for $f(x)$. The paved dual problem is then to find the best upper bound:

$$W(\mathcal{P}) = \min_{p(x) \in \mathcal{P}} \max_{x \in \{0,1\}^n} p(x),$$

where $p(x)$ takes the form of (10.1.4) with λ_k^j's satisfying (10.1.1)–(10.1.3). Let $f^* = \max_{x\in\{0,1\}^n} f(x)$. Then $W(\mathcal{P}) \geq f^*$. Let

$$u_i = \max\{0, c_i + \sum_{k\in S^{-1}(i)} \lambda_k^i\}, \quad i = 1, \ldots, n.$$

Then $W(\mathcal{P})$ can be expressed as a linear program:

$$\begin{aligned}
(LPF) \qquad & \min \sum_{k\in N} \lambda_k^0 + \sum_{i=1}^{n} u_i \\
& \text{s.t. } u_i - \sum_{k\in S^{-1}(i)} \lambda_k^i \geq c_i, \ i = 1, \ldots, n, \\
& \lambda_k^0 + \sum_{j\in S_k} \lambda_k^j \geq q_k, \ k \in N, \\
& \lambda_k^0 + \sum_{j\in S_k^i} \lambda_k^j \geq 0, \ i = 2, \ldots, 2^{s_k} - 1, k \in N, \\
& \lambda_k^0 \geq 0, \ k \in N, \\
& u_i \geq 0, \ i = 1, \ldots, n.
\end{aligned}$$

DEFINITION 10.3 *A tile of the nonlinear term $f_k(x)$ is the upper plane that minimizes the sum of the differences between $p_k(x)$ and $f_k(x)$ over all $x \in \{0,1\}^{s_k}$, or equivalently the slacks of all the inequalities in (10.1.1)–(10.1.3). A paved plane with all local upper planes being tiles is called a roof of $f(x)$.*

To characterize the conditions for a tile and therefore a roof, we need to rewrite $f(x)$ such that the coefficients of the nonlinear terms are all positive. This can be accomplished by introducing a complementary variable $\bar{x}_i = 1 - x_i$ for a primal variable x_i, when necessary. Suppose that $q_k < 0$ and $S_k = \{j_1, \ldots, j_m\}$. Then we have

$$\begin{aligned}
f_k(x) &= q_k x_{j_1} x_{j_2} \cdots x_{j_m} \\
&= q_k(1 - \bar{x}_{j_1}) x_{j_2} \cdots x_{j_m} \\
&= -q_k \bar{x}_{j_1} x_{j_2} \cdots x_{j_m} + q_k(1 - \bar{x}_{j_2}) x_{j_3} \cdots x_{j_m} \\
&= -q_k \sum_{i=1}^{m-1} \bar{x}_{j_i} \prod_{t=i+1}^{m} x_{j_t} + q_k x_{j_m}.
\end{aligned}$$

Denote

$$N^+ = \{k \in N \mid q_k > 0\}, \ N^- = \{k \in N \mid q_k < 0\}.$$

Then, we can express $f(x)$ in the following form:

$$f(x) = \sum_{i=1}^{n} \gamma_i x_i + \sum_{k \in N^+} d_k \prod_{j \in Q_k} x_j + \sum_{k \in N^-} e_k \bar{x}_{t_k} \prod_{j \in R_k} x_j, \quad (10.1.5)$$

where (i) $d_k, e_k > 0$, (ii) $Q_k \subseteq I$ for $k \in N^+$, and (iii) $R_k \subseteq I$ and $t_k \in I \setminus R_k$ for $k \in N^-$.

THEOREM 10.1 ([146]) *Let $p_k(x)$ be a tile of the nonlinear term in polynomial $f(x)$ in the form (10.1.5). Then*

$$p_k(x) = \begin{cases} \sum_{j \in Q_k} \lambda_k^j x_j, & k \in N^+, \\ v_k \bar{x}_{t_k} + \sum_{j \in R_k} \mu_k^j x_j, & k \in N^-, \end{cases} \quad (10.1.6)$$

where

$$\sum_{j \in Q_k} \lambda_k^j = d_k, \quad (10.1.7)$$

$$v_k + \sum_{j \in R_k} \mu_k^j = e_k, \quad (10.1.8)$$

$$(\lambda, \mu, v) \geq 0. \quad (10.1.9)$$

Therefore, a roof of $f(x)$ takes the following form:

$$\begin{aligned} p(x) &= \sum_{i=1}^{n} \gamma_i x_i + \sum_{k \in N^+} \sum_{i \in Q_k} \lambda_k^i x_i + \sum_{k \in N^-} (v_k(1 - x_{t_k}) + \sum_{i \in R_k} \mu_k^i x_i) \\ &= \sum_{k \in N^-} v_k + \sum_{i=1}^{n} (\gamma_i + \sum_{k \in Q^{-1}(i)} \lambda_k^i - \sum_{k \in T^{-1}(i)} v_k + \sum_{k \in R^{-1}(i)} \mu_k^i) x_i, \end{aligned} \quad (10.1.10)$$

where $T^{-1}(i) = \{k \in N^- \mid t_k = i\}$, $Q^{-1}(i) = \{k \in N^+ \mid i \in Q_k\}$, $R^{-1}(i) = \{k \in N^- \mid i \in R_k\}$, and (λ, μ, v) satisfies (10.1.7)–(10.1.9). Let $\mathcal{R}$ denote the set of roofs of $f(x)$. Since a roof is also an upper plane of $f(x)$, it holds $\mathcal{R} \subseteq \mathcal{P}$. Define the roof dual of (0-1UPP) as

$$W(\mathcal{R}) = \min_{p(x) \in \mathcal{R}} \max_{x \in \{0,1\}^n} p(x), \quad (10.1.11)$$

where $p(x)$ is defined by (10.1.10). Let

$$u_i = \max\{0, \gamma_i + \sum_{k \in Q^{-1}(i)} \lambda_k^i - \sum_{k \in T^{-1}(i)} v_k + \sum_{k \in R^{-1}(i)} \mu_k^i\}.$$

Similar to the paved dual problem, we can express the roof dual as a linear programming problem:

$$
\begin{aligned}
(LRF) \qquad & \min \sum_{k \in N^-} v_k + \sum_{i=1}^{n} u_i \\
& \text{s.t. } u_i - \sum_{k \in Q^{-1}(i)} \lambda_k^i + \sum_{k \in T^{-1}(i)} v_k - \sum_{k \in R^{-1}(i)} \mu_k^i \geq \gamma_i, \\
& \qquad\qquad\qquad\qquad\qquad\qquad\qquad i = 1, \ldots, n, \\
& \sum_{i \in Q_k} \lambda_k^i = d_k, \ k \in N^+, \\
& v_k + \sum_{i \in R_k} \mu_k^i = e_k, \ k \in N^-, \\
& (u, \lambda, \mu, v) \geq 0.
\end{aligned}
$$

It is clear that $f^* \leq W(\mathcal{P}) \leq W(\mathcal{R})$, where f^* is the optimal value of $(0\text{-}1UPP)$. It will be shown in a later subsection that $W(\mathcal{R}) = W(\mathcal{P})$ for quadratic case of $(0\text{-}1UPP)$. There exist non-quadratic instances with $W(\mathcal{P}) < W(\mathcal{R})$ (see [145]).

10.1.2 Relation to other linearization formulations

Consider expression (10.1.5) of $f(x)$. Let

$$
\begin{aligned}
y_k &= \prod_{j \in Q_k} x_j, \quad k \in N^+, \\
w_k &= \bar{x}_{t_k} \prod_{j \in R_k} x_j, \quad k \in N^-.
\end{aligned}
$$

Since $d_k > 0$ and $e_k > 0$, we can rewrite $(0\text{-}1UPP)$ as the following equivalent 0-1 linear programming problem,

$$
\begin{aligned}
(DRF) \qquad & \max \sum_{i=1}^{n} \gamma_i x_i + \sum_{k \in N^+} d_k y_k + \sum_{k \in N^-} e_k w_k \\
& \text{s.t. } y_k \leq x_i, \quad i \in Q_k, \ k \in N^+, && (10.1.12) \\
& \quad w_k \leq 1 - x_{t_k}, \quad k \in N^-, && (10.1.13) \\
& \quad w_k \leq x_i, \quad i \in R_k, \ k \in N^-, && (10.1.14) \\
& \quad x_i, y_k, w_k \in \{0, 1\}. && (10.1.15)
\end{aligned}
$$

The above problem is called *discrete Rhys form*. Let (CRF) denote the linear programming relaxation by relaxing the constraint (10.1.15) to $0 \leq x_i \leq 1$, $y_k \geq 0$ and $w_k \geq 0$. By associating constraint (10.1.12) with a dual variable

λ_k^i, (10.1.13) with v_k , (10.1.14) with μ_k^i and $x_i \le 1$ with u_i, we obtain a dual problem of (CRF) which is exactly problem (LRF), the linear programming expression of the roof duality. Therefore we have the following result.

THEOREM 10.2 $v(CRF) = v(LRF) = W(\mathcal{R})$.

Next, let us consider another linearization formulation. Let $y_k = \prod_{i\in S_k} x_i$ in (0-1UPP). Then, we have

$$y_k = \min_{i\in S_k} x_i = \max\{0, \sum_{i\in S_k} x_i - s_k + 1\},$$

where $s_k = |S_k|$. Rewrite the objective function $f(x)$ in (0-1UPP) in the following form

$$f(x) = \sum_{i=1}^{n} c_i x_i + \sum_{k\in N^+} q_k \prod_{i\in S_k} x_i + \sum_{k\in N^-} q_k \prod_{i\in S_k} x_i.$$

Substituting in the right-hand side

$$\prod_{i\in S_k} x_i = \min_{i\in S_k} x_i, \quad k \in N^+,$$
$$\prod_{i\in S_k} x_i = \max\{0, \sum_{i\in S_k} x_i - s_k + 1\}, \quad k \in N^-,$$

and relaxing the integrality restriction on x_j, we obtain a piecewise linear concave maximization problem:

$$\max_{x\in[0,1]^n} \sum_{i=1}^{n} c_i x_i + \sum_{k\in N^+} q_k \min_{i\in S_k} x_i + \sum_{k\in N^-} q_k \max\{0, \sum_{i\in S_k} x_i - s_k + 1\}. \tag{10.1.16}$$

It is easy to see that this problem is equivalent to (0-1UPP) if x_i's are restricted to 0 or 1. Therefore, the optimal value of (10.1.16) provides an upper bound for (0-1UPP). Introducing a new variable y_k, problem (10.1.16) is equivalent to the following *standard linear form (SLF)* of (0-1UPP):

$$\begin{aligned} (SLF) \qquad & \max \sum_{i=1}^{n} c_i x_i + \sum_{k\in N} q_k y_k \\ & \text{s.t. } y_k \le x_i,\ i \in S_k,\ k \in N^+, \\ & \quad \sum_{i\in S_k} x_i - y_k \le s_k - 1,\ k \in N^-, \\ & \quad 0 \le x_i \le 1,\ i = 1, \ldots, n, \\ & \quad 0 \le y_k,\ k \in N. \end{aligned}$$

Since (SLF) is a relaxation of (0-1UPP), $v(SLF)$ provides an upper bound of (0-1UPP). The following result shows that this upper bound coincides with the paved duality upper bound.

THEOREM 10.3 ([96]) $v(SLF) = W(\mathcal{P})$.

Theorems 10.2 and 10.3 together imply

$$v(SLF) = W(\mathcal{P}) \leq W(\mathcal{R}) = v(CRF). \tag{10.1.17}$$

10.1.3 Quadratic case

Now, consider the quadratic case of (0-1UPP):

$$(\text{0-1}UQP) \quad \max_{\{0,1\}^n} Q(x) = \sum_{i=1}^{n} c_i x_i + \sum_{1 \leq i < j \leq n} q_{ij} x_i x_j.$$

We have the following result:

THEOREM 10.4 ([90]) *For unconstrained quadratic 0-1 optimization problem (0-1UQP), it holds* $v(CRF) = v(SLF)$.

Proof. Let $I^+ = \{(i,j) \mid q_{ij} > 0\}$, $I^- = \{(i,j) \mid q_{ij} < 0\}$. The function $Q(x)$ can be rewritten as

$$Q(x) = \sum_{i=1}^{n} c_i x_i + \sum_{(i,j) \in I^+} q_{ij} x_i x_j - \sum_{(i,j) \in I^-} q_{ij} \bar{x}_i x_j + \sum_{(i,j) \in I^-} q_{ij} x_j.$$

Then, the continuous relaxation of problem (CRF) for problem (0-1UQP) has the following form:

$$\begin{aligned}
(CRF) \quad & \max \sum_{i=1}^{n} c_i x_i + \sum_{(i,j) \in I^+} q_{ij} y_{ij} - \sum_{(i,j) \in I^-} q_{ij} y_{ij} + \sum_{(i,j) \in I^-} q_{ij} x_j \\
& \text{s.t. } y_{ij} \leq x_i, \ y_{ij} \leq x_j, \ (i,j) \in I^+, \\
& \quad\ \ y_{ij} \leq 1 - x_i, \ y_{ij} \leq x_j, \ (i,j) \in I^-, \\
& \quad\ \ 0 \leq x_i \leq 1, \ i = 1, \ldots, n, \\
& \quad\ \ 0 \leq y_{ij}, \ 1 \leq i < j \leq n.
\end{aligned}$$

On the other hand, problem (SLF) for (0-1UQP) has the following form:

$$\begin{aligned}(SLF) \qquad \max \; & \sum_{i=1}^{n} c_i x_i + \sum_{(i,j)\in I^+} q_{ij} y_{ij} + \sum_{(i,j)\in I^-} q_{ij} y_{ij} \\ \text{s.t. } & y_{ij} \le x_i, \; y_{ij} \le x_j, \; (i,j) \in I^+, \\ & y_{ij} \ge x_i + x_j - 1, \; (i,j) \in I^-, \\ & 0 \le x_i \le 1, \; i = 1, \ldots, n, \\ & 0 \le y_{ij}, \; 1 \le i < j \le n.\end{aligned}$$

Now, for any $(i,j) \in I^-$, we have

$$\begin{aligned} & q_{ij}x_j + \max\{-q_{ij}y_{ij} \mid y_{ij} \le 1 - x_i, \; y_{ij} \le x_j\} \\ &= q_{ij}x_j - q_{ij}\max\{y_{ij} \mid y_{ij} \le 1 - x_i, \; y_{ij} \le x_j\} \\ &= q_{ij}x_j - q_{ij}\min\{1 - x_i, x_j\} \\ &= q_{ij}\max\{x_i + x_j - 1, 0\} \\ &= q_{ij}\min\{y_{ij} \mid y_{ij} \ge x_i + x_j - 1, \; y_{ij} \ge 0\} \\ &= \max\{q_{ij}y_{ij} \mid y_{ij} \ge x_i + x_j - 1, \; y_{ij} \ge 0\}.\end{aligned}$$

Thus, (CRF) and (SLF) are equivalent and $v(CRF) = v(SLF)$. □

In view of (10.1.17), Theorem 10.4 implies the following corollary.

COROLLARY 10.1 *For unconstrained quadratic 0-1 optimization (0-1UQP), it holds*

$$v(SLF) = W(\mathcal{P}) = W(\mathcal{R}) = v(CRF). \tag{10.1.18}$$

10.2 Local Search

Let $f(x)$ be defined in problem (0-1UPP). Denote by $\Delta_i(x)$ the i-th derivative of f at x,

$$\begin{aligned}\Delta_i(x) &= \frac{\partial f}{\partial x_i} \\ &= f(x_1, \ldots, x_{i-1}, 1, x_{i+1}, \ldots, x_n) - f(x_1, \ldots, x_{i-1}, 0, x_{i+1}, \ldots, x_n).\end{aligned}$$

Denote by $\Theta_i(x)$ the i-th residual

$$\begin{aligned}\Theta_i(x) &= f(x_1, \ldots, x_{i-1}, 0, x_{i+1}, \ldots, x_n) \\ &= f(x) - x_i\Delta_i(x).\end{aligned}$$

Both $\Delta_i(x)$ and $\Theta_i(x)$ are, in general, functions of $x_1, \ldots, x_{i-1}, x_{i+1}, \ldots, x_n$. Moreover, f can be expressed as

$$f(x) = x_i\Delta_i(x) + \Theta_i(x). \tag{10.2.1}$$

DEFINITION 10.4 *The m-neighborhood of $x \in \{0,1\}^n$ is defined as*

$$N_m(x) = \{y \mid \rho_H(x,y) \le m\}, \tag{10.2.2}$$

where $\rho_H(x,y)$ is the number of different components between x and y. A point $x \in \{0,1\}^n$ is called an N_m local maximizer *of f if*

$$f(y) \le f(x), \quad \forall y \in N_m(x).$$

Obviously, an N_n local maximizer is a global maximizer of f and hence an optimal solution to (0-1UPP). The following result gives an optimality criterion for a local maximizer.

THEOREM 10.5 *A point $x \in \{0,1\}^n$ is an N_1 local maximizer of $f(x)$ if and only if for all $i = 1, \ldots, n$,*

$$x_i = \begin{cases} 1, & \text{if } \Delta_i(x) > 0, \\ 0, & \text{otherwise.} \end{cases} \tag{10.2.3}$$

Proof. It is clear that $N_1(x) = \{y^1, \ldots, y^n\}$, where y^i is different from x only at the i-th component. By the definition, we have

$$\begin{aligned} f(y^i) &= y_i^i \Delta_i(y^i) + \Theta_i(y^i) \\ &= (1 - x_i)\Delta_i(x) + \Theta_i(x) \\ &= f(x) + (1 - 2x_i)\Delta_i(x). \end{aligned}$$

Therefore, $f(y^i) \le f(x)$ for $i = 1, \ldots, n$ if and only if (10.2.3) holds. □

Since the number of points in N_m increases exponentially as m increases, the cost of computing an N_m local maximizer becomes prohibitive for problem of a realistic size of m.

PROCEDURE 10.1 (LOCAL SEARCH FOR (0-1UPP))

Step 0. Choose $x^0 \in \{0,1\}^n$.

Step 1. If there exists $y \in N_m(x)$ such that $f(y) > f(x)$, set $x := y$, repeat Step 1. Otherwise, x is an N_m local maximizer of f.

10.3 Basic Algorithm

Let $f(x)$ be defined as in (0-1UPP). From (10.2.1), we have $f(x) = x_n\Delta_n(x) + \Theta_n(x)$. Since $\Delta_n(x)$ and $\Theta_n(x)$ do not depend on x_n, we can express them as functions of $x_1, \ldots, x_{n-1}$, $g_n(x_1, \ldots, x_{n-1})$ and $h_n(x_1, \ldots, x_{n-1})$, respectively. Thus

$$f(x) = x_n g_n(x_1, \ldots, x_{n-1}) + h_n(x_1, \ldots, x_{n-1}). \tag{10.3.1}$$

From the optimal condition (10.2.3), the global maximizer of f satisfies

$$x_n = \begin{cases} 1, & \text{if } g_n(x_1, \ldots, x_{n-1}) > 0, \\ 0, & \text{otherwise.} \end{cases} \quad (10.3.2)$$

Therefore, if we can express x_n defined in (10.3.2) as a polynomial of x_1, $\ldots$,x_{n-1}, $\phi_n(x_1, \ldots, x_{n-1})$, then we can eliminate x_n from the expression of $f(x)$ in (10.3.1),

$$f_{n-1}(x_1, \ldots, x_{n-1}) = \phi_n(x_1, \ldots, x_{n-1}) g_n(x_1, \ldots, x_{n-1}) + h_n(x_1, \ldots, x_{n-1}).$$

Performing the same elimination process for f_{n-1}, we will get a function f_{n-2} of $x_1, \ldots, x_{n-2}$ and this process continues recursively until we obtain $f_1(x_1)$. Let x^* denote the optimal solution of (0-1UPP). Notice that $x_1^* = 1$ if $f_1(1) > f_1(0)$ and $x_1^* = 0$ otherwise. Then $x_2^*, \ldots, x_n^*$ can be obtained by using $x_{i+1}^* = \phi_{i+1}(x_1^*, \ldots, x_i^*)$ recursively for $i = 1, \ldots, n-1$.

The basic algorithm can then be described as follows.

ALGORITHM 10.1 (BASIC ALGORITHM FOR (0-1UPP))

Step 0. Set $f_n(x) = f(x)$ and $k = n$.

Step 1. Calculate

$$g_k(x_1, \ldots, x_{k-1}) = \frac{\partial f_k}{\partial x_k},$$
$$h_k(x_1, \ldots, x_{k-1}) = f_k(x_1, \ldots, x_{k-1}, 0).$$

Determine the polynomial expression of ϕ_k defined by

$$\phi_k(x_1, \ldots, x_{k-1}) = \begin{cases} 1, & \text{if } g_k(x_1, \ldots, x_{k-1}) > 0, \\ 0, & \text{otherwise.} \end{cases} \quad (10.3.3)$$

Step 2. Compute

$$f_{k-1}(x_1, \ldots, x_{k-1}) = \phi_k(x_1, \ldots, x_{k-1}) g_k(x_1, \ldots, x_{k-1}) + h_k(x_1, \ldots, x_{k-1}$$

Step 3. If $k > 1$, then set $k := k - 1$ and go to Step 1. Otherwise, set $x_1^* = 1$ if $f_1(1) > f_1(0)$ and $x_1^* = 0$ if $f_1(1) \leq f_1(0)$. Calculate x_k^* by $x_k^* = g_k(x_1^*, \ldots, x_{k-1}^*)$ for $k = 2, \ldots, n$.

It is proved in [92] that the basic algorithm produces an optimal solution x^* to (0-1UPP). The following small-size example illustrates the algorithm.

EXAMPLE 10.1

$$\max_{x \in \{0,1\}^3} f(x) = 4x_1x_2x_3 - x_1x_2 - x_1x_3 - x_2x_3.$$

By the algorithm, we have $g_3(x_1, x_2) = 4x_1x_2 - x_1 - x_2$ and thus

$$\phi_3(x_1, x_2) = \left\{ \begin{array}{ll} 1, & \text{if } g_3(x_1, x_2) > 0 \\ 0, & \text{otherwise} \end{array} \right\} = x_1x_2.$$

Hence we get

$$\begin{aligned} f_2(x_1, x_2) &= \phi_3(x_1, x_2)g_3(x_1, x_2) + h_3(x_1, x_2) \\ &= x_1x_2(4x_1x_2 - x_1 - x_2) - x_1x_2 \\ &= x_1x_2. \end{aligned}$$

Since $g_2(x_1) = x_1$, we get

$$\phi_2(x_1) = \left\{ \begin{array}{ll} 1, & \text{if } g_2(x_1) > 0, \\ 0, & \text{otherwise} \end{array} \right\} = x_1.$$

Thus

$$f_1(x_1) = \phi_2(x_1)g_2(x_1) + h_2(x_1) = x_1.$$

Therefore, $x_1^* = 1$, $x_2^* = \phi_2(x_1^*) = x_1^* = 1$ and $x_3^* = \phi_3(x_1^*, x_2^*) = x_1^*x_2^* = 1$. The optimal solution to the example is $x^* = (1, 1, 1)^T$ with $f(x^*) = 1$.

The key task in using the basic algorithm is how to identify the polynomial expression of ϕ_k defied in (10.3.3). In principle, ϕ_k can be always constructed systematically. Let's consider the following instance, $g_4(x_1, x_2, x_3) = 4x_1x_2 - x_1 - x_2 + 3x_2x_3$. The first step is to find the mapping from all possible combinations of x_1, x_2 and x_3 to the value of g_4 which is given in the following table.

Table 10.1. Illustrative example of mapping g_k.

x_1	x_2	x_3	$g_4(x_1, x_2, x_3)$
0	0	0	0
1	0	0	−1
0	1	0	−1
0	0	1	0
1	1	0	2
1	0	1	−1
0	1	1	2
1	1	1	5

Using Boolean algebra and noticing that all possible combinations of x_1, x_2 and x_3 are mutually exclusive, we can get

$$\begin{aligned} \phi_4(x_1, x_2, x_3) &= x_1x_2(1 - x_3) + (1 - x_1)x_2x_3 + x_1x_2x_3 \\ &= x_1x_2 + x_2x_3 - x_1x_2x_3. \end{aligned}$$

Note that if g_k involves s variables, then we need to examine 2^s combinations. In the worst case, if g_n involves $n-1$ variables, calculating ϕ_n is more than enumerating 2^{n-1} possible solutions. The basic algorithm could become very powerful for (0-1UPP) where interactions between variables are weak, i.e., situations where each variable interacts with at most s other variables in some cross terms and $s \ll n$.

Techniques to obtain the polynomial expression ϕ_k are discussed in [46][95].

10.4 Continuous Relaxation and its Convexification

Consider the continuous relaxation of (0-1UPP)

$$(\overline{0\text{-}1UPP}) \qquad \max_{x\in[0,1]^n} f(x) = \sum_{i=1}^{n} c_i x_i + \sum_{k\in N} q_k \prod_{i\in S_k} x_i.$$

It always holds $v(0\text{-}1UPP) \le v(\overline{0\text{-}1UPP})$. The following interesting result shows that (0-1UPP) can actually be reduced to $(\overline{0\text{-}1UPP})$ since at least one of the global solutions of $(\overline{0\text{-}1UPP})$ is attained at a vertex of $[0,1]^n$.

THEOREM 10.6 *Let $U = [0,1]^n$ and $f(x)$ be any multi-linear polynomial function defined on U. Suppose that $x^* \in U$ is a maximizer of f over U. Define $U(x^*) = \{x \in U \mid x_i = x_i^*,\ i \in J\}$, where $J = \{i \mid x_i^* = 0 \text{ or } x_i^* = 1\}$. Then $f(x) = f(x^*)$ for all $x \in U(x^*)$.*

Proof. Without loss of generality, let $J = \{1, 2, \dots, n-k\}$ with $k \ge 1$. From (10.3.1), we have

$$f(x^*) = x_n^* g_n(x_1^*, \dots, x_{n-1}^*) + h_n(x_1^*, \dots, x_{n-1}^*).$$

Since $0 < x_n^* < 1$, we must have $g_n(x_1^*, \dots, x_{n-1}^*) = 0$, otherwise, we will get a contradiction, i.e., we are able to increase $f(x^*)$ by changing x_n^*. We then proceed to write

$$\begin{aligned} f(x^*) &= h_n(x_1^*, \dots, x_{n-1}^*) \\ &= x_{n-1}^* g_{n-1}(x_1^*, \dots, x_{n-2}^*) + h_{n-1}(x_1^*, \dots, x_{n-2}^*), \end{aligned}$$

which leads to $g_{n-1}(x_1^*, \dots, x_{n-2}^*) = 0$. Repeating the same process, we can finally conclude

$$f(x^*) = h_{n-k+1}(x_1^*, \dots, x_{n-k}^*).$$

Therefore $f(x)$ is a constant over $U(x^*)$. □

The above theorem implies that at least one maximizer of $(\overline{0\text{-}1UPP})$ is located at a vertex of $[0,1]^n$, thus a maximizer of (0-1UPP) at the same time and

$$v(\overline{0\text{-}1UPP}) = v(0\text{-}1UPP).$$

However, problem $(\overline{0\text{-}1UPP})$ is still a difficult global optimization problem since $f(x)$ is neither a convex nor a concave function on $[0,1]^n$. A promising approach is to transform $(\overline{0\text{-}1UPP})$ into a convex maximization or concave minimization problem so that the solution methods developed for solving concave minimization in global optimization literature can be applied.

The classical convexification transformation of $(0\text{-}1UPP)$ makes use of the relation $x_j = x_j^2$ for $x_j \in \{0,1\}$. Thus, adding a penalty term $(p/2)\sum_{j=1}^n (x_j^2 - x_j)$ to $f(x)$ does not change the optimal solution of $(0\text{-}1UPP)$. This leads to the following function

$$f_p(x) = f(x) + \frac{p}{2}\sum_{j=1}^n x_j^2 - \frac{p}{2}\sum_{j=1}^n x_j. \tag{10.4.1}$$

Clearly, $f_p(x)$ takes the same values as $f(x)$ on $\{0,1\}^n$. Moreover, if p is large enough then $f_p(x)$ becomes a convex function. Since a convex function always attends its maximum at a vertex of $[0,1]^n$, problem $(0\text{-}1UPP)$ can be reduced to a convex maximization problem on $[0,1]^n$. However, the threshold value ρ with which $p \geq \rho$ implies a convexity of $f_p(x)$ is difficult to determine. In the following, we will discuss some alternative ways to convexify $f(x)$.

Define

$$f^1(x) = f(x) + \frac{1}{2}\sum_{j=1}^n \Big(\sum_{k\in S^{-1}(j)} |q_k|\Big)x_j^2 - \frac{1}{2}\sum_{j=1}^n \Big(\sum_{k\in S^{-1}(j)} |q_k|\Big)x_j,$$

where $S^{-1}(j) = \{k \in N \mid j \in S_k\}$. It is easy to see that $f^1(x) = f(x)$ for all $x \in \{0,1\}^n$ and $\nabla^2 f^1(x)$ is a diagonally dominant matrix and thus is semi-definite for any $x \in [0,1]^n$. Thus $f^1(x)$ is a convex function on $[0,1]^n$. Define

$$(CP_1) \qquad \max_{x\in[0,1]^n} f^1(x).$$

Then, (CP_1) is a convex maximization problem and is equivalent to $(0\text{-}1UPP)$.

We can express the nonlinear polynomial term of $f(x)$ by

$$\prod_{j\in S_k} x_j = \min_{j\in S_k} x_j \tag{10.4.2}$$

for all $x \in \{0,1\}^{s_k}$. Letting $y_k = \min_{j\in S_k} x_j$ for $k \in N^+$, $f(x)$ becomes

$$f^2(x,y) = \sum_{i=1}^n c_i x_i + \sum_{k\in N^+} q_k y_k + \sum_{k\in N^-} q_k \min_{j\in S_k} x_j.$$

It is clear that $f^2(x, y)$ is a convex function of (x, y) since $q_k \min_{j \in S_k} x_j$ is a convex function when $q_k < 0$. Define

$$(CP_2) \qquad \begin{aligned} &\max\ f^2(x, y) \\ &\text{s.t. } y_k \leq x_j, \quad j \in S_k,\ k \in N^+, \\ &\qquad x \in [0, 1]^n. \end{aligned}$$

Then, (CP_2) is a linearly constrained nonsmooth convex maximization problem and is equivalent to (0-1UPP). Problem (CP_2), however, introduces additional variables and constraints which, in cases with a large number of terms with positive coefficients, may make the problem itself difficult to solve. An alternative way to replace (10.4.2) is to use the relationship

$$\prod_{j \in S_k} x_j = \max\{0, \sum_{j \in S_k} x_j - s_k + 1\},\ k \in N^+,$$
$$y_k = \min_{j \in S_k} x_j,\ k \in N^-,$$

where $s_k = |S_k|$. Then $f(x)$ becomes

$$f^3(x) = \sum_{i=1}^n c_i x_i + \sum_{k \in N^+} q_k \max\{0, \sum_{j \in S_k} x_j - s_k + 1\} + \sum_{k \in N^-} q_k \min_{j \in S_k} x_j.$$

Define

$$(CP_3) \qquad \max_{x \in [0,1]^n} f^3(x).$$

Since $f^3(x)$ is a convex function of x, problem (CP_3) is also a convex maximization problem and is equivalent to (0-1UPP).

Denote by $S^*(\cdot)$ the set of integer local optimal solutions to problem $(\cdot)$. We have the following result.

THEOREM 10.7 ([160]) *It holds* $S^*(CP_2) \subset S^*(CP_3) \subset S^*(CP_1)$.

Examples can be easily constructed to show that the above set inclusions can be strict (see [160]). From a computational point of view, the reformulation (CP_2) is the best in terms of the number of local integer optimal solutions.

10.5 Unconstrained Quadratic 0-1 Optimization

As a special class of unconstrained polynomial 0-1 problems, the unconstrained quadratic 0-1 optimization problem is of the following form:

$$(0\text{-}1UQP) \qquad \max_{\{0,1\}^n} Q(x) = \sum_{i=1}^n c_i x_i + \sum_{1 \leq i < j \leq n} q_{ij} x_i x_j.$$

Note here that $x_j = x_j^2$. We can express $Q(x)$ in a compact form: $Q(x) = x^T Qx$, where $Q = (a_{ij})_{n\times n}$ with $a_{ii} = c_i$ for each i, $a_{ij} = \frac{1}{2}q_{ij}$ for $i < j$, and $a_{ij} = a_{ji}$ for $i > j$.

Applications of (0-1UQP) include financial analysis [156], molecular conformation problem [177] and cellular radio channel assignment [40]. It is well-known that (0-1UQP) is an NP-hard problem. Exact solution methods for solving (0-1UQP) include branch-and-bound algorithms based on different bounding approaches and preprocessing [24][175], linear programming and cutting plane generation techniques [12][99][172], and concave minimization method [110].

10.5.1 A polynomially solvable case

If $q_{ij} \geq 0$ for $1 \leq i < j \leq n$, then (0-1UQP) can be reduced to a linear programming problem and thus can be solved in polynomial time. Let $z_{ij} = x_i x_j = \min(x_i, x_j)$. Then (0-1$UQP$) is equivalent to the following linear integer programming problem:

$$\max \sum_{i=1}^{n} c_i x_i + \sum_{1\leq i<j\leq n} q_{ij} z_{ij} \tag{10.5.1}$$

$$\text{s.t. } z_{ij} \leq x_i,\ 1 \leq i < j \leq n, \tag{10.5.2}$$

$$z_{ij} \leq x_j,\ 1 \leq i < j \leq n, \tag{10.5.3}$$

$$x_i,\ x_j,\ z_{ij} \in \{0, 1\},\ 1 \leq i < j \leq n. \tag{10.5.4}$$

Consider the linear programming relaxation of the above problem by replacing constraint (10.5.4) with

$$x_i,\ x_j,\ z_{ij} \in [0, 1],\ 1 \leq i < j \leq n. \tag{10.5.5}$$

It can be verified that the constraint matrix corresponding to (10.5.2), (10.5.3) and (10.5.5) is totally unimodular. Thus, the linear program has an integer optimal solution which also solves linear integer programming problem (10.5.1)–(10.5.4). Therefore, quadratic 0-1 program with nonnegative coefficients for all the cross terms can be solved in polynomial time.

It has been shown in [178] that problem (0-1UQP) with $q_{ij} \geq 0$ for $1 \leq i < j \leq n$ can be reduced to a minimum-cut in a graph with positive arc capacities, which is polynomially solvable. We will discuss this reduction in details in the next chapter in the context of the Lagrangian relaxation for quadratic knapsack problems.

10.5.2 Equivalence to maximum-cut problem

For any $x_i \in \{0, 1\}$, let $s_i = 2x_i - 1$. Then $s_i \in \{+1, -1\}$. Function $Q(x)$ can be rewritten as

$$\begin{aligned} P(s) &= \sum_{1 \le i < j \le n} \frac{1}{4} q_{ij} s_i s_j + \sum_{1 \le i < j \le n} \frac{1}{4} q_{ij} s_i + \sum_{1 \le i < j \le n} \frac{1}{4} q_{ij} s_j \\ &\qquad + \sum_{i=1}^{n} \frac{1}{2} c_i s_i + C_1 \\ &= \sum_{1 \le i < j \le n} \frac{1}{4} q_{ij} s_i s_j + \sum_{i=1}^{n} [\frac{1}{4} (\sum_{j=1}^{i-1} q_{ji} + \sum_{j=i+1}^{n} q_{ij}) + \frac{1}{2} c_i] s_i + C_1, \end{aligned}$$

where $C_1 = \frac{1}{4} \sum_{1 \le i < j \le n} q_{ij} + \frac{1}{2} \sum_{i=1}^{n} c_i$ is a constant. Let $w_{ij} = -\frac{1}{4} q_{ij}$ for $1 \le i < j \le n$ and

$$w_{0i} = -\frac{1}{4} (\sum_{j=1}^{i-1} q_{ji} + \sum_{j=i+1}^{n} q_{ij}) - \frac{1}{2} c_i$$

for $i = 1, \ldots, n$. Then, we have

$$P(s) = \sum_{0 \le i < j \le n} (-w_{ij}) s_i s_j + C_1,$$

where $s_0 \equiv 1$.

Now, define a graph $G = (E, V)$ with vertex set $V = \{0, 1, \ldots, n\}$ and edge set $E = \{ij \mid 1 \le i < j \le n\}$. The weight w_{ij} is associated to edge $ij \in E$. Each $s \in \{+1, -1\}^{n+1}$ corresponds to a partition of V into $V^+ = \{i \in V \mid s_i = +1\}$ and $V^- = \{i \in V \mid s_i = -1\}$. The set of edges $\delta(W) = \{ij \in E \mid i \in W, \ j \in E \setminus W\}$ is called a *cut* of G. The function value $P(s)$ can be expressed in terms of V^+ and V^- as

$$\begin{aligned} P(s) &= \sum_{i,j \in V^+} (-w_{ij}) + \sum_{i,j \in V^-} (-w_{ij}) + \sum_{ij \in \delta(V^+)} w_{ij} + C_1 \\ &= 2 \sum_{ij \in \delta(V^+)} w_{ij} + C_1 + C_2, \end{aligned}$$

where $C_2 = \sum_{ij \in E} (-w_{ij})$ is a constant. Therefore, problem (0-1UQP) is equivalent to the following maximum-cut problem:

$$(MC) \qquad \max_{W \subseteq V} \sum_{ij \in \delta(W)} w_{ij}.$$

A cut (V^+, V^-) in G is linked to x in problem (0-1UQP) via

$$x_i = \begin{cases} 1, & \text{if } i \in V^+, \\ 0, & \text{if } i \in V^-. \end{cases}$$

Therefore, algorithms for maximum-cut problems can be applied to (0-1UQP) via solving problem (MC). Barahona et al [12] proposed a branch-and-cut algorithm based on solving problem (MC) using cutting planes derived for the maximum-cut problem (see also [172] [199]).

10.5.3 Variable fixation

For convenience, assume that $q_{ij} = q_{ji}$ for $i > j$ in problem (0-1UQP).

LEMMA 10.1 *Let x^* denote the optimal solution of (0-1UQP). Let*

$$a_i := c_i + \sum_{j \neq i} \min(0, q_{ij}), \tag{10.5.6}$$

$$b_i := c_i + \sum_{j \neq i} \max(0, q_{ij}). \tag{10.5.7}$$

Then,
(i) $x_i^* = 1$, *if* $a_i \geq 0$;
(ii) $x_i^* = 0$, *if* $b_i < 0$.

Proof. (i) Notice that the term containing x_i in $Q(x)$ is $(c_i + \sum_{j \neq i} q_{ij} x_j) x_i$. Since

$$c_i + \sum_{j \neq i} q_{ij} x_j \geq a_i \geq 0$$

for any $x \in \{0, 1\}^n$, x_i must take 1 in the optimal solution x^*. Part (ii) can be proved in a similar way. □

It is easy to see that a_i and b_i define the range of the gradient of $Q(x)$ over $[0, 1]^n$, i.e.,

$$a_i \leq \frac{\partial Q(x)}{\partial x_i} \leq b_i, \quad i = 1, \ldots, n.$$

Thus, Lemma 10.1 can be interpreted as the fixation of a variable to 0 or 1 if the corresponding partial derivative does not change sign over $[0, 1]^n$. The lower bound a_i and the upper bound b_i can be further tightened after a variable is fixed. Let's consider an example to show how to exploit this property.

EXAMPLE 10.2 Consider the following example:

$$\max_{x \in \{0,1\}^3} Q(x) = 2x_1 + 3x_2 + 6x_3 - 2x_1x_2 - x_1x_3 - 4x_2x_3.$$

We have $a = (-1, -3, 1)^T$, $b = (2, 3, 6)^T$. Since $a_3 > 0$, we can fix $x_3 = 1$. Substituting $x_3 = 1$ in $Q(x)$, we obtain $Q_1(x_1, x_2) = 6 + x_1 - x_2 - 2x_1x_2$ for which $a_{1,2} = (-1, -3)^T$ and $b_{1,2} = (1, -1)^T$. Since $b_2 < 0$, we can fix $x_2 = 0$. Substituting $x_2 = 0$ in $Q_1(x_1, x_2)$, we obtain $Q_2(x_1) = 6 + x_1$. Obviously, x_1 can be fixed to 1 in $Q_2(x_1)$. Therefore, all the variables are fixed and the optimal solution to the example is $x^* = (1, 0, 1)^T$ with $Q(x^*) = 7$.

Note that Lemma 10.1 does not always guarantee a predetermination of variables, as shown in another example where $Q(x) = 2x_1 + 3x_2 - 4x_1x_2$. For this quadratic function, since $a = (-2, -1)^T$ and $b = (2, 3)^T$, no variable can be fixed.

The above variable fixation can be integrated into a branch-and-bound method using the standard depth-first binary search. A node in the binary tree can be represented by (lev, a, b, UB), where lev denotes the number of levels in the binary tree, a and b the gradient bounds of Q on the free variables, respectively, and UB the upper bound. We also denote by p_i the index of the ith fixed variable in the algorithm.

ALGORITHM 10.2 (BRANCH-AND-BOUND METHOD FOR (0-1UQP))

Step 0 (Initialization). Choose an initial feasible solution x by some heuristic method. Set the incumbent $x_{opt} = x$ and the lower bound $f_{opt} = Q(x)$. Compute an initial upper bound of $Q(x)$ over $\{0, 1\}^n$:

$$UB = \sum_{i=1}^{n} \max(0, c_i) + \sum_{1 \le i < j \le n} \max(0, q_{ij}).$$

Set $lev = 0$, $I_{fix} = \emptyset$, $I_{free} = \{1, 2, \ldots, n\}$, $L = \emptyset$.

Step 1 (Gradient bounds). For each $i = 1, \ldots, n$, compute a_i and b_i by equations (10.5.6) and (10.5.7), respectively.

Step 2 (Variable fixation). If $a_i \ge 0$ or $b_i < 0$ for some $i \in I_{free}$, then fix $x_i = 1$ if $a_i \ge 0$ or fix $x_i = 0$ if $b_i < 0$, set $lev := lev + 1$, $p_{lev} = i$, $I_{fix} := I_{fix} \cup \{i\}$, $I_{free} := I_{free} \setminus \{i\}$. Update the upper bound UB after fixing x_i. Repeat Step 2 until there is no $i \in I_{free}$ such that $a_i \ge 0$ or $b_i < 0$. Go to Step 4.

Step 3 (Branching). If no variable is fixed at Step 2, then choose j such that

$$j = \arg \max_{i \in I_{free}} \min(-a_i, b_i).$$

Let UB^0 and UB^1 be the updated upper bounds by letting $x_j = 0$ and $x_j = 1$, respectively. If $UB^0 > UB^1$, then set $x_j = 0$ and $UB = UB^0$; Otherwise, set $x_j = 1$ and $UB = UB^1$. If $\min(UB^0, UB^1) > f_{opt}$, save node $(lev+1, a, b, UB)$ to L. Set $lev := lev + 1$, $p_{lev} = j$, $I_{fix} := I_{fix} \cup \{j\}$, $I_{free} := I_{free} \setminus \{j\}$.

Step 4. If $UB > f_{opt}$ and $lev < n$, go to Step 5. Otherwise, the current node is fathomed. If $lev = n$, update the incumbent x_{opt} and the lower bound f_{opt} if $UB > f_{opt}$. If $L = \emptyset$, then stop, the incumbent x_{opt} is an optimal solution to (0-1UQP). Otherwise, select the last node in L. Set $x_{p_{lev}} := 1 - x_{p_{lev}}$. Update the upper bound UB of the selected node.

Step 5. Update the gradient bound a and b for free variables. For each $i \in I_{free}$:

(i) If $x_{p_{lev}} = 1$, then

$$a_i := a_i + \max(0, q_{i,p_{lev}}),$$
$$b_i := b_i + \min(0, q_{i,p_{lev}}).$$

(ii) If $x_{p_{lev}} = 0$, then

$$a_i := a_i - \min(0, q_{i,p_{lev}}),$$
$$b_i := b_i - \max(0, q_{i,p_{lev}}).$$

Go to Step 2.

The following greedy heuristic uses the gradient information of Q at the center point $x_c = (1/2, \ldots, 1/2)^T$ and the point $x_0 = (0, \ldots, 0)^T$ to search for a "good" point in $\{0,1\}^n$. For each i, we have

$$\frac{\partial Q}{\partial x_i}(x_c) = c_i + \frac{1}{2}\sum_{j \neq i} q_{ij},$$
$$\frac{\partial Q}{\partial x_i}(x_0) = c_i.$$

PROCEDURE 10.2 (HEURISTIC FOR (0-1UQP))

Step 1. (Initialization). Set $I = \{1, \ldots, n\}$. Set $v_i = c_i$ and calculate $w_i = \sum_{j \neq i} q_{ij}$ for $i = 1, \ldots, n$.

Step 2. (Variable selection). For each i, let $u_i = v_i + \frac{1}{2}w_i$. Set $I^+ = \{i \mid u_i > 0,\ i \in I\}$ and $I^0 = \{i \mid u_i = 0,\ v_i \geq 0,\ i \in I\}$. If $I^+ \neq \emptyset$, choose j such that $u_j = \max\{u_i \mid i \in I^+\}$ and set $x_j = 1$; If $I^+ = \emptyset$ and $I^0 \neq \emptyset$, then choose j such that $v_j = \max\{v_i \mid i \in I^0\}$ and set $x_j = 1$; Otherwise choose any $j \in I$ and set $x_j = 0$.

Step 3. (Updating). Set $I := I \setminus \{j\}$. If $I = \emptyset$, then stop and x is a solution. Otherwise, for each $i \in I$, set $w_i := w_i - q_{ij}$ for $i < j$ and $w_i := w_i - q_{ji}$

for $i > j$; If $x_j = 1$, then set $v_i := v_i + q_{ij}$ for $i < j$ and $v_i := v_i + q_{ji}$ for $i > j$; Return to Step 2.

The following example is used to illustrate the above algorithm and heuristics.

EXAMPLE 10.3 Consider the following example:

$$\max_{x\in\{0,1\}^6} Q(x) = x_1 + 3x_2 - 2x_3 + 4x_4 + 2x_5 + x_6 - 8x_1x_2 - 3x_1x_4$$
$$+3x_2x_3 - 4x_3x_5 - 2x_4x_6.$$

Procedure 10.2 starts by calculating $u = (-4.5, 0.5, -2.5, 1.5, 0, 0)$, $v = (1, 3, -2, 4, 2, 1)$. We have $I^+ = \{2, 4\}$. Then, set $x_4 = 1$ and update $u_{\{1,2,3,5,6\}} = (-6, 0.5, -2.5, 0, -1)$, $v_{\{1,2,3,5,6\}} = (-2, 3, -2, 2, -1)$, and $I^+ = \{2\}$. Set $x_2 = 1$ and update $u_{\{1,3,5,6\}} = (-10, -1, 0, -1)$, $v_{\{1,3,5,6\}} = (-10, 1, 2, -1)$. We have $I^+ = \emptyset$, $I^0 = \{5\}$. So, we set $x_5 = 1$ and update $u_{\{1,3,6\}} = (-10, -3, -1)$ and $v_{\{1,3,6\}} = (-10, -3, -1)$. We have $I^+ = I^0 = \emptyset$. Set $x_1 = 0$ and update $u_{\{3,6\}} = (-3, -1)$, $v_{\{3,6\}} = (-3, -1)$. Again, $I^+ = I^0 = \emptyset$ and we set $x_3 = x_6 = 0$. Finally, we have a feasible solution $x = (0, 1, 0, 1, 1, 0)^T$.

The iteration process of Algorithm 10.2 can be described as follows:

Step 0. Apply Procedure 10.2 to find an incumbent $x_{opt} = (0, 1, 0, 1, 1, 0)^T$ with a lower bound $f_{opt} = Q(x_{opt}) = 9$. Compute an upper bound $UB = 14$. Set $lev = 0$, $I_{fix} = \emptyset$, $I_{free} = \{1, 2, 3, 4, 5, 6\}$, $L = \emptyset$.

Step 1. Compute the gradient bounds:

$$a = (-10, -5, -6, -1, -2, -1), \; b = (1, 6, 1, 4, 2, 1).$$

Step 2. No variable can be fixed.

Step 3. $-a_2 = 5 = \max_{i\in I_{free}} \min(-a_i, b_i)$. Setting $x_2 = 0$ and 1, respectively, yields the corresponding upper bounds $UB^0 = 8$ and $UB^1 = 11$. Set $UB = 11$. Set $x_2 = 1$, $lev = 1$, $p_1 = 2$, $I_{fix} = \{2\}$ and $I_{free} = \{1, 3, 4, 5, 6\}$.

Step 4. $UB > 9 = f_{opt}$, $lev < 6$.

Step 5. Update the gradient bounds: $a_{1,3,4,5,6} = (-10, -3, -1, -2, -1)$, $b_{1,3,4,5,6} = (-7, 1, 4, 2, 1)$.

Step 2. Since $b_1 = -7 < 0$, set $x_1 = 0$, $lev = 2$, $p_2 = 1$, $I_{fix} = \{2, 1\}$ and $I_{free} = \{3, 4, 5, 6\}$. Update the upper bound $UB = 11$.

Step 4. $UB > 9 = f_{opt}$, $lev < 6$.

Step 5. Update the gradient bounds: $a_{3,4,5,6} = (-3, 2, -2, -1)$, $b_{3,4,5,6} = (1, 4, 2, 1)$.

Step 2. Since $a_4 = 2 > 0$, fix $x_4 = 1$. Set $p_3 = 4$, $lev = 3$, $I_{fix} = \{2, 1, 4\}$ and $I_{free} = \{3, 5, 6\}$. Update the upper bound $UB = 10$.

Step 4. $UB > 9 = f_{opt}$, $lev < 6$.

Step 5. Update the gradient bounds: $a_{3,5,6} = (-3,-2,-1)$, $b_{3,5,6} = (1,2,-1)$.

Step 2. Since $b_6 = -1 < 0$, fix $x_6 = 0$. Set $p_4 = 6$, $lev = 4$, $I_{fix} = \{2,1,4,6\}$ and $I_{free} = \{3,5\}$. Update the upper bound $UB = 10$.

Step 4. $UB > 9 = f_{opt}$, $lev < 6$.

Step 5. Update the gradient bounds: $a_{3,5} = (-3,-2)$, $b_{3,5} = (1,2)$.

Step 2. No variable can be fixed.

Step 3. $-a_5 = 2 = \max_{i=3,5} \min(-a_i, b_i)$. Setting $x_5 = 0$ and 1, respectively, the corresponding upper bounds are $UB^0 = 8$ and $UB^1 = 9$. Set $UB = 9$. Set $x_5 = 1$, $lev = 5$, $p_5 = 5$, $I_{fix} = \{2,1,4,6,5\}$, $I_{free} = \{3\}$.

Step 4. $UB = f_{opt}$, the current node is fathomed. Since $L = \emptyset$, the algorithm stops and the incumbent solution $x_{opt} = (0,1,0,1,1,0)^T$ is the optimal solution.

10.6 Notes

More materials about the theory of nonlinear 0-1 programming or pseudo-Boolean optimization can be found in [31][95][92].

The concept of roof duality was first introduced by Hammer, Hansen and Simeone in their pioneering paper [90] for unconstrained quadratic 0-1 optimization. Roof duality theory was later extended to polynomial 0-1 programming in [146] and its relations to other linearization approaches and Lagrangian duality were discussed in [1][96].

The basic algorithm was presented in [92] and was investigated in [31][46]. The relationship between the problems of maximizing a multilinear function on $[0,1]^n$ and $\{0,1\}^n$ was established in [184]. The concave minimization formulation for unconstrained polynomial 0-1 optimization was presented in [160]. The equivalence between the unconstrained quadratic optimization and the maximum-cut problem was shown in Hammer [88]. The branch-and-bound method based on variable fixation for unconstrained quadratic 0-1 optimization problems was proposed in [175].

Chapter 11

CONSTRAINED POLYNOMIAL 0-1 PROGRAMMING

In this chapter, we consider the following constrained polynomial 0-1 programming problem:

$$
\begin{aligned}
(0\text{-}1PP) \qquad & \max\ f(x) = \sum_{k=1}^{p} c_k \prod_{j \in S_k} x_j \\
& \text{s.t. } g_i(x) = \sum_{k=1}^{p_i} a_{ik} \prod_{j \in S_{ik}} x_j \le b_i,\ i = 1, \ldots, m, \\
& x \in X = \{0,1\}^n,
\end{aligned}
$$

where S_k and S_{ik} are subsets of $\{1, \ldots, n\}$.

This chapter is organized as follows. We discuss in Section 11.1 how to convert problem $(0\text{-}1PP)$ into an unconstrained polynomial 0-1 programming problem by an exact penalty method. We then explore in Section 11.2 how to transform problem $(0\text{-}1PP)$ into an equivalent 0-1 linear programming problem. In Section 11.3, we investigate how to improve the upper bound of $(0\text{-}1PP)$ under a branch-and-bound framework. In Section 11.4, we study cutting plane methods to replace the nonlinear constraints by linear constraints without introducing additional variables and constraints. Finally, we examine in Section 11.5 quadratic 0-1 knapsack problems in details.

11.1 Reduction to Unconstrained Problem

We can apply the results on exact penalty functions in Section 2.5 to convert constrained polynomial 0-1 programming problem $(0\text{-}1PP)$ into a corresponding unconstrained polynomial 0-1 programming problem.

We assume in this section that all a_{ik}'s and b_i's in $(0\text{-}1PP)$ are integers. We convert the inequality constraints in $(0\text{-}1PP)$ into equality constraints via introducing slack variables. Let $s_i = b_i - g_i(x)$. Then $g_i(x) \leq b_i$ is equivalent to $s_i \geq 0$. Note all s_i's are also integers. Let $\underline{g}_i = \min_{x \in X} g_i(x)$, where $X = \{0,1\}^n$. Since $s_i \leq b_i - \underline{g}_i$, we can express s_i as $s_i = \sum_{j=1}^{q_i} y_{ij} 2^{j-1}$, where $q_i = \lfloor \log_2(b_i - \underline{g}_i) \rfloor + 1$, $y_{ij} \in \{0,1\}, i = 1, \ldots, m, j = 1, \ldots, q_i$. The inequality constraint $g_i(x) \leq b_i$ is equivalent to $g_i(x) + s_i = b_i$ or

$$G_i(x, y_i) := g_i(x) + \sum_{j=1}^{q_i} y_{ij} 2^{j-1} - b_i = 0. \tag{11.1.1}$$

Applying Corollary 2.2 leads to the following result.

COROLLARY 11.1 *Suppose that $g_i(x)$ $(i = 1, \ldots, m)$ are integer-valued in problem (0-1PP). Let $c^+ = \sum_{j=1}^{p} \max(0, c_j)$ and $c^- = \sum_{j=1}^{p} \min(c_j, 0)$. Then, for any $\mu \geq \mu_0 = c^+ - c^- + 1$, any solution x^* that solves*

$$\begin{aligned} \max\ & f(x) - \mu \sum_{i=1}^{m} [G_i(x, y_i)]^2 \\ \text{s.t. }\ & x \in \{0,1\}^n, \\ & y_i = (y_{i,1}, \ldots, y_{i,q_i}), \\ & y_{i,j} \in \{0,1\},\ j = 1, \ldots, q_i,\ i = 1, \ldots, m, \end{aligned} \tag{11.1.2}$$

also solves (0-1PP), where $G_i(x, y_i)$ is defined as in (11.1.1).

EXAMPLE 11.1

$$\begin{aligned} \max\ & f(x) = -2x_1x_2 + x_1x_3 - 3x_2x_3 + 4x_1x_2x_3 \\ \text{s.t. }\ & g_1(x) = x_1 + 2x_1x_2 \leq 2, \\ & g_2(x) = 2x_3 - x_1x_3 \leq 1, \\ & x \in \{0,1\}^3. \end{aligned}$$

The optimal solution of this problem is $x^* = (1,0,1)^T$ with $f(x^*) = 1$. Since $\underline{g}_1 = 0$ and $\underline{g}_2 = -1$, so $q_1 = \lfloor \log_2(2-0) \rfloor + 1 = 2$, $q_2 = \lfloor \log_2[0-(-1)] \rfloor + 1 = 1$ and hence

$$\begin{aligned} G_1(x, y_1) &= x_1 + 2x_1x_2 + y_{11} + 2y_{12} - 2 = 0, \\ G_2(x, y_2) &= 2x_3 - x_1x_3 + y_{21} - 1 = 0. \end{aligned}$$

By Corollary 11.1, $\mu_0 = 5 - (-5) + 1 = 11$. Taking $\mu = 11$, we obtain the following exact penalty problem:

$$\begin{aligned} \max\ T(x, y; 11) = & -2x_1x_2 + x_1x_3 - 3x_2x_3 + 4x_1x_2x_3 \\ & - 11[(x_1 + 2x_1x_2 + y_{11} + 2y_{12} - 2)^2 \\ & + (2x_3 - x_1x_3 + y_{21} - 1)^2] \\ \text{s.t. }\ x \in \{0,1\}^3,\ & y_{11},\ y_{12},\ y_{21} \in \{0,1\}. \end{aligned}$$

The optimal solution of the above unconstrained polynomial problem is $(x^*, y^*) = (1, 0, 1, 1, 0, 0)^T$ with $T(x^*, y^*, 11) = 1$. Thus, solving the exact penalty problem yields an optimal solution $x^* = (1, 0, 1)^T$ of the original problem.

11.2 Linearization Methods

Consider a general polynomial term:

$$y = \prod_{j \in S} x_j, \tag{11.2.1}$$

where $S \subseteq \{1, \ldots, n\}$ and $x_j \in \{0, 1\}, j \in S$. Since x_j's are binary variables, y is also a binary variable. The following theorem shows that the nonlinear equation (11.2.1) is equivalent to two linear inequalities.

THEOREM 11.1 *Let $s = |S|$. Equation (11.2.1) holds if and only if*

$$\sum_{j \in S} x_j - y \leq s - 1, \tag{11.2.2}$$

$$-\sum_{j \in S} x_j + sy \leq 0, \tag{11.2.3}$$

$$x_j \in \{0, 1\},\ j \in S,\ y \in \{0, 1\}. \tag{11.2.4}$$

Proof. If any $x_j = 0$ then $y = 0$. In such a case, (11.2.2) is redundant and (11.2.3) becomes $y \leq \sum_{j \in S} x_j / s < 1$ which implies $y = 0$ by (11.2.4). If all $x_j = 1$, then $y = 1$. In this situation, (11.2.2) becomes $y \geq 1$ which implies $y = 1$ by (11.2.4), and (11.2.3) is redundant. □

Now consider problem (0-1PP). Let

$$y_k = \prod_{j \in S_k} x_j,\ k = 1, \ldots, p, \tag{11.2.5}$$

$$y_{ik} = \prod_{j \in S_{ik}} x_j,\ k = 1, \ldots, p_i,\ i = 1, \ldots, m. \tag{11.2.6}$$

Substituting (11.2.5) and (11.2.6) into (0-1PP) and adding constraints as defined in (11.2.2)–(11.2.4) lead to a 0-1 linear programming:

$$
\begin{aligned}
(0\text{-}1LP) \qquad & \max \sum_{k=1}^{p} c_k y_k \\
& \text{s.t.} \sum_{k=1}^{p_i} a_{ik} y_{ik} \le b_i, \ i = 1, \ldots, m, \\
& \sum_{j \in S_k} x_j - y_k \le s_k - 1, \ k = 1, \ldots, p, \\
& - \sum_{j \in S_k} x_j + s_k y_k \le 0, \ k = 1, \ldots, p, \\
& \sum_{j \in S_{ik}} x_j - y_{ik} \le s_{ik} - 1, \ k = 1, \ldots, p_i, \ i = 1, \ldots, m, \\
& - \sum_{j \in S_{ik}} x_j + s_{ik} y_{ik} \le 0, \ \ k = 1, \ldots, p_i, \ i = 1, \ldots, m, \\
& x \in \{0,1\}^n, \ y_k \in \{0,1\}, \ k = 1, \ldots, p, \\
& y_{ik} \in \{0,1\}, \ k = 1, \ldots, p_i, \ i = 1, \ldots, m,
\end{aligned}
$$

where $s_k = |S_k|$, $s_{ik} = |S_{ik}|$.

Consider Example 11.1 again. Let $y_1 = x_1 x_2$, $y_2 = x_1 x_3$, $y_3 = x_2 x_3$, $y_4 = x_1 x_2 x_3$. Then, by (11.2.2)–(11.2.4), we have the following equivalent 0-1 linear program:

$$
\begin{aligned}
\max \ & -2y_1 + y_2 - 3y_3 + 4y_4 \\
\text{s.t.} \ & x_1 + 2y_1 \le 2, \\
& 2x_3 - y_2 \le 1, \\
& x_1 + x_2 - y_1 \le 1, \\
& -x_1 - x_2 + 2y_1 \le 0, \\
& x_1 + x_3 - y_2 \le 1, \\
& -x_1 - x_3 + 2y_2 \le 0, \\
& x_2 + x_3 - y_3 \le 1, \\
& -x_2 - x_3 + 2y_3 \le 0, \\
& x_1 + x_2 + x_3 - y_4 \le 2, \\
& -x_1 - x_2 - x_3 + 3y_4 \le 0, \\
& x \in \{0,1\}^3, y \in \{0,1\}^4.
\end{aligned}
$$

As we can see from the example, for each nonlinear term, the linearization method introduces one new 0-1 variable and two inequality constraints. For a problem with large number of nonlinear terms, the linearized problem may become prohibitive due to a huge number of additionally introduced variables and constraints.

11.3 Branch-and-Bound Method

11.3.1 Upper bounds and penalties

Upper bounds for polynomial functions can be utilized in a branch-and-bound method for solving (0-1PP). Obviously, the simplest upper bound for the objective function, $f(x) = \sum_{k=1}^{p} c_k \prod_{j \in S_k} x_j$, is

$$\bar{z}_1 = \sum_{k=1}^{p} \max(0, c_k).$$

A *penalty* is defined as an increment p_j^0 or p_j^1 which may be subtracted from an upper bound when fixing a free variable x_j at 0 or 1, respectively. Penalties can be used to lower upper bounds or even be used to fix some free variables. When a free variable x_j is fixed at 0, all terms containing x_j vanish. An increment $\sum_{k \in T^{-1}(j)} \max(0, c_k)$ can be subtracted from $\bar{z}_1$, where $T^{-1}(j) = \{k \mid j \in S_k\}$. Consider now another case when a free variable x_j is fixed at 1. If there exists k such that $|S_k| = 1$ and $S_k = \{j\}$, then we can subtract an increment $\max(0, c_k) - c_k = -\min(0, c_k)$ from $\bar{z}_1$. If $f(x)$ contains term: $c_k x_l + c_{k'} x_l x_j$ with $c_k c_{k'} < 0$, then fixing x_j at 1 yields a reduced term $(c_k + c_{k'})x_l$ and hence

$$\max(0, c_k) + \max(0, c_{k'}) - \max(0, c_k + c_{k'}) = \min_{c_k c_{k'} < 0}(|c_k|, |c_{k'}|)$$

can be subtracted from $\bar{z}_1$.

Define the following for $j \in \{1, 2, \ldots, n\}$:

$$p_j^0 = \sum_{k \in T^{-1}(j)} \max(0, c_k),$$

$$p_j^1 = -\min_{S_k = \{j\}}(0, c_k) + \sum_{c_k c_{k'} < 0} \min(|c_k|, |c_{k'}|),$$

where (k, k') is such that $S_k = \{l\}$ and $S_{k'} = \{l, j\}$ for some $l \in \{1, 2, \ldots, n\}$. An improved upper bound can be obtained by considering a fixation of variables.

PROPOSITION 11.1 *$\bar{z}_2 = \bar{z}_1 - \max_{j=1,\ldots,n}[\min(p_j^0, p_j^1)]$ is an improved upper bound of $f(x)$.*

More sophisticated upper bounds can be derived by using different types of additive penalties (see [95]). We point out that roof dual can be used to derive upper bounds for a polynomial function.

11.3.2 Branch-and-bound method

The branch-and-bound method for constrained 0-1 programming is based on the following three main steps: (i) Computing upper bounds of the objective function $f(x)$ or lower bound of the constraint functions; (ii) Computing penalties associated with the upper bound of the current subproblem. The penalties can be used to improved the upper bound if a variable is fixed at 0 or 1 and to fathom the subproblems; (iii) Standard binary search or its variants can be used to branch a subproblem into two subproblems with $x_j = 0$ and $x_j = 1$, respectively.

ALGORITHM 11.1 (BRANCH-AND-BOUND METHOD FOR $(0\text{-}1PP)$)

Step 1 (Initialization). Compute a feasible solution x to (0-1PP) by certain heuristics and set the incumbent $x_{opt} = x$. Set $f_{opt} = f(x_{opt})$.

Step 2 (Upper bound). Compute an upper bound $\bar{f}$ of the current subproblem (node). If $\bar{f} \leq f_{opt}$, then the current subproblem is fathomed and go to Step 3. Otherwise, go to Step 4.

Step 3 (Backtracking). If all variables have been fixed, stop and the incumbent x_{opt} is the optimal solution. Otherwise, select a subproblem with a free variable by certain backtracking rule and return to Step 2.

Step 4 (Lower bound). If a better feasible solution $\tilde{x}$ can be found during the bounding procedure, then update x_{opt} and $f_{opt} = f(\tilde{x})$.

Step 5 (Feasibility check). Compute a lower bound $\bar{g}_i$ of the constraint function $g_i(x)$ in the current subproblem. If $\bar{g}_i > b_i$ for some i, go to Step 3.

Step 6 (Variable fixation for objective). For each unfixed variable x_j in the current subproblem, compute penalties p_j^0 and p_j^1 associated with the upper bound $\bar{f}$. If $\bar{f} - p_j^0 \leq f_{opt}$, set $x_j = 1$; if $\bar{f} - p_j^1 \leq f_{opt}$, set $x_i = 0$. If at least one variable can be fixed, return to Step 2.

Step 7 (Variable fixation for constraints). For each constraint g_i and each unfixed variable x_j, compute penalties p_{ij}^0 and p_{ij}^1 associated with the lower bound $\bar{g}_i$. If $\bar{g}_i + p_{ij}^0 > b_i$, set $x_i = 1$; if $\bar{g}_i + p_{ij}^1 > b_i$, set $x_i = 0$. If at least one variable is fixed, return to Step 2.

Step 8 (Branching). Generate two new subproblems by setting an unfixed variable $x_j = 0$ and $x_j = 1$, respectively. Choose one of the two subproblems to be explored first. Return to Step 2.

Two typical backtracking strategies can be adopted in Step 3: the depth first rule and the best first rule. In the depth first rule, the last generated subproblem is chosen. In the best first rule, the subproblem with the maximum upper bound is selected.

11.4 Cutting Plane Methods

Consider the 0-1 constrained polynomial programming with a linear objective function:

$$(0\text{-}1PP_1) \qquad \max\ f(x) = \sum_{k=1}^{n} c_k x_k$$
$$\text{s.t. } g_i(x) = \sum_{k=1}^{p_i} a_{ik} \prod_{j \in S_{ik}} x_j \le b_i,\ i = 1, \ldots, m,$$
$$x \in X = \{0, 1\}^n,$$

where S_{ik} are subsets of $\{1, \ldots, n\}$. Notice that a nonlinear objective function can be always reduced to a linear function after introducing new 0-1 variables and new constraints as discussed in Section 11.2.

The main idea of the cutting plane method is to replace the nonlinear constraints by linear constraints without introducing additional variables and constraints. The resulting problem is a generalized set covering problem which can be solved by a 0-1 linear programming algorithm.

11.4.1 Generalized covering relaxation

Consider now a general polynomial constraint function:

$$g(x) = \sum_{k \in N} a_k \prod_{j \in S_k} x_j \le b, \tag{11.4.1}$$

where N and S_k are nonempty index sets, and $\cup_{k \in N} S_k = \{1, \ldots, n\}$.

Denote $N^+ = \{k \in N \mid a_k > 0\}$, $N^- = \{k \in N \mid a_k < 0\}$. In the sequel, we assume that $\sum_{k \in N^+} a_k > b$; Otherwise, $g(x) \le b$ holds for any $\{0, 1\}^n$. Define

$$g^+(x) = \sum_{k \in N^+} a_k (\prod_{j \in S_k} x_j),$$
$$g^-(x) = \sum_{k \in N^-} a_k (\prod_{j \in S_k} x_j).$$

A set $M \subseteq N$ is said to be a *cover* for the inequality (11.4.1) if

$$\sum_{k \in M} |a_k| > b - \sum_{k \in N^-} a_k. \tag{11.4.2}$$

It is easy to see that N is a cover for (11.4.1) since $\sum_{k \in N^+} a_k > b$ from the assumption. A cover M is said to be *minimal* if no strict subset of it is a cover. If $M \subseteq N^+$, then M is a cover of $g^+(x) \le b$ if and only if $\sum_{k \in M} a_k > b$.

Let φ be a mapping that associates an index $j \in S_k$ with each $k \in N^-$. Let Φ^- denote the set of all such mappings. For any $M \subseteq N$, let $S_M = \cup_{k \in M \cap N^+} S_k$, and $S_\varphi = \{j = \varphi(k) \mid k \in M \cap N^-\}$. For $x \in \{0,1\}$, denote by $\bar{x}_j$ the *complement* of x_j, $\bar{x}_j = 1 - x_j$.

THEOREM 11.2 *If (11.4.1) is satisfied, then*

$$\sum_{j \in S_M} \bar{x}_j + \sum_{j \in S_\varphi} x_j \geq 1 \tag{11.4.3}$$

for any cover $M \subseteq N$ and $\varphi \in \Phi^-$.

Proof. Since $\prod_{j \in S_k} x_j \leq x_{\varphi(k)}$ for any $k \in N^-$, (11.4.1) implies

$$b \geq g(x) = g^+(x) + g^-(x) \geq g^+(x) + \sum_{k \in N^-} a_k x_{\varphi(k)}.$$

Notice that $-a_k = |a_k|$ for $k \in N^-$. Thus

$$g^+(x) + \sum_{k \in N^-} |a_k| \bar{x}_{\varphi(k)} \leq b + \sum_{k \in N^-} |a_k|. \tag{11.4.4}$$

Let $y_k = \prod_{j \in S_k} x_j$ for $k \in N^+$ and $y_k = \bar{x}_{\varphi(k)}$ for $k \in N^-$. For any cover $M \subseteq N$, if $y_k = 1$ for all $k \in M$, then, by (11.4.4),

$$\sum_{k \in M} |a_k| \leq \sum_{k \in N} |a_k| y_k \leq b + \sum_{k \in N^-} (-a_k),$$

which contradicts that M is a cover. Thus, $\prod_{k \in M} y_k = 0$, i.e.,

$$\prod_{k \in M \cap N^+} (\prod_{j \in S_k} x_j) \times \prod_{k \in M \cap N^-} \bar{x}_{\varphi(k)} = 0,$$

which is in turn equivalent to the following generalized covering constraint:

$$\sum_{j \in S_M} \bar{x}_j + \sum_{j \in S_\varphi} x_j \geq 1.$$

□

Suppose now that $\hat{x} \in \{0,1\}^n$ does not satisfy (11.4.1). Define the following index set:

$$G^1(\hat{x}) = \{k \in N^+ \mid \prod_{j \in S_k} \hat{x}_j = 1\},$$

$$G^0(\hat{x}) = \{k \in N^- \mid \prod_{j \in S_k} \hat{x}_j = 0\}.$$

Dropping all the terms in (11.4.1) with zero value at $\hat{x}$ and negative coefficient and letting $\varphi(k) \in S_k$ be such that $\hat{x}_{\varphi(k)} = 0$ for $k \in G^0(\hat{x})$, we have the following,

$$\begin{aligned} g(x) &\geq \sum_{k \in G^1(\hat{x})} a_k \prod_{j \in S_k} x_j + \sum_{k \in G^0(\hat{x})} a_k x_{\varphi(k)} + \sum_{k \in N^- \setminus G^0(\hat{x})} a_k \\ &= \sum_{k \in G^1(\hat{x})} a_k \prod_{j \in S_k} x_j + \sum_{k \in G^0(\hat{x})} (-a_k) \bar{x}_{\varphi(k)} + \sum_{k \in N^-} a_k. \end{aligned}$$

Let $\tilde{g}(x)$ denote the right-hand side of the above inequality. Then $\tilde{g}(x) \leq b$ is a valid inequality for (11.4.1) in the sense that for any $\tilde{x}$ that satisfies $g(\tilde{x}) \leq b$, $\tilde{x}$ also satisfies $\tilde{g}(\tilde{x}) \leq b$. Furthermore, if let $G(\hat{x}) = G^1(\hat{x}) \cup G^0(\hat{x})$, we then have

$$\tilde{g}(\hat{x}) = \sum_{k \in G(\hat{x})} |a_k| + \sum_{k \in N^-} a_k = g(\hat{x}) > b,$$

which implies that $G(\hat{x})$ is a cover for (11.4.1).

Let $M \subseteq G(\hat{x})$ be any cover for (11.4.1). For $k \in G^0(\hat{x})$, let $\varphi(k)$ be such that $\hat{x}_{\varphi(k)} = 0$. Let

$$\begin{aligned} G_M &= \cup_{k \in M \cap G^1(\hat{x})} S_k, \\ G_\varphi &= \{j = \varphi(k) \mid k \in M \cap G^0(\hat{x})\}. \end{aligned}$$

Then, by Theorem 11.2, we have the generalized covering inequality

$$\sum_{j \in G_M} \bar{x}_j + \sum_{j \in G_\varphi} x_j \geq 1. \tag{11.4.5}$$

Observe that the inequality (11.4.5) is valid for (11.4.1), but it is violated by $\hat{x}$ since $\hat{y}_k = \prod_{j \in S_k} \hat{x}_j = 1$ for $k \in G^1(\hat{x})$, and $\hat{y}_k = 1 - \hat{x}_{\varphi(k)} = 1$ for $k \in G^0(\hat{x})$.

Of course, a minimal cover M results in a generalized covering inequality with less variables. A simple way to determine a minimal cover $M \subseteq G(\hat{x})$ is as follows: (i) ranking $|a_k|$ for $k \in G(\hat{x})$ in a decreasing order, (ii) determining the smallest subset $\tilde{G}(\hat{x}) \subseteq G(\hat{x})$ such that

$$\sum_{k \in \tilde{G}(\hat{x})} |a_k| > b - \sum_{k \in N^-} a_k.$$

The index $\varphi(k)$ can be chosen as the first index $j \in S_k$ such that $\hat{x}_j = 0$.

EXAMPLE 11.2 Consider a polynomial constraint:

$$g(x) = 6x_1x_2x_4 - 3x_4x_5 - x_1x_3 + 2x_2x_6 \leq 4.$$

Solution $\hat{x} = (1,1,1,1,0,0)^T$ is infeasible with $g(\hat{x}) = 5 > 4$. We have $N = \{1,2,3,4\}$, $N^+ = \{1,4\}$, $N^- = \{2,3\}$, $G^1(\hat{x}) = \{1\}$, $G^0(\hat{x}) = \{2\}$ and $G(\hat{x}) = \{1,2\}$. It is easy to see that $G(\hat{x})$ is a minimal cover since $6 + 3 > 4 + (3 + 1)$. By (11.4.5), the generalized covering inequality is $\bar{x}_1 + \bar{x}_2 + \bar{x}_4 + x_5 \geq 1$.

Now, consider the constrained nonlinear 0-1 programming $(0\text{-}1PP_1)$. Suppose that a point $\hat{x}$ violates one of the nonlinear inequalities in problem $(0\text{-}1PP_1)$, then (11.4.5) cuts off point $\hat{x}$ while (11.4.5) is satisfied by all feasible solutions of $(0\text{-}1PP_1)$. A cutting plane method (see [82]) can then be proposed to approximate the feasible region of problem $(0\text{-}1PP_1)$ by generating generalized covering inequality successively.

A *generalized covering relaxation* (GCR) of $(0\text{-}1PP_1)$ can be formed by replacing the nonlinear constraints $g_i(x) \leq b_i$, $i = 1, \ldots, m$, by a group of generalized covering inequalities defined by (11.4.3) or (11.4.5). A GCR can be then solved by any 0-1 linear programming algorithm. An efficient heuristic method for solving GCR was proposed by Balas and Martin [8].

ALGORITHM 11.2 (CUTTING PLANE METHOD FOR $(0\text{-}1PP_1)$)

Step 0. Generate a group of generalized covering inequalities and form an initial GCR problem (GCR_0). Set $k = 0$.

Step 1. Solve (GCR_k) by certain 0-1 linear programming algorithm. Let x^k be the optimal solution of (GCR_k). If x^k is feasible to $(0\text{-}1PP_1)$, then x^k is an optimal solution to $(0\text{-}1PP_1)$.

Step 2. For each polynomial constraint $g_i(x) \leq b_i$ which is violated at x^k, generate a generalized covering inequality defined by (11.4.5). Add all such newly generated generalized covering inequalities to (GCR_k). Denote by (GCR_{k+1}) the new problem. Set $k := k + 1$, go to Step 1.

The finite convergence of Algorithm 11.2 is evident by observing that the total number of binary solution is 2^n and at least one solution x^k is eliminated at each iteration.

We will discuss in the next two subsections how to derive more compact linear inequalities than the generalized covering inequalities.

11.4.2 Lower bounding linear function

Let $s_k = |S_k|$ for $k \in N$. For any $x \in \{0,1\}^n$, denote $Q(x) = \{i \in \{1, \ldots, n\} \mid x_i = 1\}$. For any $x \in \{0,1\}^n$, let $N^0(x)$ denote the set of all the index k with $x_j = 1$ for every $j \in S_k$, and $N^1(x)$ the set of all index k with at most one $x_j = 0$ for some $j \in S_k$, i.e.,

$$N^0(x) = \{k \in N \mid |S_k \setminus Q(x)| = 0\},$$
$$N^1(x) = \{k \in N \mid |S_k \setminus Q(x)| \leq 1\}.$$

For every $M \subseteq N^+$, define

$$g_M(x) = \sum_{j \in S_M} \Big(\sum_{k \in S^{-1}(j)} a_k \Big) x_j - \sum_{k \in M} (s_k - 1) a_k, \tag{11.4.6}$$

where $S_M = \cup_{k \in M} S_k$ and $S^{-1}(j) = \{k \in M \mid j \in S_k\}$. Then $g_M(x)$ is a lower bounding linear function of $g^+(x)$ as stated in the following Lemma.

LEMMA 11.1 *Let $\emptyset \neq M \subset N^+$. For any $x \in \{0,1\}^n$, it holds*

$$g^+(x) \geq g_M(x),$$

where $g_M(x)$ is defined in (11.4.6). The equality holds if and only if $N^0(x) \cap N^+ \subseteq M \subseteq N^1(x) \cap N^+$.

Proof. Let $T_k(x) = \prod_{j \in S_k} x_j$, $W_k(x) = \sum_{j \in S_k} x_j - s_k + 1$. Obviously, $T_k(x) \geq 0$ for any $x \in \{0,1\}^n$ and $T_k(x) = 0$ if and only if at least one x_j with $j \in S_k$, is equal to zero, i.e., $k \notin N^0(x)$. It is easy to see that $T_k(x) \geq W_k(x)$ for any $x \in \{0,1\}^n$ and $T_k(x) = W_k(x)$ if and only if $k \in N^1(x)$. Thus, we have

$$g^+(x) = \sum_{k \in N^+} a_k T_k(x) \geq \sum_{k \in M} a_k T_k(x) \geq \sum_{k \in M} a_k W_k(x) = g_M(x).$$

Moreover, $g^+(x) = g_M(x)$ if and only if $T_k(x) = W_k(x)$ for every $k \in M$ and $T_k(x) = 0$ for every $k \in N^+ \setminus M$. This is, $k \in N^1(x)$ for every $k \in M$ and $k \notin N^0(x)$ for every $k \in N^+ \setminus M$, i.e., $N^0(x) \cap N^+ \subseteq M \subseteq N^1(x) \cap N^+$. □

For each $\varphi \in \Phi^-$, define

$$h_\varphi(x) = \sum_{k \in N^-} a_k x_{\varphi(k)}. \tag{11.4.7}$$

The following lemma shows that $h_\varphi(x)$ is a lower bounding linear function of $g^-(x)$.

LEMMA 11.2 *Let $\varphi \in \Phi^-$. For any $x \in \{0,1\}^n$, it holds*

$$g^-(x) \geq h_\varphi(x), \tag{11.4.8}$$

where $h_\varphi(x)$ is defined in (11.4.7). The equality holds if and only if $\varphi(k) \in S_k \setminus Q(x)$ for all $k \in N^-$ such that $S_k \setminus Q(x) \neq \emptyset$.

Proof. For any $x \in \{0,1\}^n$, let $M = \{k \in N^- \mid S_k \subseteq Q(x)\}$. Since $x_{\varphi(k)} = 1$ for $k \in M$ and $a_k < 0$ for $k \in N^-$, we have

$$g^-(x) = \sum_{k \in M} a_k \geq \sum_{k \in M} a_k x_{\varphi(k)} + \sum_{k \in N^- \setminus M} a_k x_{\varphi(k)} = h_\varphi(x). \quad (11.4.9)$$

If $\varphi(k) \in S_k \setminus Q(x) \neq \emptyset$, then $x_{\varphi(k)} = 0$ for any $k \in N^- \setminus M$. Thus the inequality in (11.4.9) holds as equality. Conversely, suppose $g^-(x) = h_\varphi(x)$ for some $x \in \{0,1\}^n$ and $\varphi \in \Phi^-$. Notice that $S_k \setminus Q(x) = \emptyset$ for $k \in M$. If $\varphi(k) \notin S_k \setminus Q(x)$ for some $k \in N^- \setminus M$, then $\varphi(k) \in Q(x) \cap S_k$ and hence $x_{\varphi(k)} = 1$. Since $a_k < 0$, it follows from (11.4.9) that $g^-(x) > h_\varphi(x)$, a contradiction. □

Combining Theorems 11.1 and 11.2, we obtain the following theorem.

THEOREM 11.3 *For any $M \subseteq N^+$ and $\varphi \in \Phi^-$, it holds*

$$g(x) \geq g_M(x) + h_\varphi(x) \quad (11.4.10)$$

for any $x \in \{0,1\}^n$. Moreover, the inequality holds as an equality if and only if $N^0(x) \cap N^+ \subseteq M \subseteq N^1(x) \cap N^+$ and $\varphi(k) \in S_k \setminus Q(x)$ for all $k \in N^-$ such that $S_k \setminus Q(x) \neq \emptyset$.

11.4.3 Linearization of polynomial inequality

The following theorem shows that polynomial inequality (11.4.1) is equivalent to a linear inequality by replacing $g(x)$ with the lower bounding linear function derived in Theorem 11.3. Let $\mathbb{M}$ denote the set of all covers for $g^+(x) \leq b$, i.e.,

$$\mathbb{M} = \{M \subseteq N^+ \mid \sum_{k \in M} a_k > b\}.$$

THEOREM 11.4 *The inequality (11.4.1) is satisfied for all $x \in \{0,1\}^n$ if and only if*

$$g_M(x) + h_\varphi(x) \leq b \quad (11.4.11)$$

for all $M \in \mathbb{M}$ and $\varphi \in \Phi^-$, where g_M and φ are defined in (11.4.6) and (11.4.7), respectively.

Proof. If (11.4.1) holds, then by Theorem 11.3, (11.4.11) holds. Conversely, suppose (11.4.11) is satisfied and there is some $x_0 \in \{0,1\}^n$ such that $g(x_0) >$

b. From Theorem 11.3, there exist M_0 with $N^0(x_0) \cap N^+ \subseteq M_0 \subseteq N^1(x_0) \cap N^+$ and $\varphi_0 \in \Phi^-$ with $\varphi_0(k) \in S_k \setminus Q(x_0) \neq \emptyset$ for $k \in N^-$, such that

$$g_{M_0}(x_0) + h_{\varphi_0}(x) = g(x_0) > b.$$

Thus, inequality (11.4.11) is not satisfied. Since $g_{M_0}(x_0) = \sum_{k \in M_0} a_k W_k(x_0)$ and $W_k(x_0) \leq 1$ for $k \in M^0 \subseteq N^1(x^0) \cap N^+$, we have

$$\sum_{k \in M_0} a_k \geq g_{M_0}(x_0) > b - h_{\varphi_0}(x_0) \geq b.$$

Therefore, $M_0 \in \mathbb{M}$. This contradicts that inequality (11.4.11) is satisfied for all $M \in \mathbb{M}$ and $\varphi \in \Phi^-$. □

THEOREM 11.5 *The inequality (11.4.1) is satisfied for all $x \in \{0,1\}^n$ if and only if*

$$\sum_{j \in S_M} \Big(\sum_{k \in S^{-1}(j)} a_k \Big) \bar{x}_j + \sum_{j \in S_\varphi} \Big(\sum_{k \in \varphi^{-1}(j)} |a_k| \Big) x_j \geq \sum_{k \in M} a_k - b \quad (11.4.12)$$

for all $M \in \mathbb{M}$ and $\varphi \in \Phi^-$, where $S_M = \cup_{k \in M} S_k$, $S^{-1}(j) = \{k \in N^+ \mid j \in S_k\}$, $S_\varphi = \{j = \varphi(k) \mid k \in N^-\}$ and $\varphi^{-1}(j) = \{k \in N^- \mid j = \varphi(k)\}$.

Proof. By using expressions (11.4.6) and (11.4.7), inequality (11.4.11) is equivalent to

$$\sum_{j \in S_M} \Big(\sum_{k \in S^{-1}(j)} a_k \Big) x_j + \sum_{k \in N^-} a_k x_{\varphi(k)} \leq b + \sum_{k \in M} (s_k - 1) a_k. \quad (11.4.13)$$

Substituting $\bar{x}_j = 1 - x_j$ in (11.4.13) for $j \in S_M$, and noting that

$$\sum_{j \in S_M} \Big(\sum_{k \in S^{-1}(j)} a_k \Big) = \sum_{k \in M} s_k a_k,$$

the inequality (11.4.13) gives rise to (11.4.12). □

For $M^+ \subseteq N^+$, let $M = M^+ \cup N^-$ be a minimal cover for (11.4.1). It was shown in [9] that inequality (11.4.12) reduces to the generalized covering inequality (11.4.3) when M is a minimal cover for (11.4.1) and $S_M \cap S_\varphi = \emptyset$.

EXAMPLE 11.3 Consider a polynomial constraint:

$$g(x) = 5x_1x_2 + 3x_2x_5 - x_1x_3 - 2x_3x_4 \leq 4.$$

Note that $N = \{1,2,3,4\}$, $N^+ = \{1,2\}$, $N^- = \{3,4\}$. Choose a minimal cover $M = \{1,2\}$ and $\varphi(3) = \varphi(4) = 3$. Then the linear inequality of the form (11.4.12) is

$$5\bar{x}_1 + 8\bar{x}_2 + 3\bar{x}_5 + 3x_3 \geq 4.$$

Now, let $M^+ = \{1\}$ and $M = M^+ \cup N^- = \{1,3,4\}$. Then M is a minimal cover for $g(x) \leq 4$. Again, choose $\varphi(3) = \varphi(4) = 3$. Since $S_M = \cup_{k \in M \cap N^+} S_k = S_1 = \{1,2\}$ and $S_\varphi = \{j = \varphi(k) \mid k \in M \cap N^-\} = \{j = \varphi(k) \mid k = 3,4\} = \{3\}$, it holds $S_M \cap S_\varphi = \emptyset$. The generalized covering inequality in the form of (11.4.3) is

$$\bar{x}_1 + \bar{x}_2 + x_3 \geq 1.$$

When $M \subseteq N$ is a cover that is not minimal, (11.4.12) gives rise to a linear inequality that is not of the generalized covering type. This kind of inequalities is usually more compact than the family of generalized covering inequalities (see [9]). Dominance relations between various linear inequalities were investigated in [10].

11.5 Quadratic 0-1 Knapsack Problems

The quadratic 0-1 knapsack problem can be expressed as follows:

$$(QKP) \qquad \max\ Q(x) = \sum_{j=1}^{n} q_{jj}x_j + \sum_{1 \leq i < j \leq n} q_{ij}x_ix_j$$

$$\text{s.t.}\ \sum_{i=1}^{n} a_ix_i \leq b,$$

$$x \in \{0,1\}^n,$$

where $q_{ij} \geq 0$ for $1 \leq i \leq j \leq n$, $a_i \geq 0$, $i = 1,\ldots,n$ and $0 < b < \sum_{i=1}^{n} a_i$.

Problem (QKP) is a special case of problem $(0\text{-}1PP)$. In this section, we will derive special properties of problem (QKP) and investigate solution methods for solving (QKP).

11.5.1 Lagrangian dual of (QKP)

Due to the special structure of the quadratic function, the dual function of the quadratic 0-1 knapsack problem possesses some special properties that can be exploited in designing efficient dual search procedures for (QKP). We will first discuss the dual function for general 0-1 knapsack problems. Characterizations and computation of the dual functions for supermodular knapsack problems and quadratic 0-1 knapsack problems will be investigated next.

11.5.1.1 Dual function of general 0-1 knapsack problem

Consider the following general singly-constrained nonlinear knapsack problem:

$$(GNKP) \qquad \begin{array}{ll} \max & f(x) \\ \text{s.t.} & g(x) \le b, \\ & x \in \{0,1\}^n, \end{array}$$

where $g(x)$ is a strictly increasing function of each x_i and $0 < b < g(e)$, where $e = (1, \ldots, 1)^T$. Assume also that $f(0) = 0$, $g(0) = 0$ and e is the unique maximizer of $f(x)$ over $\{0,1\}^n$. Note that $(GNKP)$ is more general than the knapsack problem which we discussed before, since f is not assumed to possess a monotonicity in problem $(GNKP)$.

The Lagrangian function of $(GNKP)$ is

$$L(x, \lambda) = f(x) - \lambda(g(x) - b), \tag{11.5.1}$$

where $\lambda \ge 0$. The Lagrangian relaxation problem of $(GNKP)$ is

$$(L_\lambda) \quad d(\lambda) = \max\{L(x, \lambda) \mid x \in \{0,1\}^n\}. \tag{11.5.2}$$

The Lagrangian dual is then defined as

$$(D) \quad \min_{\lambda \ge 0} d(\lambda). \tag{11.5.3}$$

Since the dual function $d(\lambda)$ is a piecewise linear function on $\mathbb{R}_+$, it is characterized by its breakpoints. Let $x^0 = (0, \ldots, 0)^T$. Define recursively

$$\begin{aligned} \lambda_k &= \max\{\frac{f(x) - f(x^{k-1})}{g(x) - g(x^{k-1})} \mid x \in \{0,1\}^n,\ g(x) > g(x^{k-1})\} \\ &= \frac{f(x^k) - f(x^{k-1})}{g(x^k) - g(x^{k-1})}. \end{aligned} \tag{11.5.4}$$

In the case where there exist multiple solutions achieving the maximum in (11.5.4), we choose x^k to be the one with maximum value of $g(x)$.

Since $\{g(x^k)\}$ is strictly increasing, there exists an index $p > 0$ such that $x^p = e$. We can easily show that λ_k $(k = 1, \ldots, p)$ corresponds to the slopes of the concave envelope of the perturbation function of $(GNKP)$. In fact, the envelope function ϕ of the perturbation function $w(y)$ of $(GNKP)$ can be expressed as

$$\phi(y) = \begin{cases} f_1 + \xi_1(y - c_1), & c_1 \le y < c_2 \\ f_2 + \xi_2(y - c_2), & c_2 \le y < c_3 \\ \cdots & \cdots \\ f_{K-1} + \xi_{K-1}(y - c_{K-1}), & c_{K-1} \le y < c_K \\ f_K, & c_K \le y < \infty \end{cases} \tag{11.5.5}$$

where (c_i, f_i), $i = 1, \ldots, K$, are the corner points of $w(y)$, and

$$\xi_i = \frac{f_{i+1} - f_i}{c_{i+1} - c_i} > 0, \quad 1 \leq i < K. \tag{11.5.6}$$

Since $f(0) = 0$ and $f(e) \geq f(x)$ for all $x \in \{0,1\}^n$, we imply that $c_1 = 0$ and $c_K = g(e)$. Note that the envelope function ϕ, in general, is not necessarily concave.

Let ψ be the concave envelope function of the perturbation function w. Then, ψ is a piecewise linear function with decreasing slopes η_i, $i = 1, \ldots, q(\leq K)$. The slope η_i can be determined recursively by

$$\begin{aligned} \eta_i &= \max\{\frac{f_j - f_{k_{i-1}}}{c_j - c_{k_{i-1}}} \mid j > k_{i-1}\} \\ &= \frac{f_{k_i} - f_{k_{i-1}}}{c_{k_i} - c_{k_{i-1}}} \end{aligned} \tag{11.5.7}$$

for $1 \leq i \leq q$, where $k_0 = 1$ and k_i is the maximum index that satisfies $k_i > k_{i-1}$ and that achieves the maximum of (11.5.7). By the definition of c_i and f_i, we imply that

$$p = q, \ \lambda_i = \eta_i, \ g(x^i) = c_{k_i}, \ f(x^i) = f_{k_i}, i = 1, \ldots, p.$$

Thus, by the concavity of ψ, we must have

$$\lambda_1 > \lambda_2 > \cdots > \lambda_p > 0. \tag{11.5.8}$$

Moveover, since $f(x) \leq f(e)$ for all $x \in \{0,1\}^n$, and $g(x) > 0$ implies that $g(x) \geq \min_{j=1,\ldots,n} g(e_j)$, where e_j is the j-th unit vector in $\mathbb{R}^n$, we have

$$\max_{j=1,\ldots,n} f(e)/g(e_j) \geq \max\{f(x)/g(x) \mid x \in \{0,1\}^n, \ g(x) > 0\} = \lambda_1.$$

By the perturbation theory in Chapter 3, we have the following results.

THEOREM 11.6 *The solution x^k solves the Lagrangian relaxation problem (L_{λ_k}), $k = 1, \ldots, p$.*

THEOREM 11.7 (i) *The points $\lambda_1, \ldots, \lambda_p$ are the breakpoints of $d(\lambda)$ on $\mathbb{R}_+$ and the slope of $d(\lambda)$ on interval $[\lambda_{k+1}, \lambda_k]$ is $b - g(x^k)$, $k = 1, \ldots, p-1$.*

(ii) *Let r be the maximum index k such that $g(x^k) \leq b$. Then, λ_{r+1} solves the dual problem (D) with optimal value $d(\lambda_{r+1}) = f(x^{r+1}) - \lambda_{r+1}(g(x^{r+1}) - b)$.*

Proof. (i) We prove that $d(\lambda)$ is linear on the interval $[\lambda_{k+1}, \lambda_k]$. Let $\lambda = \mu\lambda_{k+1} + (1-\mu)\lambda_k$ with $\mu \in [0,1]$. For any $x \in \{0,1\}^n$, by Theorem 11.6, we have

$$f(x) - \lambda_k(g(x) - b) \le f(x^k) - \lambda_k(g(x^k) - b), \quad (11.5.9)$$

$$f(x) - \lambda_{k+1}(g(x) - b) \le f(x^{k+1}) - \lambda_{k+1}(g(x^{k+1}) - b). \quad (11.5.10)$$

Using the relation $\lambda_{k+1} = (f(x^{k+1}) - f(x^k))/(g(x^{k+1}) - g(x^k))$, we obtain from (11.5.10) that

$$f(x) - \lambda_{k+1}(g(x) - b) \le f(x^k) - \lambda_{k+1}(g(x^k) - b). \quad (11.5.11)$$

Multiplying both sides of (11.5.9) by $(1 - \mu)$ and both sides of (11.5.11) by μ yields

$$f(x) - \lambda(g(x) - b) \le f(x^k) - \lambda(g(x^k) - b),$$

which implies that $d(\lambda) = f(x^k) - \lambda(g(x^k) - b)$, and hence $d(\lambda)$ is linear on $[\lambda_{k+1}, \lambda_k]$ with a slope of $b - g(x^k)$.

(ii) By the definition of λ_k and Theorem 11.6, we have

$$\begin{aligned} d(\lambda_{k+1}) &= f(x^{k+1}) - \lambda_{k+1}(g(x^{k+1}) - b) \\ &= f(x^{k+1}) - \frac{f(x^{k+1}) - f(x^k)}{g(x^{k+1}) - g(x^k)}[(g(x^{k+1}) - g(x^k)) + (g(x^k) - b)] \\ &= f(x^k) - \lambda_k(g(x^k) - b) + (\lambda_k - \lambda_{k+1})(g(x^k) - b) \\ &= d(\lambda_k) + (\lambda_k - \lambda_{k+1})(g(x^k) - b). \end{aligned}$$

Thus, $d(\lambda_{k+1}) \le d(\lambda_k)$ for $k \le r$ and $d(\lambda_{k+1}) \ge d(\lambda_k)$ for $k > r$. Therefore, λ_{r+1} solves (D).

11.5.1.2 Dual function of supermodular knapsack problem

Function $f(x)$ is said to be *supermodular* if it satisfies the following conditions:

(i) $f(0) = 0$,

(ii) $e = (1, \dots, 1)^T$ is the unique maximizer of $f(x)$ over $\{0, 1\}^n$,

(iii) $f(x \wedge y) + f(x \vee y) \ge f(x) + f(y)$ for all $x, y \in \{0, 1\}^n$, where $x \wedge y = (\min(x_1, y_1), \dots, \min(x_n, y_n))^T$ and $x \vee y = (\max(x_1, y_1), \dots, \max(x_n, y_n))^T$.

PROPOSITION 11.2 *Let $f(x)$ be a polynomial defined by*

$$f(x) = \sum_{i=1}^{n} c_i x_i + \sum_{k \in N} d_k \prod_{j \in S_k} x_j, \quad (11.5.12)$$

where $c_i \ge 0$, $i = 1, \dots, n$, $d_k \ge 0$ and $S_k \subseteq \{1, \dots, n\}$, $k \in N$. Then $f(x)$ is supermodular.

Proof. By the definition of supermodularity, it suffices to show that $p(x) = \prod_{j\in S} x_j$ is supermodular for any $S \subseteq \{1,\ldots,n\}$. We prove this by induction. The conclusion is obviously true if $|S| = 1$. Suppose that $p(x)$ is supermodular when $|S| = k-1$. Let $|S| = k$ and $i \in S$. Suppose that $x_i \leq y_i$, then

$$\begin{aligned}
&p(x \wedge y) + p(x \vee y) - p(x) - p(y) \\
&= x_i\Big(\prod_{j\in S\setminus i} x_j \wedge y_j - \prod_{j\in S\setminus i} x_j\Big) + y_i\Big(\prod_{j\in S\setminus i} x_j \vee y_j - \prod_{j\in S\setminus i} y_j\Big) \\
&\geq x_i\Big(\prod_{j\in S\setminus i} x_j \wedge y_j + \prod_{j\in S\setminus i} x_j \vee y_j - \prod_{j\in S\setminus i} x_j - \prod_{j\in S\setminus i} y_j\Big) \\
&\geq 0.
\end{aligned}$$

The above inequality can be also proved in a similar way for the case where $x_i \geq y_i$. □

The supermodular knapsack problem can be expressed as

$$\begin{aligned}
(SKP) \qquad & \max\ f(x) \\
& \text{s.t. } g(x) = \sum_{i=1}^{n} a_i x_i \leq b, \\
& x \in \{0,1\}^n,
\end{aligned}$$

where $f(x)$ is a supermodular function on $\{0,1\}^n$, $a_i > 0$ and $\sum_{i=1}^n a_i > b$.

The following result shows that for suppermodular knapsack problems, the computation of λ_k can be simplified. Denote $x \leq y$ if $x_i \leq y_i$ for all i and $x < y$ if $x \leq y$ and at least one strict inequality $x_i < y_i$ holds.

THEOREM 11.8 *For problem (SKP), let λ_k and x^k be defined by (11.5.4). Then, for $k = 1,\ldots,p$, λ_k can be calculated by the following formula,*

$$\lambda_k = \max\{\frac{f(x) - f(x^{k-1})}{a^T(x - x^{k-1})} \mid x \in \{0,1\}^n,\ x > x^{k-1}\}, \qquad (11.5.13)$$

where $x^0 = (0,\ldots,0)^T$.

Proof. Let w_k denote the right-hand side of (11.5.13). Since $a_i > 0$ for each i, $x > x^{k-1}$ implies $a^T x > a^T x^{k-1}$, i.e., $g(x) > g(x^{k-1})$. Thus, λ_k defined in (11.5.4) is greater than or equal to w_k for each k. We prove in the following $w_k \geq \lambda_k$ for each k. Since $a^T(x^k - x^{k-1}) > 0$, there exist $x_i^k = 1$ and $x_i^{k-1} = 0$. Hence $x^k \vee x^{k-1} > x^{k-1}$ and $x^k \wedge x^{k-1} < x^k$. Thus

$$w_k \geq \frac{f(x^k \vee x^{k-1}) - f(x^{k-1})}{a^T(x^k \vee x^{k-1} - x^{k-1})}. \qquad (11.5.14)$$

Notice that $x^k \vee x^{k-1} + x^k \wedge x^{k-1} = x^k + x^{k-1}$. It follows from (11.5.14) and the supermodular property of f that

$$w_k \geq \frac{f(x^k) - f(x^k \wedge x^{k-1})}{a^T(x^k - x^k \wedge x^{k-1})}. \tag{11.5.15}$$

By Theorem 11.6, we have

$$f(x^k) - \lambda_k(a^T x^k - b) \geq f(x^k \wedge x^{k-1}) - \lambda_k[a^T(x^k \wedge x^{k-1}) - b].$$

Thus,

$$\frac{f(x^k) - f(x^k \wedge x^{k-1})}{a^T(x^k - x^k \wedge x^{k-1})} \geq \lambda_k,$$

which, combined with (11.5.15), implies $w_k \geq \lambda_k$. □

The set $\{x \in \{0,1\}^n \mid x > x^{k-1}\}$ is a subset of $\{x \in \{0,1\}^n \mid g(x) > g(x^{k-1})\}$ where g is a strictly increasing function. That is why searching for λ_k using (11.5.13) for supermodular knapsack problems will be easier than searching for λ_k using (11.5.4) for quadratic 0-1 knapsack problems.

Theorem 11.8 immediately implies that $x^p > x^{p-1} > \cdots > x^1 > 0$, which in turn implies $p \leq n$. Notice that $x^0 = 0$ solves problem (L_λ) with $\lambda = 0$. In summary, we have the following corollary.

COROLLARY 11.2 *For problem (SKP),*

(i) *The number of breakpoints of $d(\lambda)$ is at most n;*

(ii) *There exist at most $n+1$ solutions $0 = x^0 < x^1 < \cdots < x^p$ such that for every $\lambda \geq 0$, one of x^k, k = 0, . . ., p, solves (L_λ).*

Using the outer Lagrangian linearization method (Procedure 3.3) for singly constrained integer program, we can find the optimal solution of the dual problem (D) of (SKP) by evaluating $d(\lambda)$ for at most $n+1$ times. It can be shown that $d(\lambda)$ can be computed in polynomial time (see [67]). Therefore, (D) can be solved in polynomial time.

We note from Proposition 11.2 that linear function $f(x) = \sum_{i=1}^n c_i x_i$ with each $c_i \geq 0$ and quadratic function $Q(x)$ defined in (QKP) are supermodular. Therefore, Corollary 11.2 is applicable to problem (QKP).

11.5.1.3 Lagrangian relaxation and minimum-cut in quadratic case

The Lagrangian relaxation problem (L_λ) of (QKP) can be expressed as

$$\begin{aligned} d(\lambda) &= \max_{x \in \{0,1\}^n} Q(x) - \lambda(a^T x - b) \\ &= \lambda b + \max_{x \in \{0,1\}^n} \{Q(x) - \lambda \sum_{j=1}^n a_j x_j\}. \end{aligned} \tag{11.5.16}$$

Consider a directed graph $G = (V, E)$ with $V = (s, 1, 2, \ldots, n, t)$, where s denotes the source and t the sink, and with $E = E_s \cup E_Q \cup E_t$, where

$$E_s = \{(s, j) \mid j = 1, \ldots, n\},$$
$$E_Q = \{(i, j) \mid q_{ij} > 0,\ 1 \le i < j \le n\},$$
$$E_t = \{(j, t) \mid j = 1, \ldots, n\}.$$

The capacities of the arcs in E are defined as follows:

$$c_{sj}(\lambda) = \max(0, \sum_{i=j}^{n} q_{ji} - \lambda a_j), \quad (s, j) \in E_s, \tag{11.5.17}$$

$$c_{ij}(\lambda) = q_{ij}, \quad (i, j) \in E_Q, \tag{11.5.18}$$

$$c_{jt}(\lambda) = \max(0, \lambda a_j - \sum_{i=j}^{n} q_{ji}), \quad (j, t) \in E_t. \tag{11.5.19}$$

Let $(U, \overline{U})$ be a partition of G with $s \in U$ and $t \in \overline{U}$. The set of arcs $\delta^+(U) = \{(i, j) \mid i \in U,\ j \in \overline{U}\}$ is called an $s - t$ cut. The capacity of $\delta^+(U)$ is $\sum_{(i,j)\in\delta^+(U)} c_{ij}(\lambda)$. The *minimum-cut* problem is to find a cut with minimum capacity. Let $\Psi(\lambda)$ be the capacity of the minimum-cut of G. Then $\Psi(\lambda) = \min_U \sum_{(i,j)\in\delta^+(U)} c_{ij}(\lambda)$. Associate each cut $\delta^+(U)$ of G with a 0-1 vector $(1, x_1, \ldots, x_n, 0)$ satisfying $x_i = 1$ if $i \in U$ and $x_i = 0$ otherwise. The following result shows that the Lagrangian relaxation problem (11.5.16) can be solved by computing the minimum-cut of the graph $G = (V, E)$.

THEOREM 11.9 $d(\lambda) = \sum_{j=1}^{n} c_{sj}(\lambda) + \lambda b - \Psi(\lambda)$.

Proof. By (11.5.17)–(11.5.19), we have

$$\begin{aligned}
&\Psi(\lambda)\\
&= \min_{x\in\{0,1\}^n} \{\sum_{j=1}^{n} c_{sj}(1 - x_j) + \sum_{1\le i<j\le n} c_{ij} x_i (1 - x_j) + \sum_{j=1}^{n} c_{jt} x_j\}\\
&= \sum_{j=1}^{n} c_{sj}(\lambda) + \min_{x\in\{0,1\}^n} \{\sum_{j=1}^{n} \min(0, \lambda a_j - \sum_{i=j}^{n} q_{ji}) x_j\\
&\quad + \sum_{i=1}^{n-1} \sum_{j=i+1}^{n} q_{ij} x_i - \sum_{1\le i<j\le n} q_{ij} x_i x_j + \sum_{j=1}^{n} \max(0, \lambda a_j - \sum_{i=j}^{n} q_{ji}) x_j\}
\end{aligned}$$

$$
\begin{aligned}
&= \sum_{j=1}^{n} c_{sj}(\lambda) + \min_{x\in\{0,1\}^n} \{\sum_{j=1}^{n}(\lambda a_j - \sum_{i=j}^{n} q_{ji})x_j + \sum_{i=1}^{n-1}\sum_{j=i+1}^{n} q_{ij}x_i \\
&\quad + \sum_{j=1}^{n} q_{jj}x_j - Q(x)\} \\
&= \sum_{j=1}^{n} c_{sj}(\lambda) + \min_{x\in\{0,1\}^n} \{\sum_{j=1}^{n}(\lambda a_j - \sum_{i=j}^{n} q_{ji})x_j + \sum_{j=1}^{n}\sum_{i=j}^{n} q_{ji}x_j - Q(x)\} \\
&= \sum_{j=1}^{n} c_{sj}(\lambda) + \min_{x\in\{0,1\}^n} \{\sum_{j=1}^{n} \lambda a_j x_j - Q(x)\} \\
&= \sum_{j=1}^{n} c_{sj}(\lambda) + \lambda b - d(\lambda).
\end{aligned}
$$

This proves the theorem. □

It is well-known that the minimum-cut problem is equivalent to the maximum-flow problem which can be solved in polynomial time (see [168]). Therefore, the dual function $d(\lambda)$ can be evaluated by computing the maximum-flow of a graph with $n+2$ vertices and $2n + n(n-1)/2$ arcs. Algorithms with different complexity bounds have been proposed for finding a maximum-flow in G (see e.g., [65][78][168]). For example, using the $O(n^3)$ maximum-flow algorithms proposed in [78] or [65], Procedure 3.3 finds an optimal solution of the dual problem (D) for quadratic 0-1 knapsack problems in $O(n^4)$ time.

11.5.2 Heuristics for finding feasible solutions

To obtain a tight initial lower bound in branch-and-bound methods, different heuristics can be used to find a good feasible solution of (QKP).

Define $q_{ij} = q_{ji}$ for $i > j$. The quadratic function can be rewritten as

$$Q(x) = \sum_{i=1}^{n}(q_{ii} + \frac{1}{2}\sum_{j\neq i} q_{ij}x_j)x_i.$$

Define $l(x) = \sum_{i=1} c_i x_i$, where c_i is given by

$$c_i = q_{ii} + \frac{1}{2}\sum_{j\neq i} q_{ij}. \tag{11.5.20}$$

Then $l(x) \geq Q(x)$ for all $x \in \{0,1\}^n$.

Another way to derive the linear approximation function $l(x)$ is via the best L_2-approximation. The best L_2-approximation is defined as the unique linear function l_Q such that

$$\sum_{x\in\{0,1\}^n} |Q(x) - l_Q(x)|^2 = \min_{l \text{ linear}} \sum_{x\in\{0,1\}^n} |Q(x) - l(x)|^2.$$

LEMMA 11.3 ([91]) *The best linear L_2-approximation of $Q(x)$ is given by $l_Q(x) = c_0 + \sum_{i=1}^n c_i x_i$, where*

$$c_0 = -\frac{1}{4} \sum_{1\le i<j\le n} q_{ij},$$

$$c_i = q_{ii} + \frac{1}{2} \sum_{j\ne i} q_{ij}, \quad i = 1, \ldots, n. \tag{11.5.21}$$

We see that c_i defined in (11.5.21) agrees with that defined in (11.5.20). Now, consider the linear approximation problem:

$$\begin{aligned} \max \ & \sum_{j=1}^n c_i x_i \\ \text{s.t.} \ & \sum_{i=1}^n a_j x_j \le b, \\ & x \in \{0,1\}^n. \end{aligned}$$

A greedy method for the above 0-1 linear knapsack problem may produce a good feasible solution of (QKP).

PROCEDURE 11.1 (HEURISTIC A FOR FINDING A FEASIBLE SOLUTION OF (QKP))

Step 1. Calculate c_i and $\rho_i = c_i/a_i$, $i = 1, \ldots, n$, by (11.5.20). Set $K_1 = \emptyset$, $K_0 = \{1, \ldots, n\}$, $I = K_0$ and $s = b$.

Step 2. Compute $k = \arg\max\{\rho_i \mid i \in I\}$. If $\sum_{i\in K_1\cup\{k\}} a_i > b$, set $I := I \setminus \{k\}$. If $I = \emptyset$, go to Step 4. Otherwise, repeat Step 2. If $\sum_{i\in K_1\cup\{k\}} a_i \le b$ set $K_1 := K_1 \cup \{k\}$ and $K_0 := K_0 \setminus \{k\}$, $s := s - a_k$.

Step 3. If $s < \min\{a_i \mid i \in K_0\}$, go to Step 4. Otherwise, update ρ_i for $i \in K_0$: $\rho_i := \rho_i - (1/2)q_{ki}/a_i$. Set $I = K_0$, return to Step 2.

Step 4. Set $x_i = 1$ for $i \in K_1$ and $x_i = 0$ for $i \in K_0$, x is a feasible solution of (QKP). Stop.

The above procedure starts from $x = (0, \ldots, 0)^T$ and improves the solution by adding 1 to some component. Alternatively, we can start from point

$x = (1, \ldots, 1)^T$ and decrease the values of the components of x in the order determined by ranking the ratios c_i/a_i.

PROCEDURE 11.2 (HEURISTIC B FOR FINDING A FEASIBLE SOLUTION OF (QKP))

Step 1. Set $K_1 = \{1, \ldots, n\}$, $K_0 = \emptyset$. Calculate $\rho_i = c_i/a_i$, $i = 1, \ldots, n$.

Step 2. Compute $k = \arg\min\{\rho_i \mid i \in K_1\}$, set $K_1 := K_1 \setminus \{k\}$ and $K_0 := K_0 \cup \{k\}$. If $\sum_{i \in K_1} a_i \leq b$, set $x_i = 1$ for $i \in K_1$ and $x_i = 0$ for $i \in K_0$, x is a feasible solution. Stop.

Step 3. Update ρ_i for $i \in K_1$: $\rho_i := \rho_i - (1/2)q_{ki}/a_i$. Return to Step 2.

The feasible solution found in Procedure 11.1 or Procedure 11.2 can be further improved by using fill-up and exchange ([66]). The derivative $\Delta_i(x)$ of the quadratic function $Q(x)$ can be written as

$$\Delta_i(x) = q_{ii} + \sum_{j \neq i} q_{ij} x_j.$$

The "second-order derivative" of $Q(x)$ (see [89]) is defined by

$$\begin{aligned} \Delta_{ij}(x) &= Q(x \mid x_i = 1, x_j = 0) - Q(x \mid x_i = 0, x_j = 1) \\ &= \Delta_i(x) - \Delta_j(x) + q_{ij}(x_i - x_j) \\ &= q_{ii} - q_{jj} + \sum_{k \neq i,j} (q_{ik} - q_{jk}) x_k. \end{aligned}$$

PROCEDURE 11.3 (HEURISTIC C FOR IMPROVING A FEASIBLE SOLUTION OF (QKP))

Given a feasible solution x. Let $K_1 = \{i \mid x_i = 1\}$, $K_0 = \{i \mid x_i = 0\}$.

Step 1. (Fill-Up). Find $k \in \arg\max\{\Delta_i(x) \mid i \in K_0\}$. If $\sum_{i \in K_1} a_i + a_k \leq b$, then set $x := x + e_k$, where e_k is the k-th unit vector. Set $K_0 := K_0 \setminus \{k\}$. Repeat Step 1 until $K_0 = \emptyset$.

Step 2. (Exchange). Reset $K_1 = \{i \mid x_i = 1\}$ and $K_0 = \{i \mid x_i = 0\}$. Find $(k, l) \in \arg\min\{\Delta_{ij}(x) \mid i \in K_1, j \in K_0\}$. If $\sum_{i \in K_1} a_i - a_k + a_l \leq b$, then, set $x := x - e_k + e_l$. Set $K_1 := K_1 \setminus \{k\}$, $K_0 := K_0 \setminus \{l\}$. Repeat Step 1 until $K_1 = \emptyset$ or $K_0 = \emptyset$.

EXAMPLE 11.4 *Consider the following problem:*

$$\begin{aligned} \max\ Q(x) = {} & x_1 + 4x_2 + x_3 + 2x_4 + 6x_1x_2 + 4x_1x_3 + 10x_1x_4 \\ & + x_2x_3 + 5x_2x_4 + 4x_3x_4 \\ \text{s.t. } & 7x_1 + 5x_2 + 4x_3 + 2x_4 \leq 13, \\ & x \in \{0, 1\}^4. \end{aligned}$$

We first apply Procedure 11.1 to the example.

Step 1. Using (11.5.20), we have $c_1 = 11$, $c_2 = 10$, $c_3 = 5.5$, $c_4 = 11.5$. Calculate the ratios: $\rho_1 = c_1/a_1 = 11/7$, $\rho_2 = c_2/a_2 = 10/5$, $\rho_3 = c_3/a_3 = 5.5/4$, $\rho_4 = c_4/a_4 = 11.5/2$. Set $K_1 = \emptyset$ and $K_0 = \{1, 2, 3, 4\}$, $I = K_0$, $s = 13$.

Step 2. Since $\rho_4 = \max\{\rho_i \mid i \in I\}$ and $a_4 = 2 < b$, set $K_1 = \{4\}$, $K_0 = \{1, 2, 3\}$, $s = 13 - 2 = 11$.

Step 3. Update ρ_i: $\rho_1 = 11/7 - (1/2) \times (10/7) = 6/7$, $\rho_2 = 2 - (1/2) \times (5/5) = 3/2$, $\rho_3 = 5.5/4 - (1/2) \times (4/4) = 7/8$. Set $I = K_0$.

Step 2. Since $\rho_2 = \max\{\rho_i \mid i \in I\}$ and $a_2 + a_4 = 7 < b$, set $K_1 = \{2, 4\}$, $K_0 = \{1, 3\}$. Set $I = K_0$, $s = 11 - 5 = 6$.

Step 3. Update ρ_i: $\rho_1 = 6/7 - (1/2) \times (6/7) = 3/7$, $\rho_3 = 7/8 - (1/2) \times (1/4) = 3/4$. Set $I = K_0$.

Step 2. Since $\rho_3 = \max\{\rho_i \mid i \in I\}$ and $a_2 + a_4 + a_3 = 11 < b$, set $K_1 = \{2, 4, 3\}$, $K_0 = \{1\}$, $s = 6 - 4 = 2$.

Step 3. $s = 2 < a_1$.

Step 4. The feasible solution is $x = (0, 1, 1, 1)^T$ with $Q(x) = 17$.

Next, we apply Procedure 11.3 to improve the feasible solution $x = (0, 1, 1, 1)^T$. We have $K_1 = \{2, 3, 4\}$ and $K_0 = \{1\}$. No fill-up occurs in Step 1. In the exchange step, the only feasible exchange is $(k, l) = (2, 1)$ with

$$
\begin{aligned}
\Delta_{21}(x) &= q_{22} - q_{11} + (q_{23} - q_{13})x_3 + (q_{24} - q_{14})x_4 \\
&= 4 - 1 + (1 - 4) + (5 - 10) \\
&= -5 < 0.
\end{aligned}
$$

The new feasible solution is $x := x - e_2 + e_1 = (1, 0, 1, 1)^T$ with $Q(x) = 22$.

Now, we consider to apply Procedure 11.2 to the example.

Step 1. $K_1 = \{1, 2, 3, 4\}$, $K_0 = \emptyset$. Compute $\rho_1 = 11/7$, $\rho_2 = 10/5$, $\rho_3 = 5.5/4$, $\rho_4 = 11.5/2$.

Step 2. Since $\rho_3 = \min\{\rho_i \mid i \in K_1\}$, set $K_1 = \{1, 2, 4\}$, $K_0 = \{3\}$. $a_1 + a_2 + a_4 = 14 > b$.

Step 3. Update ρ_i: $\rho_1 = 11/7 - (1/2) \times (4/7) = 9/7$, $\rho_2 = 10/5 - (1/2) \times (1/5) = 19/10$, $\rho_4 = 11.5/2 - (1/2) \times 4/2 = 19/4$.

Step 2. Since $\rho_1 = \min\{\rho_i \mid i \in K_1\}$, set $K_1 = \{2, 4\}$, $K_0 = \{1, 3\}$. Since $a_2 + a_4 = 7 < b$, we obtain a feasible solution $x = (0, 1, 0, 1)^T$ with $Q(x) = 11$.

Again, we can use Procedure 11.3 to improve the feasible solution $x = (0, 1, 0, 1)^T$. In the Fill-Up step, since $x + e_3 = (0, 1, 1, 1)^T$ is feasible, we set $x = (0, 1, 1, 1)^T$. The Exchange step then produces the feasible solution $x = (1, 0, 1, 1)^T$ which is the same as we obtained by applying Procedures 11.1 and 11.3.

11.5.3 Branch-and-bound method based on Lagrangian relaxation

The following algorithm consists of three main steps: (i) Finding an initial feasible solution and a lower bound of (QKP) by the heuristics described in the previous subsection; (ii) fixing certain variables to 0 or 1 by Lagrangian bound; and (iii) searching for the exact optimal solution by a back-track scheme (see Section 2.2.2).

ALGORITHM 11.3 (BRANCH-AND-BOUND METHOD FOR (QKP))

Main Step I. (Initial feasible solution). Let $I = \{1, 2, \ldots, n\}$. Compute an initial feasible solution x^0 by certain heuristic procedure. Set $x_{opt} = x^0$ and $f_{opt} = Q(x^0)$.

Main Step II. (Variable fixation)

Step 1. Compute an optimal solution λ^* to problem (D). Let x^* be the optimal solution to the corresponding Lagrangian problem (11.5.16). If $a^T x^* = b$, then the strong duality holds, stop and x^* is the optimal solution to (QKP). If x^* is feasible to (QKP) and $Q(x^*) > f_{opt}$, set $x_{opt} = x^*$ and $f_{opt} = Q(x^*)$.

Step 2. Set $J = \emptyset$. Set $j = 1$.

Step 3. Add $\{j\}$ to J if $1 - x_j^* = 1$, or add $\{-j\}$ to J if $1 - x_j^* = 0$. Add $\{\underline{-k}\}$ to J for all $k \in I \setminus J$ such that $a_k > b - \sum_{i \in J} a_i$. Solve the subproblem with x_i being fixed at 0 if $-i \in J$ and x_i being fixed at 1 if $i \in J$. Let d_j be the Lagrangian bound of the subproblem. If $d_j \leq f_{opt}$, then change $\{j\}$ in J to $\{\underline{-j}\}$ or change $\{-j\}$ in J to $\{\underline{j}\}$ and remove all underlined indices to its right out from J.

Step 4. If $j < n$, set $j := j + 1$ and go to Step 3. Otherwise go to Main Step III.

Main Step III. (Branch-and-bound).

Step 1. Compute the slack $s = b - \sum_{j \in J} a_j$. If $s < 0$, go to Step 6.

Step 2. For each $j \in I \setminus J$, if $a_j > s$, add $\underline{-j}$ to J.

Step 3. Compute the Lagrangian bound $d(\lambda^*)$ on the subproblem with x_i being fixed at 0 if $-i \in J$ and x_i being fixed at 1 if $i \in J$. If $d(\lambda^*) \leq f_{opt}$, go to Step 6.

Step 4. Let x^* be the optimal solution to the Lagrangian relaxation problem (11.5.16) corresponding to the optimal Lagrangian multiplier. If x^* is feasible to (QKP) and $Q(x^*) > f_{opt}$, set $x_{opt} = x^*$ and $f_{opt} = Q(x^*)$. If $a^T x^* = b$, go to Step 6.

Step 5. For each $j \in I \setminus J$, calculate the pseudo-cost $\rho_j = L(x^*, \lambda^*) - L(y^j, \lambda^*)$, where $L(x, \lambda) = Q(x) - \lambda(a^T x - b)$ and $y_i^j = x_i^*$ for $i \neq j$, $y_j^j = 1 - x_j^*$. Choose $j = \arg\min_{j \in I \setminus J} \rho_j$. Add j to J if $x_j^* = 0$ or add $-j$ to J if $x_j^* = 1$, go to Step 1.

Step 6. Seek from right to left the first index j or $-j$ in J that is not underlined. If no such index exists, stop and x_{opt} is the optimal solution. Otherwise, move all indexes to the right of j (or $-j$) out from J and change $\{j\}$ in J to $\{\underline{-j}\}$ or change $\{-j\}$ in J to $\{\underline{j}\}$. Go to Step 1.

EXAMPLE 11.5 Let's apply Algorithm 11.3 to Example 11.4.

Main Step I. (Initial feasible solution) A feasible solution $x^0 = (1, 0, 1, 1)^T$ is obtained by using Procedures 11.1 and 11.3. Set $x_{opt} = (1, 0, 1, 1)^T$ and $f_{opt} = Q(x^0) = 22$.

Main Step II. (Variable fixation)

Step 1. Solving dual problem (D), we obtain $\lambda^* = 2.1111$, $d(\lambda^*) = 27.4444$, $x^* = (0, 0, 0, 0)^T$.

Steps 2-4. Let $J = \emptyset$. Add $\{1\}$ to J, we get the Lagrangian bound $d_1 = 25 > f_{opt}$. Similarly, we have $d_2 = 25.7273 > f_{opt}$, $d_3 = 25.0833 > f_{opt}$, $d_4 = 26.7500 > f_{opt}$. Thus, no variable can be fixed by the Lagrangian bound.

Main Step III. (Branch-and-bound)

Step 1. $s = 13$.

Step 3. Solving the subproblem associated with $J = \emptyset$, we obtain $\lambda^* = 2.1111$, $d(\lambda^*) = 27.4444 > f_{opt}$, $x^* = (0, 0, 0, 0)^T$.

Step 5. The pseudo-costs are $\rho_1 = 27.4444 - 13.6667 = 13.7777$, $\rho_2 = 27.4444 - 20.8889 = 6.5555$, $\rho_3 = 27.4444 - 20 = 7.4444$, $\rho_4 = 27.4444 - 25.2222 = 2.2222$. So $j = 4$. Update J to $\{4\}$.

Step 1. $s = 13 - 2 = 11$.

Step 3. Solving the subproblem associated with $J = \{4\}$, we obtain $\lambda^* = 2.2500$, $d(\lambda^*) = 26.7500 > f_{opt}$, $x^* = (0, 0, 0, 1)^T$.

Step 5. The pseudo-costs are $\rho_1 = 26.75 - 22 = 4.7500$, $\rho_2 = 26.75 - 24.5 = 2.25$, $\rho_3 = 26.75 - 22.75 = 4.00$. So $j = 2$. Update J to $\{4, 2\}$.

Step 1. $s = 13 - 5 - 2 = 6$.

Step 2. Since $a_1 = 7 > 6 = s$, set $J = \{4, 2, \underline{-1}\}$.

Step 4. Solving the subproblem associated with $J = \{4, 2, \underline{-1}\}$, we obtain $\lambda^* = 0$, $d(\lambda^*) = 17 \leq 22 = f_{opt}$, $x^* = (0, 1, 1, 1)^T$.

Step 6. Back track to get an updated $J = \{4, \underline{-2}\}$.

Step 2. $s = 13 - 2 = 11$.

Step 4. Solving the subproblem associated with $J = \{4, \underline{-2}\}$, we obtain $\lambda^* = 0$, $d(\lambda^*) = 22 = f_{opt}$, $x^* = (1, 0, 1, 1)^T$.

Step 7. Back track to get an updated $J = \{\underline{-4}\}$.

Step 2. $s = 13$.

Step 4. Solving the subproblem associated with $J = \{\underline{-4}\}$, we obtain $\lambda^* = 1.0625$, $d(\lambda^*) = 13.8125 < 22 = f_{opt}$, $x^* = (0, 0, 0, 0)^T$.

Step 7. There is no index in J that is not underlined, stop and $x_{opt} = (1, 0, 1, 1)^T$ is an optimal solution to the example.

11.5.4 Alternative upper bounds

In this subsection, we investigate alternative upper bounding techniques for (QKP): Lagrangian decomposition, upper planes and linearization. A general branch-and-bound method will be also presented.

11.5.4.1 Lagrangian decomposition of (QKP)

The Lagrangian decomposition method discussed in Subsection 3.6.2 can be used to generate an upper bound of (QKP) better than the classical Lagrangian bound. Now we apply the decomposition scheme (DQ_2) in Subsection 3.6.2 to problem (QKP). Rewrite (QKP) as

$$\begin{aligned} \max \ & \sum_{j=1}^{n} q_{jj} x_j + \sum_{1 \leq i < j \leq n} q_{ij} x_i x_j \\ \text{s.t.} \ & \sum_{i=1}^{n} a_i y_i \leq b, \\ & x = y, \\ & x \in \{0, 1\}^n, \\ & y \in \{0, 1\}^n. \end{aligned}$$

Note that, different from the Lagrangian decomposition method discussed in Chapter 3, both x and y are integer vectors in the above formulation. Dualizing the equality constraints $x = y$ gives rise to the Lagrangian decomposition function:

$$\begin{aligned}\ell(\mu) &= \max\{\sum_{i=1}^{n}(q_{ii}-\mu_i)x_i + \sum_{1\le i<j\le n} q_{ij}x_ix_j \mid x\in\{0,1\}^n\} \\ &\quad + \max\{\sum_{i=1}^{n}\mu_i y_i \mid \sum_{i=1}^{n} a_i y_i \le b,\ y\in\{0,1\}^n\} \\ &= \ell_1(\mu)+\ell_2(\mu).\end{aligned}$$

Since $q_{ij} \geq 0$ for $1 \leq i < j \leq n$, the first part $\ell_1(\mu)$ can be reduced to a minimum-cut problem and thus is polynomially solvable (see Subsection 11.5.1.3). The second part $\ell_2(\mu)$ is a 0-1 linear knapsack problem which can be solved by efficient methods (see Subsection 6.2.2).

The Lagrangian decomposition dual problem is

$$(DD) \quad \min\ \{\ell(\mu) \mid \mu \in \mathbb{R}^n\}.$$

By Theorem 3.21, problem (DD) generates an upper bound of (QKP) at least as good as the classical Lagrangian bound, i.e.,

$$v(D) \geq v(DD) \geq v(QKP),$$

where (D) is defined in (11.5.3). Moreover, solving the 0-1 linear knapsack problem $\ell_2(\mu)$ also provides us a feasible solution and thus a lower bound for (QKP). We notice, however, the complexity of evaluating $\ell_2(u)$ is NP-complete while problem (D) defined in (11.5.3) is polynomial solvable. Therefore, we have to compromise between the tightness of the upper bound and the computation effort to obtain it in a branch-and-bound algorithm based on a Lagrangian decomposition bound.

LEMMA 11.4 ([163]) *There exists an optimal solution μ^* of (DD) such that $\ell_1(\mu^*) = 0$.*

The above result suggests that the dual search for problem (DD) can be designed to decrease the optimal value of $\ell_1(u)$ while updating the multiplier vector. The following *μ-updating process* for general unconstrained quadratic 0-1 problem is useful for constructing such a procedure.

Let x^* denote the optimal solution to the first part $\ell^1(\mu)$. By Lemma 10.1, if $q_{ii} > \mu_i$, then $x_i^* = 1$. Thus, increasing the value of μ_i to q_{ii} will not modify the optimal solution x^* but will decrease the value of $\ell_1(\mu)$ by $q_{ii} - \mu_i$. Meanwhile, this will increase the value of $\ell_2(\mu)$ by at most $q_{ii} - \mu_i$. Therefore, such a modification of μ will not increase the value of $\ell(\mu)$. After setting $x_i = 1$, the modified quadratic subproblem has a linear term: $\sum_{j\neq i}(q_{jj} + q_{ij} - \mu_j)x_j$. Again, if $\mu_j < q_{jj} + q_{ij}$ for some $j \neq i$, then we deduce that $x_j^* = 1$ and changing μ_j to $q_{jj} + q_{ij}$ will not increase the value of $\ell(\mu)$. The above μ-updating process repeats until no such a j exists and terminates with a new multiplier vector μ'.

By (11.5.16), the initial μ can be taken to be $\lambda^* a$, where λ^* is the optimal solution to (D). Furthermore, we have

$$\begin{aligned}\ell(\lambda^* a) &= \ell_1(\lambda^* a) + \ell_2(\lambda^* a) \\ &= \max_{x\in\{0,1\}^n} \{\sum_{i=1}^n q_{ii}x_i + \sum_{1\le i<j\le n} q_{ij}x_i x_j - \lambda^* \sum_{i=1}^n a_i x_i\} \\ &\quad + \max\{\lambda^* \sum_{i=1}^n a_i y_i \mid \sum_{i=1}^n a_i y_i \le b,\ y \in \{0,1\}^n\} \\ &\le \max_{x\in\{0,1\}^n} \{\sum_{i=1}^n q_{ii}x_i + \sum_{1\le i<j\le n} q_{ij}x_i x_j - \lambda^* (\sum_{i=1}^n a_i x_i - b)\} \\ &= d(\lambda^*).\end{aligned}$$

Then, we have

$$v(D) = d(\lambda^*) \ge \ell(\lambda^* a) \ge \ell(\mu') \ge v(DD) \ge v(QKP).$$

The following heuristic procedure is devised to find an improved upper bound better than $v(D)$.

PROCEDURE 11.4 (HEURISTIC FOR SOLVING (DD))

Step 1. Solve the Lagrangian dual problem (D) and obtain an optimal multiplier λ^*. Set $UB = d(\lambda^*)$. Set $k = 0$.

Step 2. Compute a μ' by the μ-updating process with initial $\mu = \lambda^* a$. Set $\mu = \mu'$.

Step 3. Solve the linear knapsack problem $\ell_2(\mu)$ and set $v_2 = \ell_2(\mu)$.

Step 4. Solve the quadratic problem $\ell_1(\mu)$ and set $v_1 = \ell_1(\mu)$. Let x be the optimal solution to $\ell_1(\mu)$.

Step 5. If $v_1 + v_2 < UB$, set $UB = v_1 + v_2$.

Step 6. If $v_1 > 0$, then use the μ-updating process to modify those μ_i with $x_i = 1$. If $v_1 = 0$ or k exceeds a given maximum number, stop. Otherwsie, set $k := k + 1$, return to Step 3.

11.5.4.2 Upper planes of $Q(x)$

Let $S = \{x \in \{0,1\}^n \mid a^T x \leq b\}$. An *upper plane* of the quadratic function $Q(x)$ defined in (QKP) is any linear function $l(x)$ such that $l(x) \geq Q(x)$ for all $x \in S$. Let f^* denote the optimal value of (QKP). It is clear that if $l(x)$ is an upper plane of $Q(x)$, then an upper bound of f^* can be obtained by solving the linear approximation problem:

$$\max\ \{l(x) \mid x \in \tilde{S}\}, \tag{11.5.22}$$

where $\tilde{S} \supseteq S$. Two typical choices of $\tilde{S}$ are: S and $\bar{S} := \{x \in [0,1]^n \mid a^T x \leq b\}$. If $\tilde{S} = S$, then problem (11.5.22) is a 0-1 linear knapsack problem, which is relatively easy to solve (see [153]). If $\tilde{S} = \bar{S}$, then (11.5.22) becomes a continuous linear knapsack problem that can be solved by greedy methods discussed in Subsection 6.2.2.

In the following, we describe several ways of deriving upper planes for (QKP). Let $h_{ii} = q_{ii}$ and $h_{ij} = (1/2)q_{ij}$ for all i and j. Define $H = (h_{ij})_{n\times n}$. Then $Q(x) = x^T H x$ for all $x \in \{0,1\}^n$ and the quadratic function $Q(x)$ can be rewritten as

$$Q(x) = x^T H x = \sum_{j=1}^n (\sum_{i=1}^n h_{ij} x_i) x_j.$$

Let $p_j(x) = \sum_{i=1}^n h_{ij} x_i$. Let v_j be an upper bound of $p_j(x)$ over S. Then $l(x) = \sum_{j=1}^n v_j x_j$ gives rise to an upper plane of $Q(x)$.

Since $h_{ij} \geq 0$ for all i, j, the simplest bound of $p_j(x)$ is

$$v_j^1 = \sum_{i=1}^n h_{ij} = q_{jj} + (1/2) \sum_{i\neq j}^n q_{ij}. \tag{11.5.23}$$

Let m be the largest possible number of 1's in a feasible solution of (QKP). Let I_j be the set of indexes of the m largest elements of h_{ij}, $j = 1,\ldots,n$. Then, an improved bound is given by

$$v_j^2 = \sum_{i\in I_j} h_{ij}. \tag{11.5.24}$$

Other more tighter bounds are given by

$$v_j^3 = \max\{q_{jj}x_j + (1/2) \sum_{i\neq j}^n q_{ij} x_i \mid x \in \bar{S}\}. \tag{11.5.25}$$

$$v_j^4 = \max\{q_{jj}x_j + (1/2) \sum_{i\neq j}^n q_{ij} x_i \mid x \in S\}. \tag{11.5.26}$$

Obviously, v_j^4 provides the tightest upper bound for $l_j(x)$ and

$$v_j^1 \geq v_j^2 \geq v_j^4,$$
$$v_j^1 \geq v_j^3 \geq v_j^4.$$

Since a tighter upper bound often requires more computational efforts to obtain, a good trade-off needs to be found out in order to design an efficient branch-and-bound algorithm for (QKP). It was shown in [66] that the most efficient upper plane is given by $l(x) = \sum_{j=1}^n v_j^3 x_j$.

Now, consider upper planes for Example 11.4. The optimal solution of Example 11.4 is $x^* = (1,0,1,1)^T$ with $Q(x^*) = 22$. The upper plane determined by v_j^1 can be determined by using (11.5.23): $l^1(x) = 11x_1 + 10x_2 + 5.5x_3 + 11.5x_4$. The corresponding linear knapsack problem

$$\max \ \{l^1(x) \mid 7x_1 + 5x_2 + 4x_3 + 2x_4 \leq 13, \ x \in \{0,1\}^n\}$$

has an optimal solution $\bar{x} = (1,0,1,1)^T$. So the upper bound is $UB_1 = l^1(\bar{x}) = 28$.

Consider the upper plane determined by v_j^2. Since the largest number of 1's in the knapsack is 3, we calculate $v_1^2 = 3+2+5 = 10$, $v_2^2 = 4+3+2.5 = 9.5$, $v_3^2 = 1+2+2 = 5$, $v_4^2 = 2+5+2.5 = 9.5$. So, $l^2(x) = 10x_1 + 9.5x_2 + 5x_3 + 9.5x_4$. The corresponding linear knapsack problem

$$\max \ \{l^2(x) \mid 7x_1 + 5x_2 + 4x_3 + 2x_4 \leq 13, \ x \in \{0,1\}^n\}$$

has an optimal solution $\bar{x} = (1,0,1,1)^T$ which yields the upper bound $UB_2 = l^2(\bar{x}) = 24.5$.

The upper plane determined by v_j^3 is $l^3(x) = 10.2857x_1 + 9.0714x_2 + 5x_3 + 9x_4$. Solving

$$\max \ \{l^3(x) \mid 7x_1 + 5x_2 + 4x_3 + 2x_4 \leq 13, \ x \in \{0,1\}^n\}$$

yields an optimal solution $\bar{x} = (1,0,1,1)^T$ which gives out the upper bound $UB_3 = l^3(\bar{x}) = 24.2857$. Finally, the upper plane determined by v_j^4 is $l^4(x) = 10x_1 + 7x_2 + 5x_3 + 9x_4$. The corresponding linear knapsack problem

$$\max \ \{l^4(x) \mid 7x_1 + 5x_2 + 4x_3 + 2x_4 \leq 13, \ x \in \{0,1\}^n\}$$

has an optimal solution $\bar{x} = (1,0,1,1)^T$ which yields the upper bound $UB_4 = l^4(\bar{x}) = 24$. In this example, we see that

$$UB_1 > UB_2 > UB_3 > UB_4 > Q(x^*).$$

11.5.4.3 Linearization

By replacing each quadratic term $x_i x_j$ with a new 0-1 x_{ij}, (QKP) can be converted into an equivalent 0-1 linear integer programming:

$$
\begin{aligned}
(ILP) \qquad & \max \sum_{i=1}^{n} q_{ii} x_i + \sum_{1 \le i < j \le n} q_{ij} x_{ij} \\
& \text{s.t.} \sum_{i=1}^{n} a_i x_i \le b, && (11.5.27) \\
& x_{ij} \le x_i, \ 1 \le i < j \le n, && (11.5.28) \\
& x_{ij} \le x_j, \ 1 \le i < j \le n, && (11.5.29) \\
& x_i + x_j - 1 \le x_{ij}, \ 1 \le i < j \le n, && (11.5.30) \\
& x_i \in \{0, 1\}, \ i = 1, \ldots, n, && (11.5.31) \\
& x_{ij} \in \{0, 1\}, \ 1 \le i < j \le n. && (11.5.32)
\end{aligned}
$$

We notice that constraint (11.5.30) is redundant in (ILP) because $q_{ij} \ge 0$ for all i, j. The continuous relaxation of (ILP)) then provides an upper bound of the optimal value of (QKP). However, the quality of the upper bound provided by the continuous relaxation could be very poor. Some valid inequality techniques can be used to tighten this upper bound. Multiplying both sides of (11.5.27) by x_j and using the fact $x_j^2 = x_j$, we obtain the following constraints (see [2] [22]) :

$$\sum_{i<j} a_i x_{ij} + \sum_{i>j} a_i x_{ij} \le (b - a_j) x_j, \quad j = 1, \ldots, n. \qquad (11.5.33)$$

The above constraints are redundant in (ILP) when x_j and x_{ij} are 0-1 variables and hence are valid constraints. Similarly, another set of constraints which involve six variables can be derived as follows:

$$x_i + x_j + x_k - x_{ij} - x_{ik} - x_{jk} \le 1, \quad 1 \le i < j < k \le n. \qquad (11.5.34)$$

The resulting linear programming can be expressed as:

$$(LP) \qquad \max \sum_{i=1}^{n} q_{ii}x_i + \sum_{1\le i<j\le n} q_{ij}x_{ij}$$

$$\text{s.t.} \sum_{i=1}^{n} a_i x_i \le b, \tag{11.5.35}$$

$$x_{ij} \le x_i,\ 1 \le i < j \le n, \tag{11.5.36}$$

$$x_{ij} \le x_j,\ 1 \le i < j \le n, \tag{11.5.37}$$

$$x_i + x_j - 1 \le x_{ij},\ 1 \le i < j \le n, \tag{11.5.38}$$

$$0 \le x_i \le 1,\ i = 1, \ldots, n, \tag{11.5.39}$$

$$x_{ij} \ge 0,\ 1 \le i < j \le n, \tag{11.5.40}$$

$$\sum_{i<j} a_i x_{ij} + \sum_{i>j} a_i x_j \le (b - a_j)x_j,\ j = 1, \ldots, n, \tag{11.5.41}$$

$$x_i + x_j + x_k - x_{ij} - x_{ik} - x_{jk} \le 1,\ \ 1 \le i < j < k \le n. \tag{11.5.42}$$

Numerical test shows that the upper bound computed from solving the above linear programming gives a much better upper bound than that of the direct continuous relaxation of (ILP) (see [22]). However, the number of constraints in (LP) becomes prohibitive as n increases. For example, (LP) has more than 1500 constraints when $n = 20$. One way to overcome this difficulty is to generate the constraints (11.5.36), (11.5.37), (11.5.38) and (11.5.42) sequentially during the progress of solving the linear programming. A linear programming with constraints (11.5.35), (11.5.39), (11.5.40) and (11.5.41) is first solved. If the optimal solution does not satisfy constraints in (11.5.36), (11.5.37), (11.5.38) and (11.5.42), then the corresponding constraint is generated one by one and the resulting linear programming is solved by the dual simplex method.

11.5.4.4 A general branch-and-bound method

We now describe a general framework of a branch-and-bound method for (QKP). The branch-and-bound method consists of three main steps: (i) Computing an initial lower bound and a feasible solution and improving feasible solutions by certain heuristics; (ii) Fixing certain variables by Lagrangian dual methods; (iii) Performing a standard binary search for the unfixed variables. An upper bound at each individual node can be computed by various methods: (a) classical Lagrangian method, (b) Lagrangian decomposition method, (c) upper planes and (d) linearization method. Heuristics described in Section 11.5.2 can be used to generate a lower bound and feasible solutions. Let $\tilde{x}$ be a feasible solution obtained. For each variable x_i, an upper bound is computed for the problem where x_i is fixed at $1 - \tilde{x}_i$. If the upper bound is less than or equal to the objective value of the incumbent, then set $x_i = \tilde{x}_i$ in the optimal solution of (QKP).

ALGORITHM 11.4 (BRANCH-AND-BOUND ALGORITHM FOR (QKP))

Step 1. Compute a feasible solution $\tilde{x}$ to (QKP) by certain heuristics. Let LB be the corresponding lower bound of (QKP).

Step 2. For each i, compute an upper bound ub_i for problem (QKP) with x_i fixed at $1-\tilde{x}_i$. If $ub_i \leq LB$, then, set $x_i = \tilde{x}_i$. Update the lower bound LB if a better feasible solution is found during the upper bounding procedure.

Step 3. At each node, an upper bound ub of the corresponding subproblem is computed. If $ub \leq LB$, then the node is fathomed. Otherwise, the node is branched into two nodes by setting $x_i = 1$ and $x_i = 0$, respectively.

11.6 Notes

Excellent surveys of the methods for constrained nonlinear 0-1 programming problems can be found in [94] and [95].

The reduction of problem $(0\text{-}1PP)$ to an unconstrained 0-1 optimization problem was discussed in [111][113][200]. The linearization method was first proposed by Dantzig [48] and Fortet [63][64] (see also [224]). Various branch-and-bound methods or implicit enumeration methods were proposed in, for example, [95] and the references therein. The cutting-plane method for $(0\text{-}1PP)$ with a linear objective function was originated from [83] and was extensively studied in [9][10].

The quadratic 0-1 knapsack problem was first introduced in [66] and was studied by many authors (see [22][23][38][39][89][163]).

Chapter 12

TWO LEVEL METHODS FOR CONSTRAINED POLYNOMIAL 0-1 PROGRAMMING

Consider constrained polynomial 0-1 programming problems in the following form:

$$(0\text{-}1PP_2) \qquad \min\ f(x) = \sum_{k=1}^{q} c_k \prod_{j\in Q_k} x_j$$
$$\text{s.t.}\ g_i(x) = \sum_{k=1}^{q} a_{ik} \prod_{j\in Q_k} x_j \le b_i, \quad i = 1, 2, \ldots, m,$$
$$x \in X = \{0, 1\}^n,$$

where $Q_k \subseteq \{1, \ldots, n\}$ for $k = 1, \ldots, q$. Note that any constrained polynomial 0-1 programming problem in the general form of $(0\text{-}1PP)$ can be represented in the form of $(0\text{-}1PP_2)$.

This chapter consists of a set of three solution methods for $(0\text{-}1PP_2)$. The first one is a revised version of a two-level method proposed by Taha in [211]. A systematic solution framework is established to achieve an efficiency in searching for an exact solution. The second method is to apply the revised Taha's method to an equivalent singly-constrained formulation of $(0\text{-}1PP_2)$ resulted from applying the p-norm surrogate constraint method discussed in Chapter 4. The last method is an integration of the revised Taha's method with the convergent Lagrangian and objective level cut method discussed in Chapter 7.

12.1 Revised Taha's Method

As Taha suggested in [211], the problem $(0\text{-}1PP_2)$ can be transformed into an equivalent two-level problem that consists of a 0-1 linear *master problem* with positive coefficients in the objective function,

$$\begin{aligned} \min \ \tilde{f}(y) &= \sum_{k=1}^{q} \tilde{c}_k y_k \\ \text{s.t. } \tilde{g}_i(y) &= \sum_{k=1}^{q} \tilde{a}_{ik} y_k \le \tilde{b}_i, \quad i = 1, 2, \ldots, m, \\ y &\in \{0,1\}^q, \end{aligned} \tag{12.1.1}$$

and a set of *secondary constraints*,

$$y_k = \begin{cases} \prod_{j \in Q_k} x_j, & k \in J^+ = \{k \mid c_k \ge 0\}, \\ 1 - \prod_{j \in Q_k} x_j, & k \in J^- = \{k \mid c_k < 0\}, \end{cases} \tag{12.1.2}$$

where $\tilde{c}_k = c_k$ and $\tilde{a}_{ik} = a_{ik}$ for $k \in J^+$, $\tilde{c}_k = -c_k$ and $\tilde{a}_{ik} = -a_{ik}$ for $k \in J^-$, $\tilde{b}_i = b_i - \sum_{k \in J^-} a_{ik}$. Note that $f(x) = \tilde{f}(y) + \sum_{k \in J^-} c_k$. We call $x_1, \ldots, x_n$ the *decision variables* and $y_1, \ldots, y_q$ the *decision terms.*

12.1.1 Definitions and notations

Let $N = \{1, \ldots, n\}$, $M = \{1, \ldots, m\}$ and $Q = \{1, \ldots, q\}$. Let $I_t \subseteq Q$ denote the index set of y_k's determined at iteration t. Define a signed index set

$$J_t = \{\xi \mid \xi = k \text{ if } y_k = 1,\ k \in I_t;\ \xi = -k, \text{ if } y_k = 0,\ k \in I_t\}.$$

Then, J_t represents a *partial solution* determined at iteration t. A decision term y_k with $k \in \bar{I}_t = Q \setminus I_t$ is said to be a *free* term of the partial solution J_t. Assigning binary values to all free decision terms of J_t yields a *completion* of J_t. Note that if J_t has l elements, it can determine 2^{q-l} different completions. Among all completions of J_t, the *typical* completion y^t is the completion with all the free y_k's set to be zero. Since all $\tilde{c}_k$'s in the master problem (12.1.1) are nonnegative, the typical completion of J_t has the minimum value of the objective function among all completions of J_t.

A partial solution J_t is said to be *feasible* (*infeasible*) if its typical completion constitutes a feasible (*infeasible*) solution y to the master problem (12.1.1). A partial solution J_t can be also used to partially determine some decision variables x_j's consistently via the secondary constraints (12.1.2) or can lead to an inconsistent J_{t+1}. When an inconsistency occurs, J_t is said to be an *inconsistent partial solution.* Otherwise, it is a *consistent partial solution.* It is clear an inconsistency of J_t implies that all completions of J_t are inconsistent to the secondary constraints.

When J_t is consistent, the decision variables x_j's determined by the second constraints (12.1.2) form the *converted solution* of J_t. The converted solution can be represented by the signed index set:

$$D_t = \{\xi \mid \xi = j \text{ if } x_j = 1,\ j \in d_t;\ \xi = -j, \text{ if } x_j = 0,\ j \in d_t\},$$

where d_t is the index set of all x_j's in the converted solution. The converted solution D_t could further determine some free decision terms y_k's by the secondary constraints. These determined decision terms constitute an *augmented solution* of J_t which can be represented by the signed index set:

$$B_t = \{\xi \mid \xi = k \text{ if } y_k = 1,\ j \in b_t;\ \xi = -k, \text{ if } y_k = 0,\ k \in b_t\},$$

where b_t is the index set of all y_k's determined by the converted solution D_t. If B_t is uniquely determined by D_t, then the complement of any element in B_t must lead to an inconsistency and thus all decision terms in the augmented solution can be fixed. We underline a signed index in B_t to denote that this decision term is fixed in the augmented solution. It is clear that a new partial solution $J_{t+1} = J_t \cup B_t$ must be consistent. In the case of $B_t = \emptyset$, J_t itself is consistent.

The following example illustrates the concepts introduced above.

EXAMPLE 12.1

$$\begin{aligned}
\min\ & 3x_1 + 5x_1x_2x_3 + 3x_1x_4x_5 + 8x_2x_3x_5 - 4x_3x_4x_5 \\
\text{s.t.}\ & 3x_1 - x_1x_4x_5 - x_2x_3x_5 + x_3x_4x_5 \le 2, \\
& 2x_1 - 4x_1x_2x_3 - 7x_1x_4x_5 - 3x_2x_3x_5 - x_3x_4x_5 \le -3, \\
& -6x_1 - 3x_1x_2x_3 + 5x_1x_4x_5 - 3x_2x_3x_5 + 6x_3x_4x_5 \le 5, \\
& x_1, x_2, x_3, x_4, x_5 \in \{0, 1\}.
\end{aligned}$$

The above example can be converted into a two-level formulation with a master program,

$$\begin{aligned}
\min\ & 3y_1 + 5y_2 + 3y_3 + 8y_4 + 4y_5 - 4 \\
\text{s.t.}\ & 3y_1 - y_3 - y_4 - y_5 \le 1, \\
& 2y_1 - 4y_2 - 7y_3 - 3y_4 + y_5 \le -2, \\
& -6y_1 - 3y_2 + 5y_3 - 3y_4 - 6y_5 \le -1, \\
& y_1, y_2, y_3, y_4, y_5 \in \{0, 1\},
\end{aligned} \tag{12.1.3}$$

and a set of secondary constraints,

$$\left\{ \begin{array}{lcl}
y_1 & = & x_1, \\
y_2 & = & x_1x_2x_3, \\
y_3 & = & x_1x_4x_5, \\
y_4 & = & x_2x_3x_5, \\
y_5 & = & 1 - x_3x_4x_5.
\end{array} \right. \tag{12.1.4}$$

Consider a partial solution $J_0 = \emptyset$, in which all decision terms are free and there exist in total $2^5 = 32$ completions of J_0. It is clear that the typical completion of J_0, $y^0 = (0,0,0,0,0)^T$, is an infeasible partial solution.

Assigning value one to the free term y_2 of J_0 leads to a new partial solution $J_1 = \{2\}$. Now J_1 is a feasible partial solution because its typical completion, $y^0 = (0,1,0,0,0)^T$, is feasible in the master problem (12.1.3). Also J_1 is a consistent partial solution, since it can determine $x_1 = x_2 = x_3 = 1$ via the secondary constraints (12.1.4). Thus, the converted solution D_1 is found to be $\{1,2,3\}$. From D_1 and the secondary constraints of $y_1 = x_1$, y_1 can be further fixed at one. Thus, the augmented solution of J_1 is identified to be $B_1 = \{\underline{1}\}$. We can get a new partial solution J_2 via augmenting J_1 by B_1 on the right, i.e., $J_2 = \{2, \underline{1}\}$. It is easy to check that J_2 is a consistent partial solution.

Suppose at iteration t we have a feasible partial solution $J_t = \{2, \underline{3}, \underline{5}\}$. It can be verified that J_t is an inconsistent partial solution since applying the secondary constraints leads to a contradiction ($y_2 = y_3 = 1$ implies $x_3 = x_4 = x_5 = 1$, while at the same time, $y_5 = 1$ requires at least one of x_3, x_4, x_5 equal to zero). Furthermore, all completions of J_t must be also inconsistent.

12.1.2 Fathoming, consistency and augmentation

Due to its flexibility and its generality, the backtrack scheme [73] discussed in Chapter 2 can be used as a solution concept to solve the polynomial 0-1 problems. Especially, the fathoming and augmenting techniques are suitable to be adopted for the two-level formulation (12.1.1)–(12.1.2).

Observe that a feasible solution to $(0\text{-}1PP_2)$ implies that the corresponding solution y derived from (12.1.2) is feasible to the master problem (12.1.1). Thus, the optimal solution of $(0\text{-}1PP_2)$ can be sought from among the feasible solutions to the master problem (12.1.1) that satisfy the secondary constraints (12.1.2). Since the master problem (12.1.1) is a linear 0-1 programming problem, at each iteration, the additive algorithm described in Chapter 2 can be modified to search for a feasible solution which is better than the incumbent.

The algorithm for $(0\text{-}1PP_2)$ consists of three main sub-procedures: *fathoming, consistency check and B_t recognition.*

A partial solution J_t is fathomed if there is no need to investigate further the completions of J_t. Let $\tilde{f}^t = \tilde{f}(y^t)$ and $\tilde{g}_i^t = \tilde{g}_i(y^t)$, where y^t is the typical completion of J_t. We use y^* to record the incumbent solution and let $f_{opt} = \tilde{f}(y^*)$.

LEMMA 12.1 *Let J_t be a partial solution at iteration t. J_t can be fathomed if one of the following conditions holds:*

(i) *$\tilde{f}^t \geq f_{opt}$ (domination);*

(ii) *J_t has no feasible completion (feasibility);*

(iii) *J_t is feasible and inconsistent (consistency);*

(iv) *J_t is feasible and consistent with $B_t = \emptyset$ (optimality).*

Proof. (i) Since y^t is the typical completion with the minimum objective value among all completions of J_t, no completion of J_t can have a smaller objective value than f_{opt} when $\tilde{f}^t \geq f_{opt}$.

(ii) If J_t has no feasible completion, all completions of J_t are infeasible to the original problem.

(iii) If J_t is inconsistent, all of its completions are also inconsistent to the secondary constraints. This implies that no completion of J_t can lead to a feasible solution to the original problem.

(iv) In this case, J_t is both feasible and consistent. So J_t and its converted solution D_t must satisfy the master problem and the secondary constraints simultaneously. $B_t = \emptyset$ implies that there is no need to augment J_t. Since all $\tilde{c}_k \geq 0$, the typical completion of J_t is the best feasible solution among all its completions. After setting $y^* = y^t$ and $f_{opt} = \tilde{f}^t$ if $\tilde{f}^t < f_{opt}$, no optimal solution will be lost. □

Based on Lemma 12.1, a procedure can be devised to fathom certain partial solutions at each iteration by using domination, feasibility, consistency or optimality. Let (MP_t) denote the master problem (12.1.1) with y_k, $k \in I_t$, being fixed at zero or one according to J_t. Notice that we only apply one iteration of the additive algorithm to (MP_t). We leave the iterative loop of the additive algorithm either when a feasible completion of J_t is found with an objective value less than f_{opt}, or a conclusion is reached that no feasible completion of J_t can have an objective value less than f_{opt}. In the latter case, J_t is fathomed by feasibility.

At the t-th iteration, the fathoming process starts by applying the additive algorithm to search for a feasible solution to (MP_t) which is better than the incumbent solution y^*. If a better feasible partial solution J_t is found, we check its consistency. Based on the relationship between the consistency and the converted solution, the consistency check is equivalent to solving the system of the secondary constraints (12.1.2) by assigning the values to x_j's according to J_t. If the secondary constraints (12.1.2) cannot be satisfied, J_t is inconsistent and it is fathomed by consistency; otherwise, a converted solution, D_t, can be found in the process of consistency check. The consistency check is designed as a two-phase procedure. In the first phase, the procedure deals with the case when $k \in J_t$ and $k \in J^+$ (or when $-k \in J_t$ and $k \in J^-$). More specifically, we need to specify the values to some x_j's via the following types of secondary constraints:

$$1 = \prod_{j \in Q_k} x_j, \; k \in J^+,$$

$$0 = 1 - \prod_{j \in Q_k} x_j, \; k \in J^-.$$

It is clear that the consistency check leads to fixation of all x_j with $j \in Q_k$ at 1. In the second phase, the procedure deals with the case when $k \in J_t$ and $k \in J^-$ (or when $-k \in J_t$ and $k \in J^+$). More specifically, we need to specify the values to some x_j's via the following types of secondary constraints:

$$0 = \prod_{j \in Q_k} x_j, \; k \in J^-,$$

$$1 = 1 - \prod_{j \in Q_k} x_j, \; k \in J^+.$$

In general, we are not able to determine how many of x_j with $j \in Q_k$ are zero and which one is zero. In particular, if one x_j with $j \in Q_k$ has been already fixed at zero by some previous consistency check, then we are not able to specify the remaining x_j's with $j \in Q_k$. We are able, however, to draw a conclusion in the following two cases. First, if all x_j's with $j \in Q_k$ have been already fixed at one by some previous consistency check, an inconsistency occurs. Second, if all but one x_j have been already fixed at one by some previous consistency check, the remaining x_j needs to be fixed at zero.

The consistency check procedure discussed above can be now summarized as follows.

PROCEDURE 12.1 (CONSISTENCY CHECK)

Given a nonempty partial solution J_t.

Phase 1

Step 1.0. Set $D_t = \emptyset$, $J = J_t$.

Step 1.1. If $J = \emptyset$, exit.

Step 1.2. Find k such that $k \in J$ and $k \in J^+$, or $-k \in J$ and $k \in J^-$. If no such a k exists, go to Phase 2.

Step 1.3. Set $D_t := D_t \cup Q_k$. If D_t has n elements, go to Phase 2.

Step 1.4. If $k \in J^+$, set $J := J \setminus \{k\}$; Otherwise, set $J := J \setminus \{-k\}$. Return to Step 1.1.

Phase 2

Step 2.1. Find k such that $k \in J$ and $k \in J^-$, or $-k \in J$ and $k \in J^+$. If no such a k exists, exit.

Step 2.2. If there is a $j \in Q_k$ such that $-j \in D_t$, go to Step 2.5.

Step 2.3. If $Q_k \subseteq D_t$, exit and report an inconsistency.

Step 2.4. If there is a unique $j \in Q_k$ such that $j \notin D_t$, set $D_t := D_t \cup \{-j\}$.

Step 2.5. If $k \in J^-$, set $J := J \setminus \{k\}$; Otherwise, set $J := J \setminus \{-k\}$. Return to Step 2.1.

After finite iterations, Procedure 12.1 either finds a converted solution D_t or reports an inconsistency of the partial solution J_t.

Let us consider a partial solution $J_t = \{3, 5\}$ for Example 12.1. Since $3 \in J^+ = \{1, 2, 3, 4\}$, we apply Phase 1 and obtain $D_t = Q_3 = \{1, 4, 5\}$. As $5 \in J^- = \{5\}$ and $Q_5 = \{3, 4, 5\}$ in which y_4 and y_5 have been already fixed at one by D_t, applying Phase 2 expands D_t to $\{1, 4, 5, -3\}$. Consider another partial solution $J_t = \{1, -3, -5\}$ for Example 12.1. Applying Phase 1 gives $D_t = Q_1 \cup Q_5 = \{1, 3, 4, 5\}$. Since $Q_3 = \{1, 4, 5\} \subset D_t$, applying Phase 2 leads to an inconsistency of J_t.

From D_t, the augmented solution B_t can be identified by the secondary constraints (12.1.2). The B_t recognition consists of three phases. Phase 1 deals with the following two situations: i) For a k which is not included in I_t, the index set of J_t, if all elements x_j's with $j \in Q_k$ are assigned to be one by D_t, then y_k has to be equal to one if $k \in J^+$, or equal to zero if $k \in J^-$. ii) For a k which is not included in I_t, if there exists a $j \in Q_k$ such that $-j \in D_t$, i.e., one element in Q_k is assigned to be zero by D_t, then y_k has to be equal to zero if $k \in J^+$, or equal to one, if $k \in J^-$. All elements in B_t generated in Phase 1 have to be underlined.

When performing consistency check, information of J_t, in many cases, is not enough to determine the decision variables in Q_k when $k \in J_t$ and $k \in J^-$ or when $-k \in J_t$ and $k \in J^+$. Consider $J_t = \{1, -3\}$ for Example 12.1. Consistency check only gives $D_t = \{1\}$. Decision variables x_4 and x_5 in Q_3 are left undetermined which one should be zero. Phase 2 of B_t recognition fixes one undetermined decision variable to zero for every such a decision term. If x_4 is set to zero in Example 12.1 when $J_t = \{1, -3\}$, then y_5 will be fixed at 1 further. Thus, Phase 2 of B_t recognition may generate new members in B_t, for example, $\{5\}$ in the above example. All elements in B_t generated in Phase 2 are not underlined. One point to emphasize here is that adding a non-underlined element to the right of a partial solution does not eliminate any completion to be checked.

Phase 3 of B_t recognition deals with a tricky situation. Let us consider Example 12.1 with a partial solution $J_t = \{1\}$. The corresponding D_t is $\{1\}$. Note that both $y_3 = x_1x_4x_5$ and $y_5 = 1 - x_3x_4x_5$ are free terms since x_3, x_4 and x_5 are not fixed by D_t. We can verify that y_3 and y_5 cannot be zero at the same time. Without loss of generality, we can set $B_t = B_t \cup \{-5, \underline{3}\}$ to avoid a partial solution with both y_3 and y_5 being zero. Note that adding an underlined element to the right of a partial solution of l elements eliminates

2^{q-l-1} possible completions. Phase 2 is devised to avoid possible inconsistency resulted from applying Phase 3 alone. Consider again $J_t = \{1, -3\}$ for Example 12.1. Without performing Phase 2 in advance, Phase 3 will give $B_t = \{-5, \underline{3}\}$ which contradicts the fact that $-3 \in J_t$.

The three-phase B_t recognition procedure is now described as follows.

PROCEDURE 12.2 (B_t RECOGNITION)

Given a nonempty partial solution J_t, its index set I_t and an augmented solution D_t.

Phase 1

Step 1.0. Set $B_t = \emptyset$, $\bar{I}_t = Q \setminus I_t$ and $\tilde{B} = \emptyset$.

Step 1.1. Find $k \in \bar{I}_t$ such that $j \in Q_k$ and $-j \in D_t$. Set $\bar{I}_t := \bar{I}_t \setminus \{k\}$. If $k \in J^+$, set $B_t := B_t \cup \{-\underline{k}\}$; otherwise, set $B_t := B_t \cup \{\underline{k}\}$.

Step 1.2. Find $k \in \bar{I}_t$ such that $Q_k \subseteq D_t$. Set $\bar{I}_t := \bar{I}_t \setminus \{k\}$. If $k \in J^+$, set $B_t := B_t \cup \{\underline{k}\}$; Otherwise, set $B_t := B_t \cup \{-\underline{k}\}$.

Step 1.3. If there exists a $k \in B_t$, set $\tilde{B} = B_t$.

Phase 2

Step 2.0. Set $J = J_t$.

Step 2.1. Find k such that $k \in J$ and $k \in J^-$, or $-k \in J$ and $k \in J^+$. If no such a k exists, go to Step 2.4.

Step 2.2. If there is a $j \in Q_k$ such that $-j \in D_t$, set $J := J \setminus \{k \text{ or } -k\}$, return to Step 2.1.

Step 2.3. Find $j \in Q_k$ such that $j \notin D_t$ and $-j \notin D_t$, set $D_t := D_t \cup \{-j\}$, $J := J \setminus \{k \text{ or } -k\}$, return to Step 2.1.

Step 2.4. If $J = J_t$, go to Phase 3.

Step 2.5. Find $k \in \bar{I}_t$ such that $j \in Q_k$ and $-j \in D_t$, set $\bar{I}_t := \bar{I}_t \setminus \{k\}$. If $k \in J^+$, set $B_t := B_t \cup \{-k\}$; otherwise, set $B_t := B_t \cup \{k\}$.

Step 2.6. If there exists a $j \in B_t$, set $\tilde{B} = B_t$.

Phase 3

Step 3.1. For $k \in \bar{I}_t \cap J^-$, set $\bar{I}_t := \bar{I}_t \setminus \{k\}$, $D_t := D_t \cup Q_k$ and $B_t := B_t \cup \{-k\}$. If no such a k exists, exit.

Step 3.2. Find $k \in \bar{I}_t$ such that $Q_k \subseteq D_t$, set $B_t := B_t \cup \{\underline{k}\}$. If no such a k exists, set $B_t = \tilde{B}$, exit.

Let us consider the following instance for Example 12.1 to illustrate how to construct B_t. Suppose we have $J_t = \{1, 4\}$. A converted solution $D_t =$

$\{1, 2, 3, 5\}$ is derived. Phase 1 of Procedure 12.2 identifies $B_t = \{\underline{2}\}$. Phase 3 of Procedure 12.2 generates $y_5 = 0$ and $y_3 = 1$. Finally, $B_t = \{\underline{2}, -5, \underline{3}\}$.

The above B_t recognition procedure finds an augmented solution B_t within finite iterations. When B_t only contains elements of $\{-k\}$, B_t is also considered as an empty set. An empty B_t implies that no free term need to be fixed at one by B_t, i.e., there is no need to augment J_t any further. Thus, y^t is a feasible and consistent solution. The incumbent can then be replaced by $y^* = y^t$ and $f_{opt} = \tilde{f}^t$, and J_t is fathomed by optimality. If B_t is not empty, we augment J_t by adding B_t on the right, i.e., $J_t = J_t \cup B_t$. We then re-calculate the values of $\tilde{f}^t$ and $\tilde{g}_i^t$. If $\tilde{f}^t$ is greater than or equal to the incumbent value f_{opt}, J_t is fathomed by domination. If the inequality $\tilde{f}^t < f_{opt}$ holds and also J_t is still a feasible partial solution, we update the incumbent with $y^* = y^t$ and $f_{opt} = \tilde{f}^t$. If J_t is infeasible, let $J_{t+1} = J_t$ and apply the modified additive algorithm to (MP_{t+1}) for a feasible partial solution of J_{t+1} with an objective value less than f_{opt}.

As in the backtrack technique discussed in Chapter 2, when a partial solution J_t is fathomed at iteration t, we locate the rightmost element in J_t which is not underlined. If none exists, we could claim that all 2^q solutions are implicitly enumerated and the algorithm terminates. Otherwise, we replace it by its underlined complement and delete all elements to its right. A non-redundant partial solution J_{t+1} is then generated for the next iteration.

12.1.3 Solution algorithm

The solution algorithm is presented as follows, while a flow diagram is given in Figure 12.1.

ALGORITHM 12.1 (REVISED TAHA'S METHOD)

Step 0. Set $J_0 = \emptyset$, $t = 0$, and $f_{opt} = \infty$.

Step 1. Apply the modified additive algorithm to (MP_t) to find a feasible partial solution J_t^* whose objective value is strictly less than f_{opt}. If such feasible partial solution is found, let $J_t = J_t \cup J_t^*$, where $J_t^* \subseteq \bar{I}_t$. Otherwise, fathom J_t by feasibility and go to the Step 6 of backtracking.

Step 2 (Consistency check). If J_t is inconsistent, fathom J_t by consistency and go to Step 6. Otherwise, construct D_t.

Step 3 (B_t recognition). If $B_t = \emptyset$, set $y^* = y^t$, $f_{opt} = \tilde{f}^t$, fathom J_t by optimality and go to Step 6. Otherwise, augment J_t with B_t on the right.

Step 4. If $\tilde{f}^t \geq f_{opt}$, fathom J_t by domination and go to Step 6.

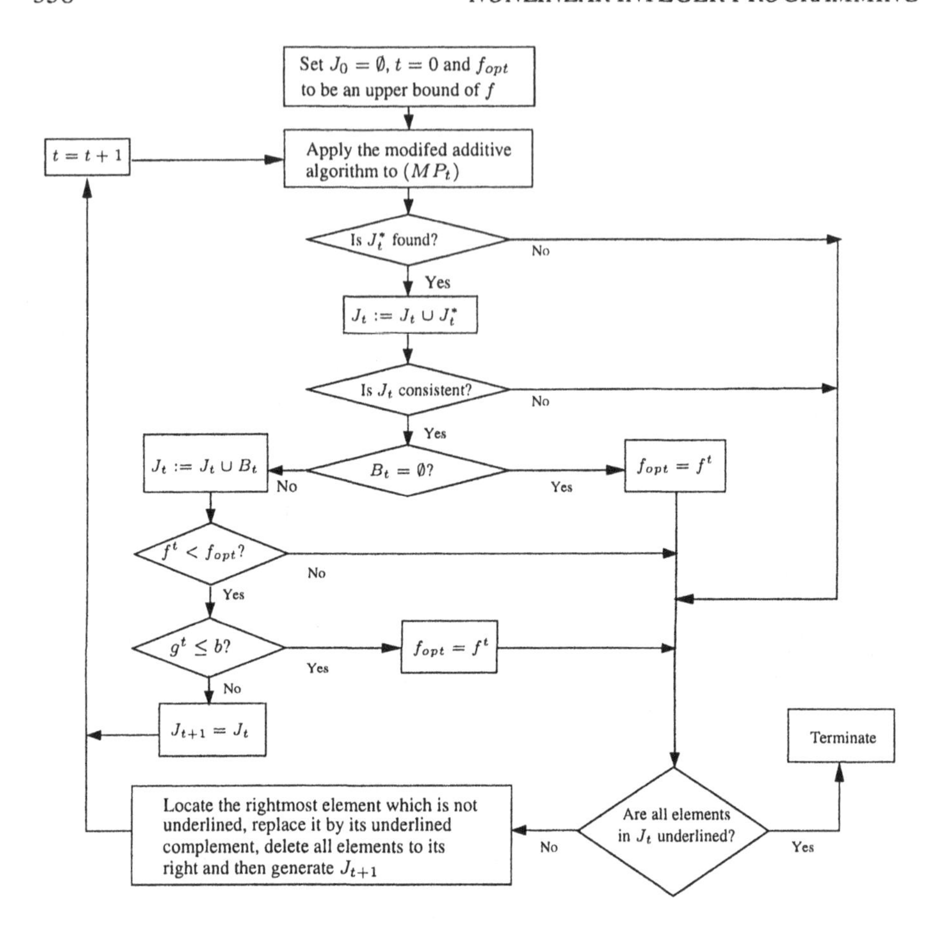

Figure 12.1. Diagram of revised Taha's method.

Step 5. If $\tilde{g}_i^t = \tilde{g}_i^t(y^t) \leq b_i$ for all $i \in M$, set $y^* = y^t$ and $f_{opt} = \tilde{f}^t$, fathom J_t by optimality and go to Step 6. Otherwise, set $J_{t+1} = J_t$, and $t = t + 1$, go to Step 1.

Step 6 (Backtracking). If all elements in J_t are underlined, stop. Otherwise, generate J_{t+1} by replacing the rightmost element of J_t which is not underlined by its underlined complement, delete all elements to its right. Set $t = t + 1$, go to Step 1.

The following theorem is straightforward.

THEOREM 12.1 *Algorithm 12.1 terminates in finite iterations either at an optimal solution x_{opt} to $(0\text{-}1PP_2)$ with $f_{opt} < \infty$ or reporting an infeasibility of $(0\text{-}1PP_2)$ with $f_{opt} = \infty$.*

We now apply Algorithm 12.1 to solve Example 12.1 step by step.

Initial Iteration

Step 0. Set $J_0 = \emptyset$ and $f_{opt} = \infty$.

Iteration 1 ($t = 0$)

Step 1. Applying the modified additive algorithm to (MP_0) with $f_{opt} = \infty$ finds $J_0 = J_0 \cup J_0^* = \{2\}$.

Step 2. Consistency check. J_0 is consistent with $D_0 = \{1, 2, 3\}$.

Step 3. B_t recognition. $B_0 = \{\underline{1}, -5, \underline{3}, \underline{4}\}$ and $J_0 = \{2, \underline{1}, -5, \underline{3}, \underline{4}\}$.

Step 4. $\tilde{f}^0 = 15 < f_{opt} = \infty$.

Step 5. $\tilde{g}_1^0 = 1 \leq b_1 = 1$; $\tilde{g}_2^0 = -12 \leq b_2 = -2$; $\tilde{g}_3^0 = -7 \leq b_3 = -1$. Set incumbent $y^* = (1, 1, 1, 1, 0)^T$ and $f_{opt} = 15$.

Step 6. Backtrack. $J_1 = \{2, \underline{1}, \underline{5}\}$.

Iteration 2 ($t = 1$)

Step 1. Applying the modified additive algorithm to (MP_1) with $f_{opt} = 15$ finds $J_1 = J_1 \cup J_1^* = \{2, \underline{1}, \underline{5}, 3\}$.

Step 2. Consistency check. J_1 is inconsistent.

Step 6. Backtrack. $J_2 = \{2, \underline{1}, \underline{5}, -\underline{3}\}$.

Iteration 3 ($t = 2$)

Step 1. Applying the modified additive algorithm to (MP_2) with $f_{opt} = 15$ reports an infeasibility.

Step 6. Backtrack. $J_3 = \{-\underline{2}\}$.

Iteration 4 ($t = 3$)

Step 1. Applying the modified additive algorithm to (MP_3) with $f_{opt} = 15$ finds $J_3 = J_3 \cup J_3^* = \{-\underline{2}, 4\}$.

Step 2. Consistency check. J_3 is consistent with $D_3 = \{2, 3, 5, -1\}$.

Step 3. B_t recognition. $B_3 = \emptyset$. Update incumbent by $y^* = (0, 0, 0, 1, 0)^T$ with $f_{opt} = 4$.

Step 6. Backtrack. $J_4 = \{-\underline{2}, -\underline{4}\}$.

Iteration 5 ($t = 4$)

Step 1. Applying the modified additive algorithm to (MP_4) with $f_{opt} = 4$ finds $J_4 = J_4 \cup J_4^* = \{-\underline{2}, -\underline{4}, 5, 3\}$.

Step 2. Consistency check. J_4 is consistent with $D_4 = \{1, 4, 5, -3\}$.

Step 3. B_t recognition. $B_4 = \{\underline{1}\}$ and $J_4 = \{-\underline{2}, -\underline{4}, 5, 3, \underline{1}\}$.

Step 4. $\tilde{f}^4 = 6 > f_{opt} = 4$.

Step 6. Backtrack. $J_5 = \{-\underline{2}, -\underline{4}, 5, -\underline{3}\}$.

Iteration 6 ($t = 5$)

Step 1. Applying the modified additive algorithm to (MP_5) with $f_{opt} = 4$ reports an infeasibility.

Step 6. Backtrack. $J_6 = \{-\underline{2}, -\underline{4}, -\underline{5}\}$.

Iteration 7 ($t = 6$)

Step 1. Applying the modified additive algorithm to (MP_6) with $f_{opt} = 4$ reports an infeasibility.

Step 6. Backtrack. All elements in J_6 are underlined and the procedure terminates at an optimal term $y^* = (0,0,0,1,0)^T$ with $f_{opt} = 4$. The corresponding optimal solution to Example 12.1 is $(0,1,1,1,1)^T$.

12.2 Two-Level Method for p-Norm Surrogate Constraint Formulation

The efficiency of the revised Taha's method developed in the previous section depends on the efficiency in carrying out two major tasks: seeking feasibility and checking consistency. It is observed that seeking feasibility in the master problem (12.1.1) will become much easier if the problem is singly constrained, i.e., $m = 1$. Adopting the p-norm surrogate constraint method discussed in Chapter 4, a multiply constrained polynomial 0-1 programming problem can be converted into an equivalent singly constrained polynomial 0-1 programming problem if the positive parameter p is selected to be large enough.

Let $v(PG_i)$ denote the optimal value of the following unconstrained polynomial 0-1 problem for $i = 1, 2, \ldots, m$,

$$(PG_i) \qquad \min\ g_i(x) = \sum_{j=1}^{q} a_{ij} \prod_{j \in Q_k} x_j$$
$$\text{s.t. } x \in \{0,1\}^n.$$

Problem (PG_i) can be solved by any solution method for unconstrained polynomial 0-1 integer programming problems. We assume that $b_i \geq v(PG_i)$, $i = 1, \ldots, m$, otherwise problem (0-1PP_2) is infeasible. Let $s_i = -v(PG_i) + 1$, $i = 1, \ldots, m$. For a positive integer p, we consider the following p-norm surrogate constraint formulation of (0-1PP_2):

$$\min\ f(x) = \sum_{k=1}^{q} c_k \prod_{j \in Q_k} x_j \tag{12.2.1}$$
$$\text{s.t. } g_s(x) := \sum_{i=1}^{m} [\mu_i (\sum_{k=1}^{q} a_{ik} \prod_{j \in Q_k} x_j + s_i)]^p \leq \sum_{i=1}^{m} [\mu_i (b_i + s_i)]^p = b^s,$$
$$x \in \{0,1\}^n,$$

where μ_i's are determined by the following equations

$$\mu_1(b_1 + s_1) = \mu_2(b_2 + s_2) = \ldots = \mu_m(b_m + s_m), \tag{12.2.2}$$
$$\sum_{i=1}^{m} \mu_i = 1,\ \mu_i > 0, i = 1, \ldots, m. \tag{12.2.3}$$

Note that for problems with $g_i(x) \geq 0$ for all $x \in \{0,1\}^n$ (e.g. all a_{ik}'s are nonnegative), s_i can be set to 0. Assume that all a_{ik}'s in (12.2.1) are integers.

From Chapter 4, we know that problems (0-1PP_2) and (12.2.1) are equivalent if

$$p \geq \left\lceil \ln(m)/\ln\left(\min_{1\leq i\leq m} \frac{b_i+s_i+1}{b_i+s_i}\right)\right\rceil, \qquad (12.2.4)$$

where $\lceil \alpha \rceil$ denotes the minimum integer that is greater than or equal to α.

Note that $x_j^p = x_j$ if $x_j \in \{0,1\}$. Thus $g_s(x)$ is still a polynomial after expanding and combining similar terms. One problem is how to calculate the coefficients of the expanded polynomial of $g_s(x)$. Consider a linear function $h(z) = \alpha_1 z_1 + \alpha_2 z_2 + \cdots + \alpha_n z_n$, where $z_i \in \{0,1\}$ for $i = 1, 2, \ldots, n$. By the multinomial theorem, we have

$$\begin{aligned} h^p(z) &= (\alpha_1 z_1 + \alpha_2 z_2 + \cdots + \alpha_n z_n)^p \\ &= \sum_{t_1+t_2+\cdots+t_n=p} \frac{p!}{t_1!\cdots t_n!}(\alpha_1 z_1)^{t_1}\cdots(\alpha_n z_n)^{t_n}. \end{aligned}$$

Thus, for any combination $\{i_1, \ldots, i_k\} \subseteq \{1, \ldots, n\}$, the coefficient of the item $z_{i_1}\cdots z_{i_k}$ is

$$\beta_{i_1\ldots i_k} = \sum_{\substack{t_1+\cdots+t_k=p \\ t_j\geq 1, j=1,\ldots,k}} \frac{p!}{t_1!\cdots t_k!}\alpha_{i_1}^{t_1}\cdots\alpha_{i_k}^{t_k}. \qquad (12.2.5)$$

Notice that computing $\beta_{i_1\ldots i_k}$ by applying (12.2.5) directly could be very time-consuming when p is large. The following proposition greatly simplifies the calculation of $\beta_{i_1\ldots i_k}$.

PROPOSITION 12.1 *Let $\beta_{i_1 i_2\ldots i_k}$ be the coefficient of $z_{i_1}\cdots z_{i_k}$ in the expansion of $h^p(z)$. Then*

$$\beta_{i_1\ldots i_k} = \sum_{j=1}^{k}(-1)^{k-j} \sum_{N^j\subseteq\{i_1,\ldots,i_k\}} (\sum_{i\in N^j} \alpha_i)^p, \qquad (12.2.6)$$

where $1 \leq k \leq p$ and N^j is an index set with cardinality $|N^j| = j$.

Proof. When $k = 1$, the theorem is valid as $\beta_{i_1} = (\alpha_{i_1})^p$. Suppose that (12.2.6) holds true for $k \geq 1$. Then, we have the following from the definition of $\beta_{i_1 i_2\ldots i_k i_{k+1}}$,

$$\begin{aligned} \beta_{i_1\ldots i_k i_{k+1}} &= (\sum_{j=1}^{k+1} \alpha_{i_j})^p - \sum_{\{i'_1,\ldots,i'_k\}\subset\{i_1,\ldots,i_k,i_{k+1}\}} \beta_{i'_1\ldots i'_k} \\ &\quad - \cdots - \sum_{\{i'_1,\ldots,i'_l\}\subset\{i_1,\ldots,i_k,i_{k+1}\}} \beta_{i'_1\ldots i'_l} - \cdots - \sum_{j=1}^{k+1}\beta_{i_j}. \end{aligned}$$

Equation (12.2.6) further yields the following based on the induction assumption,

$$\begin{aligned}
&\beta_{i_1\ldots i_k i_{k+1}} \\
&= (\sum_{j=1}^{k+1} \alpha_{i_j})^p - \sum_{\{i'_1,\ldots,i'_k\}\subset\{i_1,\ldots,i_k,i_{k+1}\}} \sum_{j=1}^{k}(-1)^{k-j} \sum_{N^j\subseteq\{i'_1,\ldots,i'_k\}} (\sum_{i\in N^j} \alpha_i)^p \\
&- \cdots - \sum_{\{i'_1,\ldots,i'_l\}\subset\{i_1,\cdots,i_k,i_{k+1}\}} \sum_{j=1}^{l}(-1)^{l-j} \sum_{N^j\subseteq\{i'_1,\ldots,i'_l\}} (\sum_{i\in N^j} \alpha_i)^p \\
&- \cdots - \sum_{\{i'_1\}\subset\{i_1,\ldots,i_k,i_{k+1}\}} \sum_{j=1}^{1}(-1)^{1-j} \sum_{N^j\subseteq\{i'_1\}} (\sum_{i\in N^j} \alpha_i)^p. \qquad (12.2.7)
\end{aligned}$$

Note that

$$\begin{aligned}
&\sum_{\{i'_1,\ldots,i'_l\}\subset\{i_1,\ldots,i_k,i_{k+1}\}} \sum_{j=1}^{l}(-1)^{l-j} \sum_{N^j\subseteq\{i'_1,\ldots,i'_l\}} (\sum_{i\in N^j} \alpha_i)^p \\
&= \sum_{j=1}^{l}(-1)^{l-j} \sum_{N^j\subseteq\{i_1,\ldots,i_l,\ldots,i_k,i_{k+1}\}} C_{k+1-j}^{l-j}(\sum_{i\in N^j} \alpha_i)^p \qquad (12.2.8)
\end{aligned}$$

for $l = 1,\ldots,k$. Equations (12.2.7) and (12.2.8) lead to the following,

$$\begin{aligned}
&\beta_{i_1\ldots i_k i_{k+1}} \\
&= (\sum_{i=1}^{k+1} \alpha_{i_j})^p - \sum_{j=1}^{k}(-1)^{k-j} \sum_{N^j\subseteq\{i_1,\ldots,i_k,i_{k+1}\}} C_{k+1-j}^{k-j}(\sum_{i\in N^j} \alpha_i)^p \\
&- \ldots - \sum_{j=1}^{l}(-1)^{l-j} \sum_{N^j\subseteq\{i'_1,\ldots,i_k,i_{k+1}\}} C_{k+1-j}^{l-j}(\sum_{i\in N^j} \alpha_i)^p \\
&- \cdots - \sum_{j=1}^{1}(-1)^{1-j} \sum_{N^j\subseteq\{i_1,\ldots,i_k,i_{k+1}\}} C_{k+1-j}^{1-j}(\sum_{i\in N^j} \alpha_i)^p.
\end{aligned}$$

Combining the similar terms in the above equation gives

$$\beta_{i_1 \ldots i_k i_{k+1}} = (\sum_i^{k+1} \alpha_{i_j})^p - C_1^0 \sum_{N^k \subseteq \{i_1, \ldots, i_k, i_{k+1}\}} (\sum_{i \in N^k} \alpha_i)^p$$

$$-(-C_2^1 + C_2^0) \sum_{N^{k-1} \subseteq \{i_1, \ldots, i_k, i_{k+1}\}} (\sum_{i \in N^{k-1}} \alpha_i)^p - \cdots$$

$$-[\sum_{j=1}^{k+1-l} (-1)^{k+1-l-j} C_{k+1-l}^{k+1-l-j}] \cdot \sum_{N^{k+1-l} \subseteq \{i_1, \ldots, i_k, i_{k+1}\}} (\sum_{i \in N^{k+1-l}} \alpha_i)^p - \cdots$$

$$-[\sum_{j=1}^{k} (-1)^{k-j} C_k^{k-j}] \cdot \sum_{N^1 \subseteq \{i_1, \ldots, i_k, i_{k+1}\}} (\sum_{i \in N^1} \alpha_i)^p. \qquad (12.2.9)$$

It is easy to verify that

$$-\sum_{j=1}^{k+1-l} (-1)^{k+1-l-j} C_{k+1-l}^{k+1-l-j} = (-1)^{k+1-l}, \quad l = 1, \ldots, k.$$

Therefore, (12.2.9) implies that (12.2.6) holds true for $k+1$. □

In the following, we focus on solving the singly constrained polynomial 0-1 problem (12.2.1). After expanding $g_s(x)$ and rearranging the cross terms, (12.2.1) can be expressed as

$$\min \ f(x) = \sum_{k=1}^{T} c_k \prod_{j \in Q_k} x_j \qquad (12.2.10)$$

$$\text{s.t. } \ g_s(x) = \sum_{k=1}^{T} a_k \prod_{j \in Q_k} x_j \le b_s,$$

$$x \in \{0,1\}^n,$$

where $q \le T \le 2^q$, $Q_k \subseteq N$ and $c_k := 0$ for all newly generated cross terms.

Consider the two-level formulation of problem (12.2.10):

$$\min \ \tilde{f}(y) = \sum_{k=1}^{T} \tilde{c}_k y_k \qquad (12.2.11)$$

$$\text{s.t. } \ \tilde{g}_s(y) = \sum_{k=1}^{T} \tilde{a}_k y_k \le \tilde{b}_s,$$

$$y_k \in \{0,1\}, \quad k = 1, 2, \ldots, T,$$

with a set of nonlinear secondary constraints

$$y_k = \begin{cases} \prod_{j \in Q_k} x_j, & k \in J_s^+ = \{k \mid c_k \geq 0\}, \\ 1 - \prod_{j \in Q_k} x_j, & k \in J_s^- = \{k \mid c_k < 0\}, \end{cases} \tag{12.2.12}$$

where $\tilde{c}_k = c_k$ and $\tilde{a}_k = a_k$ for $k \in J_s^+$, $\tilde{c}_k = -c_k$ and $\tilde{a}_k = -a_k$ for $k \in J_s^-$, $\tilde{b}_s = b_s - \sum_{k \in J_s^-} a_k$.

Taking the advantage of the single constraint in (12.2.11), a simple procedure can be derived to search for a feasible partial solution of (12.2.11) rather than to apply the additive algorithm.

Suppose J_t is a partial solution at iteration t. Let I_t be the index of J_t and y^t the typical completion of J_t. Denote by (MP_s^t) the master problem (12.2.11) with y_k, $k \in I_t$, being fixed at zero or one according to J_t. When $\tilde{g}_s(y^t) > \tilde{b}_s$, J_t is an infeasible partial solution. If

$$\tilde{g}_s(y^t) + \sum_{k \in \bar{I}_t} \min(0, \tilde{a}_k) > \tilde{b}_s, \tag{12.2.13}$$

then, it is impossible to augment J_t to obtain a feasible completion. Thus, J_t can be fathomed. Otherwise, there must exist at least one feasible completion of J_t. The following procedure can be used to find a feasible completion of J_t.

PROCEDURE 12.3 (SEARCH FOR A FEASIBLE PARTIAL SOLUTION)

Given a partial solution J_t and its index set I_t.

Step 0. If (12.2.13) holds, exit and there is no feasible completion of J_t. Otherwise, calculate $\alpha = \tilde{g}_s(y^t)$. Set $I = \{1, \ldots, T\} \setminus I_t$.

Step 1. Calculate $i = \arg\min_{k \in I} \tilde{a}_k$.

Step 2. Set $J_t := J_t \cup \{i\}$. If $\alpha := \alpha + \tilde{a}_i \leq \tilde{b}_s$, exit and J_t is a feasible partial solution. Otherwise, set $I := I \setminus \{i\}$, return to Step 1.

Procedure 12.3 either finds a feasible partial solution or reports that no feasible completion of J_t can be found. Replacing the additive algorithm (Algorithm 12.1) with Procedure 12.3 yields the following two-level solution method) for problem (12.2.10). Denote $\tilde{g}_s(y^t)$ by $\tilde{g}_s^t$ and $\tilde{f}(y^t)$ by $\tilde{f}_s^t$. The flow diagram of the algorithm is given in Figure 12.2.

ALGORITHM 12.2 (TWO-LEVEL SOLUTION METHOD FOR THE p-NORM SURROGATE CONSTRAINT FORMULATION PROBLEM)

Step 0. Apply the p-norm surrogate constraint method to convert the multiply constrained polynomial 0-1 problem into a singly constrained one. Set $J_0 = \emptyset$, $t = 0$, and $f_{opt} = \infty$.

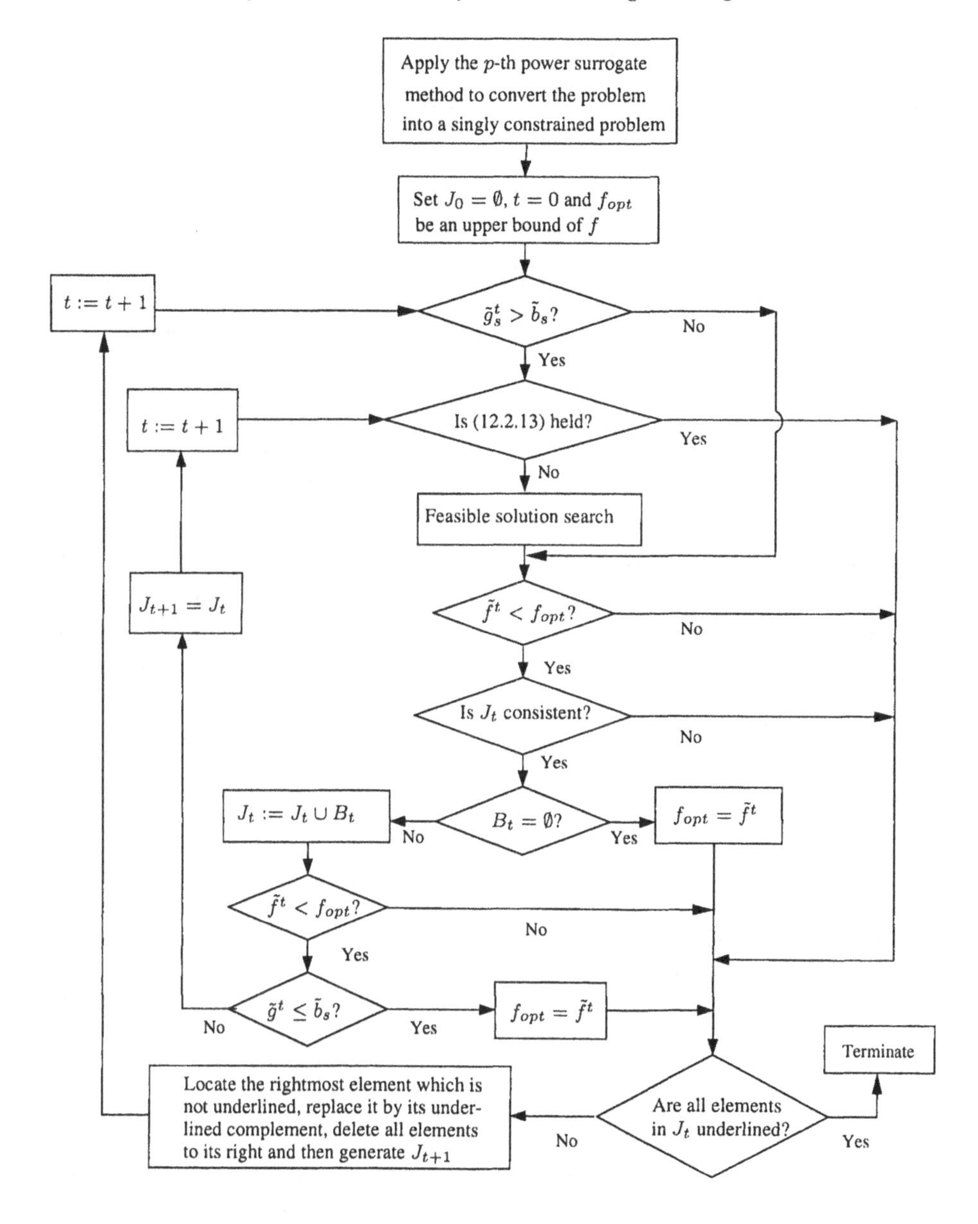

Figure 12.2. Diagram of the p-norm surrogate-constraint algorithm.

Step 1. If $\tilde{g}_s^t \leq \tilde{b}_s$, go to Step 4.

Step 2. If (12.2.13) holds, fathom J_t by feasibility and go to Step 9.

Step 3. Apply Procedure 12.3 to search for a feasible J_t.

Step 4. If $\tilde{f}^t \geq f_{opt}$, fathom J_t by domination and go to Step 9.

Step 5. Consistency check. If J_t is inconsistent, fathom J_t by consistency and go to Step 9. Otherwise, obtain D_t.

Step 6. B_t recognition. If $B_t = \emptyset$, set $y^* = y^t$ and $f_{opt} = \tilde{f}^t$, fathom J_t by optimality and go to Step 9. Otherwise, augment J_t with B_t on the right.

Step 7. If $\tilde{f}^t \geq f_{opt}$, fathom J_t by domination and go to Step 9.

Step 8. If $\tilde{g}_s^t \leq b_s$, set $y^* = y^t$ and $f_{opt} = f^t$, fathom J_t by optimality and go to Step 9. Otherwise, set $J_{t+1} = J_t$, $t = t + 1$ and go to Step 2.

Step 9. Backtrack. If all elements in J_t are underlined, terminate the algorithm. Otherwise, generate J_{t+1} by replacing the rightmost element of J_t which is not underlined by its underlined complement and delete all elements to its right. Set $t = t + 1$ and go to Step 1.

The finite termination of the algorithm directly follows from Theorem 12.1.

Now we apply Algorithm 12.2 to solve Example 12.1 again. To apply the p-norm surrogate constraint method, we need to make all constraints of Example 12.1 to take strictly positive values by adding proper constants, $s_i's$, as follows,

$$\begin{aligned}
\min\ & 3x_1 + 5x_1x_2x_3 + 3x_1x_4x_5 + 8x_2x_3x_5 - 4x_3x_4x_5 \\
\text{s.t. } & 3x_1 - x_1x_4x_5 - x_2x_3x_5 + x_3x_4x_5 + 2 \leq 4, \\
& 2x_1 - 4x_1x_2x_3 - 7x_1x_4x_5 - 3x_2x_3x_5 - x_3x_4x_5 + 14 \leq 11, \\
& - 6x_1 - 3x_1x_2x_3 + 5x_1x_4x_5 - 3x_2x_3x_5 + 6x_3x_4x_5 + 13 \leq 18, \\
& x_1, x_2, x_3, x_4, x_5 \in \{0, 1\}.
\end{aligned}$$

Initial Iteration

Step 0. Applying the p-norm surrogate constraint method to Example 12.1 with $p = 21$ from (12.2.4) and $\mu_1 = 0.6306$, $\mu_2 = 0.2293$ and $\mu_3 = 0.1401$ from (12.2.2) yields the following surrogate constraint of Example 12.1,

$$\begin{aligned}
& [0.6306 \times (3x_1 - x_1x_4x_5 - x_2x_3x_5 + x_3x_4x_5 + 2)]^{21} \\
& +[0.2293 \times (2x_1 - 4x_1x_2x_3 - 7x_1x_4x_5 - 3x_2x_3x_5 - x_3x_4x_5 + 14)]^{21} \\
& +[0.1401 \times (-6x_1 - 3x_1x_2x_3 + 5x_1x_4x_5 - 3x_2x_3x_5 + 6x_3x_4x_5 + 13)]^{21} \\
& \leq [0.6306 \times 4]^{21} + [0.2293 \times 11]^{21} + [0.1401 \times 18]^{21}.
\end{aligned}$$

After expanding the above surrogate constraint, combining the similar terms, and dividing both sides by 10^{10}, Example 12.1 can be transformed into the following equivalent singly constrained master problem,

$$\begin{aligned}
\min\ & \tilde{f}(y) = 3y_1 + 5y_2 + 3y_3 + 8y_4 + 4y_5 - 4 \\
\text{s.t. } & \tilde{g}(y) = 70.24y_1 - 71.44y_2 - 74.55y_3 - 4.31y_4 + 3.34y_5, \\
& + 6.31y_6 + 3.32y_7 + 1.20y_8 + 68.27y_9 \leq -0.91,
\end{aligned}$$

with the secondary constraints,

$$\begin{cases} y_1 &= x_1, \\ y_2 &= x_1x_2x_3, \\ y_3 &= x_1x_4x_5, \\ y_4 &= x_2x_3x_5, \\ y_5 &= 1 - x_3x_4x_5, \\ y_6 &= x_1x_3x_4x_5, \\ y_7 &= x_2x_3x_4x_5, \\ y_8 &= x_1x_2x_3x_5, \\ y_9 &= x_1x_2x_3x_4x_5. \end{cases}$$

Note that four more decision terms, y_6, y_7, y_8 and y_9, are introduced. Set $J_0 = \emptyset$ and $f_{opt} = \sum_{j=1}^{9} c_j - 4 = 19$.

Iteration 1 ($t = 0$)
Step 1. $\tilde{g}_s^0 = 0 > \tilde{b}_s = -0.91$.
Step 2. $\tilde{g}_s^0 + \sum_{j \in \bar{I}_0} \min(0, \tilde{a}_j) = -150.30 < \tilde{b}_s = -0.91$.
Step 3. Feasible solution search. $J_0 = \{3\}$.
Step 4. $\tilde{f}^0 = -1 < f_{opt} = 19$.
Step 5. Consistency check. J_0 is consistent and $D_0 = \{1, 4, 5\}$.
Step 6. B_t recognition. $B_0 = \{\underline{1}, -5, \underline{6}\}$ and $J_0 = \{3, \underline{1}, -5, \underline{6}\}$.
Step 7. $\tilde{f}^0 = 2 < f_{opt} = 19$.
Step 8. $\tilde{g}_s^1 = 2.00 > \tilde{b}_s = -0.91$. $J_1 = \{3, \underline{1}, -5, \underline{6}\}$ and go to Step 2.

Iteration 2 ($t = 1$)
Step 2. $\tilde{g}_s^1 + \sum_{j \in \bar{I}_1} \min(0, \tilde{a}_j) = -73.75 < b_s = -0.91$.
Step 3. Feasible solution search. $J_1 = \{3, \underline{1}, -5, \underline{6}, 2\}$.
Step 4. $\tilde{f}^1 = 7 < f_{opt} = 19$.
Step 5. Consistency check. J_1 is consistent and $D_1 = \{1, 2, 3, 4, 5\}$.
Step 6. B_t recognition. $B_1 = \{\underline{4}, \underline{7}, \underline{8}, \underline{9}\}$ and $J_1 = \{3, \underline{1}, -5, \underline{6}, 2, \underline{4}, \underline{7}, \underline{8}, \underline{9}\}$.
Step 7. $\tilde{f}^1 = 15 < f_{opt} = 19$.
Step 8. $\tilde{g}_s^1 = -0.96 < \tilde{b}_s = -0.91$. $f_{opt} = 15$ and go to Step 9.
Step 9. Backtrack. $J_2 = \{3, \underline{1}, -5, \underline{6}, -\underline{2}\}$.

Iteration 3 ($t = 2$)
Step 1. $\tilde{g}_s^2 = 2.00 > b_s = -0.91$.
Step 2. $\tilde{g}_s^2 + \sum_{j \in \bar{I}_2} \min(0, a_j) = -2.31 < b_s = -0.91$.
Step 3. Feasible solution search. $J_2 = \{3, \underline{1}, -5, \underline{6}, -\underline{2}, 4\}$.
Step 4. $\tilde{f}^2 = 10 < f_{opt} = 15$.
Step 5. Consistency check. J_2 is inconsistent.
Step 9. Backtrack. $J_3 = \{3, \underline{1}, -5, \underline{6}, -\underline{2}, -\underline{4}\}$.

Iteration 4 ($t = 3$)
Step 1. $\tilde{g}_s^3 = 2.00 > \tilde{b}_s = -0.91$.
Step 2. $\tilde{g}_s^3 + \sum_{j \in \bar{I}_3} \min(0, \tilde{a}_j) = 2.00 > \tilde{b}_s = -0.91$, and go to Step 9.

Step 9. Backtrack. $J_4 = \{3, \underline{1}, \underline{5}\}$.

Iteration 5 ($t = 4$)

Step 1. $\tilde{g}_s^4 = -0.97 \leq \tilde{b}_s = -0.91$ and go to Step 4.

Step 4. $f^4 = 6 < f_{opt} = 15$.

Step 2. Consistency check. J_4 is consistent and $D_4 = \{1, 4, 5, -3\}$.

Step 3. B_t recognition. $B_4 = \emptyset$, $f_{opt} = 6$ and go to Step 9.

Step 9. Backtrack. $J_5 = \{-\underline{3}\}$.

The details of the next few iterations are omitted. The algorithms stops at Iteration 13 with $y^* = (0, 0, 0, 1, 0)^T$ and $f_{opt} = 4$. The corresponding optimal solution is $x_{opt} = (0, 1, 1, 1, 1)^T$.

12.3 Convergent Lagrangian Method Using Objective Level Cut

Adopting the p-norm surrogate constraint method reduces a multiply constrained polynomial 0-1 programming problem into an equivalent singly constrained polynomial 0-1 programming problem. While it significantly simplifies the task of seeking feasibility in the implicit enumeration algorithm, the p-norm transformation, at the same time, largely increases the number of decision terms, thus increasing the computation amount for checking consistency in the implicit enumeration algorithm. This section studies a convergent Lagrangian dual search method for multiply constrained polynomial 0-1 programming problem. Using the solution concept of the objective level cut discussed in Chapter 7, the developed Lagrangian dual search method is guaranteed to find an optimal solution of the primal problem within a finite iterations. Furthermore, the resulting Lagrangian relaxation problem is a singly constrained polynomial 0-1 programming problem which can be efficiently solved by the implicit enumeration algorithm discussed in the previous section.

We assume in this section that in $(0\text{-}1PP_2)$ all coefficients of the objective function are integers. This assumption can be relaxed to situations where all coefficients of the objective function are rational numbers. We consider now the Lagrangian relaxation of problem $(0\text{-}1PP_2)$,

$$d(\lambda) = \min_{x \in \{0,1\}^n} \sum_{k=1}^{q} c_k \prod_{j \in Q_k} x_j + \sum_{i=1}^{m} \lambda_i [\sum_{k=1}^{q} a_{ik} \prod_{j \in Q_k} x_j - b_i],$$

where $\lambda \in \mathbb{R}_+^m$ is a Lagrangian multiplier vector. The conventional Lagrangian dual approach searches for an optimal Lagrangian multiplier vector that maximizes $d(\lambda)$ over all $\lambda \in \mathbb{R}_+^m$. It is often the case that the conventional Lagrangian dual approach does not identify an optimal solution to $(0\text{-}1PP_2)$. Adopting the solution concept discussed in Chapter 7, the following convergent Lagrangian dual method using objective level cut can be developed for polynomial 0-1 programming problems.

For a given lower bound l of the optimal value f^*, we consider a revised version of $(0\text{-}1PP_2)$ by imposing an objective cut:

$$\min\ f(x) = \sum_{k=1}^{q} c_k \prod_{k\in Q_k} x_j \tag{12.3.1}$$

$$\text{s.t. } g_i(x) = \sum_{k=1}^{q} a_{ik} \prod_{j\in Q_k} x_j \le b_i, \quad i = 1, 2, \ldots, m,$$

$$l \le f(x) = \sum_{k=1}^{q} c_k \prod_{j\in Q_k} x_j,$$

$$x \in \{0,1\}^n.$$

Obviously, problem (12.3.1) is equivalent to $(0\text{-}1PP_2)$ if $l \le f^*$. Define the Lagrangian relaxation of problem (12.3.1) for a given $\lambda \in \mathbb{R}^m_+$ as follows,

$$(L^l_\lambda) \qquad d^l(\lambda) = \min_{x\in\{0,1\}^n} \sum_{k=1}^{q} c_k \prod_{j\in Q_k} x_j + \sum_{i=1}^{m} \lambda_i [\sum_{k=1}^{q} a_{ik} \prod_{j\in Q_k} x_j - b_i],$$

$$\text{s.t. } l \le f(x) = \sum_{k=1}^{q} c_k \prod_{j\in Q_k} x_j,$$

$$x \in \{0,1\}^n.$$

The corresponding dual problem then is

$$(D^l) \quad \max_{\lambda\in\mathbb{R}^m_+} d^l(\lambda).$$

Similar to what shown in Chapter 7, the lower bound l can be adjusted such that $f^* - l \to 0$. This leads to the following convergent Lagrangian solution algorithm.

ALGORITHM 12.3 (CONVERGENT LAGRANGIAN AND OBJECTIVE LEVEL CUT ALGORITHM FOR $(0\text{-}1PP_2)$)

Step 0 (Initialization). Compute a lower bound l_0 of f^*. Set $t = 0$ and $f_{opt} = \infty$.

Step 1. If $l_t \ge f_{opt}$, stop.

Step 2 (Dual search with objective cut). Solve (D^{l_t}) by some dual search procedure, while the Lagrangian relaxation problem $(L^{l_t}_\lambda)$ is solved by using Algorithm 12.2. The dual search method terminates when the algorithm is not able to increase the dual value after a given number of iterations. Let

λ^t be the dual vector that generates the highest dual value in the dual search process. Set $d^t = d^{l_t}(\lambda^t)$.

Step 3. If $d^t > l_t$, set $l_{t+1} = \lceil d^t \rceil$ and let $t := t + 1$. If a feasible solution $\tilde{x}$ with $f(\tilde{x}) < f_{opt}$ is found during the dual search process, set $x_{opt} = \tilde{x}$, set $f_{opt} = f(\tilde{x})$. Go to Step 1.

Step 4. If $d^t = l_t$, solve the following problem using Algorithm 12.2 without considering constraints:

$$\min\ f(x) = \sum_{k=1}^{q} c_j \prod_{j \in Q_k} x_j \tag{12.3.2}$$

$$\text{s.t. } l_t \leq f(x) = \sum_{k=1}^{q} c_k \prod_{j \in Q_k} x_j,$$

$$x \in \{0,1\}^n.$$

If there is a feasible optimal solution x^t to (12.3.2), stop and x^t is the optimal solution to (0-1PP_2). Otherwise, set $l_{k+1} = f(x^t) + 1$, where x^t is an optimal solution to (12.3.2). Set $t := t + 1$ and go to Step 1.

The algorithm enters Step 4 only when the algorithm is not able to raise the dual value at Step 3. Step 4 is corresponding to the Lagrangian relaxation problem with $\lambda = 0$. When Step 4 identifies a feasible solution, it will be optimal to the primal problem. When Step 4 is not able to find feasible solutions, it can still help to raise the lower objective cut.

Now we apply Algorithm 12.3 to solve Example 12.1 again.

Iteration 0. Set $l^0 = -4$, the incumbent $x_{opt} = \emptyset$ and $f_{opt} = \infty$.

Iteration 1. The dual search terminates with an optimal multiplier $\lambda^0 = (0, 1.225, 0.612)^T$, a feasible solution $(1,1,1,1,1)^T$ and dual value -1.67. Set $x_{opt} = (1,1,1,1,1)^T$, $f_{opt} = f(x_{opt}) = 15$, and $l_1 = -1$.

Iteration 2. The dual search terminates with an optimal multiplier $\lambda^1 = (0.264, 0, 0)^T$, a feasible solution $(1,0,0,1,1)^T$ and dual value 3.26. Set $x_{opt} = (1,0,0,1,1)^T$, $f_{opt} = f(x_{opt}) = 6$, and $l_1 = 4$.

Iteration 3. The dual search terminates with an optimal multiplier $\lambda^3 = (0,0,0)^T$, a feasible solution $(0,1,1,1,1)^T$ and dual value 4. Set the incumbent $x_{opt} = (0,1,1,1,1)^T$, $f_{opt} = f(x_{opt}) = 4$, stop and x_{opt} is an optimal solution.

12.4 Computational Results

In this section, we report some numerical results for Algorithms 12.1, 12.2 and 12.3 for constrained polynomial 0-1 programming problems in the form of (0-1PP_2).

The test problems are randomly generated using the following ranges of the coefficients: $c_k \in [-10, 20]$, $a_{ik} \in [-5, 15]$ and the right-hand side is taken as $b_i = (1-r)\sum_{k=1}^{q} \min(0, a_{ik}) + r\sum_{k=1}^{q} \max(0, a_{ik})$ where $r \in (0,1)$ is an adjustable ratio of the right-hand side. The density number $D \in (0,1]$ is also adjustable in controlling the ratio between the number of nonzero coefficients (c_k's and a_{ik}'s) and q, the maximum number of coefficients in the objective function or in each individual constraint.

Tables 12.1 and 12.2 summarize the numerical results of the revised Taha's algorithm and the objective level cut method, respectively, for different sizes of test problems and densities of coefficients. Table 12.3 presents the numerical results for the p-norm surrogate algorithm for different sizes of test problems with density 0.25, where the computational time is divided into T_1, the CPU time to convert the problem into the p-norm surrogate problem, and T_2, the CPU time used in solving the resulting singly constrained problem. The average CPU time in all the three tables is measured by running the respective algorithm for 20 times on a SUN Workstation (Blade 2000). The comparison clearly reveals that Algorithm 12.1 performs the best among the three algorithms. Algorithm 12.2 seems to suffer from the computation effort needed in forming the surrogate constraint and from the expanding number of decision terms.

Table 12.1. Numerical results with the revised Taha's algorithm ($r = 0.5$).

q	n	m	Average CPU Time (seconds)			
			$D = 0.25$	$D = 0.50$	$D = 0.75$	$D = 1.0$
50	50	20	0.31	0.37	0.36	0.36
50	100	20	19.3	19.1	17.7	14.8
50	150	20	165.2	231.9	185.5	160.5
50	200	20	1011.1	812.9	1084.4	910.6
50	80	20	5.0	4.8	4.2	3.5
100	80	20	38.8	50.9	32.0	30.3
150	80	20	199.3	88.8	205.2	74.3
200	80	20	610.5	396.8	178.2	99.6

12.5 Notes

Following the backtrack concept of Geoffrion [73], Taha [211] extended the additive algorithm of Balas [7] for linear 0-1 programming to constrained polynomial 0-1 programming by designing a two-level solution scheme. Note that Taha's original results [211] can only deal with problem ($0\text{-}1PP_2$) with all c_j's nonnegative. Wang et al. [223] further developed a revised version of [211]

Table 12.2. Numerical results with the objective level cut algorithm ($r = 0.5$).

q	n	m	Average CPU Time (seconds)			
			$D = 0.25$	$D = 0.50$	$D = 0.75$	$D = 1.0$
50	50	20	9.0	2.6	1.5	0.84
50	60	20	14.2	21.2	8.1	2.5
50	70	20	92.8	17.5	8.1	9.8
50	80	20	240.5	59.0	30.1	17.5
60	80	20	560.1	80.4	53.0	29.2
70	80	20	1351.5	101.6	79.7	39.8
80	80	20	944.1	288.2	100.1	68.2

Table 12.3. Numerical results with the p-th power surrogate algorithm ($r = 0.5$).

q	n	m	D	T_1	T_2
15	30	20	0.25	1.3	2.9
20	30	20	0.25	2.5	4.2
25	30	25	0.25	4.0	16.1
30	30	20	0.25	6.7	31.6

which is applicable to all types of constrained polynomial 0-1 programming problems in $(0\text{-}1PP_2)$.

Chapter 13

MIXED-INTEGER NONLINEAR PROGRAMMING

This chapter discusses algorithms for solving mixed-integer nonlinear programming (MINLP) problems. The decision variables in this class of integer programming problems include both integer variables and continuous variables. Optimization models of an MINLP structure arise in a variety of fields, including chemical engineering, reliability networks and optimization of core reload patterns for nuclear reactors.

13.1 Introduction

The general formulation of mixed-integer nonlinear programming problems is of the following form:

$$
\begin{aligned}
(MINLP) \qquad & \min \ f(x,y) \\
& \text{s.t. } g_i(x,y) \le 0, \ i = 1, \ldots, q, \\
& \qquad h_i(x,y) = 0, \ i = 1, \ldots, l, \\
& \qquad x \in X \subseteq \mathbb{R}^n, \ y \in Y \subset \mathbb{Z}^m,
\end{aligned}
$$

where $f : X \times Y \to \mathbb{R}$, $g_i : X \times Y \to \mathbb{R}$ $(i = 1, \ldots, q)$, $h_i : X \times Y \to \mathbb{R}$ $(i = 1, \ldots, l)$, and $\mathbb{Z}^m$ denotes the set of integer vectors in $\mathbb{R}^m$. We assume that X is a nonempty convex set in $\mathbb{R}^n$ and Y is a finite integer set in $\mathbb{Z}^m$, e.g., $Y = \{0,1\}^m$. Let $g = (g_1, \ldots, g_q)^T$ and $h = (h_1, \ldots, h_l)^T$.

In many real-world applications, problem $(MINLP)$ often possesses certain special structures. One important instance is the convex mixed-integer programming problem where f and g are convex in (x,y), and the equality constraints are absent:

$$(MINLP_1) \qquad \begin{array}{ll} \min & f(x,y) \\ \text{s.t.} & g(x,y) \le 0, \\ & x \in X \subseteq \mathbb{R}^n, \; y \in Y \subset \mathbb{Z}^m. \end{array}$$

Another prominent mixed-integer programming problem arises from chemical engineering ([53]) where the equality constraints are absent, the continuous variable vector x and the integer variable vector y are separable in $(MINLP)$, and f and g_i's are both convex in x and linear in y. Problem $(MINLP)$, in this instance, becomes

$$(MINLP_2) \qquad \begin{array}{ll} \min & f(x) + c^T y \\ \text{s.t.} & g_i(x) + b_i^T y \le 0, \; i = 1, \ldots, q, \\ & x \in X \subseteq \mathbb{R}^n, \; y \in Y \subset \mathbb{Z}^m. \end{array}$$

The difficulty of developing an efficient method for $(MINLP)$ lies not only on the nonlinearity of the functions involved, but also on the simultaneous presence of both discrete and continuous variables. Let us consider the following small-size illustrative example.

EXAMPLE 13.1

$$\begin{array}{ll} \min & f(x,y) = 5y - 2\ln(x+1) \\ \text{s.t.} & g_1(x,y) = \mathrm{e}^{x/2} - (1/2)\sqrt{y} - 1 \le 0, \\ & g_2(x,y) = -2\ln(x+1) - y + 2.5 \le 0, \\ & g_3(x,y) = x + y - 4 \le 0, \\ & x \in [0,2], \; y \in [1,3] \text{ integer}. \end{array}$$

As shown in Figure13.1, the feasible region of this example consists of two isolated line segments. The optimal solution of the example is achieved at $(x^*, y^*) = (1.07, 2)^T$ with $f(x^*, y^*) = 8.5453$.

As we can see from this example, the feasible region of problem $(MINLP)$ is non-connected. A simple way to overcome this difficulty is to fix or to relax the integrality of the discrete variables so as to obtain a continuous relaxation of problem $(MINLP)$ with a convex feasible region. This strategy turns out to be one of the basic strategies in various solution methods for solving $(MINLP)$. Another basic idea underlying the solution methods for $(MINLP)$ is to separate the nonlinearity from the mixed-integer model so that the primal problem can be reduced to relatively easier subproblems that can be solved by existing solution methods. The basic strategies to derive subproblems are summarized as follows.

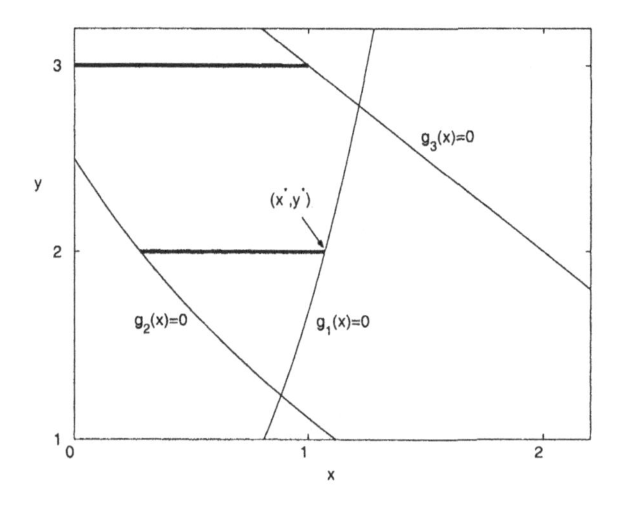

Figure 13.1. Example 13.1 of $(MINLP)$.

- Relaxing the integrality restriction on y results in a nonlinear programming (NLP) subproblem of continuous variables (x, y) which provides a lower bound to $(MINLP)$;
- Fixing a value for the integer variable y results in an NLP subproblem of continuous variable x which provides an upper bound to $(MINLP)$;
- Constructing linear or convex underestimation of f and g_i's at certain known points results in a mixed-integer linear or convex program which provides a lower bound to $(MINLP)$.

Algorithms based on the above solution strategies include branch-and-bound (BB) method, generalized Benders decomposition (GBD) method, and outer approximation (OA) method. In Sections 13.2-13.4, we will focus on methods for convex $(MINLP)$ problems. Global optimization methods for nonconvex cases of $(MINLP)$ will be discussed in Section 13.5.

13.2 Branch-and-Bound Method

Branch-and-bound method for problem $(MINLP)$ is based on the continuous relaxation of $(MINLP)$. By relaxing the integrality of variable y, we obtain the following nonlinear programming problem:

$$
\begin{array}{lll}
(NLP) & \min & f(x,y) \\
& \text{s.t.} & g(x,y) \le 0, \\
& & h(x,y) = 0, \\
& & x \in X \subseteq \mathbb{R}^n, \ y \in conv(Y),
\end{array}
$$

where α and β are the lower bound and upper bound of y, respectively. We need the following assumptions for $(MINLP)$:

ASSUMPTION 13.1 (i) $X \subseteq \mathbb{R}^n$ *is a compact convex set and* Y *is a finite integer set;*

(ii) f *and* g_i $(i = 1, \ldots, q)$ *are convex and differentiable functions of* (x, y), *and* h_i $(i = 1, \ldots, l)$ *are linear functions of* (x, y)*;*

(iii) *Certain constraint qualification of* (NLP) *is satisfied.*

Assumption 13.1 (i)-(iii) ensure that any local solution of (NLP) is a global solution and this solution can be identified by applying the KKT conditions directly. A typical sufficient condition for Assumption 13.1 (iii) is that the optimal solution of every feasible subproblem of (NLP) is a regular point, i.e., the gradient vectors of the active constraints are linearly independent.

The branch-and-bound procedure for $(MINLP)$ is similar to the one described in Chapter 2 for pure nonlinear integer programming problems. The subproblems are derived by relaxing the integrality of the integer variable y and imposing the lower bound and upper bound on y_j for each j. Let Z^k denote the lower bound obtained from solving the subproblem at node k, and UB the current best upper bound.

EXAMPLE 13.2 Applying the branch-and-bound method to Example 13.1, we find the optimal solution $(x^*, y^*) = (1.07, 2)$ after solving three subproblems. Figure 13.2 shows the search tree of the branch-and-bound method for Example 13.1.

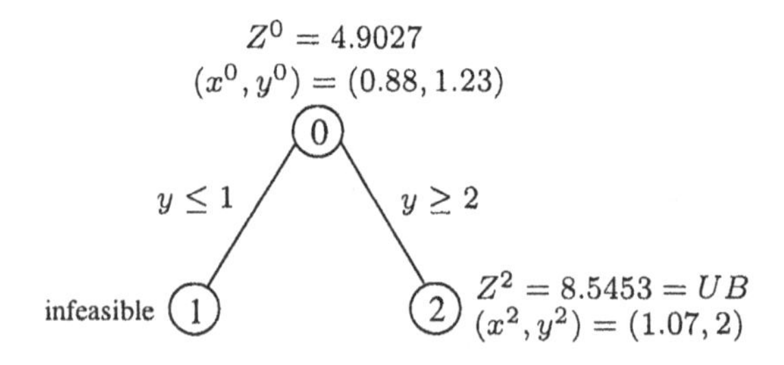

Figure 13.2. Branch-and-bound search tree for Example 13.1.

EXAMPLE 13.3 Let us consider the following example arising from process synthesis ([53]),

$$\begin{aligned}
\min\ & 10x_1 - 7x_3 - 18\ln(x_2+1) - 19.2\ln(x_1 - x_2 + 1) + 10 \\
& \qquad + 5y_1 + 6y_2 + 8y_3 \\
\text{s.t.}\ & -0.8\ln(x_2+1) - 0.96\ln(x_1 - x_2 + 1) + 0.8x_3 \le 0, \\
& -x_1 + x_2 \le 0, \\
& x_2 - 2y_1 \le 0, \\
& x_1 - x_2 - 2y_2 \le 0, \\
& -\ln(x_2+1) - 1.2\ln(x_1 - x_2 + 1) + x_3 + 2y_3 - 2 \le 0, \\
& y_1 + y_2 \le 1, \\
& y \in \{0,1\}^3,\ a \le x \le b,\ x = (x_1, x_2, x_3), \\
& a = (0,0,0),\ b = (2,2,1).
\end{aligned}$$

The optimal solution of this example is $(x^*, y^*) = (1.3009, 0, 1, 0, 1, 0)^T$ with $f(x^*, y^*) = 6.0097$. Note that the objective function and the inequality constraint functions of the problem are convex. The branch-and-bound solution process using depth-first with backtracking to the best node is summarized in Table 13.1 and the search tree is illustrated in Figure 13.3.

Table 13.1. Summary of the branch-and-bound method for Example 13.3.

Node	x^i	y^i	Z^i	UB
0	$(1.1465, 0.5466, 1)^T$	$(0.2732, 0.3, 0)^T$	0.7593	∞
1	$(1, 1, 0.6931)^T$	$(0.5, 0, 0)^T$	5.1713	∞
2	$(0, 0, 0)^T$	$(0, 0, 0)^T$	10	10
3	$(1.3009, 0, 1)^T$	$(0, 1, 0)^T$	6.0097	6.0097
4	$(1.5, 1.5, 0.9162)^T$	$(1, 0, 0)^T$	7.0927	6.0097

13.3 Generalized Benders Decomposition

The generalized Benders decomposition (GBD) has been a popular technique in solving mixed-integer linear programming problem ([74]). In this section, we discuss an extension of GBD method for solving the inequality constrained convex mixed-integer programming problem $(MINLP_1)$. The methods developed in this section and the next section can be easily extended to deal with problems with additional linear equality constraints.

Let

$$S = \{(x, y) \in X \times Y \mid g(x, y) \le 0\} \tag{13.3.1}$$

and

$$V = \{y \in Y \mid \text{ there exists } x \in X \text{ such that } g(x, y) \le 0\}. \tag{13.3.2}$$

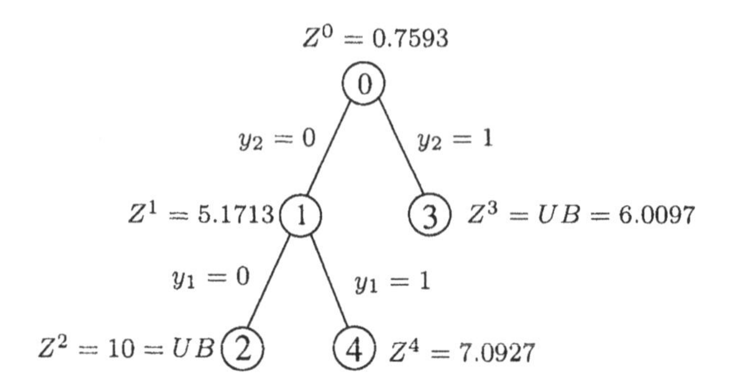

Figure 13.3. Branch-and-bound search tree for Example 13.3.

For any $y \in V$, consider the following nonlinear programming subproblem

$$(NLP(y)) \qquad \begin{aligned} \min\ & f(x,y) \\ \text{s.t. } & g(x,y) \le 0, \\ & x \in X. \end{aligned}$$

Since the optimal solution of $(NLP(y))$ is a feasible solution to $(MINLP_1)$, the optimal value $v(NLP(y))$ provides an upper bound to $(MINLP_1)$. We need the following assumption to ensure that $(NLP(y))$ can be solved correctly.

ASSUMPTION 13.2 *For any $y \in V$, the optimal solution of $(NLP(y))$ is a regular point, i.e., the gradient vectors of the active constraints at the optimal solution are linear independent.*

The Lagrangian relaxation of $(NLP(y))$ is

$$d_y(\lambda) = \min_{x \in X} L(x,y,\lambda) = f(x,y) + \lambda^T g(x,y),$$

where $\lambda \in \mathbb{R}^q_+$. Then, the Lagrangian dual problem of $(NLP(y))$ is

$$(D_y) \qquad \max_{\lambda \in \mathbb{R}^q_+} d_y(\lambda).$$

Under Assumption 13.1 (i)-(ii) and Assumption 13.2, there is no duality gap between $(NLP(y))$ and (D_y). Therefore,

$$\begin{aligned} \min_{(x,y)\in S} f(x,y) &= \min_{y \in V} v(NLP(y)) \\ &= \min_{y\in V}(\max_{\lambda \in \mathbb{R}^q_+} \min_{x \in X} L(x,y,\lambda)) \\ &= \min\ \alpha \qquad\qquad (13.3.3) \\ &\quad \text{s.t. } \alpha \ge \min_{x\in X} L(x,y,\lambda),\ \forall \lambda \ge 0, \\ &\quad\quad y \in V. \end{aligned}$$

Since set V is only known implicitly, we need to find a way to represent it explicitly by certain inequality constraints. For any $y \in Y$, consider the following feasibility-check problem:

$$\min_{x \in X} \max\{g_1(x,y), \ldots, g_q(x,y)\},$$

which is equivalent to

$$(NLPF(y)) \qquad \begin{aligned} &\min \ \beta \\ &\text{s.t.} \ \ \beta \geq g_i(x,y), \quad i = 1, \ldots, q, \\ &\qquad x \in X. \end{aligned}$$

It is easy to see that for any $y \in Y$, $(NLP(y))$ is infeasible if and only if $(NLPF(y))$ has a positive optimal value $\beta^* > 0$. The Lagrangian dual of $(NLPF(y))$ is

$$(DF(y)) \qquad \begin{aligned} &\max \ \min_{x \in X} \mu^T g(x,y) \\ &\text{s.t.} \ \ \mu \in \Lambda = \{\sum_{i=1}^{q} \mu_i = 1, \ \mu_i \geq 0, \ i = 1, \ldots, q\}. \end{aligned}$$

Thus, $y \in V$ can be characterized by the inequality constraints:

$$0 \geq \min_{x \in X} \mu^T g(x,y), \quad \forall \mu \in \Lambda. \tag{13.3.4}$$

Incorporating (13.3.4) into (13.3.3) leads to the following *master* problem

$$(MGBD) \qquad \begin{aligned} &\min \ \alpha \\ &\text{s.t.} \ \ \alpha \geq \min_{x \in X} L(x,y,\lambda), \quad \forall \lambda \geq 0, \\ &\qquad 0 \geq \min_{x \in X} \mu^T g(x,y), \quad \forall \mu \in \Lambda, \\ &\qquad y \in Y. \end{aligned}$$

The following is clear from the above discussion.

THEOREM 13.1 *Problem $(MGBD)$ is equivalent to $(MINLP_1)$.*

Notice that $(MGBD)$ has infinite constraints and the constraint functions are value functions. In order to get a solvable mixed-integer linear integer programming problem, consider the following relaxation of $(MGBD)$:

$$(MGBD_k) \qquad \begin{aligned} &\min \ \alpha \\ &\text{s.t.} \ \ \alpha \geq L(x^i,y^i,\lambda^i) + \nabla_y^T L(x^i,y^i,\lambda^i)(y - y^i), \ i \in I^k, \\ &\qquad 0 \geq (\mu^i)^T[g(x^i,y^i) + \nabla_y^T g(x^i,y^i)(y - y^i)], \ i \in J^k, \\ &\qquad y \in Y, \end{aligned}$$

where (x^i, λ^i) is the optimal primal-dual pair to $(NLP(y^i))$ if $(NLP(y^i))$ is feasible, (x^i, μ^i) is the optimal primal-dual pair to $(NLPF(y^i))$ if $(NLP(y^i))$ is infeasible, $i = 1, \ldots, k$, and $|I^k \cup J^k| = k$. By the convexity of $f(x, \cdot)$ and $g(x, \cdot)$, $(MGBD_k)$ is a relaxation of problem $(MGBD)$ and it thus provides a lower bound to $(MGBD)$ and the solution y^{k+1} to $(MGBD_k)$ can then be used to generate problem $(NLP(y^{k+1}))$ in the next iteration.

An iterative scheme can now be developed as follows.

ALGORITHM 13.1 (GENERALIZED BENDERS DECOMPOSITION ALGORITHM FOR $(MINLP_1)$)

Step 0. Choose $y^1 \in Y$. Set $LB^0 = -\infty$, $UB^0 = +\infty$, $I^0 = J^0 = \emptyset$, $k = 1$.

Step 1. Solve $(NLP(y^k))$.

(i) If $(NLP(y^k))$ is feasible, we obtain an optimal solution x^k and an optimal multiplier vector λ^k. Set $I^k = I^{k-1} \cup \{k\}$ and $J^k = J^{k-1}$. Set $UB^k = \min\{UB^{k-1}, f(x^k, y^k)\}$. If $UB^k = f(x^k, y^k)$, set $(x^*, y^*) = (x^k, y^k)$.

(ii) If $(NLP(y^k))$ is infeasible, solve $(NLPF(y^k))$ and obtain an optimal solution x^k and an optimal multiplier vector μ^k, set $J^k = J^{k-1} \cup \{k\}$ and $I^k = I^{k-1}$.

Step 2. Solve the master problem $(MGBD_k)$ and obtain an optimal solution (α^k, y^{k+1}). Set $LB^k = \alpha^k$. If $LB^k \geq UB^k$, stop and (x^*, y^*) is the optimal solution to $(MINLP_1)$. Otherwise, set $k := k + 1$ and go to Step 1.

THEOREM 13.2 *Algorithm 13.1 stops at an optimal solution (x^*, y^*) to problem $(MINLP_1)$ within a finite number of iterations.*

Proof. Let f^* denote the optimal value of $(MINLP_1)$. It is clear that $\alpha^{k-1} \leq \alpha^k \leq f^* \leq UB^k \leq UB^{k-1}$ for each $k \geq 1$. If the algorithm stops at the k-th iteration, then $LB^k = f^* = UB^k$, i.e., (x^*, y^*) is an optimal solution to $(MINLP_1)$. We prove in the following that if the algorithm does not stop at the k-th iteration, then the optimal solution y^{k+1} of $(MGBD_k)$ does not repeat any previous solutions $y^1, \ldots, y^k$. If $(NLP(y^i))$ is feasible, then $i \in I^k$. Since (x^i, λ^i) is an optimal primal-dual pair of $(NLP(y^i))$, the KKT conditions give $(\lambda^i)^T g(x^i, y^i) = 0$. Thus

$$L(x^i, y^i, \lambda^i) = f(x^i, y^i) \geq UB^i \geq UB^k > LB^k = \alpha^k. \qquad (13.3.5)$$

The optimal solution y^{k+1} must not be equal to y^i, otherwise, the first constraint in $(MGBD_k)$ becomes,

$$\alpha^k \geq L(x^i, y^i, \lambda^i) + \nabla_y^T L(x^i, y^i, \lambda^i)(y - y^i) = L(x^i, y^i, \lambda^i),$$

which contradicts (13.3.5). If $(NLP(y^i))$ is infeasible, then $i \in J^k$. Since the optimal value β^i of problem $(NLPF(y^i))$ is positive, it follows from the duality theorem that $(\mu^i)^T g(x^i, y^i) = \beta^i > 0$. Thus, y^i violates the constraint

$$0 \geq (\mu^i)^T [g(x^i, y^i) + \nabla_y^T g(x^i, y^i)(y - y^i)].$$

Therefore, in either case, y^{k+1} will not repeat any of the previous solutions $y^1, \ldots, y^k$. The finite termination of the algorithm then follows from the finiteness of integer set Y. □

EXAMPLE 13.4 To illustrate the GBD algorithm, let us apply Algorithm 13.1 to Example 13.1.

Iteration 0

Step 0. Choose $y^1 = 3$. Set $LB^0 = -\infty$, $UB^0 = +\infty$, $I^0 = J^0 = \emptyset$, $k = 1$.

Iteration 1

Step 1. Solve $(NLP(y^1))$:

$$\begin{aligned} &\min\ 15 - 2\ln(x+1) \\ &\text{s.t. } 0 \geq e^{x/2} - \sqrt{3}/2 - 1, \\ &\qquad 0 \geq -2\ln(x+1) - 0.5, \\ &\qquad 0 \geq x - 1, \\ &\qquad x \in [0, 2]. \end{aligned}$$

We obtain: $x^1 = 1$, $\lambda^1 = (0, 0, 1)^T$. Set $UB^1 = 13.6137$, $I^1 = \{1\}$ and $J^1 = \emptyset$.

Step 2. The master problem $(MGBD_1)$ is

$$\begin{aligned} &\min\ \alpha \\ &\text{s.t. } \alpha \geq 15 - 2\ln(2) + 6(y - 3), \\ &\qquad y \in [1, 3], \text{ integer}. \end{aligned}$$

We obtain: $y^2 = 1$, $LB^1 = 1.6137$.

Iteration 2

Step 1. The primal problem $(NLP(y^2))$ is infeasible. The feasibility-check problem is

$$\begin{aligned} &\min\ \beta \\ &\text{s.t. } \beta \geq e^{x/2} - 3/2, \\ &\qquad \beta \geq -2\ln(x+1) + 1.5, \\ &\qquad \beta \geq x - 3, \\ &\qquad x \in [0, 2]. \end{aligned}$$

We have $x^2 = 0.9808$ and $\mu^2 = (0.5528, 0.4471, 0)^T$. Set $J^2 = \{2\}$ and $I^2 = I^1$.

Step 2. The new master problem $(MGBD_2)$ is:

$$\begin{aligned}
&\min\ \alpha \\
&\text{s.t. } \alpha \geq 15 - 2\ln(2) + 6(y-3), \\
&\quad\ \ 0 \geq 0.5528 \times (0.3830 - 0.25y) + 0.4471 \times (-11.9971 - y), \\
&\quad\ \ y \in [1,3], \text{ integer.}
\end{aligned}$$

We have $y^3 = 2$ and $LB^2 = 7.6137$.

Iteration 3

Step 1 The primal problem $(NLP(y^3))$ is

$$\begin{aligned}
&\min\ 10 - 2\ln(x+1) \\
&\text{s.t. } 0 \geq e^{x/2} - \sqrt{2}/2 - 1, \\
&\quad\ \ 0 \geq -2\ln(x+1) + 0.5, \\
&\quad\ \ 0 \geq x - 2, \\
&\quad\ \ x \in [0,2].
\end{aligned}$$

We have $x^3 = 1.0696$ and $\lambda^3 = (1.1322, 0, 0)^T$. Set $UB^3 = 8.5453$, $I^3 = \{1,3\}$ and $J^3 = \{2\}$.

Step 2. The new master problem $(MGBD_3)$ is

$$\begin{aligned}
&\min\ \alpha \\
&\text{s.t. } \alpha \geq 15 - 2\ln(2) + 6(y-3), \\
&\quad\ \ 0 \geq 0.5528 \times (0.3830 - 0.25y) + 0.4471 \times (-11.9971 - y), \\
&\quad\ \ \alpha \geq 10 - 2\ln(2.0696) + 4.7998(y-2), \\
&\quad\ \ y \in [1,3], \text{ integer.}
\end{aligned}$$

We obtain $y^4 = 2$ and $LB^3 = 8.5453 = UB^3$. The algorithm terminates with $(1.0696, 2)$ as the optimal solution.

EXAMPLE 13.5 Applying Algorithm 13.1 to Example 13.3 yields an optimal solution $(x^*, y^*) = (1.3009, 0, 1, 0, 1, 0)^T$ after solving two master problems and two nonlinear programming subproblems. The solution process is summarized in Table 13.2.

13.4 Outer Approximation Method

The basic idea underlying the outer approximation method (OA) is similar to the GBD method. The method alternates between solving a nonlinear programming subproblem and solving a mixed-integer linear programming master

Table 13.2. Solution process of the GBD method for Example 13.3.

Iteration	y^k	x^k	LB^k	UB^k
1	$(1,0,1)^T$	$(1.5,1.5,0.9163)^T$	0	15.0927
2	$(0,1,0)^T$	$(1.3009,0,1)^T$	6.0097	6.0097

problem. The major difference between the OA method and the GBD method lies in the different derivations of the mixed-integer linear programming master problem.

Consider inequality constrained problem $(MINLP_1)$. We assume in this section that conditions (i) and (ii) of Assumption 13.1 and Assumption 13.2 hold for problem $(MINLP_1)$. Let S and V be defined the same as in (13.3.1) and (13.3.2) of Section 13.3. For any $y^i \in V$, let x^i be the optimal solution to $(NLP(y^i))$. By (i) and (ii) of Assumption 13.1 and Assumption 13.2, we have

$$
\begin{aligned}
\min_{(x,y)\in S} f(x,y) &= \min_{y^i\in V}\min_{x\in X}\{f(x,y^i) \mid g(x,y^i)\le 0\} \\
&= \min_{y^i\in V}\min\; f(x^i,y^i)+\nabla^T f(x^i,y^i)\begin{pmatrix} x-x^i \\ 0 \end{pmatrix} \\
&\qquad \text{s.t. } g(x^i,y^i)+\nabla g(x^i,y^i)\begin{pmatrix} x-x^i \\ 0 \end{pmatrix}\le 0 \\
&\qquad x\in X \\
&= \min_{y^i\in V}\min\; \alpha \qquad\qquad (13.4.1) \\
&\qquad \text{s.t. } \alpha \ge f(x^i,y^i)+\nabla^T f(x^i,y^i)\begin{pmatrix} x-x^i \\ 0 \end{pmatrix} \\
&\qquad 0\ge g(x^i,y^i)+\nabla g(x^i,y^i)\begin{pmatrix} x-x^i \\ 0 \end{pmatrix} \\
&\qquad x\in X, \alpha\in \mathbb{R}^1,
\end{aligned}
$$

where the second equation is due to the fact that the KKT conditions of $(NLP(y^i))$ and its linearization at x^i are identical. Let

$$T=\{i \mid y^i\in V \text{ and } x^i \text{ solves } (NLP(y^i))\}.$$

Consider the following MILP master problem:

$$
\begin{aligned}
(MOAV) \qquad & \min \alpha \\
& \text{s.t. } \alpha \geq f(x^i, y^i) + \nabla^T f(x^i, y^i) \begin{pmatrix} x - x^i \\ y - y^i \end{pmatrix}, \ i \in T, \\
& 0 \geq g(x^i, y^i) + \nabla^T g(x^i, y^i) \begin{pmatrix} x - x^i \\ y - y^i \end{pmatrix}, \ i \in T, \\
& x \in X, y \in V, \alpha \in \mathbb{R}^1.
\end{aligned}
$$

Let (x^*, y^*) be an optimal solution to $(MINLP_1)$, then (α^*, x^*, y^*) is an optimal solution to (13.4.1) with $\alpha^* = f(x^*, y^*)$. By the convexity of $f(x, y)$ and $g(x, y)$, for any $i \in T, \alpha \geq f(x^i, y^i)$ and $0 \geq g(x^i, y^i)$ imply that (α, x^i, y^i) is feasible to $(MOAV)$. Thus $\hat{\alpha} = v(MOAV) \leq \alpha^*$. On the other hand, since there exists i such that $(x^i, y^i) = (x^*, y^*)$, it follows from the first constraint in problem (13.4.1) that $\hat{\alpha} \geq f(x^*, y^*) = \alpha^*$. Therefore, we have the following theorem.

THEOREM 13.3 *The master problem $(MOAV)$ is equivalent to problem $(MINLP_1)$.*

In order to derive a solvable MILP from $(MOAV)$, we have to represent V by a set of inequality constraints of (x, y) and to relax the index set T by iteratively generating (x^i, y^i). For any $y \in Y$, consider the feasibility-check problem $(NLPF(y))$. We have the following lemma.

LEMMA 13.1 *Let $y^i \in Y$ be such that $(NLP(y^i))$ is infeasible. Let x^i be the optimal solution to the feasibility check problem $(NLPF(y^i))$. Then y^i is infeasible to the following inequality system:*

$$
0 \geq g_j(x^i, y^i) + \nabla^T g_j(x^i, y^i) \begin{pmatrix} x - x^i \\ y - y^i \end{pmatrix}, \ j = 1, \ldots, q, \quad (13.4.2)
$$

for all $x \in X$.

Proof. Suppose on the contrary, y^i is feasible for $\tilde{x} \in X$ to (13.4.2). Then

$$
0 \geq g_j(x^i, y^i) + \nabla_x^T g_j(x^i, y^i)(\tilde{x} - x^i), \ j = 1, \ldots, q. \quad (13.4.3)
$$

Since x^i is the optimal solution to $(NLPF(y^i))$, by the KKT conditions, there exist optimal multipliers μ_j, $j = 1, \ldots, q$, such that

$$
\sum_{j=1}^{q} \mu_j \nabla_x g_j(x^i, y^i) = 0, \ \sum_{j=1}^{q} \mu_j = 1, \ \mu_j \geq 0, \ \forall j = 1, \ldots, q. \quad (13.4.4)
$$

Multiplying (13.4.3) by μ_j and summing up for all $j = 1, \ldots, q$, we obtain by using (13.4.4) that

$$0 \geq \sum_{j=1}^{q} \mu_j g_j(x^i, y^i). \tag{13.4.5}$$

On the other hand, since x^i is the optimal solution to $(NLPF(y^i))$ and $(\mu_1, \ldots, \mu_q)^T$ is the optimal solution to the dual problem $(DF(y^i))$, it follows from the strong duality theorem that

$$\alpha^* = \sum_{j=1}^{q} \mu_j g_j(x^i, y^i),$$

where α^* is the optimal value of $(NLPF(y^i))$. Thus, (13.4.5) implies $\alpha^* \leq 0$, which contradicts the infeasibility of $(NLP(y^i))$. □

Let F denote the index set of all $y^i \in Y$ such that $(NLP(y^i))$ is infeasible. Then, by Lemma 13.1, constraint (13.4.2) excludes all $y^i \in F$. Therefore, incorporating (13.4.2) into problem $(MOAV)$ and replacing V by Y give rise to an equivalent master problem

$$
\begin{aligned}
(MOA) \qquad & \min \ \alpha \\
& \text{s.t. } \alpha \geq f(x^i, y^i) + \nabla^T f(x^i, y^i) \begin{pmatrix} x - x^i \\ y - y^i \end{pmatrix}, \ i \in T, \\
& \quad 0 \geq g(x^i, y^i) + \nabla^T g(x^i, y^i) \begin{pmatrix} x - x^i \\ y - y^i \end{pmatrix}, \ i \in T, \\
& \quad 0 \geq g(x^i, y^i) + \nabla^T g(x^i, y^i) \begin{pmatrix} x - x^i \\ y - y^i \end{pmatrix}, \ i \in F, \\
& \quad x \in X, y \in Y, \alpha \in \mathbb{R}^1.
\end{aligned}
$$

Replacing the points x^i, $i \in T$ and $i \in F$ in (MOA), by the points obtained in the previous k iterations yields

$$
\begin{aligned}
(MOA_k) \qquad & \min \ \alpha \\
& \text{s.t. } \alpha \geq f(x^i, y^i) + \nabla^T f(x^i, y^i) \begin{pmatrix} x - x^i \\ y - y^i \end{pmatrix}, \ i \in T^k, \\
& \quad 0 \geq g(x^i, y^i) + \nabla^T g(x^i, y^i) \begin{pmatrix} x - x^i \\ y - y^i \end{pmatrix}, \ i \in T^k, \\
& \quad 0 \geq g(x^i, y^i) + \nabla^T g(x^i, y^i) \begin{pmatrix} x - x^i \\ y - y^i \end{pmatrix}, \ i \in F^k, \\
& \quad x \in X, y \in Y, \alpha \in \mathbb{R}^1,
\end{aligned}
$$

where

$$T^k = \{i \mid y^i \in V \text{ and } x^i \text{ solves } NLP(y^i),\ i = 1, \ldots, k\},$$
$$F^k = \{i \mid NLP(y^i) \text{ is infeasible and } x^i \text{ solves } NLPF(y^i),\ i = 1, \ldots, k\}.$$

Comparing the structures of $(MGBD_k)$ and (MOA_k), we can see that $(MGBD_k)$ is a relaxation of (MOA_k). In fact, the first constraint in $(MGBD_k)$ can be derived from (MOA_k) by using KKT conditions of $(NLP(y^i))$ and surrogating the first and second constraints in (MOA_k) with optimal multipliers λ_j, $j = 1, \ldots, q$. The second constraint in $(MGBD_k)$ can be obtained from (MOA_k) by using the KKT conditions of $(NLPF(y^i))$ and surrogating the third constraint with the optimal multipliers μ_j, $j = 1, \ldots, q$. Therefore, the master problem (MOA_k) can provide a lower bound better than $(MGBD_k)$, but with a price of including more constraints.

The outer approximation (OA) algorithm can be now described as follows.

ALGORITHM 13.2 (OUTER APPROXIMATION ALGORITHM FOR $(MINLP_1)$)

Step 1. Choose $y^1 \in Y$. Set $LB = -\infty$, $UB = +\infty$, $T^0 = F^0 = \emptyset$, $k = 1$.

Step 2. Solve $(NLP(y^k))$.

- **(i)** If $(NLP(y^k))$ is feasible, we obtain an optimal solution x^k and optimal multiplier vector λ^k. Set $UB^k = f(x^k, y^k)$ and $T^k = T^{k-1} \cup \{k\}$. Set $UB = \min\{UB, UB^k\}$. If $UB = UB^k$, set $(x^*, y^*) = (x^k, y^k)$.
- **(ii)** If $(NLP(y^k))$ is infeasible, solve $(NLPF(y^k))$ and obtain an optimal solution x^k, set $F^k = F^{k-1} \cup \{k\}$.

Step 3. Solve the master problem (MOA_k) and obtain an optimal solution $(\alpha^k, \bar{x}^{k+1}, y^{k+1})$. Set $LB^k = \alpha^k$. If $LB^k \geq UB$, stop and (x^*, y^*) is the optimal solution to $(MINLP_1)$. Otherwise, set $k := k + 1$ and go to Step 2.

THEOREM 13.4 *Under (i) and (ii) of Assumption 13.1 and Assumption 13.2, Algorithm 13.2 stops in a finite number of iterations either at an optimal solution to problem $(MINLP_1)$ or reporting an infeasibility of problem $(MINLP_1)$ if $UB = +\infty$.*

Proof. When the algorithm stops, the optimality of (x^*, y^*) or the correctness of infeasibility reported is obvious. We now prove the finite termination of the algorithm. From the finiteness of Y, it suffices to show that if the algorithm does not stop at the k-th iteration, then the integer optimal solution y^{k+1} of the master problem (MOA_k) does not repeat any integer point in $T^k \cup F^k = \{1, \ldots, k\}$.

For any y^i with $i \le k$, if $y^i \in F^k$, then Lemma 13.1 implies that y^i is infeasible to (MOA_k) and thus $y^{k+1} \neq y^i$. If $y^i \in T^k$, then $(NLP(y^i))$ is feasible and x^i is an optimal solution to $(NLP(y^i))$. Thus, by KKT conditions, there exist $\lambda_j \ge 0, j = 1, \dots, q$, such that

$$\nabla_x f(x^i, y^i) + \sum_{j=1}^{q} \lambda_j \nabla_x g_j(x^i, y^i) = 0, \tag{13.4.6}$$

$$g_j(x^i, y^i) \le 0, \; j = 1, \dots, q, \tag{13.4.7}$$

$$\lambda_j g_j(x^i, y^i) = 0, \; j = 1, \dots, q. \tag{13.4.8}$$

Since $(\alpha^k, \bar{x}^{k+1}, y^{k+1})$ solves (MOA_k), we have

$$\alpha^k < UB \le f(x^i, y^i), \tag{13.4.9}$$

$$\alpha^k \ge f(x^i, y^i) + \nabla f(x^i, y^i) \begin{pmatrix} \bar{x}^{k+1} - x^i \\ 0 \end{pmatrix}, \tag{13.4.10}$$

$$0 \ge g_j(x^i, y^i) + \nabla g_j(x^i, y^i) \begin{pmatrix} \bar{x}^{k+1} - x^i \\ 0 \end{pmatrix}, \; j = 1, \dots, q. \tag{13.4.11}$$

Multiplying inequality (13.4.11) by λ_j and summing up for $j = 1, \dots, q$, and then adding the resulting inequality to (13.4.10), we obtain from (13.4.6)–(13.4.8) that $\alpha^k \ge f(x^i, y^i)$ which contradicts (13.4.9). □

REMARK 13.1 For 0-1 MINLP problems, it is possible to avoid solving the feasibility-check problem $(NLPF(y^k))$ by replacing the constraints for $i \in F^k$ in (MOA_k) with the following integer cuts:

$$\sum_{j \in B^i} y_j - \sum_{j \in N^i} y_j \le |B^i| - 1, \quad i \in F^k, \tag{13.4.12}$$

where $B^i = \{j \mid y_j^i = 1\}$ and $N^i = \{j \mid y_j^i = 0\}$.

EXAMPLE 13.6 Let's apply Algorithm 13.2 to Example 13.1.

Iteration 0
Step 1. Choose $y^1 = 3$. Set $LB = -\infty$, $UB = +\infty$, $T^0 = F^0 = \emptyset$, $k = 1$.
Iteration 1
Step 2. Solving $(NLP(y^1))$ gives $x^1 = 1$, $UB^1 = 13.6137$, $T^1 = \{1\}$.

Step 3. The master problem (MOA_1) is

$$\begin{aligned}
&\min\ \alpha \\
&\text{s.t.}\ \ \alpha \geq 5y - x + 1 - 2\ln(2), \\
&\quad\ 0 \geq e^{0.5} - \sqrt{3}/2 - 1 + 0.5e^{0.5}(x-1) - (\sqrt{3}/12)(y-3), \\
&\quad\ 0 \geq -2\ln(2) - 0.5 - (x-1) - (y-3), \\
&\quad\ 0 \geq x + y - 4, \\
&\quad\ x \in [0,2],\ y \in [1,3],\ \text{integer}.
\end{aligned}$$

The optimal solution to (MOA_1) is $(\alpha^1, \bar{x}^2, y^2) = (3, 1.6138, 1)$. Set $LB^1 = \alpha^1 = 3$, $k = 2$.

Iteration 2

Step 2. Since the primal problem $(NLP(y^2))$ is infeasible, set $F^1 = \{2\}$. Solving the feasibility-check problem $(NLPF(y^2))$, we obtain $x^3 = 0.9808$.

Step 3. The master problem (MOA_2) is

$$\begin{aligned}
&\min\ \alpha \\
&\text{s.t.}\ \ \alpha \geq 5y - x + 1 - 2\ln(2), \\
&\quad\ 0 \geq e^{0.5} - \sqrt{3}/2 - 1 + 0.5e^{0.5}(x-1) - (\sqrt{3}/12)(y-3), \\
&\quad\ 0 \geq -2\ln(2) - 0.5 - (x-1) - (y-3), \\
&\quad\ 0 \geq x + y - 4, \\
&\quad\ 0 \geq e^{0.4904} - 1.5 + 0.5e^{0.4904}(x - 0.9808) - 0.25(y-1), \\
&\quad\ 0 \geq -2\ln(1.9808) + 1.5 - 1.0097(x - 0.9808) - (y-1), \\
&\quad\ x \in [0,2],\ y \in [1,3],\ \text{integer}.
\end{aligned}$$

The optimal solution of (MOA_2) is $(\alpha^2, \bar{x}^3, y^3) = (8.4896, 1.1241, 2)$. Set $LB^2 = \alpha^2 = 8.4896$, $k = 3$.

Iteration 3

Step 2. Solving $(NLP(y^3))$, we obtain $x^3 = 1.0696$, $UB^3 = 8.5453$. Set $T^2 = \{1, 3\}$.

Step 3. The master problem (MOA_3) is

$$\begin{aligned}
&\min \alpha \\
&\text{s.t. } \alpha \geq 5y - x + 1 - 2\ln(2), \\
&\quad \alpha \geq 5y - 2\ln(2.0696) - 0.9664(x - 1.0696), \\
&\quad 0 \geq \mathrm{e}^{0.5} - \sqrt{3}/2 - 1 + 0.5\mathrm{e}^{0.5}(x - 1) - (\sqrt{3}/12)(y - 3), \\
&\quad 0 \geq -2\ln(2) - 0.5 - (x - 1) - (y - 3), \\
&\quad 0 \geq x + y - 4, \\
&\quad 0 \geq \mathrm{e}^{0.4904} - 1.5 + 0.5\mathrm{e}^{0.4904}(x - 0.9808) - 0.25(y - 1), \\
&\quad 0 \geq -2\ln(1.9808) + 1.5 - 1.0097(x - 0.9808) - (y - 1), \\
&\quad 0 \geq \mathrm{e}^{0.5348} - \sqrt{2}/2 - 1 + 0.5\mathrm{e}^{0.5348}(x - 1.0696) - (\sqrt{2}/8)(y - 2), \\
&\quad 0 \geq -2\ln(2.0696) + 0.5 + 0.9664(x - 1.0696) - (y - 2), \\
&\quad x \in [0, 2],\ y \in [1, 3],\ \text{integer}.
\end{aligned}$$

The optimal solution to (MOA_3) is $(\alpha^3, \bar{x}^4, y^4) = (8.5453, 1.0696, 2)$. Set $LB^3 = \alpha^3 = 8.5453 = UB^3$. So, the algorithm terminates at an optimal solution $(1.0696, 2)$.

The solution process of the OA method for Example 13.3 is summarized in Table 13.3. Since Example 13.3 is a 0-1 nonlinear integer program, the integer cut (13.4.12) is used in the master problem.

Table 13.3. Solution process of the OA method for Example 13.3.

Iteration	y^k	x^k	LB^k	UB^k
1	$(1,0,1)^T$	$(1.5, 1.5, 0.9163)^T$	2.3927	15.0927
2	$(0,0,0)^T$	$(0,0,0)^T$	6	10
3	$(0,1,0)^T$	$(1.3009, 0, 1)^T$	6.0097	6.0097

13.5 Nonconvex Mixed-Integer Programming

In this section, we investigate global optimization methods for solving nonconvex mixed-integer problem $(MINLP_1)$. Nonconvexity often arises in real-world applications of mixed-integer nonlinear programming models such as in chemical engineering and complex reliability systems. Convexity assumptions of f and g_i, however, play a key role in guaranteeing the validness of upper bounds and lower bounds used in the branch-and-bound method, the generalized Benders decomposition method and the outer approximation method for problem $(MINLP_1)$ discussed in the previous sections. In fact, without (ii) of

Assumption 13.1, the continuous nonlinear subproblem $(NLP(y))$ may be a nonconvex problem and may have multiple local solutions. Moreover, the master problems $(MGBD)$ or (MOA) do not necessarily generate a valid lower bound.

To overcome the difficulties caused by the nonconvexity, convex approximation or convexification method can be used to construct lower bounding convex subproblems. Combined with upper bounding procedures, the nonconvex problems can then be solved by branch-and-bound methods.

13.5.1 Convex relaxation

Let $\tilde{f}$ and $\tilde{g}_j$ $(j = 1, \ldots, q)$ be convex underestimators of functions f and g_j $(j = 1, \ldots, q)$, respectively. Consider the following convex lower bounding problem:

$$(CLBP) \qquad \begin{array}{ll} \min & \tilde{f}(x, y) \\ \text{s.t.} & \tilde{g}_j(x, y) \le 0, \quad j = 1, \ldots, q, \\ & x \in X, \ y \in \widetilde{Y}, \end{array}$$

where $\widetilde{Y} \supseteq Y$. Problem $(CLBP)$ is a convex mixed-integer programming problem and its optimal value provides a valid lower bound for the original problem $(MINLP_1)$. Branch-and-bound methods based on the convex relaxation can then be developed.

Many convexification schemes to underestimate a nonconvex function have been proposed in the literature. Especially, convex piecewise linear underestimators can be derived for some special functions.

Billinear function. Let $a_{ij}x_ix_j$ be the bilinear function defined on $[x_i^l, x_i^u] \times [x_j^l, x_j^u]$. Let $z_{ij} = x_ix_j$.

Case (a). $a_{ij} \ge 0$. Since $(x_i - x_i^l)(x_j - x_j^l) \ge 0$ and $(x_i - x_i^u)(x_j - x_j^u) \ge 0$, we have

$$z_{ij} \ \ge \ x_i^l x_j + x_j^l x_i - x_i^l x_j^l, \qquad (13.5.1)$$

$$z_{ij} \ \ge \ x_i^u x_j + x_j^u x_i - x_i^u x_j^u. \qquad (13.5.2)$$

Thus, the convex underestimator of the bilinear term $a_{ij}x_ix_j$ is $a_{ij}\max(U, V)$, where U and V are the right-hand sides of (13.5.1) and (13.5.2), respectively.

Case (b). $a_{ij} < 0$. Since $(x_i - x_i^u)(x_j - x_j^l) \le 0$ and $(x_i - x_i^l)(x_j - x_j^u) \le 0$, we have

$$z_{ij} \ \le \ x_i^u x_j + x_j^l x_i - x_i^u x_j^l, \qquad (13.5.3)$$

$$z_{ij} \ \le \ x_i^l x_j + x_j^u x_i - x_i^l x_j^u. \qquad (13.5.4)$$

Thus, the convex underestimator of the bilinear term $a_{ij}x_ix_j$ is $a_{ij}\min(U, V)$ or $-a_{ij}\max(-U, -V)$, where U and V are the right-hand sides of (13.5.3) and (13.5.4), respectively.

Fractional function. Let $b_{ij}(x_i/x_j)$ be the fractional function defined on $[x_i^l, x_i^u] \times [x_j^l, x_j^u]$, where $x_i^l > 0$ and $x_j^l > 0$. Let $w_{ij} = x_i/x_j$.

Case (a). $b_{ij} \geq 0$. Note that $(x_i^u - x_i)(x_i - x_i^l) \geq 0$, $(x_i - x_i^u)(1/x_j - 1/x_j^l) \geq 0$ and $(x_i - x_i^l)(1/x_j - 1/x_j^u) \geq 0$. We have

$$w_{ij} \geq x_i/x_j^l + x_i^u/x_j - x_i^u/x_j^l, \tag{13.5.5}$$

$$w_{ij} \geq x_i/x_j^u + x_i^l/x_j - x_i^l/x_j^u, \tag{13.5.6}$$

$$w_{ij} \geq \frac{1}{x_j}\left(\frac{x_i + \sqrt{x_i^l x_i^u}}{\sqrt{x_i^l} + \sqrt{x_i^u}}\right)^2. \tag{13.5.7}$$

Thus, the convex underestimating function of $b_{ij}(x_i/x_j)$ is $b_{ij}\max(U, V, W)$, where U, V, and W are the right-hand sides of (13.5.5)–(13.5.7), respectively.

Case (b). $b_{ij} < 0$. Since $(1/x_j^l)(x_j - x_j^l)(x_i/x_j - x_i^l/x_j^u) \geq 0$ and $(1/x_j^u)(x_j - x_j^u)(x_i/x_j - x_i^u/x_j^l) \geq 0$, we have

$$w_{ij} \leq \frac{1}{x_j^l x_j^u}(x_j^u x_i - x_i^l x_j + x_i^l x_j^l), \tag{13.5.8}$$

$$w_{ij} \leq \frac{1}{x_j^l x_j^u}(x_j^l x_i - x_i^u x_j + x_i^u x_j^u). \tag{13.5.9}$$

Thus, the convex underestimating function of $b_{ij}(x_i/x_j)$ is $b_{ij}\min(U, V)$ or $-b_{ij}\max(-U, -V)$, where U and V are the right-hand sides of (13.5.8)–(13.5.9), respectively.

Univariate concave function. Let $h_i(x_i)$ be a univariate concave function on $[x_i^l, x_i^u]$. The convex underestimator of $h_i(x_i)$ is the linear function correcting $(x^l, h_i(x_i^l))$ and $(x_i^u, h_i(x_i^u))$:

$$\hat{h}_i(x_i) = h_i(x_i^l) + \frac{h_i(x_i^u) - h_i(x_i^l)}{x_i^u - x_i^l}(x_i - x_i^l). \tag{13.5.10}$$

Consider a nonconvex version of problem $(MINLP_2)$:

$$\begin{aligned} &\min\ f(x) + c^T y \\ &\text{s.t. } g(x) + By \leq 0, \\ &\qquad x \in X \subseteq \mathbb{R}^n,\ y \in Y \subset \mathbb{Z}^m, \end{aligned} \tag{13.5.11}$$

where f and $g = (g_1, \ldots, g_q)^T$ are not necessarily convex functions and $X = [x^l, x^u]$. Suppose that f and g_k's can be decomposed into sums of bilinear functions, fractional functions, univariate functions and convex functions.

$$f(x) = \sum_{(i,j)\in I} a_{ij}x_ix_j + \sum_{(i,j)\in J} b_{ij}\frac{x_i}{x_j} + \sum_{i\in K} h_i(x_i) + t(x), \tag{13.5.12}$$

$$g_k(x) = \sum_{(i,j)\in I_k} a_{ij}^k x_ix_j + \sum_{(i,j)\in J_k} b_{ij}^k\frac{x_i}{x_j} + \sum_{i\in K_k} h_i^k(x_i) + t_k(x), \quad k = 1,\ldots,q, \tag{13.5.13}$$

where h_i's and h_i^k's are univariate concave functions, K and K_k $(k = 1,\ldots,q)$ are subsets of $\{1,\ldots,n\}$, t and t_k's are convex functions.

Let's introduce the following new variables for each bilinear terms and fractional terms in f and g_k, $k = 1,\ldots,q$:

$$z_{ij} = x_ix_j, \ (i,j) \in I \cup (\cup_{k=1}^q I_k),$$
$$w_{ij} = \frac{x_i}{x_j}, \ (i,j) \in J \cup (\cup_{k=1}^q J_k).$$

Let

$$I^+ = \{(i,j) \in I \mid a_{ij} \geq 0\}, \ I^- = I \setminus I^+,$$
$$J^+ = \{(i,j) \in J \mid b_{ij} \geq 0\}, \ J^- = J \setminus J^+,$$
$$I_k^+ = \{(i,j) \in I_k \mid a_{ij}^k \geq 0\}, \ I_k^- = I_k \setminus I_k^+,$$
$$J_k^+ = \{(i,j) \in J_k \mid b_{ij}^k \geq 0\}, \ J_k^- = J_k \setminus J_k^+.$$

Then, the convex underestimating problem of (13.5.11) is:

$$\min \ \sum_{(i,j)\in I} a_{ij}z_{ij} + \sum_{(i,j)\in J} b_{ij}w_{ij} + \sum_{i\in K} \hat{h}_i(x_i) + t(x) + c^Ty \tag{13.5.14}$$

$$\text{s.t.} \ \sum_{(i,j)\in I_k} a_{ij}^k z_{ij} + \sum_{(i,j)\in J_k} b_{ij}^k w_{ij} + \sum_{i\in K_k} \hat{h}_i^k(x_i) + t_k(x) + By \leq 0, \quad k = 1,\ldots,q,$$

$$(13.5.1) - (13.5.2), \ (i,j) \in I^+ \cup (\cup_{k=1}^q I_k^+),$$
$$(13.5.3) - (13.5.4), \ (i,j) \in I^- \cup (\cup_{k=1}^q I_k^-),$$
$$(13.5.5) - (13.5.7), \ (i,j) \in J^+ \cup (\cup_{k=1}^q J_k^+),$$
$$(13.5.8) - (13.5.9), \ (i,j) \in J^- \cup (\cup_{k=1}^q J_k^-),$$
$$(13.5.10), \ i \in K \cup (\cup_{k=1}^q K_k),$$
$$x \in X, \ y \in Y,$$

where $\hat{h}_i$ and $\hat{h}_i^k$ are convex underestimators of h_i and h_i^k, respectively. Notice that problem (13.5.14) is a convex mixed-integer programming problem.

Factorable functions form another class of functions whose convex underestimators can be derived efficiently. A function $h(x)$ defined on $\mathbb{R}^n$ is said to be *factorable* if it can be expressed as recursive sums and products of univariate functions. Recursive procedures can be derived to generate a convex underestimating function for a factorable function (see [157][214]).

Finally, convex relaxation schemes for general $\mathcal{C}^2$ functions were investigated in [3][5]. Let $h(x)$ be a twice differentiable function on domain $[x^l, x^u]$. Consider the following function:

$$h_\alpha(x) = h(x) + \sum_{i=1}^{n} \alpha_i(x_i^l - x_i)(x_i^u - x_i), \qquad (13.5.15)$$

with $\alpha_i > 0$, $\forall\, i = 1, \ldots, n$. It is clear that $h(x) \geq h_\alpha(x)$ for all $x \in [x^l, x^u]$ and $h_\alpha(x)$ is a convex function when α_i's are sufficiently large. The Hessian of $h_\alpha(x)$ is:

$$\nabla^2 h_\alpha(x) = \nabla^2 h(x) + 2diag(\alpha_1, \ldots, \alpha_n).$$

Thus, $h_\alpha(x)$ is a convex function on $[x^l, x^u]$ if and only if

$$\nabla^2 h(x) + 2diag(\alpha_1, \ldots, \alpha_n)$$

is a positive semi-definite matrix for all $x \in [x^l, x^u]$. In a special choice of α where $\alpha_1 = \cdots = \alpha_n = \alpha_0$, $h_\alpha(x)$ is a convex function if and only if

$$\alpha_0 \geq \bar{\alpha} = \max\{0, -\frac{1}{2} \min_{x \in [x^l, x^u]} \lambda_{\min}(x)\}, \qquad (13.5.16)$$

where $\lambda_{\min}(x)$ is the minimum eigenvalue of the Hessian of $h(x)$. For nonconvex quadratic function h, it is easy routine work to find out $\bar{\alpha}$ defined in (13.5.16). For general nonconvex function h, however, it could be difficult to determine the value of $\bar{\alpha}$. A number of methods have been proposed to calculate an appropriate $\alpha \geq \bar{\alpha}$ (see [3][5]).

13.5.2 Convexification method

Consider the following monotone mixed-integer programming problem:

$$\begin{aligned} &\max\ f(x, y) && (13.5.17)\\ &\text{s.t. } g_i(x, y) \leq 0,\ i = 1, \ldots, q, \\ &\qquad x \in X = [a, b] \subset \mathbb{R}^n,\ y \in Y = [\alpha, \beta] \cap \mathbb{Z}^m, \end{aligned}$$

where $f : X \times Y \to \mathbb{R}$ and $g_i : X \times Y \to \mathbb{R}$, $i = 1, \ldots, q$, are continuous increasing functions, $0 < a \leq b$, $0 < \alpha \leq \beta$ and α and β are integer vectors. Problem (13.5.17) is, in general, a nonconvex mixed-integer programming problem since we do not assume the convexities of f and g_i's.

A combination of the convexification transformation and the outer approximation discussed in Section 9.3 provides a global optimization method for solving the continuous relaxed subproblem of (13.5.17). The branch-and-bound methodology can be then adopted to find the optimal solution of (13.5.17).

The branch-and-bound algorithm for problem (13.5.17) is similar to that for the pure monotone integer programming (MP) discussed in Chapter 9. The only difference is that the branching process is only applied to y variables.

A prominent example of the problem (13.5.17) is the mixed-integer reliability problem in a complex system:

$$
\begin{aligned}
(MRELI) \qquad & \max\ R_s(x,y) = f(x_1,\ldots,x_n,R_1(y_1),\ldots,R_m(y_m)) \\
& \text{s.t. } g_i(x_1,\ldots,x_n,R_1(y_1),\ldots,R_m(y_m)) \le c_i,\ i=1,\ldots,q, \\
& \quad 0 < a_j \le x_j \le b_j < 1,\ j=1,\ldots,n, \\
& \quad 1 \le \alpha_j \le y_j \le \beta_j,\ y_j \text{ integer},\ j=1,\ldots,m,
\end{aligned}
$$

where x_j is the reliability of the j-th subsystem ($j=1,\ldots,n$), y_j represents the number of redundant components in the $(n+j)$-th subsystem, $R_j(y_j) = 1-(1-r_j)^{y_j}$ is the reliability of the $(n+j)$-th parallel subsystem with $0 < r_j < 1$ ($j=1,\ldots,m$), R_s is the overall system reliability, g_i is the i-th resource consumed; c_i is the total available i-th resource, α_j and β_j are lower and upper integer bounds of y_j respectively.

An inherent property of problem $(MRELI)$ is that functions f and g_i are strictly increasing with respect to each variable. Since $R_j(y_j)$ is a strictly increasing function of y_j, the overall reliability $R_s(x,y)$ is also a strictly increasing function of each variable. Therefore, problems $(MRELI)$ is a monotone mixed-integer programming problem and can thus be solved by the convexification method discussed in Chapter 9.

Computational results for the four typical types of complex networks were reported in [137].

13.6 Notes

Discussions of various applications of mixed-integer nonlinear programming can be found in [4][53][60][85][137][181][213][219][220].

Further discussions of the branch-and-bound (BB) methods can be found in [30][87][130][201]. The generalized Benders decomposition (GBD) method was proposed in [75]. The outer approximation (OA) method was proposed in [53][59]. Global optimization methods were investigated in [4][213][214] for nonconvex cases of $(MINLP)$.

Further discussions of the convexification schemes to underestimate a nonconvex function can be found in [5][6][157][197][214][232]. Branch-and-bound methods based on the convex relaxation were developed in [4][61][187] [213][214].

Applications of the convexification methods to mixed-integer programming problems arising from complex reliability networks were investigated in [137].

Chapter 14

GLOBAL DESCENT METHODS

We consider in this chapter the following general nonlinear integer programming problem,

$$(P) \quad \min_{x \in X} f(x),$$

where $X \subset \mathbb{Z}^n$ is a finite integer set and the function f defined on X is not necessarily continuous.

It is obvious that a global minimum of problem (P) must be also a local minimum. Thus, an optimal (global) solution to (P) can be sought from among local minima of (P), by a two-level solution scheme that switches between local search and global descent (from the current local minimum to a better point with a lower objective value). When compared with global search of an optimal solution to (P), local search is much easier to perform by using local information at the current solution, for example, using an algorithm similar to the steepest descent in continuous minimization. Global descent, on the other hand, cannot only rely on local information. In order to escape from the neighborhood of the current local minimum and to land in a neighborhood of a better local minimum, global descent methods often need global information of the problem, for example, the Lipschitz constant of the problem. While the parameters, such as the Lipschitz constant, are usually unknown for problem (P), a global descent method should be devised such that the estimation of such parameters can be adjusted in the solution process.

Throughout this chapter, we assume the following conditions for X and f.

ASSUMPTION 14.1 (i) *$X \subset \mathbb{Z}^n$ is a finite integer set with at least two integer points.*

(ii) *f satisfies the following Lipschitz condition:*

$$|f(x^1) - f(x^2)| \leq L \, \|x^1 - x^2\|, \quad \forall x^1, \, x^2 \in X,$$

where $\|\cdot\|$ is the usual Euclidean norm and $0 < L < \infty$ is the Lipschitz constant.

Assumption 14.1 (i) implies that there exists a constant $K > 0$ such that

$$1 \leq \max_{x^1, x^2 \in X} \|x^1 - x^2\| \leq K < \infty.$$

14.1 Local Search and Global Descent

14.1.1 Local minima and local search

A local search can be defined under different definitions of neighborhood in discrete optimization. We first introduce the concepts of a local minimum and a local search for problem (P).

DEFINITION 14.1 *Let $x^* \in X$. The m-neighborhood of x^*, $N_m(x^*)$, is defined by*

$$N_m(x^*) = \{x \in X \mid x \text{ differs from } x^* \text{ in no more than } m \text{ components}\}.$$

The unit neighborhood of x^*, $U(x^*)$, is defined by

$$U(x^*) \doteq \{x \in X \mid x_i \in \{x_i^* - 1, x_i^*, x_i^* + 1\}, i = 1, \ldots, n\}.$$

In particular, the unit m-neighborhood is defined as

$$U_m(x^*) = U(x^*) \cap N_m(x^*).$$

DEFINITION 14.2 *A point $x^* \in X$ is called an N_m or U_m local minimizer of f over X if $f(x^*) \leq f(x)$ for all $x \in N_m(x^*)$ or for all $x \in U_m(x^*)$ and $x \neq x^*$. Furthermore, if $f(x^*) \leq f(x)$ for all $x \in X$ and $x \neq x^*$, then x^* is called a global minimizer of f over X. If the strict inequality holds in the inequality $f(x^*) \leq f(x)$ for the local (global) minimizer x^*, then x^* is called a strict local (global) minimizer of f over X.*

Let us examine again Example 1.1. Table 14.1 lists the values of the objective function on X where the objective values of U_1 local minimizers are marked by "*" and the objective values of the U_2 minimizers by "**". Observe that the problem has seven U_1 local minimizers among which two are U_2 local minimizers. Also, the global minimizer $x_{global} = (6, 5)^T$ is both a U_1 and a U_2 local minimizer of f on X.

It is easy to see that for any $2 \leq m \leq n$, $U_{m-1} \subseteq U_m$. Thus, a U_m local minimizer is also a U_{m-1} local minimizer. In particular, a global minimizer is a U_m local minimizer for any m with $1 \leq m \leq n$. However, the number of integer points in a U_m neighborhood increases exponentially with respect to m. When m is large, a U_m local optimal solution is not easy, or even

Table 14.1. U_1 and U_2 local minimizers for Example 1.1.

	$x_1 = 0$	1	2	3	4	5	6	7
$x_2 = 0$	3^{**}	28	139	334	616	982	1434	1971
1	80	6^*	18	115	297	565	918	1356
2	271	99	12	10^*	93	262	516	856
3	578	306	120	20	4^*	75	230	471
4	999	629	344	144	30	2^*	58	200
5	1535	1066	682	384	171	43	1^{**}	44
6	2185	1617	1135	738	426	200	59	3^*

computationally infeasible, to find. Therefore, U_1 local minimizer is most often used in designing solution algorithms, including the global descent method discussed in this chapter. All the discussions below on local search are confined to U_1 local minimizers.

Let e_i denote the i-th unit vector in $\mathbb{R}^n$. Define $\mathbb{D} = \{\pm e_i \mid i = 1, 2 \ldots, n\}$.

DEFINITION 14.3 *For any $x \in X$, $d \in \mathbb{D}$ is said to be a descent direction of f at x if $x + d \in X$ and $f(x + d) < f(x)$. Furthermore, $d^* \in \mathbb{D}$ is called a steepest descent direction of f at x if $f(x + d^*) \leq f(x + d)$ for any other descent direction d.*

Similar to the continuous situation, we can design a discrete version of the steepest descent method for finding a U_1 local minimizer of f over X.

PROCEDURE 14.1 (DISCRETE STEEPEST DESCENT PROCEDURE FOR FINDING U_1 LOCAL MINIMIZER)

Step 0. Choose an initial point $x \in X$.

Step 1. If x is a U_1 local minimizer of f over X, then stop. Otherwise, a steepest descent direction $d^* \in \mathbb{D}$ of f at x over X can be found.

Step 2. Set $x := x + \lambda d^*$, where $\lambda \in \mathbb{Z}_+$ is the stepsize such that f has a maximum decrease in direction d^*. Go to Step 1.

The following basic properties will be useful in the later analysis.

LEMMA 14.1 (i) *For any $\tilde{x}$, $x^* \in X$ and $d \in \mathbb{D}$, it holds*

$$\|\tilde{x} - x^*\| \neq \|\tilde{x} + d - x^*\|.$$

(ii) *For any $\tilde{x}$, $x^* \in X$, if there exists an $i \in \{1, \ldots, n\}$ such that both $\tilde{x} \pm e_i \in X$, then there exists $d \in \{\pm e_i\}$ such that $\|\tilde{x} + d - x^*\| > \|\tilde{x} - x^*\|$.*

(iii) *If x^* and $\tilde{x}$ are distinct strict U_1 local minimizers of f over X, then* $\|x^* - \tilde{x}\| > 1$.

Proof. Let $d \in \{\pm e_i\}$, then

$$\|\tilde{x} + d - x^*\|^2 - \|\tilde{x} - x^*\|^2 = 2\text{sgn}(d_i)(\tilde{x}_i - x_i^*) + 1 \neq 0,$$

where $sgn(x) = 1$ if $x > 0$ and $sgn(x) = -1$ if $x < 0$. Therefore, part (i) is true. Moreover, if $\tilde{x}_i - x_i^* \geq 0$, set $d = e_i$, otherwise, set $d = -e_i$. Thus, part (ii) is true. If $\|x^* - \tilde{x}\| = 1$, then there exists $d \in \mathbb{D}$ such that $x^* - \tilde{x} = d$. This contradicts the assumption that x^* and $\tilde{x}$ are strict local minimizers of f over X. Thus, $\|x^* - \tilde{x}\| > 1$ and part (iii) is true. □

14.1.2 Identification of global minimum from among local minima

The above discussion, especially Example 1.1 motivates us to search for a global minimizer from among local minimizers. Such a solution scheme for solving nonlinear integer programming problem (P) is termed the global descent method. The global descent method enables the algorithm moving from one local minimizer of the objective function f on X to another better one at each iteration with the help of an auxiliary function, entitled the discrete global descent function. The local minimizers of a discrete global descent function on X coincide with better local minimizers of f over X under some assumptions.

A point $x \in X$ is called a *corner point* of X if for each $d \in \mathbb{D}$, $x + d \in X$ implies $x - d \notin X$. Denote by X_c the set of corner points of X. Also, let $X(x^*) = \{x \in X \mid x \neq x^*, \ f(x) \geq f(x^*)\}$.

We now give the formal definition for *discrete global descent functions.*

DEFINITION 14.4 *Let x^* be a U_1 local minimizer of f on X. A function $G : X \to \mathbb{R}$ is said to be a discrete global descent function of f at x^* if it satisfies the following conditions:*

(D1) *x^* is a strict U_1 local maximizer of G on X.*

(D2) *G has no U_1 local minimizers on the set $X(x^*) \setminus X_c$.*

(D3) *$\tilde{x} \in X \setminus X_c$ is a U_1 local minimizer of f on X with $f(\tilde{x}) < f(x^*)$ if and only if $\tilde{x}$ is a U_1 local minimizer of G on X.*

Since only the U_1 local optimality will be used in the following development, a U_1 local minimizer (maximizer) will be called a local minimizer (maximizer) for the sake of simplicity.

14.2 A Class of Discrete Global Descent Functions

We discuss in this section a class of two-parameter discrete global descent functions. Let x^* be a local minimizer of f over X. Define

$$G_{x^*,\mu,\rho}(x) = A_\mu(f(x) - f(x^*)) - \rho\|x - x^*\|, \tag{14.2.1}$$

where $\rho > 0$, $0 < \mu < 1$,

$$A_\mu(y) = y \cdot V_\mu(y), \tag{14.2.2}$$

and $V_\mu : \mathbb{R} \to \mathbb{R}$ is a continuous function that satisfies the following conditions:

(V1) $V_\mu(y)$ is strictly decreasing when $y < 0$ and non-increasing when $y \geq 0$,

(V2) $V_\mu(-\tau) = 1$, $V_\mu(0) = \mu$, and $V_\mu(y) \geq c\mu$ for all y,

where $\tau > 0$ is a sufficiently small number and $0 < c \leq 1$. In theory, the parameter τ is required to satisfy:

$$0 < \tau < \min\{|f(x^1) - f(x^2)| \mid x^1,\, x^2 \in X,\ f(x^1) \neq f(x^2)\}. \tag{14.2.3}$$

Thus, for any x^1, $x^2 \in X$, $f(x^1) < f(x^2)$ implies $f(x^1) < f(x^2) - \tau$. If $f(x)$ is an integer-valued function on X, we can simply set τ as a positive number less than 1. However, the global descent algorithm developed in later sections is insensitive to the value of τ in numerical implementation. Thus, τ is always set to be 1 in calculation.

Some examples of V_μ that satisfy the above conditions are as follows.

EXAMPLE 14.1 *Define*

$$V_\mu(y) = \begin{cases} (1-\mu)\left(\dfrac{y}{-\tau}\right)^{k+1} + \mu, & \text{if } y \leq 0 \\ \mu, & \text{if } y \geq 0 \end{cases}$$

for $k = 0, 1, 2, \ldots$. Then, $V_\mu \in \mathbb{C}^k$ and V_μ satisfies conditions (V1) and (V2).

EXAMPLE 14.2 *Define*

$$\begin{aligned} V_\mu^1(y) &= \mu\left[(1-c)\left(\frac{1-c\mu}{\mu - c\mu}\right)^{-y/\tau} + c\right], \\ V_\mu^2(y) &= \mu\left[\sqrt{(c'y)^2 + (1-c)^2} + c - c'y\right], \end{aligned}$$

where $0 < c < 1$ *and* $c' = \dfrac{(1-\mu)(1+\mu-2c\mu)}{2\mu\tau(1-c\mu)}$.

It can be verified that both V_μ^1 and $V_\mu^2 \in \mathbb{C}^\infty$ and satisfy conditions (V1) and (V2). Figure 14.1 illustrates $V_\mu^1(y)$, $V_\mu^2(y)$, $A_\mu^1(y) = y \cdot V_\mu^1(y)$ and $A_\mu^2(y) = y \cdot V_\mu^2(y)$ with $c = \mu = 0.5$ and $\tau = 1$.

We have the following lemma.

LEMMA 14.2 (i) $\operatorname{sgn}(A_\mu(y)) = \operatorname{sgn}(y)$.

(ii) *$A_\mu(y)$ is a strictly increasing function of y for $y \leq 0$.*

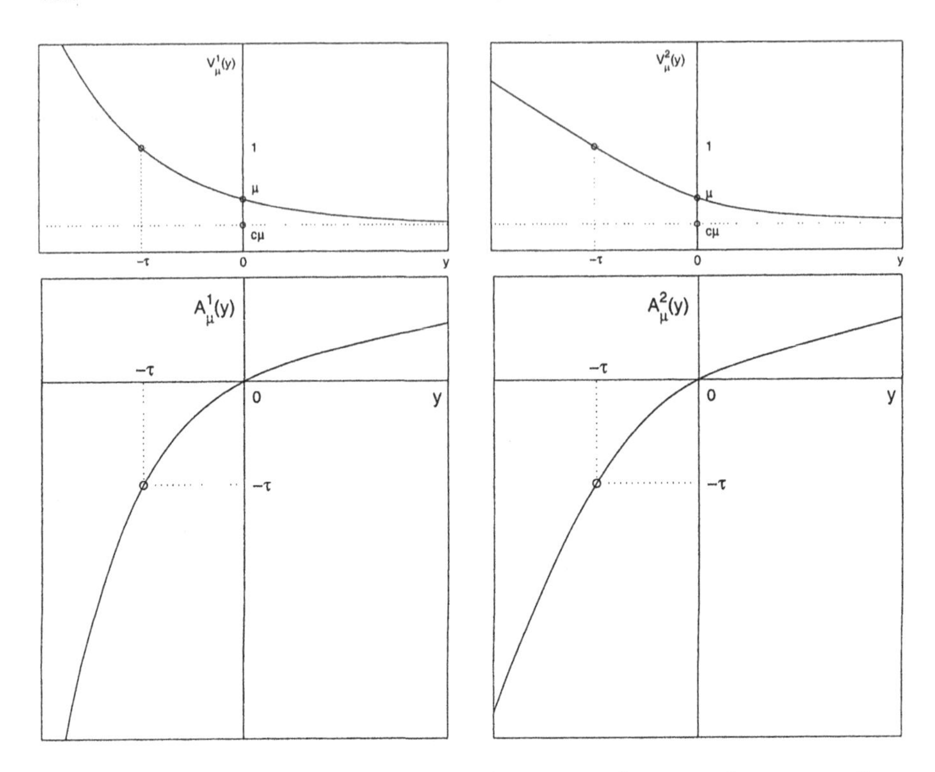

Figure 14.1. Illustrations of $V_\mu^1(y)$, $V_\mu^2(y)$, $A_\mu^1(y) = y \cdot V_\mu^1(y)$ and $A_\mu^2(y) = y \cdot V_\mu^2(y)$ in Example 14.2 with $c = \mu = 0.5$ and $\tau = 1$.

(iii) *If* $y^1 < y^2 \leq -\tau$, *then* $0 < y^2 - y^1 < A_\mu(y^2) - A_\mu(y^1)$.
(iv) *If* $y^1 < -\tau < y^2 < 0$, *then* $A_\mu(y^1) < y^1 < -\tau < y^2 < A_\mu(y^2) < 0$.

Proof. (i) From the definition of V_μ, we have $V_\mu(y) \geq c\mu > 0$, for all y, thus $\text{sgn}(A_\mu(y)) = \text{sgn}(y \cdot V_\mu(y)) = \text{sgn}(y)$.

(ii) Since $V_\mu(y)$ is strictly decreasing when $y < 0$ and $V_\mu(y) \geq c\mu > 0$, for all y, thus $V_\mu(y^1) > V_\mu(y^2) > 0$, for all $y^1 < y^2 < 0$. Therefore, $A_\mu(y^1) = y^1 \cdot V_\mu(y^1) < y^2 \cdot V_\mu(y^2) = A_\mu(y^2)$, for all $y^1 < y^2 \leq 0$.

(iii) Since $V_\mu(y)$ is strictly decreasing when $y < 0$ and $V_\mu(-\tau) = 1$, so if $y^1 < y^2 \leq -\tau$, then $V_\mu(y^1) > V_\mu(y^2) \geq 1$. Therefore, $A_\mu(y^2) - A_\mu(y^1) = y^2 \cdot V_\mu(y^2) - y^1 \cdot V_\mu(y^1) > (y^2 - y^1) \cdot V_\mu(y^2) \geq y^2 - y^1 > 0$.

(iv) Since $V_\mu(-\tau) = 1$, $A_\mu(-\tau) = -\tau \cdot V_\mu(-\tau) = -\tau$. If $y^1 < -\tau$, from part (iii), we have $A_\mu(-\tau) - A_\mu(y^1) > -\tau - y^1$. This gives $A_\mu(y^1) < y^1$. Furthermore, by the definition, $V_\mu(y)$ is strictly decreasing when $y < 0$, $V_\mu(-\tau) = 1$ and $V_\mu(y) \geq c\mu > 0$, for all y. Thus, if $-\tau < y^2 < 0$, then $1 > V_\mu(y^2) > 0$ and hence $y^2 < y^2 \cdot V_\mu(y^2) = A_\mu(y^2) < 0$. □

In the next three subsections, we will show that $G_{x^*,\mu,\rho}(\cdot)$ satisfies conditions (D1)–(D3) if the parameters μ and ρ satisfy certain conditions.

14.2.1 Condition (D1)

LEMMA 14.3 *Let x^* be a local minimizer of f on X. Suppose that $\bar{x} \in X(x^*)$. If $\rho > 0$ and $0 < \mu < \min(1, \rho/L)$, then $G_{x^*,\mu,\rho}(\bar{x}) < 0 = G_{x^*,\mu,\rho}(x^*)$.*

Proof. Since $f(\bar{x}) \geq f(x^*)$, by Assumption 14.1 (ii), we have $0 \leq f(\bar{x}) - f(x^*) \leq L\|\bar{x} - x^*\|$. Moreover, from the definition of V_μ, we have $V_\mu(y) \leq \mu$ for all $y \geq 0$. Thus, $V_\mu(f(\bar{x}) - f(x^*)) \leq \mu$. Therefore,

$$A_\mu(f(\bar{x}) - f(x^*)) = [f(\bar{x}) - f(x^*)] \cdot V_\mu(f(\bar{x}) - f(x^*)) \leq L\|\bar{x} - x^*\| \cdot \mu.$$

Since $\|\bar{x} - x^*\| > 0$, if $\rho > 0$ and $0 < \mu < \min(1, \rho/L)$, then

$$\begin{aligned} G_{x^*,\mu,\rho}(\bar{x}) &= A_\mu(f(\bar{x}) - f(x^*)) - \rho\|\bar{x} - x^*\| \\ &\leq L\mu\|\bar{x} - x^*\| - \rho\|\bar{x} - x^*\| \\ &< 0 = G_{x^*,\mu,\rho}(x^*). \end{aligned}$$

□

THEOREM 14.1 *Let x^* be a local minimizer of f on X. If $\rho > 0$ and $0 < \mu < \min(1, \rho/L)$, then x^* is a strict local maximizer of $G_{x^*,\mu,\rho}(\cdot)$ on X. If, in addition, x^* is a global minimizer of f on X, then $G_{x^*,\mu,\rho}(x) < 0$ for all $x \in X \setminus x^*$.*

Proof. Since x^* is a local minimizer of f over X, $f(x) \geq f(x^*)$ for all $x \in U_1(x^*)$. By Lemma 14.3, if $\rho > 0$ and $0 < \mu < \min(1, \rho/L)$, then $G_{x^*,\mu,\rho}(x) < 0 = G_{x^*,\mu,\rho}(x^*)$ for all $x \in U_1(x^*) \setminus x^*$. Therefore, x^* is a strict local maximizer of $G_{x^*,\mu,\rho}(\cdot)$.

If x^* is a global minimizer of f over X, then $f(x) \geq f(x^*)$ for all $x \in X$. The result then follows from Lemma 14.3. □

From Theorem 14.1, we conclude that $G_{x^*,\mu,\rho}(\cdot)$ satisfies condition (D1) if $\rho > 0$ and $0 < \mu < \min(1, \rho/L)$.

14.2.2 Condition (D2)

LEMMA 14.4 *Let x^* be a local minimizer of f on X. Suppose that x^1, $x^2 \in X(x^*)$ are two integer points such that $0 < \|x^1 - x^*\| < \|x^2 - x^*\|$. If $\rho > 0$ and $0 < \mu < \min(1, \rho/(2K^2L))$, then*

$$G_{x^*,\mu,\rho}(x^2) < G_{x^*,\mu,\rho}(x^1) < 0 = G_{x^*,\mu,\rho}(x^*). \tag{14.2.4}$$

Proof. We first show that

$$1 - \frac{\|x^1 - x^*\|}{\|x^2 - x^*\|} > \frac{1}{2K^2}. \tag{14.2.5}$$

Since x^1, x^2 and x^* are integer points and $\|x^1 - x^*\| < \|x^2 - x^*\|$, it holds

$$\|x^2 - x^*\|^2 - \|x^1 - x^*\|^2 \geq 1. \tag{14.2.6}$$

Moreover, by Assumption 14.1 (i), we have $0 < \|x^2 - x^*\| + \|x^1 - x^*\| < 2K$. It then follows from (14.2.6) that

$$\|x^2 - x^*\| - \|x^1 - x^*\| \geq \frac{1}{\|x^2 - x^*\| + \|x^1 - x^*\|} > \frac{1}{2K}.$$

Dividing both side of the above inequality by $\|x^2 - x^*\|$ and using $\|x^2 - x^*\| \leq K$ give rise to (14.2.5).

Since $f(x^2) \geq f(x^*)$, by Assumption 14.1 (ii), we have $0 \leq f(x^2) - f(x^*) \leq L\|x^2 - x^*\|$. Moreover, from the definition of V_μ, we have $V_\mu(y) \leq \mu$, for all $y \geq 0$. Thus, $V_\mu(f(x^2) - f(x^*)) \leq \mu$ and

$$\begin{aligned} A_\mu(f(x^2) - f(x^*)) &= [f(x^2) - f(x^*)] \cdot V_\mu(f(x^2) - f(x^*)) \\ &\leq L\|x^2 - x^*\| \cdot \mu. \end{aligned}$$

On the other hand, since $f(x^1) \geq f(x^*)$, by Lemma 14.2 (i), we have $A_\mu(f(x^1) - f(x^*)) \geq 0$. Therefore, by (14.2.5), if $\rho > 0$ and $0 < \mu < \min(1, \rho/(2K^2L))$, then

$$\begin{aligned} & G_{x^*,\mu,\rho}(x^2) - G_{x^*,\mu,\rho}(x^1) \\ &= [A_\mu(f(x^2) - f(x^*)) - A_\mu(f(x^1) - f(x^*))] \\ &\quad - \rho(\|x^2 - x^*\| - \|x^1 - x^*\|) \\ &\leq L\mu\|x^2 - x^*\| - \rho(\|x^2 - x^*\| - \|x^1 - x^*\|) \\ &= \|x^2 - x^*\| \cdot \left[L\mu - \rho\left(1 - \frac{\|x^1 - x^*\|}{\|x^2 - x^*\|}\right)\right] \\ &< \|x^2 - x^*\| \cdot \left(L\mu - \frac{\rho}{2K^2}\right) < 0. \end{aligned}$$

By Assumption 14.1 (i), we have $K \geq 1$, thus $0 < \mu < \min(1, \rho/(2K^2L)) \leq \min(1, \rho/L)$. The second inequality of (14.2.4) follows from Lemma 14.3. □

THEOREM 14.2 *Let x^* be a local minimizer of f over X and $\bar{d} \in \mathbb{D}$ be a feasible direction at an integer point $\bar{x} \in X(x^*)$ such that $\|\bar{x} + \bar{d} - x^*\| > \|\bar{x} - x^*\|$. If $\rho > 0$ and $0 < \mu < \min(1, \rho/(2K^2L))$, then $G_{x^*,\mu,\rho}(\bar{x} + \bar{d}) < G_{x^*,\mu,\rho}(\bar{x}) < 0 = G_{x^*,\mu,\rho}(x^*)$.*

Proof. Consider the following two cases:

Case (i): $f(\bar{x}+\bar{d}) \geq f(x^*)$. Since both $\bar{x}$ and $\bar{x}+\bar{d} \in X(x^*)$, $0 < \|\bar{x}-x^*\| < \|\bar{x}+\bar{d}-x^*\|$, $\rho > 0$ and $0 < \mu < \min(1, \rho/(2K^2L))$, it follows from Lemma 14.4 that $G_{x^*,\mu,\rho}(\bar{x}+\bar{d}) < G_{x^*,\mu,\rho}(\bar{x}) < 0 = G_{x^*,\mu,\rho}(x^*)$.

Case (ii): $f(\bar{x}+\bar{d}) < f(x^*) \leq f(\bar{x})$. From Lemma 14.2 (i), we have $A_\mu(f(\bar{x}+\bar{d}) - f(x^*)) < 0 \leq A_\mu(f(\bar{x}) - f(x^*))$. Therefore, for $\rho > 0$,

$$\begin{aligned} G_{x^*,\mu,\rho}(\bar{x}+\bar{d}) &= A_\mu(f(\bar{x}+\bar{d}) - f(x^*)) - \rho\|\bar{x}+\bar{d}-x^*\| \\ &< A_\mu(f(\bar{x}) - f(x^*)) - \rho\|\bar{x}-x^*\| \\ &= G_{x^*,\mu,\rho}(\bar{x}). \end{aligned}$$

Since $0 < \mu < \min(1, \rho/(2K^2L)) \leq \min(1, \rho/L)$, by Lemma 14.3, we have $G_{x^*,\mu,\rho}(\bar{x}+\bar{d}) < G_{x^*,\mu,\rho}(\bar{x}) < 0 = G_{x^*,\mu,\rho}(x^*)$. □

COROLLARY 14.1 *Let x^* be a local minimizer of f over X. If $\rho > 0$ and $0 < \mu < \min(1, \rho/(2K^2L))$, then $G_{x^*,\mu,\rho}(\cdot)$ satisfies condition (D2).*

Proof. For any $\bar{x} \in X(x^*) \setminus X_c$, since $\bar{x}$ is not a corner point of X, there exists i such that $\bar{x} \pm e_i \in X$. By Lemma 14.1 (ii), there exists $\bar{d} \in \{\pm e_i\}$ such that $\|\bar{x}+\bar{d}-x^*\| > \|\bar{x}-x^*\|$. By Theorem 14.2, $\bar{d}$ is a descent feasible direction of $G_{x^*,\mu,\rho}(\cdot)$ at $\bar{x}$. Thus, $\bar{x}$ is not a local minimizer of $G_{x^*,\mu,\rho}(\cdot)$. □

14.2.3 Condition (D3)

THEOREM 14.3 *Let x^* be a local minimizer of f over X. Suppose that $\tilde{x}$ is a strict local minimizer of f over X with $f(\tilde{x}) < f(x^*)$. If $\rho > 0$ is sufficiently small and $0 < \mu < 1$, then $\tilde{x}$ is a strict local minimizer of $G_{x^*,\mu,\rho}(\cdot)$ over X.*

Proof. From Lemma 14.1 (i), we have $\|\tilde{x}+d-x^*\| \neq \|\tilde{x}-x^*\|$ for all $d \in \mathbb{D}$. For any feasible direction $\bar{d} \in \mathbb{D}$ at $\tilde{x}$, we will show that

$$G_{x^*,\mu,\rho}(\tilde{x}) < G_{x^*,\mu,\rho}(\tilde{x}+\bar{d}). \tag{14.2.7}$$

Consider the following two cases:

Case (i): $\|\tilde{x}+\bar{d}-x^*\| < \|\tilde{x}-x^*\|$. If $f(\tilde{x}) < f(\tilde{x}+\bar{d}) \leq f(x^*)$, it then follows from Lemma 14.2 (ii) that

$$A_\mu(f(\tilde{x}) - f(x^*)) < A_\mu(f(\tilde{x}+\bar{d}) - f(x^*)). \tag{14.2.8}$$

Otherwise, if $f(\tilde{x}) < f(x^*) < f(\tilde{x}+\bar{d})$, from Lemma 14.2 (i), we have

$$A_\mu(f(\tilde{x}) - f(x^*)) < 0 < A_\mu(f(\tilde{x}+\bar{d}) - f(x^*)). \tag{14.2.9}$$

Inequalities (14.2.8) and (14.2.9) imply that

$$\begin{aligned} G_{x^*,\mu,\rho}(\tilde{x}) &= A_\mu(f(\tilde{x}) - f(x^*)) - \rho\|\tilde{x}-x^*\| \\ &< A_\mu(f(\tilde{x}+\bar{d}) - f(x^*)) - \rho\|\tilde{x}+\bar{d}-x^*\| \\ &= G_{x^*,\mu,\rho}(\tilde{x}+\bar{d}). \end{aligned}$$

Case (ii): $\|\tilde{x}+\bar{d}-x^*\| > \|\tilde{x}-x^*\|$. By (14.2.3), $f(\tilde{x}) < f(x^*)$ implies $f(\tilde{x}) < f(x^*) - \tau$. Consider the following three cases:

$$f(\tilde{x}) < f(\tilde{x}+\bar{d}) \le f(x^*) - \tau, \tag{14.2.10}$$

$$f(\tilde{x}) < f(x^*) - \tau < f(\tilde{x}+\bar{d}) < f(x^*), \tag{14.2.11}$$

$$f(\tilde{x}) < f(x^*) - \tau < f(x^*) \le f(\tilde{x}+\bar{d}). \tag{14.2.12}$$

If (14.2.10) holds, from Lemma 14.2 (iii), we have

$$f(\tilde{x}+\bar{d}) - f(\tilde{x}) < A_\mu(f(\tilde{x}+\bar{d}) - f(x^*)) - A_\mu(f(\tilde{x}) - f(x^*)). \tag{14.2.13}$$

Let

$$\rho_1 = \min_{d \in \mathbb{D}_0(\tilde{x})} [f(\tilde{x}+d) - f(\tilde{x})]/K, \tag{14.2.14}$$

where $\mathbb{D}_0(\tilde{x}) = \{d \in \mathbb{D} \mid \tilde{x}+d \in X\}$. Since $\tilde{x}$ is a strict local minimizer of f, we have $\rho_1 > 0$. Also, by Assumption 14.1 (i) and Lemma 14.1 (iii), $\|\tilde{x}+\bar{d}-x^*\| - \|\tilde{x}-x^*\| < K$. Therefore, if $0 < \rho \le \rho_1$, we obtain from (14.2.13) that

$$\begin{aligned} \rho &\le \rho_1 \le \frac{f(\tilde{x}+\bar{d}) - f(\tilde{x})}{K} \\ &< \frac{A_\mu(f(\tilde{x}+\bar{d}) - f(x^*)) - A_\mu(f(\tilde{x}) - f(x^*))}{\|\tilde{x}+\bar{d}-x^*\| - \|\tilde{x}-x^*\|}, \end{aligned}$$

which in turn implies (14.2.7).

If (14.2.11) holds, by Lemma 14.2 (iv), we have $A_\mu(f(\tilde{x}) - f(x^*)) < f(\tilde{x}) - f(x^*) < -\tau < f(\tilde{x}+\bar{d}) - f(x^*) < A_\mu(f(\tilde{x}+\bar{d}) - f(x^*))$. Therefore, (14.2.13) is satisfied and hence (14.2.7) holds if $0 < \rho < \rho_1$.

Finally, if (14.2.12) holds, since $f(\tilde{x}+\bar{d}) - f(x^*) \ge 0$, by Lemma 14.2 (i), we have

$$A_\mu(f(\tilde{x}+\bar{d}) - f(x^*)) \ge 0. \tag{14.2.15}$$

Moreover, since $f(\tilde{x}) - f(x^*) < -\tau$, by Lemma 14.2 (iv), we have

$$A_\mu(f(\tilde{x}) - f(x^*)) < -\tau. \tag{14.2.16}$$

Let $\rho_2 = \tau/K$. If $0 < \rho \le \rho_2$, then, by (14.2.15) and (14.2.16), we have

$$\begin{aligned} \rho &\le \frac{\tau}{K} < \frac{\tau}{\|\tilde{x}+\bar{d}-x^*\| - \|\tilde{x}-x^*\|} \\ &< \frac{A_\mu(f(\tilde{x}+\bar{d}) - f(x^*)) - A_\mu(f(\tilde{x}) - f(x^*))}{\|\tilde{x}+\bar{d}-x^*\| - \|\tilde{x}-x^*\|}. \end{aligned}$$

Thus, (14.2.7) holds.

In summary, if $0 < \rho \leq \min(\rho_1, \rho_2)$, then $\tilde{x}$ is a strict local minimizer of $G_{x^*,\mu,\rho}(\cdot)$ over X. □

It is assumed in the above theorem that the better local minimizer $\tilde{x}$ of f over X is strict. This requirement on $\tilde{x}$ can be relaxed to

$$f(\tilde{x} + \bar{d}) > f(\tilde{x}), \quad \forall \bar{d} \in \mathbb{D}_1(\tilde{x}, x^*), \tag{14.2.17}$$

where $\mathbb{D}_1(\tilde{x}, x^*) = \{d \in \mathbb{D} \mid \tilde{x} + d \in X, \; \|\tilde{x} + d - x^*\| > \|\tilde{x} - x^*\|\}$.

THEOREM 14.4 *Let x^* be a local minimizer of f over X. Suppose that $\tilde{x}$ is a local minimizer of f over X with $f(\tilde{x}) < f(x^*)$ that satisfies (14.2.17). If $\rho > 0$ is sufficiently small and $0 < \mu < 1$, then $\tilde{x}$ is a strict local minimizer of $G_{x^*,\mu,\rho}(\cdot)$ over X.*

Proof. From Lemma 14.1 (i), we have $\|\tilde{x} + d - x^*\| \neq \|\tilde{x} - x^*\|$ for all $d \in \mathbb{D}$. Let $\bar{d} \in \mathbb{D}$ be a feasible direction at $\tilde{x}$. Then $f(\tilde{x} + \bar{d}) \geq f(\tilde{x})$. If, in addition, $\|\tilde{x} + \bar{d} - x^*\| > \|\tilde{x} - x^*\|$, by (14.2.17), we have $f(\tilde{x} + \bar{d}) > f(\tilde{x})$. To prove (14.2.7), we can use the similar arguments as in the proof of Theorem 14.3 except for the following additional case: $\|\tilde{x} + \bar{d} - x^*\| < \|\tilde{x} - x^*\|$ and $f(\tilde{x}) = f(\tilde{x} + \bar{d}) < f(x^*)$. In this case, we have

$$\begin{aligned} G_{x^*,\mu,\rho}(\tilde{x}) &= A_\mu(f(\tilde{x}) - f(x^*)) - \rho\|\tilde{x} - x^*\| \\ &< A_\mu(f(\tilde{x} + \bar{d}) - f(x^*)) - \rho\|\tilde{x} + \bar{d} - x^*\| \\ &= G_{x^*,\mu,\rho}(\tilde{x} + \bar{d}). \end{aligned}$$

Thus, (14.2.7) holds. □

THEOREM 14.5 *Let x^* be a local minimizer of f over X. Suppose that $\tilde{x}$ is a local minimizer of $G_{x^*,\mu,\rho}(\cdot)$ over X. Assume further that $\rho > 0$ and $0 < \mu < \min(1, \rho/(2K^2L))$. If there exists a feasible direction $\bar{d} \in \mathbb{D}$ at $\tilde{x}$ such that $\|\tilde{x} + \bar{d} - x^*\| > \|\tilde{x} - x^*\|$, and ρ is sufficiently small, then $\tilde{x}$ is a local minimizer of f over X.*

Proof. Since x^* is a local minimizer of f over X, by Theorem 14.1, x^* is a strict local maximizer of $G_{x^*,\mu,\rho}(\cdot)$. Therefore, $\tilde{x} \neq x^*$. We claim that $f(\tilde{x}) < f(x^*)$. Suppose on the contrary that $f(\tilde{x}) \geq f(x^*)$. Then $\tilde{x} \in X(x^*)$. If there exists a feasible direction $\bar{d} \in \mathbb{D}$ at $\tilde{x}$ such that $\|\tilde{x} + \bar{d} - x^*\| > \|\tilde{x} - x^*\|$, then, by Theorem 14.2, $G_{x^*,\mu,\rho}(\tilde{x} + \bar{d}) < G_{x^*,\mu,\rho}(\tilde{x})$, a contradiction to the assumption that $\tilde{x}$ is a local minimizer of $G_{x^*,\mu,\rho}(\cdot)$ over X. Therefore, $f(\tilde{x}) < f(x^*)$ and $f(\tilde{x}) < f(x^*) - \tau$ by the definition of τ.

Now, suppose on the contrary that $\tilde{x}$ is not a local minimizer of f over X. Then there exists a descent direction $\bar{d} \in \mathbb{D}$ at $\tilde{x}$ such that $f(\tilde{x} + \bar{d}) < f(\tilde{x})$

and hence $f(\tilde{x}+\tilde{d}) - f(x^*) < f(\tilde{x}) - f(x^*) < -\tau$. By Lemma 14.2 (iii), we have

$$\begin{aligned} 0 &< f(\tilde{x}) - f(\tilde{x}+\tilde{d}) \\ &< A_\mu(f(\tilde{x}) - f(x^*)) - A_\mu(f(\tilde{x}+\tilde{d}) - f(x^*)). \end{aligned} \quad (14.2.18)$$

Since, from Lemma 14.1 (i), $\|\tilde{x}+\tilde{d}-x^*\| \neq \|\tilde{x}-x^*\|$. If $\|\tilde{x}+\tilde{d}-x^*\| > \|\tilde{x}-x^*\|$, then, by (14.2.18), we have

$$\begin{aligned} G_{x^*,\mu,\rho}(\tilde{x}+\tilde{d}) &= A_\mu(f(\tilde{x}+\tilde{d}) - f(x^*)) - \rho\|\tilde{x}+\tilde{d}-x^*\| \\ &< A_\mu(f(\tilde{x}) - f(x^*)) - \rho\|\tilde{x}-x^*\| \\ &= G_{x^*,\mu,\rho}(\tilde{x}), \end{aligned}$$

which contradicts the assumption that $\tilde{x}$ is a local minimizer of $G_{x^*,\mu,\rho}(\cdot)$. On the other hand, if $\|\tilde{x}+\tilde{d}-x^*\| < \|\tilde{x}-x^*\|$, then, by (14.2.18), we have $G_{x^*,\mu,\rho}(\tilde{x}+\tilde{d}) < G_{x^*,\mu,\rho}(\tilde{x})$, if we choose ρ such that

$$\begin{aligned} 0 < \rho &< \frac{f(\tilde{x}) - f(\tilde{x}+\tilde{d})}{K} \\ &< \frac{A_\mu(f(\tilde{x}) - f(x^*)) - A_\mu(f(\tilde{x}+\tilde{d}) - f(x^*))}{\|\tilde{x}-x^*\| - \|\tilde{x}+\tilde{d}-x^*\|}. \end{aligned}$$

Let

$$\rho_3 = \min_{d\in\mathbb{D}}[f(\tilde{x}) - f(\tilde{x}+d)]/K > 0.$$

Then, in summary, choosing ρ such that $0 < \rho < \rho_3$ leads to $G_{x^*,\mu,\rho}(\tilde{x}+\tilde{d}) < G_{x^*,\mu,\rho}(\tilde{x})$. Again, this is a contradiction. □

COROLLARY 14.2 *Let x^* be a local minimizer of f over X. Assume that every local minimizer of f over X is strict. Suppose that $\rho > 0$ is sufficiently small and $0 < \mu < \min(1, \rho/(2K^2L))$. Then, $\tilde{x} \in X \setminus X_c$ is a local minimizer of f over X with $f(\tilde{x}) < f(x^*)$ if and only if $\tilde{x}$ is a local minimizer of $G_{x^*,\mu,\rho}(\cdot)$ over X.*

Proof. The "if" part follows directly from Theorem 14.3. Now, suppose that $\tilde{x}$ is a local minimizer of $G_{x^*,\mu,\rho}(\cdot)$ over X. Since $\tilde{x} \notin X_c$, we have $\tilde{x} \pm e_i \in X$ for some i. Thus, by Lemma 14.1 (ii), there exists a feasible direction $d \in \mathbb{D}$ at $\tilde{x}$ such that $\|\tilde{x}+d-x^*\| > \|\tilde{x}-x^*\|$. If $\rho > 0$ is small enough and $0 < \mu \leq \min(1, \rho/(2K^2L))$, by Theorem 14.5, $\tilde{x}$ is a local minimizer of f over X. □

Corollary 14.2 indicates that if every local minimizer of f over X is strict, then $G_{x^*,\mu,\rho}(\cdot)$ satisfies the condition (D3) for suitable parameters μ and ρ.

We consider now the following illustrative example.

EXAMPLE 14.3 (3-HUMPBACK CAMEL FUNCTION)

$$\min_x f(x) = 2(\frac{x_1}{1000})^2 - 1.05(\frac{x_1}{1000})^4 + \frac{1}{6}(\frac{x_1}{1000})^6 - (\frac{x_1}{1000})(\frac{x_2}{1000}) + (\frac{x_2}{1000})^2,$$
$$\text{s.t. } x \in X = \{x \in \mathbb{Z}^2 \mid -2000 \le x_1 \le 2000,\ -1500 \le x_2 \le 1500\}.$$

This problem has three local minima: $x_1^* = (-1748, -874)^T$ with $f(x_1^*) = 0.2986$, $x_2^* = (1748, 874)^T$ with $f(x_2^*) = 0.2986$ and $x_3^* = (0,0)^T$ with $f(x_3^*) = 0$, among which x_3^* is the global optimal solution. We construct a global descent function $G_{x_2^*,\mu,\rho}(x)$ at the local minimum $x_2^* = (1748, 874)^T$ with $\mu = \rho = 0.01$. Figure 14.2 shows the contours of $f(x)$ and $G_{x_2^*,\mu,\rho}(x)$, and the figures of $f(x)$ and $G_{x_2^*,\mu,\rho}(x)$.

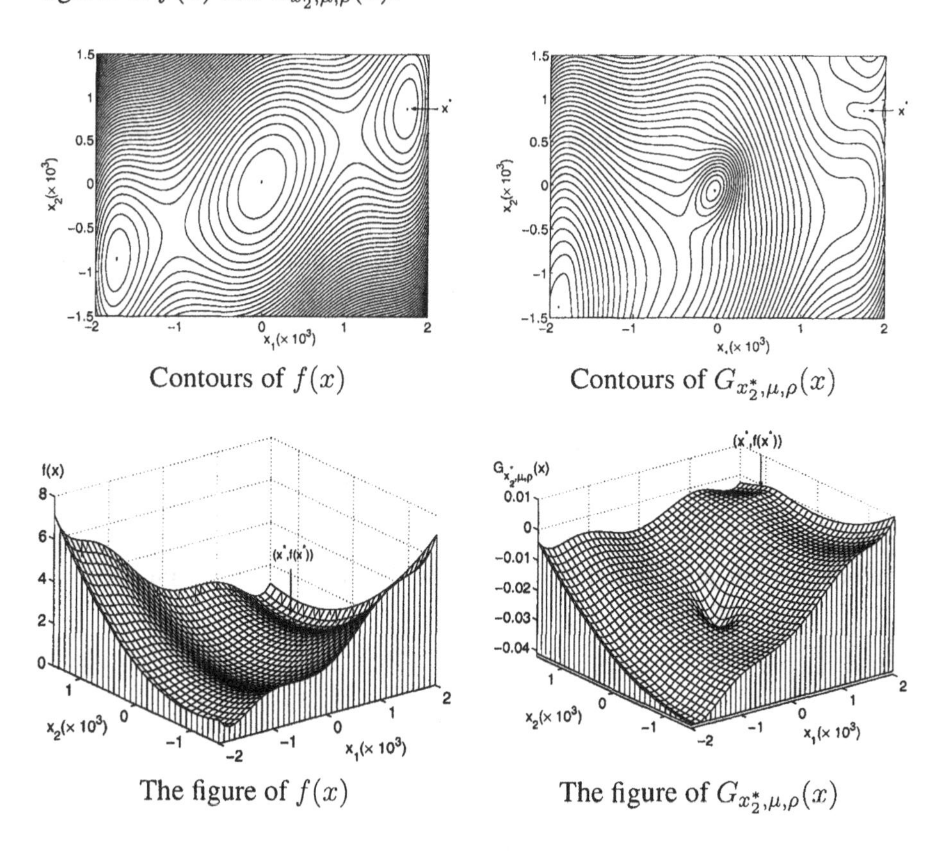

Contours of $f(x)$ | Contours of $G_{x_2^*,\mu,\rho}(x)$

The figure of $f(x)$ | The figure of $G_{x_2^*,\mu,\rho}(x)$

Figure 14.2. Illustration of the discrete global descent function.

14.3 The Discrete Global Descent Method

Based on the theoretical results in the previous section, the discrete global descent method for (P) is described now as follows.

ALGORITHM 14.1 (DISCRETE GLOBAL DESCENT METHOD FOR (P))

Step 0 (Initialization).

(i) Choose a function V_μ satisfying conditions (V1) and (V2).

(ii) Choose an initial point $x_{ini} \in X$, two fractions: $\hat{\rho}$ $(0 < \hat{\rho} < 1)$ and $\hat{\mu}$ $(0 < \hat{\mu} < 1)$, and a lower bound of ρ: $\rho_L > 0$.

(iii) Starting from x_{ini}, apply Procedure 14.1 to obtain a local minimizer x^* of f over X. Set $k = 0$.

Step 1. Generate a set of m initial points: $\{x_{ini}^{(i)} \in X \backslash \{x^*\} \mid i = 1, 2, \ldots, m\}$. Set $i = 1$.

Step 2. Set the current point $x_{cur} := x_{ini}^{(i)}$.

Step 3. If $f(x_{cur}) < f(x^*)$, then starting from x_{cur}, apply Procedure 14.1 to find a local minimizer $\tilde{x}$ such that $f(\tilde{x}) < f(x^*)$. Set $x^* := \tilde{x}$, $k := k + 1$. Go to Step 1.

Step 4. Let $D_0 := \{d \in \mathbb{D} : x_{cur} + d \in X\}$. If there exists $d \in D_0$ such that $f(x_{cur} + d) < f(x^*)$, then starting from $x_{cur} + d^*$, where $d^* = \arg\min\{f(x_{cur} + d) \mid d \in D_0\}$, apply Procedure 14.1 to find a local minimizer $\tilde{x}$ such that $f(\tilde{x}) < f(x^*)$. Set $x^* := \tilde{x}$, $k := k + 1$. Go to Step 1.

Step 5. If x_{cur} is a local minimizer of $G_{x^*,\mu,\rho}(\cdot)$ and the set $D_1 := \{d \in D_0 \mid \|x_{cur} + d - x^*\| > \|x_{cur} - x^*\|\}$ is empty, then go to Step 8.

Step 6. If x_{cur} is a local minimizer of $G_{x^*,\mu,\rho}(\cdot)$, then, set $\mu_0 = \mu$ and choose a positive integer l such that $\mu = \hat{\mu}^l \mu_0$ and there exists a descent direction of $G_{x^*,\mu,\rho}(\cdot)$ at x_{cur}.

Step 7. Let $D_2 := \{d \in D_0 : G_{x^*,\mu,\rho}(x_{cur} + d) < G_{x^*,\mu,\rho}(x_{cur}),\ f(x_{cur} + d) < f(x_{cur})\}$. If $D_2 \neq \emptyset$, then set $d^* := \arg\min\{f(x_{cur} + d) + G_{x^*,\mu,\rho}(x_{cur} + d) \mid d \in D_2\}$. Otherwise set $d^* := \arg\min\{G_{x^*,\mu,\rho}(x_{cur} + d) \mid d \in D_0\}$. Set $x_{cur} := x_{cur} + d^*$. Go to Step 4.

Step 8. Set $i := i + 1$. If $i \leq m$, go to Step 2.

Step 9. Set $\rho := \hat{\rho}\rho$. If $\rho \geq \rho_L$, then go to Step 1. Otherwise, the algorithm is incapable of finding a better local minimizer starting from the initial points, $\{x_{ini}^{(i)} \mid i = 1, 2, \ldots, m\}$. The algorithm stops and x^* is taken as an approximate global minimizer.

The motivation and mechanism behind the algorithm are explained below.

A set of m initial points is generated in Step 1 to minimize $G_{x^*,\mu,\rho}(\cdot)$. If no additional information about the objective function is provided, we set the initial points symmetric about the current local minimizer. For example, we can set $m = 2n$ and choose $x^* \pm e_i$, for $i = 1, 2, \ldots, n$, as initial points for the discrete global descent method.

Step 3 represents the situation where the current initial point, $x_{ini}^{(i)}$, satisfies $f(x_{ini}^{(i)}) < f(x^*)$. Therefore, we can further minimize the objective function f by any discrete local minimization method starting from $x_{ini}^{(i)}$. Note that Step 3 is necessary only if we choose some initial points outside the U_1 neighborhood of x^*.

Recall from Theorem 14.2 that if $f(x_{cur}) \geq f(x^*)$ and μ is sufficiently small, then x_{cur} cannot be a local minimizer of $G_{x^*,\mu,\rho}(\cdot)$. In determining whether the current point x_{cur} is a local minimizer of $G_{x^*,\mu,\rho}(\cdot)$, we compare $G_{x^*,\mu,\rho}(x_{cur})$ with $G_{x^*,\mu,\rho}(x)$ for all $x \in U_1(x_{cur}) \setminus \{x_{cur}\}$. Step 4 represents the situation where one of the neighboring points of x_{cur}, namely $x_{cur} + d^*$ with $d^* \in \mathbb{D}$, has a smaller objective function value than the current local minimum. We can then further minimize $f(\cdot)$ by any discrete local minimization method starting from $x_{cur} + d^*$.

If it is found that x_{cur} is a local minimizer of $G_{x^*,\mu,\rho}(\cdot)$ with $f(x_{cur}) \geq f(x^*)$, this implies that μ is not small enough. Step 5 represents the situation when it is impossible to move further away from x^* than x_{cur} and thus x_{cur} must be a corner point of X. Then, we give up the point x_{cur} without reducing the value of μ and try another initial point generated in Step 1. On the other hand, if x_{cur} is not a corner point of X, then Step 6 reduces the value of μ to a preselected fraction recursively until there exists a descent direction of $G_{x^*,\mu,\rho}(\cdot)$ at x_{cur}.

Step 7 aims at selecting a more promising successor point. Note that if the algorithm goes from Step 6 to Step 7, $G_{x^*,\mu,\rho}(\cdot)$ has at least one descent direction at x_{cur}. If there exists a descent direction of both f and $G_{x^*,\mu,\rho}(\cdot)$ at x_{cur}, we then reduce both $f(\cdot)$ and $G_{x^*,\mu,\rho}(\cdot)$ at the same time in order to take advantages of their reductions. On the other hand, if every descent direction of $G_{x^*,\mu,\rho}(\cdot)$ at x_{cur} is an increasing direction of $f(\cdot)$ at x_{cur}, we reduce $G_{x^*,\mu,\rho}(\cdot)$ alone.

Recall from Corollary 14.2 that the value of ρ should be selected small enough. Otherwise, there could not exist a local minimizer of $G_{x^*,\mu,\rho}(\cdot)$, even there exists a better $\tilde{x}$ with $f(\tilde{x}) < f(x^*)$. Thus, the value of ρ is reduced successively in the solution process in Step 9 if no better solution is found when minimizing the discrete global descent function. If the value of ρ reaches its lower bound ρ_L and no better solution is found, the current local minimizer is taken as a global minimizer.

14.4 Computational Results

The developed discrete global descent method is programmed in MATLAB and run on a Pentium IV system with 3.2GHz CPU. An illustrative example is given first in the following to show the solution procedure of the algorithm described in the previous section. The computational results in solving several test problems are then reported.

Throughout the tests, $V_\mu^1(y)$ is selected as the discrete global descent function with $\tau = 1$ and $c = 0.5$. Procedure 14.1 is used to perform the local searches. Suppose that a local minimizer $\tilde{x}$ of f over X is obtained using $x^* + e_j$ as the initial point, the neighboring points of $\tilde{x}$ are then arranged in the following order as the initial points in minimizing the discrete global descent function:

$$\begin{gathered}\tilde{x} + e_j,\\ \tilde{x} + e_{j+1}, \tilde{x} - e_{j+1}, \ldots, \tilde{x} + e_n, \tilde{x} - e_n,\\ \tilde{x} + e_1, \tilde{x} - e_1, \ldots, \tilde{x} + e_{j-1}, \tilde{x} - e_{j-1},\\ \tilde{x} - e_j.\end{gathered}$$

Notice that, if the current local minimizer of f is on the boundary of X, then there are less than $2n$ initial points. In addition, $\rho = \rho_L = 0.1$ is set in all the tests. In other words, if the algorithm could not find a local minimizer of $G_{x^*,\mu,\rho}(\cdot)$ using all initial points, the algorithm stops immediately. Besides these, $\mu = 0.1$ is set at the beginning of the algorithm. Once the current μ is classified as not sufficiently small, μ is reduced to $\mu/10$.

EXAMPLE 14.4 (see [71][233])

$$\begin{aligned}\min\ & f(x) = x_1 + 10x_2\\ \text{s.t. } & 66x_1 + 14x_2 \geq 1430,\\ & -82x_1 + 28x_2 \geq 1306,\\ & 0 \leq x_1 \leq 15, \quad 68 \leq x_2 \leq 102, \quad x_1, x_2 \text{ integers.}\end{aligned}$$

This problem is a linear integer programming problem. There are 314 feasible points among which seven are local minimizers and one is the global minimum solution: $x^*_{global} = (7, 70)^T$ with $f(x^*_{global}) = 707$.

The algorithm starts from a feasible point $x_{ini} = (15, 102)^T$ with $f(x_{ini}) = 1035$ and uses the discrete steepest descent method to minimize $f(x)$. After 30 function evaluations, an initial local minimizer $x^* = (3, 88)^T$ is obtained with $f(x^*) = 883$.

In the first iteration of the algorithm, $\mu = 0.1$ is found to be not small enough. When $\mu = 0.01$, the algorithm starts from $x^1_{ini} = (4, 88)^T$ and reaches

$\bar{x} = (4, 87)^T$ with $f(\bar{x}) = 874 < f(x^*)$. Then, the algorithm switches to the local search again and obtains $\tilde{x} = (4, 84)^T$ with $f(\tilde{x}) = 844$. The cumulative number of function evaluations is 42.

In the second iteration of the algorithm, the algorithm sets $x^* = (4, 84)^T$ and starts from $x_{ini}^1 = (5, 84)^T$ and reaches $\bar{x} = (5, 83)^T$ with $f(\bar{x}) = 835 < f(x^*)$. Then, the algorithm switches to the local search and obtains $\tilde{x} = (5, 79)^T$ with $f(\tilde{x}) = 795$. The cumulative number of function evaluations is 55.

In the same fashion, the algorithm generates $x^* = (5, 79)^T$, $x_{ini}^1 = (6, 79)^T$, $\bar{x} = (6, 78)^T$ with $f(\bar{x}) = 786 < f(x^*)$, $\tilde{x} = (6, 74)^T$ with $f(\tilde{x}) = 746$ and the cumulative number of function evaluations is 68 in the third iteration. Similarly, the algorithm generates $x^* = (6, 74)^T$, $x_{ini}^1 = (7, 74)^T$, $\bar{x} = (7, 73)^T$ with $f(\bar{x}) = 737 < f(x^*)$, $\tilde{x} = (7, 70)^T$ with $f(\tilde{x}) = 707$ and the cumulative number of function evaluations is 79 in the fourth iteration.

In the fifth iteration of the algorithm, three starting points, $(8, 70)^T$, $(6, 70)^T$ and $(7, 69)^T$, are infeasible. Besides these, the algorithm cannot find a feasible point with function value less than 707 using the remaining starting point $(7, 71)^T$. The cumulative number of function evaluations is 193.

In general, ρ should be reduced by a fraction and continue the process until $\rho < \rho_L$. Since $\rho = \rho_L = 0.1$ is selected in the numerical tests, and thus the algorithm is terminated. Therefore, $N_{iter} = 4$, $x_{global}^* = (7, 70)^T$, $f(x_{global}^*) = 707$, $N_{tfval} = 79$ and $N_{sfval} = 193$. The ratio of the number of function evaluations to reach the global minimum to the number of feasible points is $79/314 \approx 0.2516$.

The following test problems are used in computational experiments in testing the discrete global descent method.

PROBLEM 14.1 (see [226][166][170])

$$\begin{aligned}
\min\ & f(x) = x_1^2 + x_2^2 + 3x_3^2 + 4x_4^2 + 2x_5^2 - 8x_1 - 2x_2 - 3x_3 - x_4 - 2x_5, \\
\text{s.t. }\ & x_1 + x_2 + x_3 + x_4 + x_5 \leq 400, \\
& x_1 + 2x_2 + 2x_3 + x_4 + 6x_5 \leq 800, \\
& 2x_1 + x_2 + 6x_3 \leq 200, \\
& x_3 + x_4 + 5x_5 \leq 200, \\
& x_1 + x_2 + x_3 + x_4 + x_5 \geq 55, \\
& x_1 + x_2 + x_3 + x_4 \geq 48, \\
& x_2 + x_4 + x_5 \geq 34, \\
& 6x_1 + 7x_5 \geq 104, \\
& 0 \leq x_i \leq 99, \quad x_i \text{ integer}, \quad i = 1, 2, 3, 4, 5.
\end{aligned}$$

This problem is a quadratic integer programming problem. It has 251401581 feasible points. The optimal solution to the problem as given in [226] is

$x^*_{global} = (17, 18, 7, 7, 9)^T$ with $f(x^*_{global}) = 900$. Five initial points are used in the test experiment: $x_{ini} = (17, 18, 7, 7, 9)^T$, $(21, 34, 0, 0, 0)^T$, $(0, 0, 0, 48, 15)^T$, $(100, 0, 0, 0, 40)^T$ and $(0, 8, 32, 8, 32)^T$. For every experiment, the discrete global descent method succeeded in identifying a better minimum solution $x^*_{global} = (16, 22, 5, 5, 7)^T$ with $f(x^*_{global}) = 807$, which is the same as the optimal solution given in [166]. Moreover, the maximum numbers of function evaluations to reach the global minimum and to stop were only 5883 and 9792, respectively. They were much smaller than the average number of function evaluation (187794) reported in [166]. The average CPU time to reach the global minimum was about 2.87 seconds. The ratio of the average number of function evaluations to reach the global minimum to the number of feasible points was about 1.41×10^{-5}.

PROBLEM 14.2 (Goldstein and Price's function, see [236])

$$\begin{aligned} &\min_x f(y) = g(y)h(y), \\ &\text{s.t. } y_j = 0.001x_j, \quad -2000 \le x_j \le 2000, \quad x_j \text{ integer}, \quad j = 1, 2, \end{aligned}$$

where

$$\begin{aligned} g(y) &= 1 + (y_1 + y_2 + 1)^2(19 - 14y_1 + 3y_1^2 - 14y_2 + 6y_1y_2 + 3y_2), \\ h(y) &= 30 + (2x_1 - 3x_2)^2(18 - 32y_1 + 12y_1^2 + 48y_2 - 36y_1y_2 + 27y_2^2). \end{aligned}$$

This problem is a discrete counterpart of the Goldstein and Price's function in [79]. It is a box constrained/unconstrained nonlinear integer programming problem. It has $4001^2 \approx 1.60 \times 10^7$ feasible points and many local minimizers. More precisely, it has 207 and 2 local minimizers in the interior and on the boundary of the box $-2000 \le x_i \le 2000$, $i = 1, 2$, respectively. Nevertheless, it has only one global minimum solution: $x^*_{global} = (0, -1000)^T$ with $f(x^*_{global}) = 3$. Seven initial points are used in the test experiment: $x_{ini} = (\alpha, \alpha)^T$ for $\alpha = -2000, -1000, 0, 1000, 2000$, and $x_{ini} = (\beta, -\beta)^T$ for $\beta = -2000, 2000$. For every experiment, the global descent method succeeded in identifying the global minimum solution. The average CPU time to reach the global minimum was about 4.11 seconds. The ratio of the average number of function evaluations to reach the global minimum to the number of feasible points was about 5.32×10^{-4}.

PROBLEM 14.3

$$\begin{aligned} &\min\ f(x) = (x_1 - 1)^2 + (x_n - 1)^2 + n\sum_{i=1}^{n-1}(n - i)(x_i^2 - x_{i+1})^2, \\ &\text{s.t. } -5 \le x_i \le 5, \quad x_i \text{ integer}, \quad i = 1, 2, \ldots, n. \end{aligned}$$

This problem is a generalization of the problem 282 in [190]. It is a box constrained/unconstrained nonlinear integer programming problem. It has 11^n feasible points and many local minimizers (4, 6, 7, 10 and 12 local minimizers for $n = 2, 3, 4, 5$ and 6, respectively), but only one global minimum solution: $x^*_{global} = (1, \ldots, 1)^T$ with $f(x^*_{global}) = 0$, for all n. Three sizes of the problem are considered: $n = 25, 50$ and 100, and there are about 1.08×10^{26}, 1.17×10^{52} and 1.38×10^{104} feasible points, for $n = 25$, 50 and 100, respectively. For all problems with different sizes, nine initial points are used in the test experiment: $x_{ini} = (\alpha, \ldots, \alpha)^T$ for $\alpha = -5, -3, 0, 3, 5$, and $x_{init} = (\beta, \ldots, \beta, \beta, -\beta, \ldots, -\beta)^T$ and $(\beta, -\beta, \beta, -\beta, \ldots, -\beta)^T$ for $\beta = -5, 5$. For every experiment, the global descent method succeeded in identifying the global minimum solution. The average CPU times to reach the global minima were about 0.85 seconds, 2.89 seconds and 10.83 seconds, for $n = 25$, 50 and 100, respectively. The ratios of the average numbers of function evaluations to reach the global minima to the numbers of feasible points were about 2.30×10^{-23}, 8.27×10^{-49} and 2.74×10^{-100}, for $n = 25$, 50 and 100, respectively.

PROBLEM 14.4 (Rosenbrock's function)

$$\min\ f(x) = \sum_{i=1}^{n-1} [100(x_{i+1} - x_i^2)^2 + (1 - x_i)^2],$$
$$\text{s.t.} \quad -5 \le x_i \le 5, \quad x_i \text{ integer}, \quad i = 1, 2, \ldots, n.$$

This problem is a generalization of the problems 294–299 in [190]. It is a box constrained/unconstrained nonlinear integer programming problem. It has 11^n feasible points and many local minimizers (5, 6, 7, 9 and 11 local minimizers for $n = 2, 3, 4, 5$ and 6, respectively), but only one global minimum solution: $x^*_{global} = (1, \ldots, 1)^T$ with $f(x^*_{global}) = 0$, for all n. Three sizes of the problem are considered: $n = 25$, 50 and 100, and there are about 1.08×10^{26}, 1.17×10^{52} and 1.38×10^{104} feasible points, for $n = 25$, 50 and 100, respectively. For all problems with different sizes, nine initial points are used in the test experiment: $x_{ini} = (\alpha, \ldots, \alpha)^T$ for $\alpha = -5, -3, 0, 3, 5$, and $x_{init} = (\beta, \ldots, \beta, \beta, -\beta, \ldots, -\beta)^T$ and $(\beta, -\beta, \beta, -\beta, \ldots, -\beta)^T$ for $\beta = -5, 5$. For every experiment, the discrete global descent method succeeded in identifying the global minimum solution. The average CPU times to reach the global minima were about 51.78 seconds, 6.72 minutes and 54.76 minutes for $n =$ 25, 50 and 100, respectively. The ratios of the average numbers of function evaluations to reach the global minima to the numbers of feasible points were about 8.31×10^{-22}, 6.21×10^{-47} and 4.27×10^{-98}, for $n = 25$, 50 and 100, respectively.

PROBLEM 14.5

$$\min \ f(x) = \sum_{i=1}^{n} x_i^4 + \left(\sum_{i=1}^{n} x_i\right)^2$$
$$\text{s.t.} \ -5 \le x_i \le 5, \quad x_i \text{ integer}, \quad i = 1, 2, \ldots, n.$$

This problem is a box constrained/unconstrained nonlinear integer programming problem. It has 11^n feasible points and many local minimizers (3, 7, 19, 51 and 141 local minimizers for $n = 2, 3, 4, 5$ and 6, respectively), but only one global minimum solution: $x^*_{global} = (0, \ldots, 0)^T$ with $f(x^*_{global}) = 0$, for all n. Three sizes of the problem are considered: $n = 25$, 50 and 100, and there are about 1.08×10^{26}, 1.17×10^{52} and 1.38×10^{104} feasible points, for $n = 25$, 50 and 100, respectively. For all problems with different sizes, ten initial points are used in the test experiment: $x_{ini} = (\alpha, \ldots, \alpha)^T$ for $\alpha = -5, -3, -1, 1, 3, 5$, and $x_{init} = (\beta, \ldots, \beta, \beta, -\beta, \ldots, -\beta)^T$ and $(\beta, -\beta, \beta, -\beta, \ldots, -\beta)^T$ for $\beta = -5, 5$. For every experiment, the global descent method succeeded in identifying the global minimum solution. The average CPU times to reach the global minima were about 50.59 seconds, 1.96 minutes, and 7.73 minutes, for $n = 25$, 50 and 100, respectively. The ratios of the average numbers of function evaluations to reach the global minima to the numbers of feasible points were about 8.90×10^{-22}, 2.00×10^{-47} and 6.60×10^{-99}, for $n = 25$, 50 and 100, respectively.

The performance of the algorithm for Problems 14.1–14.5 is summarized in Table 14.2, where

- n= the number of the integer variables;
- N_{test}=the number of runs of the algorithm;
- N_{iter}=the average number of iterations;
- T_{final}=the average CPU time in seconds to obtain the final results;
- T_{stop}= the average CPU time in seconds for the algorithm to stop at Step 9
- N_{tfval}= the average numbers of objective function evaluations to obtain the final results;
- N_{sfval}= the average numbers of objective function evaluations to stop at Step 9.

14.5 Notes

Global optimization methods were mostly developed for continuous nonconvex optimization problems. The tunneling algorithm [129] was probably the

Table 14.2. Numerical results for Problems 14.1–14.5.

Problem	n	N_{test}	N_{iter}	T_{final}	T_{stop}	N_{tfval}	N_{sfval}
Problem 14.1	5	5	28	2.87	5.52	3547	7456
Problem 14.2	2	7	33	4.11	38.04	8521	68196
Problem 14.3	25	9	1	0.85	145.71	2489	243797
Problem 14.3	50	9	1	2.89	1086.75	9707	1925497
Problem 14.3	100	9	1	10.83	8864.24	37742	15316224
Problem 14.4	25	9	1	51.78	176.49	90057	305712
Problem 14.4	50	9	1	403.13	1343.91	728415	2423847
Problem 14.4	100	9	1	3285.58	10845.75	5879747	19333797
Problem 14.5	25	10	12	50.59	183.76	96382	341632
Problem 14.5	50	10	24	117.65	1164.25	234692	2184592
Problem 14.5	100	10	49	463.98	9029.10	909960	16459760

first method developed for searching for a global minimizer from among local minimizers in continuous optimization. The concept of the filled functions was introduced by Ge in [70] for continuous global optimization. Further results on filled function methods were reported by various authors (see, e.g., [72] [93][144][231][234]). A discrete filled function method [170] was developed for nonlinear integer programming problems. The materials presented in this chapter are mainly based on [169].

References

[1] W. E. Adams, A. Billionnet, and A. Sutter. Unconstrained 0-1 optimization and Lagrangian relaxation. *Discrete Applied Mathematics*, 29:131–142, 1990.

[2] W. P. Adams and H. D. Sherali. A tight linearization and an algorithm for zero-one quadratic programming problems. *Management Science*, 32:1274–1290, 1986.

[3] C. S. Adjiman, I. P. Androulakis, and C. A. Floudas. A global optimization method, αBB, for general twice-differentiable constrained NLPs: II. Implementation and computational results. *Computers and Chemical Engineering*, 22:1159–1179, 1998.

[4] C. S. Adjiman, I. P. Androulakis, and C. A. Floudas. Global optimization of mixed-integer nonlinear problems. *AIChE Journal*, 46:1769–1797, 2000.

[5] C. S. Adjiman, S. Dallwig, C. A. Floudas, and A. Neumaier. A global optimization method, αBB, for general twice-differentiable constrained NLPs: I. Theoretical advances. *Computers and Chemical Engineering*, 22:1137–1158, 1998.

[6] F. A. Al-Khayyal and J. E. Falk. Jointly constrained biconvex programming. *Mathematics of Operations Research*, 8:273–286, 1983.

[7] E. Balas. An additive algorithm for solving linear programs with zero-one variables. *Operations Research*, 13:517–546, 1965.

[8] E. Balas and C. H. Martin. Pivot and complement–A heuristic for 0-1 programming. *Management Science*, 26:86–96, 1980.

[9] E. Balas and J. B. Mazzola. Nonlinear 0-1 programming: I. Linearization techniques. *Mathematical Programming*, 30:1–21, 1984.

[10] E. Balas and J. B. Mazzola. Nonlinear 0-1 programming: II. Dominance relations and algorithms. *Mathematical Programming*, 30:22–45, 1984.

[11] R. Baldick. Refined proximity and sensitivity results in linearly constrained convex separable integer programming. *Linear Algebra and Its Applications*, 226-228:389–407, 1995.

[12] F. Barahona, M. Jünger, and G. Reinelt. Experiments in quadratic 0-1 programming. *Mathematical Programming*, 44:127–137, 1989.

[13] M. S. Bazaraa, H. D. Sherali, and C. M. Shetty. *Nonlinear Programming: Theory and Algorithms*. Wiley, New York, 1993.

[14] J. E. Beasley, N. Meade, and T.-J. Chang. An evolutionary heuristic for the index tracking problem. *European Journal of Operational Research*, 148:621–643, 2003.

[15] A. Beck and M. Teboulle. Global optimality conditions for quadratic optimization problems with binary constraints. *SIAM Journal on Optimization*, 11:179–188, 2000.

[16] E. F. Beckenbach and R. Bellman. *Inequalities, 3rd Printing*. Springer-Verlag, New York, 1971.

[17] D. E. Bell and J. F. Shapiro. A convergent duality theory for integer programming. *Operations Research*, 25:419–434, 1977.

[18] R. E. Bellman. *Dynamic Programming*. Princeton University Press, Princeton, 1957.

[19] H. P. Benson and S. S. Erenguc. An algorithm for concave integer minimization over a polyhedron. *Naval Research Logistics*, 37:515–525, 1990.

[20] H. P. Benson, S. S. Erenguc, and R. Horst. A note on adapting methods for continuous global optimization to the discrete case. *Annals of Operations Research*, 25:243–252, 1990.

[21] D. Bienstock. Computational study of a family of mixed-integer quadratic programming problems. *Mathematical Programming*, 74:121–140, 1996.

[22] A. Billionnet and F. Calmels. Linear programming for the 0-1 quadratic knapsack problem. *European Journal of Operational Research*, 92:310–325, 1996.

[23] A. Billionnet and E. Soutif. Using a mixed integer programming tool for solving the 0-1 quadratic knapsack problem. *INFORMS Journal of Computing*, 16:188–197, 2004.

[24] A. Billionnet and A. Sutter. Minimization of a quadratic pseudo-Boolean function. *European Journal of Operational Research*, 78:106–115, 1994.

[25] A. Birolini. *Reliability Engineering: Theory and Practice*. Springer, New York, 1999.

[26] G. R. Bitran and D. Tirupati. Capacity planning in manufacturing networks with discrete options. *Annals of Operations Research*, 17:119–135, 1989.

[27] G. R. Bitran and D. Tirupati. Tradeoff curves, targeting and balancing in manufacturing queueing networks. *Operations Research*, 37:547–564, 1989.

[28] C. E. Blair and R. G. Jeroslow. The value function of a mixed integer program: II. *Discrete Mathematics*, 25:7–19, 1979.

[29] I. M. Bomze and G. Danninger. A global optimization algorithm for concave quadratic programming problems. *SIAM Journal on Optimization*, 3:826–842, 1993.

[30] B. Borchers and J. E. Mitchell. An improved branch and bound algorithm for mixed integer nonlinear programs. *Computers & Operations Research*, 21:359–367, 1994.

[31] E. Boros and P. L. Hammer. Pseudo-Boolean optimization. *Discrete Applied Mathematics*, 123:155–225, 2002.

[32] K. M. Bretthauer, A. V. Cabot, and M. A. Venkataramanan. An algorithm and new penalties for concave integer minimization over a polyhedron. *Naval Research Logistics*, 41:435–454, 1994.

[33] K. M. Bretthauer, A. Ross, and B. Shetty. Nonlinear integer programming for optimal allocation in stratified sampling. *European Journal of Operational Research*, 116:667–680, 1999.

[34] K. M. Bretthauer and B. Shetty. The nonlinear resource allocation problem. *Operations Research*, 43:670–683, 1995.

[35] K. M. Bretthauer and B. Shetty. The nonlinear knapsack problem-algorithms and applications. *European Journal of Operational Research*, 138:459–472, 2002.

[36] K. M. Bretthauer and B. Shetty. A pegging algorithm for the nonlinear resource allocation problem. *Computers & Operations Research*, 29:505–527, 2002.

[37] A. V. Cabot and S. S. Erenguc. A branch and bound algorithm for solving a class of nonlinear integer programming problems. *Naval Research Logistics*, 33:559–567, 1986.

[38] A. Caprara, D. Pisinger, and P. Toth. Exact solution of the quadratic knapsack problem. *INFORMS Journal on Computing*, 11:125–139, 1999.

[39] P. Chaillou, P. Hansen, and Y. Mahieu. Best network flow bounds for the quadratic knapsack problem. *Lecture Notes in Mathematics*, 1403:226–235, 1986.

[40] P. Chardaire and A. Sutter. A decomposition method for quadratic zero-one programming. *Management Science*, 41:704–712, 1995.

[41] P. C. Chen, P. Hansen, and B. Jaumard. On-line and off-line vertex enumeration by adjacency lists. *Operations Research Letters*, 10:403–409, 1991.

[42] W. G. Cochran. *Sampling Techniques*. Wiley, New York, 1963.

[43] W. Cook, A. M. H. Gerards, A. Schrijver, and E. Tardos. Sensitivity theorems in integer linear programming. *Mathematical Programming*, 34:251–264, 1986.

[44] M. W. Cooper. The use of dynamic programming for the solution of a class of nonlinear programming problems. *Naval Research Logistics Quarterly*, 27:89–95, 1980.

[45] M. W. Cooper. A survey of methods for pure nonlinear integer programming. *Management Science*, 27:353–361, 1981.

[46] Y. Crama, P. Hansen, and B. Jaumard. The basic algorithm for pseudo-Boolean programming revisited. *Discrete Applied Mathematics*, 29:171–185, 1990.

[47] G. Danninger and I. M. Bomze. Using copositivity for global optimality criteria in concave quadratic programming problems. *Mathematical Programming*, 62:575–580, 1993.

[48] G. B. Dantzig. On the significance of solving linear programming with some integer variables. The Rand Corporation, document P1486 (1958). Also published in *Econometrica* 28:30–40, 1960.

[49] G. B. Dantzig. Discrete-variable extremum problems. *Operations Research*, 5:266–277, 1957.

[50] E. V. Denardo. *Dynamic Programming Models and Applications*. Prentice-Hall, Englewood Cliffs, New Jersey, 1982.

[51] M. Djerdjour, K. Mathur, and H. M. Salkin. A surrogate relaxation based algorithm for a general quadratic multi-dimensional knapsack problem. *Operations Research Letters*, 7:253–258, 1988.

[52] M. Djerdjour and K. Rekab. A branch and bound algorithm for designing reliable systems at a minimum cost. *Applied Mathematics and Computation*, 118:247–259, 2001.

[53] M. A. Duran and I. E. Grossmann. An outer-approximation algorithm for a class of mixed-integer nonlinear programs. *Mathematical Programming*, 36:307–339, 1986.

[54] M. E. Dyer. Calculating surrogate constraints. *Mathematical Programming*, 19:255–278, 1980.

[55] M. E. Dyer and J. Walker. Solving the subproblem in the Lagrangian dual of separable discrete programs with linear constraints. *Mathematical Programming*, 24:107–112, 1982.

[56] M. L. Fisher. The Lagrangian relaxation method for solving integer programming problems. *Management Science*, 27:1–18, 1981.

[57] M. L. Fisher and J. F. Shapiro. Constructive duality in integer programming. *SIAM Journal on Applied Mathematics*, 27:31–52, 1974.

[58] R. Fletcher. *Practical Methods of Optimization*. Wiley-Interscience, 1987.

[59] R. Fletcher and S. Leyffer. Solving mixed integer nonlinear programs by outer approximation. *Mathematical Programming*, 66:327–349, 1994.

[60] C. A. Floudas. *Nonlinear and Mixed-Integer Optimization: Fundamentals and Applications*. Oxford University Press, 1995.

[61] C. A. Floudas. *Deterministic Global Optimization: Theory, Methods and Applications*. Kluwer Academic, Dordrecht, 2000.

[62] C. A. Floudas and V. Visweswaran. Quadratic optimization. In R. Horst and P. M. Pardalos, editors, *Handbook of Global Optimization*, pages 217–269. Kluwer Academic Publishers, 1995.

[63] R. Fortet. L'algèbre de Boole et ses applications en recherche opérationnelle. *Cahiers du Centre d'Études de Recherche Opérationnelle*, 1:5–36, 1959.

[64] R. Fortet. Applications de l'algèbre de Boole en recherche opérationnelle. *Revue Francaise d'Informatique et de Recherche Opérationnelle*, 4:17–26, 1960.

[65] G. Gallo, M. Grigoridis, and R. E. Tarjan. A fast parametric maximum flow algorithm and applications. *SIAM Journal on Computing*, 18:30–55, 1989.

[66] G. Gallo, P. L. Hammer, and B. Simeone. Quadratic knapsack problems. *Mathematical Programming*, 12:132–149, 1980.

[67] G. Gallo and B. Simeone. On the supermodular knapsack problem. *Mathematical Programming*, 45:295–309, 1988.

[68] R. S. Garfinkel and G. L. Nemhauser. *Integer Programming*. John Wiley & Sons, New York, 1972.

[69] B. Gavish, F. Glover, and H. Pirkul. Surrogate constraints in integer programming. *Journal of Information and Optimization Sciences*, 12:219–228, 1991.

[70] R. P. Ge. A filled function method for finding a global minimizer of a function of several variables. *Mathematical Programming*, 46(2):191–204, 1990.

[71] R. P. Ge and C. B. Huang. A continuous approach to nonlinear integer programming. *Applied Mathematics and Computation*, 34:39–60, 1989.

[72] R. P. Ge and Y. F. Qin. A class of filled functions for finding global minimizers of a function of several variables. *Journal of Optimization Theory and Applications*, 54:241–252, 1987.

[73] A. M. Geoffrion. Integer programming by implicit enumeration and Balas' method. *SIAM Review*, 9:178–190, 1967.

[74] A. M. Geoffrion. Generalized Benders decomposition. *Journal of Optimization Theory and Applications*, 10:237–260, 1972.

[75] A. M. Geoffrion. Lagrangean relaxation for integer programming. *Mathematical Programming Study*, 2:82–114, 1974.

[76] A. M. Geoffrion and R. E. Marsten. Integer programming algorithms: A framework and state-of-the-art survey. *Management Science*, 18:465–491, 1972.

[77] F. Glover. A multiphase-dual algorithm for the zero-one integer programming problem. *Operations Research*, 13:879–919, 1965.

[78] A. V. Goldberg and R. E. Tarjar. A new approach to the maximum flow problem. *Proceedings of the 18th Annual ACM Symposium on Theory of Computing*, pages 136–146, 1986.

[79] A. A. Goldstein and J. F. Price. On descent from local minima. *Mathematics of Computation*, 25:569–574, 1971.

[80] R. E. Gomory. Outline of an algorithm for integer solutions to linear programs. *Bulletin of the American Mathematical Society*, 64:275–278, 1958.

[81] M. Gondran. Programmation entiere non lineaire. *R. I. R. O.*, 4:107–110, 1970.

[82] D. Granot and F. Granot. Generalized covering relaxation for 0-1 programs. *Operations Research*, 28:1442–1450, 1980.

[83] F. Granot and P. L. Hammer. On the role of generalized covering problems. *Cahiers du Centre d'Études de Recherche Opérationnelle*, 17:277–289, 1975.

[84] H. J. Greenberg and W. P. Pierskalla. Surrogate mathematical programming. *Operations Research*, 18:924–939, 1970.

[85] I. E. Grossmann and Z. Kravanja. Mixed-integer nonlinear programming: A survey of algorithms and applications. In R. Conn A, T. F. Coleman L. T. Biegler, and F. N. Santosa, editors, *Large-scale optimization with applications: II. Optimization design and control*. Springer, New Year, Berlin, 1997.

[86] M. Guignard and S. Kim. Lagrangean decomposition: A model yielding stronger Lagrangean bounds. *Mathematical Programming*, 39:215–228, 1987.

[87] O. K. Gupta and A. Ravindran. Branch and bound experiments in convex nonlinear integer programming. *Management Science*, 31:1533–1546, 1985.

[88] P. L. Hammer. Some network flow problems solved with pseudo-Boolean programming. *Operations Research*, 13:388–399, 1965.

[89] P. L. Hammer and Jr. D. J. Rader. Efficient methods for solving quadratic 0-1 knapsack problems. *INFOR*, 35:170–182, 1997.

[90] P. L. Hammer, P. Hansen, and B. Simeone. Roof duality, complementation and persistency in quadratic 0-1 optimization. *Mathematical Programming*, 28:121–155, 1984.

[91] P. L. Hammer and R. Holzman. Approximations of pseudo-Boolean functions; applications to game theory. *ZOR-Methods and Models of Operations Research*, 36:3–21, 1992.

[92] P. L. Hammer and S. Rudeanu. *Boolean Methods in Operations Research and Related Areas*. Springer-Verlag, Berlin, Heidelberg, New York, 1968.

[93] Q. M. Han and J. Y. Han. Revised filled function methods for global optimization. *Applied Mathematics and Computation*, 119:217–228, 2001.

[94] P. Hansen. Methods of nonlinear 0-1 programming. *Annals of Discrete Mathematics*, 5:53–70, 1979.

[95] P. Hansen, B. Jaumard, and V. Mathon. Constrained nonlinear 0-1 programming. *ORSA Journal on Computing*, 5:97–119, 1993.

[96] P. Hansen, S. H. Lu, and B. Simeone. On the equivalence of paved-duality and standard linearization in nonlinear 0-1 optimization. *Discrete Applied Mathematics*, 29:187–193, 1990.

[97] M. Held and R. M. Karp. The traveling-salesman problem and minimum spanning trees. *Operations Research*, 18:1138–1162, 1970.

[98] R. Helgason, J. Kennington, and H. Lall. A polynomially bounded algorithm for a singly constrained quadratic program. *Mathematical Programming*, 18:338–343, 1980.

[99] C. Helmberg and F. Rendl. Solving quadratic (0,1)-problems by semidefinite programs and cutting planes. *Mathematical Programming*, 82:291–315, 1998.

[100] J. B. Hiriart-Urruty and C. Lemaréchal. *Convex Analysis and Minimization Algorithms, Volumes 1 and 2*. Springer-Verlag, Berlin, 1993.

[101] D. S. Hochbaum. A nonlinear knapsack problem. *Operations Research Letters*, 17:103–110, 1995.

[102] D. S. Hochbaum and J. G. Shanthikumar. Convex separable optimization is not much harder than linear optimization. *Journal of the Association for Computing Machinery*, 37:843–862, 1990.

[103] K. L. A. Hoffman. A method for globally minimizing concave functions over convex sets. *Mathematical Programming*, 20:22–32, 1981.

[104] R. Horst, J. de Vries, and N. V. Thoai. On finding new vertices and redundant constraints in cutting plane algorithms for global optimization. *Operations Research Letters*, 7:85–90, 1988.

[105] R. Horst and H. Tuy. *Global Optimization: Deterministic Approaches*. Springer-Verlag, Heidelberg, 1993.

[106] T. Ibaraki and N. Katoh. *Resource Allocation Problems: Algorithmic Approaches*. MIT Press, Cambridge, Mass., 1988.

[107] H. Everett III. Generalized Lagrange multiplier method for solving problems of optimum allocation of resources. *Operations Research*, 11:399–417, 1963.

[108] N. J. Jobst, M. D. Horniman, C. A. Lucas, and G. Mitra. Computational aspects of alternative portfolio selection models in the presence of discrete asset choice constraints. *Quantitive Finance*, 1:489–501, 2001.

[109] E. L. Johnson. *Integer Programming: Facets, Subadditivity, and Duality for Group and Semi-group Problems*. SIAM, Philadelphia, 1980.

[110] B. Kalantari and A. Bagchi. An algorithm for quadratic zero-one programs. *Naval Research Logistics*, 37:527–538, 1990.

[111] B. Kalantari and J. B. Rosen. Penalty for zero-one integer equivalent problem. *Mathematical Programming*, 24:229–232, 1982.

[112] B. Kalantari and J. B. Rosen. An algorithm for global minimization of linearly constrained concave quadratic functions. *Mathematics of Operations Research*, 12:544–561, 1987.

[113] B. Kalantari and J. B. Rosen. Penalty formulation for zero-one nonlinear programming. *Discrete Applied Mathematics*, 16:179–182, 1987.

[114] M. H. Karwan and R. L. Rardin. Searchability of the composite and multiple surrogate dual functions. *Operations Research*, 28:1251–1257, 1980.

[115] M. H. Karwan and R. L. Rardin. Surrogate dual multiplier search procedures in integer programming. *Operations Research*, 32:52–69, 1984.

[116] A. Kaufmann and A. Henry-Labordere. *Integer and Mixed Programming: Theory and Applications*. Academic Press, New York, 1977.

[117] H. Kellerer, U. Pferschy, and D. Pisinger. *Knapsack Problems*. Springer-Verlag, Berlin Heideberg, 2004.

[118] S. Kim and H. Ahn. Convergence of a generalized subgradient method for nondifferentiable convex optimization. *Mathematical Programming*, 50:75–80, 1991.

[119] S. Kim, H. Ahn, and S-C Cho. Variable target value subgradient method. *Mathematical Programming*, 49:359–369, 1991.

[120] S. L. Kim and S. Kim. Exact algorithm for the surrogate dual of an integer programming problem: Subgradient method approach. *Journal of Optimization Theory and Applications*, 96:363–375, 1998.

[121] K. C. Kiwiel. *Methods of Descent for Nondifferentiable Optimization*. Springer, Berlin, 1985.

[122] M. S. Kodialam and H. Luss. Algorithms for separable nonlinear resource allocation problems. *Operations Research*, 46:272–284, 1998.

[123] F. Korner. A hybrid method for solving nonlinear knapsack problems. *European Journal of Operational Research*, 38:238–241, 1989.

[124] A. H. Land and A. G. Doig. An automatic method of solving discrete programming problems. *Econometrica*, 28:497–520, 1960.

[125] J. B. Lasserre. An explicit equivalent positive semidefinite program for nonlinear 0-1 programs. *SIAM Journal on Optimization*, 12:756–769, 2002.

[126] D. J. Laughhunn. Quadratic binary programming with application to capital-budgeting problems. *Operations Research*, 18:454–461, 1970.

[127] C. Lemaréchal. Nondifferentiable optimization. In G. L. Nemhauser, A. H. G. Rinnooy Kan, and M. J. Todd, editors, *Optimization*, pages 529–572. North-Holland, Amsterdam, 1989.

[128] C. Lemaréchal and A. Renaud. A geometric study of duality gaps, with applications. *Mathematical Programming*, 90:399–427, 2001.

[129] A. V. Levy and A. Montalvo. The tunneling algorithm for the global minimization of functions. *SIAM Journal on Scientific and Statistical Computing*, 6:15–29, 1985.

[130] S. Leyffer. Integrating SQP and branch-and-bound for mixed integer nonlinear programming. *Computational Optimization and Applications*, 18:295–309, 2001.

[131] D. Li. Iterative parametric dynamic programming and its application in reliability optimization. *Journal of Mathematical Analysis and Applications*, 191:589–607, 1995.

[132] D. Li. Zero duality gap for a class of nonconvex optimization problems. *Journal of Optimization Theory and Applications*, 85:309–324, 1995.

[133] D. Li. Zero duality gap in integer programming: p-norm surrogate constraint method. *Operations Research Letters*, 25:89–96, 1999.

[134] D. Li and X. L. Sun. Success guarantee of dual search in integer programming: p-th power Lagrangian method. *Journal of Global Optimization*, 18:235–254, 2000.

[135] D. Li and X. L. Sun. Towards strong duality in integer programming. *Journal of Global Optimization*, 2005. Accepted for publication.

[136] D. Li, X. L. Sun, M. P. Biswal, and F. Gao. Convexification, concavification and monotonization in global optimization. *Annals of Operations Research*, 105:213–226, 2001.

[137] D. Li, X. L. Sun, and K. McKinnon. An exact solution method for reliability optimization in complex systems. *Annals of Operations Research*, 133:129–148, 2005.

[138] D. Li, X. L. Sun, and F. L. Wang. Convergent Lagrangian and contour-cut method for nonlinear integer programming with a quadratic objective function. Technical report, Department of Systems Engineering & Engineering Management, The Chinese University of Hong Kong, Shatin, N. T., Hong Kong, 2005. Submitted for publication.

[139] D. Li, X. L. Sun, and J. Wang. Convergent Lagrangian methods for separable nonlinear integer programming: Objective level cut and domain cut methods. In J. Karlof, editor, *Integer Programming: Theory and Practice*, pages 19–36. Taylor & Francis Group, London, 2005.

[140] D. Li, X. L. Sun, and J. Wang. Optimal lot solution to cardinality constrained mean-variance formulation for portfolio selection. *Mathematical Finance*, 16:83–101, 2006.

[141] D. Li, X. L. Sun, J. Wang, and K. McKinnon. A convergent Lagrangian and domain cut method for nonlinear knapsack problems. Technical report, Department of Systems Engineering & Engineering Management, The Chinese University of Hong Kong, Shatin, N. T., Hong Kong, 2002. Submitted for publication.

[142] D. Li, J. Wang, and X. L. Sun. A convergent Lagrangian and objective level cut method for separable nonlinear programming. Technical report, Department of Systems Engineering & Engineering Management, The Chinese University of Hong Kong, 2003. Submitted for publication.

[143] D. Li and D. J. White. p-th power Lagrangian method for integer programming. *Annals of Operations Research*, 98:151–170, 2000.

[144] X. Liu. Finding global minima with a computable filled function. *Journal of Global Optimization*, 19:151–161, 2001.

[145] S. H. Lu and A. C. Williams. Roof duality and linear relaxation for quadratic and polynomial 0-1 optimization. Technical report, Rutgers University, New Brunswick, NJ, 1987. RUTCOR Research Report #8-87.

[146] S. H. Lu and A. C. Williams. Roof duality for polynomial 0-1 optimization. *Mathematical Programming*, 37:357–360, 1987.

[147] D. G. Luenberger. Quasi-convex programming. *SIAM Journal on Applied Mathematics*, 16:1090–1095, 1968.

[148] D. G. Luenberger. *Linear and Nonlinear Programming*. Addison-Wesley, Reading, Massachusetts, 1984.

[149] H. Luss and S. K. Gupta. Allocation of effort resources among competing activities. *Operations Research*, 23:360–366, 1975.

[150] H. M. Markowitz. *Portfolio Selection: Efficient Diversification of Investment*. John Wiley & Sons, New York, 1959.

[151] R. E. Marsten and T. L. Morin. A hybrid approach to discrete mathematical programming. *Mathematical Programming*, 14:21–40, 1978.

[152] S. Martello and P. Toth. An upper bound for the zero-one knapsack problem and a branch and bound algorithm. *European Journal of Operational Research*, 1:169–175, 1977.

[153] S. Martello and P. Toth. *Knapsack Problems: Algorithms and Computer Implementations*. John Wiley & Sons, New York, 1990.

[154] K. Mathur, H. M. Salkin, and B. B. Mohanty. A note on a general nonlinear knapsack problems. *Operations Research Letters*, 5:79–81, 1986.

[155] K. Mathur, H. M. Salkin, and S. Morito. A branch and search algorithm for a class of nonlinear knapsack problems. *Operations Research Letters*, 2:155–160, 1983.

[156] R. D. Mcbride and J. S. Yormark. An implicit enumeration algorithm for quadratic integer programming. *Management Science*, 26:282–296, 1980.

[157] G. P. McCormick. Computability of global solutions to factorable nonconvex programs: I. Convex underestimating problems. *Mathematical Programming*, 10:147–175, 1976.

[158] H. Müller-Merbach. An improved upper bound for the zero-one knapsack problem; a note on the paper by Martello and Toth. *European Journal of Operational Research*, 2:212–213, 1978.

[159] A. Melman and G. Rabinowitz. An efficient method for a class of continuous nonlinear knapsack problems. *SIAM Review*, 42:440–448, 2000.

[160] P. Michelon. Unconstrained 0-1 nonlinear programming: A nondifferentiable approach. *Journal of Global Optimization*, 2:155–165, 1992.

[161] P. Michelon and N. Maculan. Lagrangian decomposition for integer nonlinear programming with linear constraints. *Mathematical Programming*, 52:303–313, 1991.

[162] P. Michelon and N. Maculan. Lagrangian methods for 0-1 quadratic problems. *Discrete Applied Mathematics*, 42:257–269, 1993.

[163] P. Michelon and L. Veilleux. Lagrangian methods for the 0-1 quadratic knapsack problem. *European Journal of Operational Research*, 92:326–341, 1996.

[164] K. B. Misra and U. Sharma. An efficient algorithm to solve integer-programming problems arising in system-reliability design. *IEEE Transaction on Reliability*, 40:81–91, 1991.

[165] G. Mitra, K. Darby-Dowman an C. A. Lucas, and J. Yadegar. Linear, integer separable and fuzzy programming problems: A unified approach towards reformulation. *Journal of Operational Research Society*, 39:161–171, 1988.

[166] C. Mohan and H. T. Nguyen. A controlled random search technique incorporating the simulated annealing concept for solving integer and mixed integer global optimization problems. *Computational Optimization and Applications*, 14:103–132, 1999.

[167] P. Neame, N. Boland, and D. Ralph. An outer approximate subdifferential method for piecewise affine optimization. *Mathematical Programming*, 87:57–86, 2000.

[168] G. L. Nemhauser and L. A. Wolsey. *Integer and Combinatorial Optimization*. John Wiley & Sons, New York, 1988.

[169] C. K. Ng, D. Li, and L. S. Zhang. Global descent method for discrete optimization. Technical report, Department of Systems Engineering & Engineering Management, The Chinese University of Hong Kong, Shatin, N. T., Hong Kong, 2005. Submitted for publication.

[170] C. K. Ng, L. S. Zhang, D. Li, and W. W. Tian. Discrete filled function method for discrete global optimization. *Computational Optimization and Applications*, 31:87–115, 2005.

[171] S. S. Nielsen and S. A. Zenios. A massively parallel algorithm for nonlinear stochastic network problems. *Operations Research*, 41:319–337, 1993.

[172] G. Palubeckis. A heuristic-based branch and bound algorithm for unconstrained quadratic zero-one programming. *Computing*, 54:283–301, 1995.

[173] P. Papalambros and D. J. Wilde. *Principles of Optimal Design–Modeling and Computation*. Cambridge University Press, Cambridge, UK, New York, 1986.

[174] P. M. Pardalos and J. B. Rosen. *Constrained Global Optimization: Algorithms and Applications*. Springer-Verlag, Berlin, 1987.

[175] P. M. Pardolas and G. P. Rodgers. Computational aspects of a branch and bound algorithm for quadratic zero-one programming. *Computing*, 45:131–144, 1990.

[176] R. G. Parker and R. L. Rardin. *Discrete Optimization*. Academic Press, Boston, 1988.

[177] A. T. Phillips and J. B. Rosen. A quadratic assignment formulation of the molecular conformation problem. *Journal of Global Optimization*, 4:229–241, 1994.

[178] J. C. Picard and H. D. Ratliff. Minimum cuts and related problems. *Networks*, 5:357–370, 1975.

[179] M. C. Pinar. Sufficient global optimality conditions for bivalent quadratic optimization. *Journal of Optimization Theory and Applications*, 122:433–440, 2004.

[180] B. T. Polyak. Minimization of nonsmooth functions. *USSR Computational Mathematics and Mathematical Physics*, 9:14–29, 1969.

[181] A. J. Quist, E. de Klerk, C. Roos, T. Terlaky, R. van Geemert, J. E. Hoogenboom, and T. Illes. Finding optimal nuclear reactor core reload patterns using nonlinear optimization and search heuristics. *Engineering Optimization*, 32:143–176, 1999.

[182] R. T. Rockafellar. *Convex Analysis*. Princeton University Press, NJ, 1970.

[183] J. B. Rosen and P. M. Pardalos. Global minimization of large-scale constrained concave quadratic problems by separable programming. *Mathematical Programming*, 34:163–174, 1986.

[184] I. G. Rosenberg. 0-1 optimization and nonlinear programming. *R.A.I.R.O*, 6:95–97, 1972.

[185] P. A. Rubin. Comment on "A nonlinear Lagrangian dual for integer programming". *Operations Research Letters*, 32:197–198, 2004.

[186] A. Rubinov, H. Tuy, and H. Mays. An algorithm for monotonic global optimization problems. *Optimization*, 49:205–221, 2001.

[187] H. S. Ryoo and N. V. Sahinidis. Global optimization of nonconvex NLPs and MINLPs with applications in process design. *Computers and Chemical Engineering*, 19:551–566, 1995.

[188] H. M. Salkin and K. Mathur. *Foundations of Integer Programming*. North-Holland, New York, 1989.

[189] S. Sarin, M. H. Karwan, and R. L. Rardin. A new surrogate dual multiplier search procedure. *Naval Research Logistics*, 34:431–450, 1987.

[190] K. Schittkowski. *More Test Examples for Nonlinear Programming Codes*. Springer-Verlag, 1987.

[191] A. Schrijver. *Theory of Linear and Integer Programming*. John Wiley & Sons, New York, 1986.

[192] J. F. Shapiro. A survey of Lagrangian techniques for discrete optimization. *Annals of Discrete Mathematics*, 5:113–138, 1979.

[193] W. F. Sharpe. A simplified model for portfolio analysis. *Management Science*, 9:277–293, 1963.

[194] W. F. Sharpe. *Portfolio Theory and Capital Markets*. McGraw Hill, New York, 1970.

[195] J. P. Shectman and N. V. Sahinidis. A finite algorithm for global minimization of separable concave programs. *Journal of Global Optimization*, 12:1–35, 1998.

[196] H. D. Sherali and W. P. Adams. *A Reformulation-Linearization Technique for Solving Discrete and Continuous Nonconvex Problems*. Kluwer Academic, Dordrecht, 1999.

[197] H. D. Sherali and H. J. Wang. Global optimization of nonconvex factorable problems. *Mathematical Programming*, 89:459–478, 2001.

[198] N. Z. Shor. *Minimization Methods for Non-differentiable functions*. Springer, Berlin, 1985.

[199] C. De Simone and G. Rinaldi. A cutting plane algorithm for the max-cut problem. *Optimization Methods and Software*, 3:195–214, 1994.

[200] M. Sinclair. An exact penalty function approach for nonlinear integer programming problems. *European Journal of Operational Research*, 27:50–56, 1986.

[201] R. A. Stubbs and S. Mehrotra. A branch-and-cut method for 0-1 mixed convex programming. *Mathematical Programming*, 86:515–532, 1999.

[202] X. L. Sun and D. Li. Asymptotic strong duality for bounded integer programming: A logarithmic-exponential dual formulation. *Mathematics of Operations Research*, 25:625–644, 2000.

[203] X. L. Sun and D. Li. New dual formulations in constrained integer programming. In X. Q. Yang, editor, *Progress in Optimization*, pages 79–91. Kluwer, 2000.

[204] X. L. Sun and D. Li. On the relationship between the integer and continuous solutions of convex programs. *Operations Research Letters*, 29:87–92, 2001.

[205] X. L. Sun and D. Li. Optimality condition and branch and bound algorithm for constrained redundancy optimization in series systems. *Optimization and Engineering*, 3:53–65, 2002.

[206] X. L. Sun, H. Z. Luo, and D. Li. Convexification of nonsmooth monotone functions. *Journal of Optimization Theory and Applications*, 2005. Accepted for publication.

[207] X. L. Sun, K. I. M. McKinnon, and D. Li. A convexification method for a class of global optimization problems with applications to reliability optimization. *Journal of Global Optimization*, 21:185–199, 2001.

[208] X. L. Sun, F. L. Wang, and D. Li. Exact algorithm for concave knapsack problems: Linear underestimation and partition method. *Journal of Global Optimization*, 33:15–30, 2005.

[209] C. Sundararajan. *Guide to Reliability Engineering: Data, Analysis, Applications, Implementation, and Management*. Van Nostrand Reinhold, New York, 1991.

[210] S. S. Syam. A dual ascent method for the portfolio selection problem with multiple constraints and linked proposals. *European Journal of Operational Research*, 108:196–207, 1998.

[211] H. A. Taha. A Balasian-based algorithm for zero-one polynomial programming. *Management Science*, 18:B328–B343, 1972.

[212] H. A. Taha. *Integer Programming, Theory, Applications, and Computations*. Academic Press, New York, 1975.

[213] M. Tawarmalani and N. V. Sahinidis. *Convexification and Global Optimization in Continuous and Mixed-Integer Nonlinear Programming: Theory, Algorithms, Software, and Applications*. Kluwer Academic Publishers, Dordrecht, 2002.

[214] M. Tawarmalani and N. V. Sahinidis. Global optimization of mixed integer nonlinear programs: A theoretical and computational study. *Mathematical Programming*, 99:563–591, 2004.

[215] N. V. Thoai. Global optimization techniques for solving the general quadratic integer programming problem. *Computational Optimization and Applications*, 10:149–163, 1998.

[216] N. V. Thoai. General quadratic programming. In C. Audet, P. Hansen, and G. Savard, editors, *Essays and Surveys in Global Optimization*, pages 107–129. Springer, 2005.

[217] F. A. Tillman, C. L. Hwuang, and W. Kuo. *Optimization of System Reliability*. Marcel Dekker, 1980.

[218] H. Tuy. Monotonic optimization: Problems and solution approaches. *SIAM Journal on Optimization*, 11:464–494, 2000.

[219] S. G. Tzafestas. Optimization of system reliability: A survey of problems and techniques. *International Journal of Systems Science*, 11:455–486, 1980.

[220] J. Viswanathan and I. E. Grossmann. Optimal feed location and number of trays for distillation columns with multiple feeds. *I& EC Research*, 32:2942–2949, 1993.

[221] V. Visweswaran and C. A. Floudas. New properties and computational improvement of the GOP algorithm for problems with quadratic objective functions and constraints. *Journal of Global Optimization*, 3:439–462, 1993.

[222] S. Walukiewicz. *Integer programming*. Kluwer Academic, Dordrecht, 1990.

[223] J. Wang and D. Li. Efficient solution schemes for polynomial zero-one programming. Technical report, Department of Systems Engineering & Engineering Management, The Chinese University of Hong Kong, Shatin, N. T., Hong Kong, 2004. Submitted for publication.

[224] L. J. Watters. Reduction of integer polynomial programming problems to zero-one linear programming problems. *Operations Research*, 15:1171–1174, 1967.

[225] M. Werman and D. Magagnosc. The relationship between integer and real solutions of constrained convex programming. *Mathematical Programming*, 51:133–135, 1991.

[226] C. William. *Computer Optimization Techniques*. Petrocelli Books Inc., New York, 1980.

[227] G. B. Wolfe and P. Wolfe. Decomposition principle for linear programs. *Operations Research*, 8:101–111, 1960.

[228] L. A. Wolsey. *Integer programming*. John Wiley & Sons, New York, 1998.

[229] Y. F. Xu and D. Li. A nonlinear Lagrangian dual for integer programming. *Operations Research Letters*, 30:401–407, 2002.

[230] Y. F. Xu, C. L. Liu, and D. Li. Generalized nonlinear Lagrangian formulation for bounded integer programming. *Journal of Global Optimization*, 33:257–272, 2005.

[231] Z. Xu, H. X. Huang, P. M. Pardalos, and C. X. Xu. Filled functions for unconstrained global optimization. *Journal of Global Optimization*, 20:49–65, 2001.

[232] J. M. Zamora and I. E. Grossmann. A branch and contract algorithm for problems with concave univariate, bilinear and linear fractional terms. *Journal Of Global Optimization*, 14:217–249, 1999.

[233] L. S. Zhang, F. Gao, and W. X. Zhu. Nonlinear integer programming and global optimization. *Journal of Computational Mathematics*, 17:179–190, 1999.

[234] L. S. Zhang, C. K. Ng, D. Li, and W. W. Tian. A new filled function method for global optimization. *Journal of Global Optimization*, 28:17–43, 2004.

[235] X. Zhao and P. B. Luh. New bundle methods for solving Lagrangian relaxation dual problems. *Journal of Optimization Theory and Applications*, 113:373–397, 2002.

[236] Q. Zheng and D. M. Zhuang. Integral global minimization: Algorithms, implementations and numerical tests. *Journal of Global Optimization*, 7:421–454, 1995.

[237] H. Ziegler. Solving certain singly constrained convex optimization problems in production planning. *Operations Research Letters*, 1:246–252, 1982.

Index

Early Titles in the

INTERNATIONAL SERIES IN OPERATIONS RESEARCH & MANAGEMENT SCIENCE

Frederick S. Hillier, Series Editor, *Stanford University*

Saigal/ *A MODERN APPROACH TO LINEAR PROGRAMMING*
Nagurney/ *PROJECTED DYNAMICAL SYSTEMS & VARIATIONAL INEQUALITIES WITH APPLICATIONS*
Padberg & Rijal/ *LOCATION, SCHEDULING, DESIGN AND INTEGER PROGRAMMING*
Vanderbei/ *LINEAR PROGRAMMING*
Jaiswal/ *MILITARY OPERATIONS RESEARCH*
Gal & Greenberg/ *ADVANCES IN SENSITIVITY ANALYSIS & PARAMETRIC PROGRAMMING*
Prabhu/ *FOUNDATIONS OF QUEUEING THEORY*
Fang, Rajasekera & Tsao/ *ENTROPY OPTIMIZATION & MATHEMATICAL PROGRAMMING*
Yu/ *OR IN THE AIRLINE INDUSTRY*
Ho & Tang/ *PRODUCT VARIETY MANAGEMENT*
El-Taha & Stidham/ *SAMPLE-PATH ANALYSIS OF QUEUEING SYSTEMS*
Miettinen/ *NONLINEAR MULTIOBJECTIVE OPTIMIZATION*
Chao & Huntington/ *DESIGNING COMPETITIVE ELECTRICITY MARKETS*
Weglarz/ *PROJECT SCHEDULING: RECENT TRENDS & RESULTS*
Sahin & Polatoglu/ *QUALITY, WARRANTY AND PREVENTIVE MAINTENANCE*
Tavares/ *ADVANCES MODELS FOR PROJECT MANAGEMENT*
Tayur, Ganeshan & Magazine/ *QUANTITATIVE MODELS FOR SUPPLY CHAIN MANAGEMENT*
Weyant, J./ *ENERGY AND ENVIRONMENTAL POLICY MODELING*
Shanthikumar, J.G. & Sumita, U./ *APPLIED PROBABILITY AND STOCHASTIC PROCESSES*
Liu, B. & Esogbue, A.O./ *DECISION CRITERIA AND OPTIMAL INVENTORY PROCESSES*
Gal, T., Stewart, T.J., Hanne, T. / *MULTICRITERIA DECISION MAKING: Advances in MCDM Models, Algorithms, Theory, and Applications*
Fox, B.L. / *STRATEGIES FOR QUASI-MONTE CARLO*
Hall, R.W. / *HANDBOOK OF TRANSPORTATION SCIENCE*
Grassman, W.K. / *COMPUTATIONAL PROBABILITY*
Pomerol, J-C. & Barba-Romero, S. / *MULTICRITERION DECISION IN MANAGEMENT*
Axsäter, S. / *INVENTORY CONTROL*
Wolkowicz, H., Saigal, R., & Vandenberghe, L. / *HANDBOOK OF SEMI-DEFINITE PROGRAMMING: Theory, Algorithms, and Applications*
Hobbs, B.F. & Meier, P. / *ENERGY DECISIONS AND THE ENVIRONMENT: A Guide to the Use of Multicriteria Methods*
Dar-El, E. / *HUMAN LEARNING: From Learning Curves to Learning Organizations*
Armstrong, J.S. / *PRINCIPLES OF FORECASTING: A Handbook for Researchers and Practitioners*
Balsamo, S., Personé, V., & Onvural, R./ *ANALYSIS OF QUEUEING NETWORKS WITH BLOCKING*
Bouyssou, D. et al. / *EVALUATION AND DECISION MODELS: A Critical Perspective*
Hanne, T. / *INTELLIGENT STRATEGIES FOR META MULTIPLE CRITERIA DECISION MAKING*
Saaty, T. & Vargas, L. / *MODELS, METHODS, CONCEPTS and APPLICATIONS OF THE ANALYTIC HIERARCHY PROCESS*
Chatterjee, K. & Samuelson, W. / *GAME THEORY AND BUSINESS APPLICATIONS*
Hobbs, B. et al. / *THE NEXT GENERATION OF ELECTRIC POWER UNIT COMMITMENT MODELS*
Vanderbei, R.J. / *LINEAR PROGRAMMING: Foundations and Extensions, 2nd Ed.*
Kimms, A. / *MATHEMATICAL PROGRAMMING AND FINANCIAL OBJECTIVES FOR SCHEDULING PROJECTS*
Baptiste, P., Le Pape, C. & Nuijten, W. / *CONSTRAINT-BASED SCHEDULING*
Feinberg, E. & Shwartz, A. / *HANDBOOK OF MARKOV DECISION PROCESSES: Methods and Applications*
Ramík, J. & Vlach, M. / *GENERALIZED CONCAVITY IN FUZZY OPTIMIZATION AND DECISION ANALYSIS*
Song, J. & Yao, D. / *SUPPLY CHAIN STRUCTURES: Coordination, Information and Optimization*
Kozan, E. & Ohuchi, A. / *OPERATIONS RESEARCH/ MANAGEMENT SCIENCE AT WORK*
Bouyssou et al. / *AIDING DECISIONS WITH MULTIPLE CRITERIA: Essays in Honor of Bernard Roy*

Early Titles in the
INTERNATIONAL SERIES IN OPERATIONS RESEARCH & MANAGEMENT SCIENCE
(Continued)

Cox, Louis Anthony, Jr. / *RISK ANALYSIS: Foundations, Models and Methods*
Dror, M., L'Ecuyer, P. & Szidarovszky, F. / *MODELING UNCERTAINTY: An Examination of Stochastic Theory, Methods, and Applications*
Dokuchaev, N. / *DYNAMIC PORTFOLIO STRATEGIES: Quantitative Methods and Empirical Rules for Incomplete Information*
Sarker, R., Mohammadian, M. & Yao, X. / *EVOLUTIONARY OPTIMIZATION*
Demeulemeester, R. & Herroelen, W. / *PROJECT SCHEDULING: A Research Handbook*
Gazis, D.C. / *TRAFFIC THEORY*
Zhu/ *QUANTITATIVE MODELS FOR PERFORMANCE EVALUATION AND BENCHMARKING*
Ehrgott & Gandibleux/ *MULTIPLE CRITERIA OPTIMIZATION: State of the Art Annotated Bibliographical Surveys*
Bienstock/ *Potential Function Methods for Approx. Solving Linear Programming Problems*
Matsatsinis & Siskos/ *INTELLIGENT SUPPORT SYSTEMS FOR MARKETING DECISIONS*
Alpern & Gal/ *THE THEORY OF SEARCH GAMES AND RENDEZVOUS*
Hall/*HANDBOOK OF TRANSPORTATION SCIENCE - 2nd Ed.*
Glover & Kochenberger/ *HANDBOOK OF METAHEURISTICS*
Graves & Ringuest/ *MODELS AND METHODS FOR PROJECT SELECTION: Concepts from Management Science, Finance and Information Technology*
Hassin & Haviv/ *TO QUEUE OR NOT TO QUEUE: Equilibrium Behavior in Queueing Systems*
Gershwin et al/ *ANALYSIS & MODELING OF MANUFACTURING SYSTEMS*

**** A list of the more recent publications in the series is at the front of the book ****